Radiationless Processes

NATO ADVANCED STUDY INSTITUTES SERIES

A series of edited volumes comprising multifaceted studies of contemporary scientific issues by some of the best scientific minds in the world, assembled in cooperation with NATO Scientific Affairs Division.

Series B. Physics

Recent Volumes in this Series

Volume 52 – Physics of Nonlinear Transport in Semiconductors
edited by David K. Ferry, J. R. Barker, and C. Jacoboni

Volume 53 – Atomic and Molecular Processes in Controlled Thermonuclear Fusion
edited by M. R. C. McDowell and A. M. Ferendeci

Volume 54 – Quantum Flavordynamics, Quantum Chromodynamics, and Unified Theories
edited by K. T. Mahanthappa and James Randa

Volume 55 – Field Theoretical Methods in Particle Physics
edited by Werner Rühl

Volume 56 – Vibrational Spectroscopy of Molecular Liquids and Solids
edited by S. Bratos and R. M. Pick

Volume 57 – Quantum Dynamics of Molecules: The New Experimental Challenge to Theorists
edited by R. G. Woolley

Volume 58 – Cosmology and Gravitation: Spin, Torsion, Rotation, and Supergravity
edited by Peter G. Bergmann and Venzo De Sabbata

Volume 59 – Recent Developments in Gauge Theories
edited by G. 't Hooft, C. Itzykson, A. Jaffe, H. Lehmann, P. K. Mitter, I. M. Singer, and R. Stora

Volume 60 – Theoretical Aspects and New Developments in Magneto-Optics
Edited by Jozef T. Devreese

Volume 61 – Quarks and Leptons: *Cargèse 1979*
edited by Maurice Lévy, Jean-Louis Basdevant, David Speiser, Jacques Weyers, Raymond Gastmans, and Maurice Jacob

Volume 62 – Radiationless Processes
edited by Baldassare Di Bartolo

This series is published by an international board of publishers in conjunction with NATO Scientific Affairs Division

A Life Sciences B Physics	Plenum Publishing Corporation London and New York
C Mathematical and Physical Sciences	D. Reidel Publishing Company Dordrecht, Boston and London
D Behavioral and Social Sciences	Sijthoff & Noordhoff International Publishers
E Applied Sciences	Alphen aan den Rijn, The Netherlands, and Germantown, U.S.A.

Radiationless Processes

Edited by

Baldassare DiBartolo
Department of Physics
Boston College

Assistant Editor

Velda Goldberg
Department of Physics
Boston College

PLENUM PRESS • NEW YORK AND LONDON
Published in cooperation with NATO Scientific Affairs Division

Library of Congress Cataloging in Publication Data

Nato Advanced Study Institute on Radiationless Processess, Ericc, Italy, 1979.
Radiationless processes.

(Nato advanced study institutes series: series B, Physics; v. 62)
Includes bibliographical references and index.
1. Radiationless transitions–Congresses. 2. Molecular theory–Congresses. I. Di Bartolo, Baldassare. II. Goldberg, Velda. III. Title. IV. Series.
QC795.8.I5N37 1979 541.2'24 80-21961
ISBN-13: 978-1-4613-3176-6 e-ISBN-13: 978-1-4613-3174-2
DOI: 10.1007/978-1-4613-3174-2

Proceedings of the NATO Advanced Study Institute on Radiationless Processes, held at the Ettore Majorana Centre for Scientific Culture, Erice, Sicily, Italy, November 18–December 1, 1979.

A Division of Plenum Publishing Corporation
227 West 17th Street, New York, N.Y. 10011
Softcover reprint of the hardcover 1st edition 1980

God loves the search for truth
by means of reason and science

Frederick II (1194-1250)
Holy Roman Emperor
King of Germany and Sicily

PREFACE

This book presents an account of the NATO Advanced Study Institute on "Radiationless Processes," held in Erice, Italy, from November 18 to December 1, 1979. This meeting was organized by the International School of Atomic and Molecular Spectroscopy of the "Ettore Majorana" Centre for Scientific Culture.

The objective of the Institute was to formulate a comprehensive treatment of the various processes by which molecules and crystals in excited electronic levels relax nonradiatively to the ground level.

A total of 83 participants came from 62 laboratories and 22 nations (Australia, Belgium, Brasil, Canada, Czechoslovakia, France, F. R. Germany, Greece, Hungary, India, Ireland, Israel, Italy, Mexico, The Netherlands, Poland, Portugal, Switzerland, Turkey, United Kingdom, United States, and U.S.S.R.). The secretaries of the Institute were: Velda Goldberg for the scientific aspects and Antonino La Francesca for the administrative aspects of the meeting.

Eleven series of lectures for a total of 36 hours were given. Nine "long" seminars and 7 "short" seminars were also presented. In addition, two informal seminars and 2 round-table discussions were held.

After an introductory overview of the theory of radiationless processes, the Institute dealt firstly with the interaction of electrons with the distribution of vibrational modes in simple molecules, then with the increasingly complex situation found in large isolated molecules, and finally with the coupling of excited electrons with the continuous phonon distribution in insulating solids.

The theory of nonradiative relaxation was also considered in two areas of current technological importance, namely that of luminescence of phosphors and that of solid-state lasers. Nonradiative transitions in semiconductors and defect structures in phosphors produced by radiationless processes were also treated.

Specific aspects of the theory were also presented in seminar form together with several applications to various areas of current interest.

I would like to acknowledge the sponsorship of the Institute by the North Atlantic Treaty Organization and the very valuable assistance received from Dr. Mario di Lullo, Head of the NATO Advanced Study Institutes Programme. I wish to acknowledge also the sponsorship by the Italian Ministry of Public Education, the Italian Ministry of Scientific and Technological Research, the Regional Sicilian Government and the Department of Physics of Boston College.

I would like to thank for their help Prof. A. Zichichi, Director of the "Ettore Majorana" Centre for Scientific Culture, Dr. A. Gariele, Ms. P. Savalli, and all the personnel of the "Ettore Majorana" organization in Erice, Prof. R. L. Carovillano, chairman of the Department of Physics at Boston College, Prof. V. Adragna, Dr. G. Denaro and Rag. M. Strazzera.

I would like to thank the members of the organizing committee of the meeting (Prof. F. Williams and Dr. F. Auzel) and Prof. A. Scharmann for their valuable help and advice.

I am especially grateful to my friends and collaborators, A. La Francesca and V. Goldberg, who helped me tremendously with the various aspects of the meeting. V. Goldberg was the assistant editor of this book: without her patient and intelligent work, these proceedings could not have been produced.

It is difficult to describe in words the interest and the enthusiasm with which everybody participated in the proceedings of this Institute. I brought with me many fond memories of the two weeks I spent in Erice directing this meeting. I feel richer for the scientific knowledge I acquired by listening to the lectures of my learned colleagues. I feel also richer from the human and spiritual points of view for having shared this experience with so many fine people. I am sure we shall meet again soon.

Erice, December, 1979

B. Di Bartolo
Editor and Director of the Institute

CONTENTS

PRESENT TRENDS IN THE THEORY OF RADIATIONLESS PROCESSES

F. Williams, D. E. Berry and J. E. Bernard

Physics Department
University of Delaware
Newark, Delaware 19711, U.S.A.

ABSTRACT

The diversity of radiationless processes, e.g. inelastic collisions, electronic-vibrational relaxation, deexcitation, energy transfer, are unified by considering a system, molecular or solid, divided into two subsystems; one characterized as "fast"; the other, as "slow". Thus, the adiabatic (Born-Oppenheimer) approximation can be applied and partially separates the eigenstates of the two subsystems. These are frequently the electronic states and vibrational levels, but for some processes are local modes (vibrons) and normal modes (phonons). Most radiationless processes are characterized by the irreversible degradation of a single large quantum of the "fast" subsystem into many small quanta of the "slow" subsystem, however, the inverse occurs in special cases, for example thermal excitation and anti-Stokes emission. The calculation of radiationless transition probabilities is difficult mainly because, in contrast to radiative transitions, the operator coupling the two subsystems is not simple, the static (Condon) approximation to the "fast" eigenfunctions is not valid, and the origin of the irreversibility is often not obvious. The methods of calculation are reviewed; and current trends, emphasized. The trends include: (a) non-Condon analyses, (b) attention to collective effects, (c) novel phenomenological theories, and (d) detailed calculations on specific materials. A glossary of terms used in publications on radiationless processes and a bibliography of theoretical studies thereon are included.

I. INTRODUCTION

Radiationless processes limit the efficiencies of lamps, electronic devices and solar energy converters; and are important in chemical reactions, for example pre-dissociation as occurs in ozone depletion; and are also of fundamental interest in molecular and solid state chemistry and physics. In some energy conversion devices, such as anti-Stokes converters and photovoltaic solar converters, the second law of thermodynamics requires radiationless processes; in others, such as electroluminescent lamps, there is no such requirement, nevertheless radiationless processes occur.

The theory of radiationless processes is in general not in the advanced stage of understanding in which the theory of radiative transitions flourishes. Part of the basis for this is the diversity of types of radiationless processes: inelastic collisions between electrons and atoms, between atoms and molecules, between electrons and impurities in crystals; relaxation with respect to the nuclear coordinates, either of molecules or of condensed matter, following electronic transitions; non-radiative de-excitation from excited electronic states; and phonon-assisted energy transfer involving partial degradation of electronic energy into thermal vibrations. Other reasons for the paucity of theoretical understanding include the facts that the operator responsible for the coupling in non-radiative transitions is rather intractable and that the origin of the irreversibility is often not transparent.

Radiationless processes are usually characterized by an irreversibility which originates from the degradation of a single large quantum of energy, which characterizes a subsystem with fast orbital motion, into many small quanta, which characterize another subsystem with relatively slow motion. In contrast, the irreversibility of radiative transitions arises from the final state involving the continuum of photon states characterizing infinite free space. Returning to radiationless processes we note that the large and small quanta are frequently associated with electronic states and vibrational levels, respectively, but for some processes are local modes (vibrons) and normal lattice modes (phonons), respectively.

Another reason that our understanding of radiationless processes has evolved more slowly than our understanding of radiative processes has been the comparative situations regarding relevant experiments. Photon spectroscopy has been, and remains, a powerful tool for studying radiative transitions; there has been no comparable high resolution spectroscopy for radiationless transitions, although techniques like photoacoustic and capacitance spectroscopy are beginning to provide useful data on radiationless processes. In addition, the time constants for radiative decay, $\sim 10^{-1}$ to 10^{-9} sec, were earlier accessible to experimental determination than were the faster non-radiative decay, i.e. $\sim 10^{-11}$ to 10^{-13} sec, via intra-band transitions or vibrational relaxation.

In subsequent sections we shall not give a detailed review of the history of theoretical studies of radiationless processes. The complexity of the history of this subject can be inferred from the titles in Appendix B. Suffice it to say that the theory for molecules and the theory for solids appear to have evolved rather independently, and in fact within these two broad fields many specific theoretical developments have occurred with negligible interactions with each other, in part because the formalisms used are often grounded in different disciplines. Our goal in the subsequent sections will be to interrelate the more important general concepts and analyses, and then focus on the trends in recent and current research on the theory of radiationless processes. In Section II we review the adiabatic approximation, which is basic to the separation of most "fast" and "slow" modes; in Section III we consider general aspects of radiationless transitions and discuss some broad classes of radiationless processes; in Section IV we review the methods of calculation; and finally in Section V we discuss the present trends.

II. THE BORN-OPPENHEIMER ADIABATIC APPROXIMATION

This approximation is well known for the separation of the many-body problem of electronic and atomic motion into two separate many-body problems. It has also been widely applied to separate other "fast" and "slow" subsystems, for example vibrational and rotational modes of molecules and local modes (vibrons) and normal modes (phonons) of crystals. In fact, a further separation into a hierarchy of subsystems is possible, for example the double adiabatic approximation in which the orbital states of the electrons are "fast"; the vibrons, "intermediate"; and the phonons "slow". In the following we shall consider a system subdivided into two subsystems: one, "fast"; the other "slow".

The system is described according to quantum mechanics by a wavefunction $\psi(r,R)$ which satisfies the time-independent Schrodinger equation:

$$H\psi(r,R) = \mathcal{E}\psi(r,R) \quad , \tag{1}$$

where H is the Hamiltonian of the system with eigenvalue $\mathcal{E}$; r specifies the collective position coordinates of the fast subsystem and R specifies the collective position coordinates of the slow subsystem. The basic idea is that the eigenstates of the fast subsystem can be obtained by taking the coordinates R as fixed. Thus, we have a time-independent Schrodinger equation for the fast subsystem:

$$H_f\phi(r;R) = E(R)\phi(r;R) , \qquad (2)$$

where $H_f = H - T_s$ and T_s is the kinetic energy operator of the slow subsystem; and in which its eigenfunction $\phi(r;R)$ and eigenvalue E(R) both have parametric dependence on the coordinates, R, of the slow subsystem. In other words these are smooth functions of R and as long as the changes in R occur slowly (adiabatically) no transitions between eigenstates of the fast subsystem occur.

The total wavefunction is then chosen of the following form:

$$\psi(r,R) = \chi(R)\phi(r;R) , \qquad (3)$$

where $\chi(R)$ is the eigenfunction for the slow subsystem. Substitution of eq. (3) into eq. (1) and making use of eq. (2) and of the explicit form of T_s, we obtain:

$$[-\frac{\hbar^2}{2M}\nabla_R^2 + E(R)]\chi(R) = \mathcal{E}\chi(R) , \qquad (4)$$

if we neglect certain terms to be discussed. The dynamics of the slow subsystem is thus governed by the effective potential E(R) which we identify as the adiabatic potential. We emphasize that E(R) is of course dependent on the state of the fast subsystem.

The terms neglected in obtaining eq. (4) are the following:

$$\frac{\hbar^2}{2M}[2\nabla_R\chi(R)\nabla_R\phi(r;R) + \chi(R)\nabla_R^2\phi(r;R)] . \qquad (5)$$

The diagonal components of the first term are zero for a system with real eigenfunctions (non-magnetic) whose particles are conserved; the diagonal component of the second term is smaller than the kinetic energy term for the fast subsystem by the ratio of the energies of the slow and fast subsystems, for example, for electron and phonon subsystems, by the ratio of electron and atomic masses. The off-diagonal components of eq. (5) are identified as the interaction between the fast and slow subsystems, for the same example, the electron-phonon interaction, and will be further discussed in the Section III.

The plot of E(R) is shown in Fig. 1 for two states of the fast subsystems and with eigenvalues of the slow subsystem also shown. Radiative transitions are shown vertically in accordance with the Franck-Condon principle and occur not far from equilibrium values of R. Thus, in computing transition matrices, the

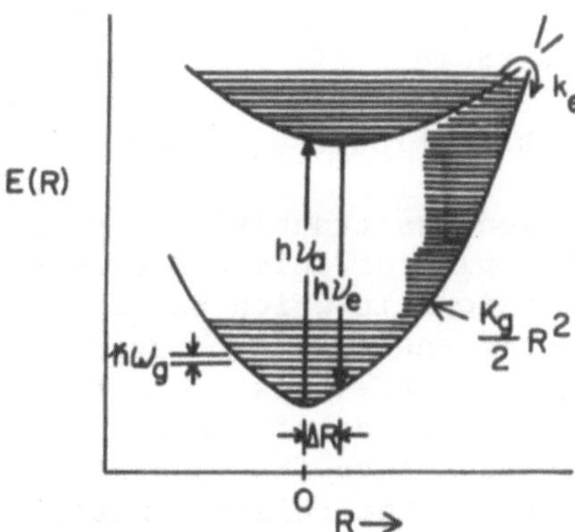

Fig. 1. Adiabatic potential curves diagram including vibrational levels; and radiative and radiationless transitions.

approximation of $\phi(r;R)$ as $\phi(r;R_o)$ where R_o is a constant has frequently been used. This is the "crude" adiabatic approximation and is sometimes referred to as the Condon approximation (see glossary, Appendix A). Radiationless transitions are shown as occurring in the region of R where crossing of the E(R)'s for the two states of the fast subsystem is imminent. This is the region where the Condon approximation is no longer a good approximation and in fact where the adiabatic approximation itself is failing. In the next section we consider the radiationless transition in more detail.

III. THE RADIATIONLESS TRANSITION

III.A. Introduction to Radiationless Transitions

The radiationless transition is any transition of a system in which energy is not exchanged with the radiation field (i.e. electromagnetic field). More precisely, we consider a system composed of the following three subsystems: (a) the electromagnetic field, (b) a subsystem of interest - designated as the fast subsystem and (c) a heat bath, which is large compared to the fast subsystem, that is, a subsystem having a more dense energy level spectrum than the fast subsystem - called the slow subsystem.

The radiationless transition is a transfer of energy between the fast subsystem of interest and the slow subsystem, in contradiction to the radiative process which is the transfer of energy between the electromagnetic field and either subsystem.

III.B. Some Types of Radiationless Transitions

A class of radiationless transitions is the class of inelastic collisions. Examples of this class of transitions are: (a) Electron-atom collisions in which the atomic state and the relative electron kinetic energy are changed. Here a portion of energy is exchanged between the relative motion of the external electron and the internal motion of the atom or vice versa. In such processes the division into fast and slow subsystems depends on the relationship between the relative motion of the scattered particles (electron and atom) and the internal motion of the atom. (b) Collision between two atoms, for example, in the case of charge exchange collision reactions. Here the relative motion is usually the slow subsystem, and the internal motion is the fast subsystem. (c) Similar to the example of atomic collisions is the case of molecular collisions and of atomic-molecular collisions. (d) Similar to the example of electron-atom collision is the case of the "collision" of an electron in the conduction band with an imperfection in the crystal lattice.

A second class of radiationless transitions consists of those processes which exchange energy between different modes of a molecule. Here examples include: (a) energy exchange between high and low energy vibrational modes of a molecule , for example, between the stretching modes and bending modes of a molecule, with the high energy mode being the fast subsystem ; or (b) exchange of energy between vibrational modes and rotational modes. Usually the rotational modes have the lower energy and are viewed as the slow subsystem.

We should note that these energy exchanges can take place in a quasi-isolated molecule if the molecule is large enough, or in molecules of any size embedded in a matrix. The matrix modes act as intermediate virtual coupling states for the molecules.

A third class of transitions is energy exchange transitions between local modes and crystal modes, for which we have the following examples: (a) In crystals there is the possibility of energy exchange between a local mode at an imperfection of the crystal and the phonon modes, in particular, the acoustic phonons. (b) In molecular crystals, energy can be exchanged between the intramolecular vibrational modes, the fast subsystem, and the crystal phonon modes - the slow system.

(c) In polar crystals, we could consider the exchange of energy between the optical modes and the acoustic modes with the optical modes being the fast subsystem.

A final class of radiationless transitions is the interchange of electronic and vibrational and/or rotational energy. Here we are concerned with the electronic energy being converted into crystal phonon energy or molecular vibrational and/or rotational energy, with the electronic subsystem usually being the fast subsystem.

III.C. Radiationless Transitions Revisited

A transition problem in general and a radiationless transition in particular have the following general format. The first quantity we must have is a Hamiltonian, H, which can be divided as follows

$$H = H_o + H' \quad , \tag{6}$$

where for some region of coordinate space and/or time H' can be neglected. The second aspect of the problem is that the coordinate space can be divided into two subspaces denoted by r and R, where r represents a complete set of coordinates for the fast subsystem and R represents a complete set of coordinates for the slow subsystem. Thirdly, when H' is neglected, it is meaningful to speak of the energy levels and corresponding states of the fast subsystem between which radiationless transitions take place. These transitions are induced by the interaction H'.

III.D. Radiationless Electronic Transitions

In this subsection, we detail radiationless transitions in which electronic energy is lost or gained. It is usual to assume that the adiabatic separation has been made in which the fast subsystem is taken to be the electronic particles of interest and the slow subsystem is the nuclear motion. The radiationless transitions shown in Figs. 2A, 2B, and 2C are single electronic particle transitions. In Figs. 2A and 2B are shown electronic deexcitations, the first being non-activated and the second being activated with activation energy E_a. Fig. 2C shows radiationless relaxations after either a radiative excitation or deexcitation. These processes, shown in Fig. 2C, are just the exchange of local vibrational energy into crystal phonon energy. Figs. 2D, 2E, 2F, 2G, and 2H are double electronic particle transitions. Figs. 2D, 2E, and 2F are impact transitions in which one particle being a hot particle loses energy which is directly given to the second electronic particle. Fig. 2D represents impact excitation, Fig. 2E represents impact ionization of a defect, and Fig. 2F represents

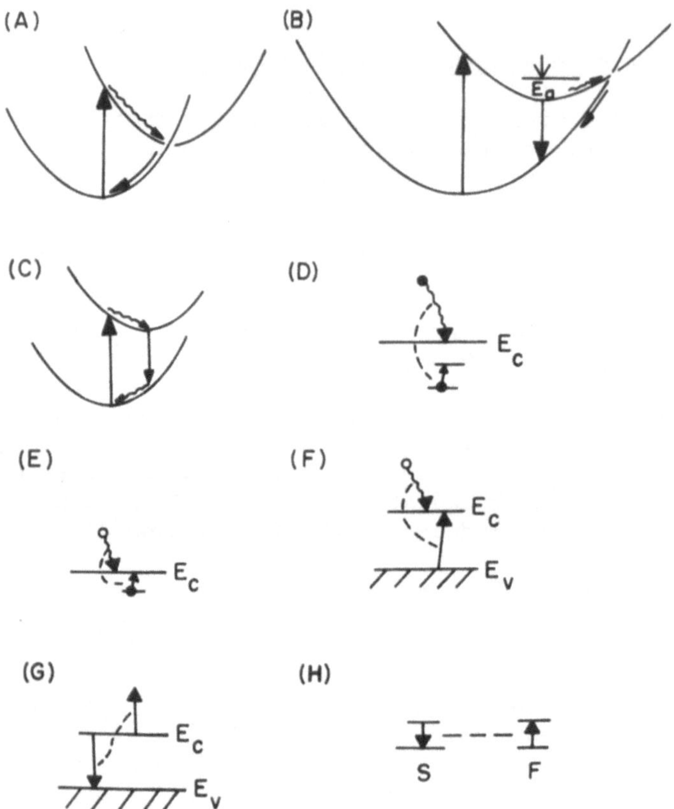

Fig. 2. Types of electron-phonon transitions.

an impact band-to-band transition. Fig. 2G represents the Auger transition which is the inverse process of the impact processes. Pictured in Fig. 2G is the case of a band-to-band deexcitation with the formation of a hot carrier. The final process shown in Fig. 2H is a resonance transfer process. Here energy is transferred from the sensitizer, S, to the fluorescer, F. In all of the processes represented in Figs. 2D to 2H the radiationless transition is a multi-electronic particle transition involving the nuclear motion only for energy conservation or for forming intermediate virtual coupling states.

III.E. The Problem of Radiationless Decay

There are several parts of this problem to be considered, when we need to make predictions of experimental results. First, what measurable properties distinguish the states of the fast subsystem? For example, in the charge exchange reaction

$$A^+ + B^- \to A + B \quad , \tag{7}$$

it is the charge of the individual atom. Without a clearly measurable characteristic, we cannot discuss, in particular, a radiationless transition or any type of transition. In this example, eq. (7), the true stationary states of H are mixtures of the states on the left and the right of eq. (7), and no transition can take place between these mixtures. However, when asking about the charge of A and/or B, the corresponding operator for this measurement does not commute with the Hamiltonian H, but only with H_o. Therefore, the state on the right or the left is a mixture of different energy states, each of which has a different time evolution. Since the measurement states, that is the states in which the system is prepared, are mixtures of different energy states, we must specify the particular preparation of the system. We can, for example, uniquely specify this preparation as that of a pure state which diagonalizes a particular operator which does not commute with H. Or, on the other hand, we can assume that it is not in a pure state, and then in this case, we can specify the preparation of the system by specifying the quantum statistical density matrix of the state. In general, if a density matrix, ρ, is specified, it is usual to assume that it can be approximated by a product of $\rho^{(f)}$ - the density matrix of the fast subsystem, and of $\rho^{(s)}$ - the density matrix of the slow subsystem. Two general cases have been studied; in both cases $\rho^{(s)}$ is taken to be the Boltzmann equilibrium distribution, and $\rho^{(f)}$ in one case is also taken to be the Boltzmann equilibrium distribution whereas in the other case, $\rho^{(f)}$ is given by a master equation defining a non-equilibrium distribution.

A third problem is the nature of H' and whether or not a model H' should be used. For example, in the adiabatic separation, we have H' given by:

$$H'\chi(R)\phi(r;R) = T_s\chi_s(R)\phi(r;R) - \phi(r;R)T_s\chi(R) , \tag{8a}$$

$$H'\chi(R)\phi(r;R) = \frac{-\hbar^2}{2m}\left[\chi(R)\frac{\partial^2}{\partial R^2}\phi(r;R) + \frac{\partial\chi(R)}{\partial R}\cdot\frac{\partial\phi(r;R)}{\partial R}\right]. \tag{8b}$$

In many calculations the first term on the right of eq. (8b) is neglected, and the second term which is just the electron-phonon interaction for the case of electronic radiationless transition, can, in some cases, be reduced to

$$\int \phi^* \frac{\partial\phi}{\partial R}\,dr\,\frac{\partial\chi}{\partial R} = AR\,\frac{\partial\chi}{\partial R} . \tag{9}$$

In this reduction the electronic part of the matrix element has been written as a power series in R. We must also note that the first term, although negligible when the adiabatic potential curves are well separated in energy, is not negligible near crossing or pseudo-crossing (places where the curves are close together). We can broaden the concept of adiabatic potential, $E_m(R)$ to a generalized adiabatic potential $V_n(R)$, where $V_n(R)$ is the nth diagonal element of the matrix

$$E_m(R)\delta_{mm'} - \frac{\hbar^2}{2m}\int dr\phi_m^*(r;R)\,\frac{\partial^2}{\partial R^2}\,\phi_{m'}(r;R) , \tag{10}$$

with $\chi_{nj}(R)$ satisfying

$$\left[-\frac{\hbar^2}{2m}\frac{\partial^2}{\partial R^2} + V_n(R)\right]\chi_{nj}(R) = E_{nj}\chi_{nj}(R) , \tag{11}$$

in the general adiabatic approximation, whose wavefunctions are

$$\psi_{nj}(r;R) = \chi_{nj}(R)\sum_m a_m^{(n)}\phi_m(r;R) . \tag{12}$$

The $\phi_m(r;R)$ would satisfy eq. (2), and the $a_m^{(n)}$ are given by diagonalization of eq. (10). It should be noted that the usual adiabatic approximation breaks down at crossings.

In the general adiabatic separation the only non-general adiabatic term is a term like the second term on the right of eq. (8b).

A final example of the problems involved in the radiationless transition is the method by which we find the transition rate. For example, for the weak coupling case, that is, H' small everywhere for all times, we can use the usual perturbative calculation. However if for some region of R-space and/or for some period of time, H' is not small, then other approximation schemes would be more appropriate, for example, a scheme like that of the Landau-Zener-Rosen theory.

III.F. Summary

We can sum up the radiationless transition problem as follows: If we let n and m index states of the fast subsystem and j and k index states of the slow subsystem, then radiationless transition is a transition from the state ψ_{nj} to the state ψ_{mk} where $E_{nj} = E_{mk}$.

IV. SOME GENERAL METHODS USED IN RADIATIONLESS DECAY CALCULATIONS

IV.A. Introduction

The problem of determining measurable quantities connected with the radiationless transition involves two aspects. The first is the problem of finding the quantum mechanical transition probabilities for a complex multi-particle system, in which part of the problem is to correctly model the interaction between the fast and slow subsystems so that the important effects are included. The second aspect is to incorporate the thermodynamics. Here again, we need to approximate because of the complexities, but we must also incorporate the important statistical effects.

It is usual, in the weak coupling limit, that is, the limit where H' is small for all values of R and for all times, to use a perturbation calculation related to Fermi's golden rule to find the transition probability. We will refer to this as the golden rule. The statistics are then incorporated using a density matrix formalism, with the slow subsystem considered in a Boltzmann distribution and the fast subsystem is either treated using equilibrium statistics, or using the Pauli master equation - nonequilibrium statistics. In addition, to the use of the golden rule, in the method of Landau-Zener-Rosen the transition probability is calculated in a non-quantum mechanical perturbative way. Here, the approximation is to the form of the interaction not the transition probability for the given form. We will discuss these methods later in this section. In addition, we will briefly mention a method which uses the correlation function and a method which uses an effective Hamiltonian for dissipative systems.

IV.B. The Landau-Zener-Rosen Method

Although this method is not limited to two interacting levels of the fast subsystem, it is both easier to understand the method in that situation and is most often used in that approximation. The starting assumption is that only adiabatic levels which cross or become very close together (that is, a psuedo-crossing) can interact strongly, and that each crossing acts independently.

The basic philosophy of this method is the following: when the Born-Oppenheimer adiabatic approximation is used to separate the motion of the fast and slow subsystems, the simple adiabatic potential curves can cross or almost cross at specific points in the R-configurational space. Furthermore, at these points non-adiabatic transitions are viewed as being caused by a perturbation consisting of the terms neglected in the adiabatic separation.

In particular, we assume an adiabatic separation of the fast and slow subsystems and in the energy range of concern that only two levels of the fast subsystem are important, say levels 1 and 2. In what follows, we will outline the procedure where the slow subsystem is treated classically. Delos, Thorson and Knudson have given derivations of the Landau-Zener-Rosen theory in a semi-classical formulation and in a momentum-space formulation. In the classical formulation the simple adiabatic states are just the electronic functions. That is,

$$\psi_1^{(a)}(r,R) = \phi_1(r;R) \text{ and } \psi_2^{(a)}(r,R) = \phi_2(r;R). \tag{13}$$

We assume that the motion in the R-space can be specified classically and is unchanged to first order. Therefore we have in the simplest case $R = vt$, where v is the constant velocity in the region of R-space where the crossing takes place.

In the following, we will use the notation:

$$H = H_o + H' , \tag{14a}$$

where H' is the non-adiabatic part of the complete Hamiltonian, H,

$$H_o\phi_i(r;R) = E_i(R)\phi_i(r;R), \tag{14b}$$

$$H\phi_i(r;R) = E_i(R)\phi_i(r;R) + E_{12}(R)\phi_j(r;R) , \tag{14c}$$

where $i,j = 1,2$ and $i \neq j$,

$$\int \phi_i(r;R)\phi_i(r;R)dr = N, \tag{14d}$$

$$\frac{\partial}{\partial t} \phi_i(r;R) = 0, \tag{14e}$$

and

and

$$H\Psi(r,R,t) = i\hbar \frac{\partial\psi(r\ R,t)}{\partial t} , \tag{14f}$$

where $\psi(r,R,t)$ is the wavefunction of the system including the effect of H', and is given by

$$\Psi(r,R,t) = C_1(t)e^{\frac{i}{\hbar}\int_{-\infty}^{t} E_1(R(t))dt}\phi_1(r;R) + C_2(t)e^{\frac{i}{\hbar}\int_{-\infty}^{t} E_2(R(t))dt}\phi_2(r;R). \tag{15}$$

We can derive, following Zener, the Landau-Zener equations which have the form:

$$\begin{aligned} i\hbar \frac{dC_1}{dt} &= -E_{12}(R(t))e^{-\frac{i}{\hbar}\int_{-\infty}^{t}[E_1 - E_2]dt}C_2 \\ i\hbar \frac{dC_2}{dt} &= -E_{12}(R(t))e^{\frac{i}{\hbar}\int_{-\infty}^{t}[E_1 - E_2]dt}C_1 . \end{aligned} \tag{16a}$$

These equations are to be solved using the initial conditions

$$C_1(-\infty) = 1 \quad \text{and} \quad C_2(-\infty) = 0 . \tag{17}$$

The Rosen-Zener theory differs only in that using a change of the time variable, allows eqs. (16a) to be rewritten in the form

$$\begin{aligned} i\frac{dC_1}{d\tau} &= -\exp[-2i\int_{-\infty}^{\tau} T(S)dS]C_2 \\ i\frac{dC_2}{d\tau} &= -\exp[2i\int_{-\infty}^{\tau} T(S)dS]C_1. \end{aligned} \tag{16b}$$

A general derivation of eqs. (16b) was given by Dinterman and Delos. The problem is to solve either eqs. (16a) or (16b) for C_1 and C_2. If we do a simple perturbation solution of these equations as Henry and Lang have done, we get the usual perturbative solution of the original problem. However, what has been done is to exactly solve eqs. (16a) and (16b) for special cases, that is, where E_{12} and $E_1 - E_2$ are taken to be

$$\begin{aligned} E_{12}(R) &= \text{constant} = \beta \\ E_1(R(t)) - E_2(R(t)) &= \alpha t , \end{aligned} \tag{18a}$$

near the crossing; or where T(S) is taken to be

$$T(S) = [\alpha \cos\beta S]^{-1}. \tag{18b}$$

In eqs. (18a) and (18b), α and β are parameters which are to be chosen so that E_{12} and $E_1 - E_2$ or T(S) are good approximations for the potentials of the problem at hand.

To sum-up, the Landau-Zener-Rosen theory assumes interaction between states of the first subsystem in those regions of R-space where the energy difference is small. It is assumed that the full time-dependent wavefunction for the system can be written as a linear combination of the interacting states. Using the time-dependent Schrodinger's equation, we can derive coupled equations for the coefficients of the linear combination of the full wavefunction. The most useful aspect of the Landau-Zener-Rosen theory is that the equations are solved exactly for an approximate interaction rather than perturbatively for an exact interaction. Note that taking the probability for a transition found in the Landau-Zener-Rosen theory and expanding it in a power series of E_{12} and keeping only the lowest order term gives the same thing as the lowest order perturbation solution for the same problem.

The Landau-Zener-Rosen theory can be classified as a pure state theory in contradistinction to a statistical theory. It is most easily applied to collision problems. The Landau-Zener-Rosen theories are important in those cases for which H' is large in some region of R-space during some period of time, for which perturbation theory would be inappropriate.

IV.C. The Density Matrix Method

Both equilibrium and non-equilibrium statistical treatments have been used. It is the thermodynamics of the problem which is focused upon. In the non-equilibrium case only the fast subsystem is taken to be in non-equilibrium and the slow subsystem is assumed to be in equilibrium and to have a Boltzmann distribution. As usual the Hamiltonian is split into two parts.

$$H = H_o + H' , \tag{19a}$$

with H' being the non-adiabatic part of the Hamiltonian. The Schrodinger equation for H_o is then

$$H_o \psi_{nj}^{(a)}(r,R) = E_{nj} \psi_{nj}^{(a)}(r,R), \tag{19b}$$

where

$$\psi_{nj}^{(a)}(r,R) = \chi_{nj}(R)\phi_n(r;R), \tag{19c}$$

and

$$\Psi(r,R,t) = \sum_{n,j} a_{nj}(t)\psi_{nj}^{(a)}(r,R) , \tag{19d}$$

where $\Psi(r,R,t)$ is the state of the system including the non-adiabatic effects. Then the density matrix is defined as

$$\rho_{nj;mk}(t) = a^*_{mk}(t)a_{nj}(t) . \tag{20}$$

We assume that the density matrix can be written as a product

$$\rho_{nj;mk}(t) = \rho^{(f)}_{nm}(t)\rho^{(s)}_{nj;mk}(t), \tag{21}$$

where $\rho^{(f)}$ is the density matrix for the fast subsystem and $\rho^{(s)}$ is the density matrix for the slow subsystem. Note that the assumption that eq. (21) is a good approximation for the density matrix is equivalent to the assumption that

$$[H_o,H'] \approx 0. \tag{22}$$

Further, the statement that the slow subsystem is in equilibrium implies that

$$\rho^{(s)}_{nj;mk} = 0 \text{ if } n \neq m \text{ and/or } k \neq j, \tag{23a}$$

and

$$\rho^{(s)}_{nj;nj} = N \exp[-(E_{nj}-E_{no})/kT], \tag{23b}$$

where

$$N = \sum_j \exp[-(E_{nj}-E_{no})/kT].$$

Now using the general theory of quantum statistical mechanics we can write an equation of motion for $\rho^{(f)}_{nn}(t)$. In the weak coupling limit this equation takes the form of the Pauli master equation, which is

$$\frac{d\rho^{(f)}_{nn}}{dt} = \sum_m k_{nm}\exp[E_{no}/kT]\{\exp[-E_{no}/kT]\,\rho^{(f)}_{mm} + - \rho^{(f)}_{nn}\exp[-E_{mo}/kT]\}, \tag{24}$$

where k_{nm} is called the rate constant for the problem. It is given by

$$k_{nm} = \frac{2\pi}{\hbar} \sum_j \sum_k \rho^{(s)}_{mk;mk}|\langle nj|H'|mk\rangle|^2\delta(E_{nj}-E_{mk}). \tag{25}$$

A large part of the work done has involved choosing a simple form for H' and working out the consequences. Some of the forms chosen have been, for example,

$$H' = Cre^{-aR}, \tag{26a}$$

which models, according to Lin, a system in which the important non-adiabatic term arises from the repulsive forces in V(r,R), or

$$H' = rF(R), \tag{26b}$$

for a system having only one excited state of the fast subsystem, and

$$H' = ArF(R) + Br^2F(R), \tag{26c}$$

for a system having two excited states of the fast subsystem, interacting with the slow subsystem and also with each other.

Let us turn to the case of equilibrium statistics. This is the case to which the radiationless problem has been reduced in the past. That is, both the fast and slow subsystems are assumed to be in equilibrium. This is applicable so long as we are concerned with times short enough so that the percent change in the number of excited states of the fast subsystem is negligible.

IV.D. Effective Hamiltonian and Correlation Function Methods

In the effective Hamiltonian and correlation function methods, we are not concerned with eliminating the complexities of the problem per se, but insuring that in the basic formulation we have included all of the complexities that have important effects on the measurable quantities.

The effective Hamiltonian method essentially assumes that we are dealing with a dissipative quantum system and uses recent developments in the theory of quantum dissipative systems. The basic idea is that the Hamiltonian for the system can be written in the form,

$$H = H_o + \Delta - \frac{i}{2}\Gamma , \tag{27}$$

where Δ is Hermitian and is the level shift operator arising from H' and Γ is Hermitian and is the damping operator arising from H'. Γ includes fundamentally the natural half-width of the decaying states.

The correlation function method arises from linear response theory, the primary aspect of which is the inclusion of coherence effects. Let N_α denote the number of particles of the fast subsystem in the state α, $\overline{N_\alpha}$ denotes the equilibrium value of N_α and < > represents the equilibrium statistical average. Then we can consider two related problems. The first is the relaxation problem characterized by a rate constant k_α, where

$$\frac{d\Delta N_\alpha}{dt} = -k_\alpha \Delta N_\alpha \quad , \tag{28}$$

with $\Delta N_\alpha = N_\alpha - \overline{N_\alpha}$. The second problem is the response of the system to an external force having a potential energy of the form

$$H_F = aN_\alpha \ , \tag{29}$$

which induces a change in N_α given by

$$\frac{d\Delta N_\alpha}{dt} = G_\alpha a , \tag{30}$$

where G is defined as the rate coefficient. It has been shown, (see for example S. Fischer), that

$$k_\alpha = kTG_\alpha(0)/\langle N_\alpha \rangle ,$$

where

$$G_\alpha(0) = \int_0^\infty dt \int_0^{1/kT} d\lambda \langle N_\alpha(-t-ih\lambda)\dot{N}_\alpha(0)\rangle .$$

In the correlation function method, we are interested in finding k_α, the rate constant, when coherence effects are included.

IV.E. Summary

At the present time there are many methods being investigated. Each method focuses primarily on one or only a few aspects of the complex problem of the radiationless decay. In this section, we have only mentioned a few of the methods being used.

V. SUMMARY OF PRESENT TRENDS

The previous discussions in this paper do not exhaustively cover the literature of the theories of radiationless processes. In order to give a flavor of the many aspects of this field we will add some further comments.

In the early literature on radiationless decay the Condon approximation was assumed. It has been since recognized that in order to do the calculation correctly a non-Condon approach is needed. Even though, in the case of radiative decay it is reasonable, but not totally correct, to use the Condon approximation, it is not reasonable in the cases of radiationless decay, since these are non-vertical transitions, that is, non-Franck-

Condon transitions. The non-Condon approximation introduces the problem of finding the wavefunctions for the fast subsystem as a function of R, the coordinates of the slow subsystem. One solution to this problem is the use of a model potential, for example, see the work of Lin, In this approach a simple form for the total potential V(r,R) is assumed in order to obtain the R-dependence of $\phi(r;R)$ in analytic form. A second approach to the non-Condon problem is to use a polaron-like approach, as used by Howgate. In this approach, the readjustment of the slow subsystem due to changes in the excitation of the fast subsystem are handled non-perturbatively by a transformation to polaron-like states. Other fast-slow subsystem interactions, which are weak, are handled perturbatively. These methods leave the problem of how to relate the model interactions to the real interactions of the system.

A third attempt is worth mentioning in this context, that is, a molecular orbital approach, in which the orbitals are used to calculate explicitly all the interactions for the system. In the cases to which this approach have been applied, the potential V(r,R) is usually given as a power series expansion of R. This type of calculation is limited to systems involving only a few atoms, for example, for the formaldehyde molecule as investigated by Lin. This method is adapted to transition metal and rare earth dopants interacting with nearest neighbor ligands in complexes or in crystals. In contradistinction to this general trend, we must mention the work of Struck and Fonger. They have pursued the general form of the statistical part of the transition rate using the Condon approximation with equilibrium Boltzmann statistics.

In spite of the difficulties involved, the general trend seems to be to include the non-Condon effects in current calculations.

A second trend is the inclusion of cooperative effects. We have mentioned two of these theories in Section IV. The first being the use of the quantum theory of dissipative systems; for example as used by Jortner. This method includes the natural line width in a fundamental way. Here, the quantum cooperative effects due to non-discrete levels are included. The second cooperative effect already mentioned is statistical correlation as used by Fischer. Here, the method of the correlation function includes the effects of non-independent relaxation processes.

In addition to these two cooperative effects, we should mention the work on superradiance and coherent states. For the superradiance, as discussed by Fong, it is assumed that because you have more than one fast subsystem these will interact and generate more slow particles than if each fast subsystem were isolated from the others. This idea is modeled on Dicke's theory of optical superradiance.

The research involving coherent states is just one of many ideas being culled from the theory of laser physics for use in understanding radiationless transitions. A recent example in this area is the work of Dodonov and Manko. In essence, the states of the system which participate in the radiationless process are assumed to have an indefinite number of slow particles. In order to deal with states with an indefinite number of particles, laser physics developed semiclassical states called coherent states, which retain some of the underlying quantum particle aspects, while featuring the classical field aspects of having an indefiniteness in the number of particles. The coherent states diagonalize the annihilation operator for the slow particles with normalization of unity. The theories which include coherence introduce aspects which are normally neglected, but which may be important in some radiationless transitions. This trend can be summed up as a search to see whether or not within the usual formulation of the radiationless problem all the important effects have been included.

A third general trend is the consideration of new types of radiationless processes which are becoming accessible to measurements because of the development of new experimental techniques. Some of the processes considered recently are radiationless recombination in semiconductors via deep centers; decay of vibrons, optical phonon and local modes; and intraband relaxation of hot carriers. The study of deep centers has developed because of the advancement in the photocapacitance technique by which radiationless transitions have been directly studied, notably by C. Henry and Lang. The other radiationless processes mentioned are now being studied because of advances in short time resolution spectroscopy. For example, the thermalization time constant for hot carriers in gallium arsenside has been measured by Shank, Fork, Leheny and Shah.

A final trend to be mentioned is a move away from the general theory of radiationless transitions to theories for particular systems. As examples we cite the following calculations: (a) rare earth dopant for particular crystal lattices, as studied by Perlin, e.g. $YA\ell O_3$:Yb,Tm; (b) consideration of large or of small molecular systems, e.g. of the latter case, the Lin and van Dijk calcuation for formaldehyde, mentioned earlier; (c) the Auger calculations for electron-hole recombination in semiconductors, for example, the Pilkuhn work for silicon; and (d) radiationless transfer from excited rare earth ions to vibrations of their ligands, considered by Ermolaev and Sveshnikova as resonance energy transfer.

Final aspects of the calcuations of radiationless decay are the use of: (a) computer techniques to get numerical results, for example by Struck and Fonger, (b) formal scattering theory techniques, e.g. the use of deflection functions by Bárány and (c) stationary state techniques, for example by B. Henry and Kasha and separately by Jortner. This last method in some sense eliminates the radiationless transition.

In some of the current research there is a trend away from strict adiabaticity. For example, the use of the sudden approximation in the regions of crossings of adiabatic surfaces. In some of these cases the perturbation causing the transition is a potential operator, not the usual interaction operator - this arises from initially making the crude adiabatic approximation and then the treating $[V(r,R)-V(r,R_o)]$ as the perturbation.

With regard to multiphonon relaxation, earlier analyses assumed equilibrium distribution among phonon levels for the initial electronic state; some current research focuses on dynamical multiphonon relaxation with non-equilibrium characterizing the initial phonon distribution.

Finally, we note that the crucial step when irreversibility is determined has not received adequate attention: for Auger processes, irreversibility occurs during the intra-band relaxation of the Auger particle; for the multiphonon relaxation of defects with local modes, during the decay of local modes into crystal modes; for cascade capture of electronic particles by defects, irreversibility occurs when occupancy is at an energy level from which re-emission into the continuum is improbable.

The preparation of this manuscript was in part supported under a grant from the U. S. Army Research Office-Durham.

APPENDIX A

Glossary of Terms

Accepting modes are those vibrational modes of the slow subsystem for which the non-adiabatic transition probability is proportional to the square of the overlap of wavefunctions of the mode for the two states between which the transition is taking place. The accepting modes act as sinks for the energy of the fast subsystem. See promoting modes definition.

Adiabatic approximation is the approximation to the state of a system composed of a fast and a slow subsystem as one element of the complete set of states having the form $\chi_{nj}^{(\alpha)}(R)\phi_n(r;R)$. That is

$$\psi_\alpha(r,R) = \chi_{nj}^{(\alpha)}(R)\phi_n(r;R).$$

See Born-Oppenheimer expansion. Note that a more complete terminology for the adiabatic approximation is the Born-Oppenheimer adiabatic approximation.

Adiabatic potential curve is a non-crossing effective potential energy curve for the motion of the slow subsystem in the adiabatic approximation. Adiabatic curves are not necessarily characterized by a single value for a given characteristic, α. In a collision, let α be the charge of molecule A, then the adiabatic curves are curves for which molecule A is charged for some regions of R-space, and uncharged for other regions. See the solid curves in Fig. 3. See diabatic curve.

Auger processes are two-electron transitions in which for the final state one of the electrons is in a continuum. Examples are auto-ionization transitions in atoms, and in semiconductors electron-hole recombination to form hot carriers. For the latter example the process is viewed as a three quasi-particle process.

Born-Oppenheimer expansion is the expansion of the state of a system composed of a fast and a slow subsystem in terms of the complete set of states having the form $\chi_{nj}^{(\alpha)}(R)\phi_n(r;R)$. That is,

$$\psi_\alpha(r,R) = \sum_{n,j} \chi_{nj}^{(\alpha)}(R)\phi_n(r;R).$$

See adiabatic approximation.

The *Condon approximation* is the approximation in which matrix elements involving wavefunctions of the fast subsystem are taken to be independent of the coordinates of the slow subsystem. In some instances, this approximation is interpreted as taking the

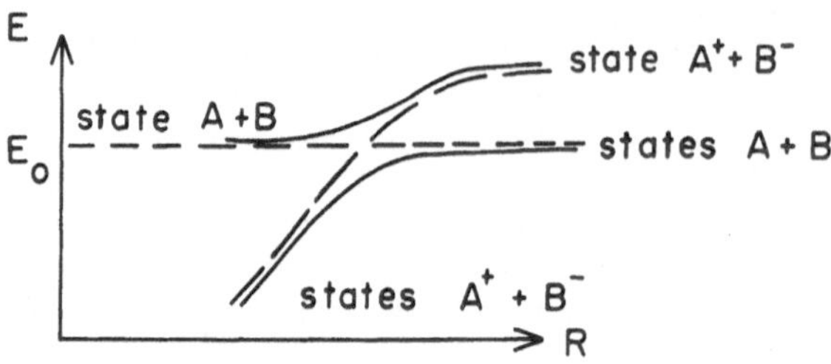

Fig. 3. Adiabatic (solid curve) and diabatic (dashed curves) curves for the collision of molecule A with molecule B at a given initial kinetic energy, E_0. R is the separation of the molecules.

fast subsystem wavefunctions themselves as independent of the coordinates of the slow subsystem. The latter has been referred to as the crude adiabatic approximation.

The *crude adiabatic approximation* is the approximation in which the wavefunctions for the fast subsystem are taken as independent of the coordinates of the slow subsystem. See the Condon approximation.

Diabatic curve is a effective potential energy curve for the motion of the slow subsystem in the adiabatic approximation. These curves can cross. Diabatic curves, in contradistinction to adiabatic curves, have a single value of a given characteristic α. The diabatic curves depend on which characteristic is chosen to define them. As an example, let the characteristic α, in a molecular collision, be the charge of molecule A, then the corresponding diabatic curves are curves for which molecule A has the same charge for all regions of R-space. See the dash curves in Fig. 3.

Double adiabatic approximation is the extension of the adiabatic approximation to systems composed of three subsystems: a fast subsystem with coordinates ρ, a intermediate subsystem with coordinates r, and a slow subsystem with coordinates R. In this approximation the wavefunction of system is assumed to be

$$\psi_{\alpha}(\rho,r,R) = \mu^{(\alpha)}_{njp}(\rho)\chi_{nj}(R;\rho)\phi_{n}(r;R,\rho).$$

Franck-Condon principle: An optical transition between states of fast subsystem is considered to occur on a time scale short compared to the period of the slow subsystem. Hence, the slow subsystem coordinates and the momentum remain fixed during the transition. This is often expressed by saying that Franck-Condon transitions are those which occur vertically on a configurational coordinate diagram.

Golden rule refers to transition rate formulas which are related to the classic Fermi golden rule of time dependent perturbation theory. We sketch a derivation here in order to emphasize the assumptions that are central to the question of the applicability of such formulas. In time dependent perturbation theory, as applied to time-independent problems, the Hamiltonian is split into two parts

$$H = H_o + H' ,$$

where H' is some small perturbation which causes transitions among the eigenstates of the zero-order Hamiltonian H_o. Then the eigenfunctions Ψ of H are written as an expansion

$$\Psi = \sum_n c_n(t) e^{-iE_n t/\hbar} \psi_n,$$

where the ψ_n are eigenfunctions of H_o. First-order perturbation theory then yields equations of motion for the expansion coefficients $c_n(t)$,

$$i\hbar \frac{dc^{(1)}_m}{dt} = \sum_n e^{i\omega_{mn}t} c^{(0)}_n H'_{mn} ,$$

where the superscripts on the c_n's denote the order of approximation,

$$\omega_{mn} = \frac{E_m - E_n}{\hbar} ,$$

and

$$H'_{mn} = \langle\psi_m|H'|\psi_n\rangle.$$

Now, if the system is in a definite initial state, say ψ_s at t = 0, then the zero-order coefficients are $c^{(0)}_k = \delta_{ks}$, so that the equations for the $c^{(1)}_n$'s can be easily integrated:

$$c_n^{(1)}(t) = -\frac{2i}{\hbar\omega_{ns}} H'_{ns}\, e^{i\frac{\omega_{ns}t}{2}} \sin\frac{\omega_{ns}t}{2} .$$

Then the transition probabilities are

$$|c_n^{(1)}(t)|^2 = \frac{2\pi}{\hbar^2} |H'_{ns}|^2\, t\left[\frac{1}{\pi}\frac{\sin^2\frac{\omega_{ns}t}{2}}{\frac{\omega_{ns}t}{2}}\right].$$

In the limit of large t, the factor in brackets becomes a δ-function. Thus, the transition rate is

$$W_{s\to n} = \frac{d}{dt}|c_n^{(1)}(t)|^2 = \frac{2\pi}{\hbar}|H'_{ns}|^2\delta(E_n-E_s). \qquad \text{(I)}$$

From here it is easy to get the usual Fermi golden rule for the total transition rate to all final states by multiplying by the density of final states $\rho_f(E_n)$ and integrating over E_n. The result is

$$W_s = \frac{2\pi}{\hbar} |H'_{ns}|^2\rho_f(E_s). \qquad \text{(II)}$$

Note that it must be assumed that the final states form a quasi-continuum in the vicinity of E_s. It is usual in applications to radiationless transitions to use eq. (I) rather than eq. (II) and to include a sum over final states and a thermal average over initial states. Hence, one usually finds the transition rate written in the form

$$W = \frac{2\pi}{\hbar} \underset{s}{\mathrm{Av}} \sum_n |H'_{ns}|^2\delta(E_n - E_s). \qquad \text{(III)}$$

We include such formulas in the class of golden rule formulas.

Note that it is necessary to assume that t is large enough so that

$$\frac{\sin^2 \cdot\frac{\omega_{ns}t}{2}}{\frac{\omega_{ns}t}{2}} \approx \pi\delta(\omega_{ns}),$$

and that t is small enough that the coefficients $c_n^{(1)}(t)$ for each term in the thermal average are not appreciably different from δ_{ns}.

Huang-Rhys Factor, $S_n^{(m)}$, is the ratio of the relaxation energy to the average vibrational energy of the slow subsystem when the system is in the nth state of the fast subsystem. More specifically, if $k_n(j)$ is the force constant for the nth state of the fast subsystem for the jth mode of slow subsystem, $\omega_n(j)$ is the angular frequency of the jth mode of nth state and if $\Delta R_{nm}(j)$ is the displacement between the minima of the nth and mth adiabatic potential curve for the jth mode then

$$S_n^{(m)} = \frac{\frac{1}{2}\sum_j k_n(j)[\Delta R_{nm}(j)]^2}{\frac{\hbar}{N}\sum_j \omega_n(j)} ,$$

where N is the total number modes involved in the transition. $S_n^{(m)}$ is Huang-Rhys factor used when the transition is from the state m to the state n.

Note that the Huang-Rhys factor at zero temperature is the average number of slow quanta for energy conservation in a Franck-Condon transition. It is not common to define S when the force constants are unequal nor to indicate the states between which the transition is taking place. Also if the force constants for the states n and m are equal then

$$S_n^{(m)} = S_m^{(n)} .$$

Internal conversion is a radiationless transition involving electronic states of the same multiplicity. See intersystem crossing.

Intersystem crossing is a radiationless transition involving electronic states of different multiplicity. See internal conversion.

The *non-adiabaticity operator* is the operator whose matrix elements are neglected in making the adiabatic approximation. If it is denoted H_{NA}, we can write

$$H_{NA}\phi(r;R)\chi(R) = -\frac{\hbar^2}{M}\vec{\nabla}_R\phi(r;R)\cdot\vec{\nabla}_R\chi(R) - \frac{\hbar^2}{2M}\chi(R)\nabla_R^2\phi(r;R) .$$

Thus, it represents the action of the kinetic energy operator for the slow subsystem on the wavefunction of the fast subsystem.

Non-crossing rule states that the adiabatic potential curves cannot cross, since the neglected non-adiabatic terms which couple together the states of the fast subsystem will cause the curves to be displaced away from one another. Note that if the adiabatic curves did cross, then in that region of R-space near the crossing the adiabatic approximation would be invalid.

Promoting modes are those vibrational modes of the slow subsystem for which the non-adiabatic transition probability is proportional to the square of the matrix element of conjugate momentum of the mode. The promoting modes are responsible for the non-Franck-Condon part of the transition of the fast subsystem.

Resonance Energy Transfer is the transfer of energy between molecules or ions by direct electromagnetic coupling, originally formulated by Förster for dipole-dipole interaction and later generalized by Dexter to include magnetic dipole and exchange interactions, clearly to be distinguished from radiative transfer. In contrast to exciton motion, resonance energy transfer occurs from the relaxed sensitizer (donor of energy) in its excited state to the relaxed fluorescer (acceptor of energy) in its ground state.

The *statistical limit* corresponds to cases for which a quasi-continuum exists for the final state of a transition.

Strong coupling case is the case when $S_n^{(m)}$ is greater than one, where $S_n^{(m)}$ is the Huang-Rhys factor.

Vibron are local vibrational modes, frequently intramolecular modes of molecular dopants.

Weak coupling case is the case when $S_n^{(m)}$ is less than one, where $S_n^{(m)}$ is the Huang-Rhys factor.

APPENDIX B. BIBLIOGRAPHY

A. Reviews

1. Robert Englman, Non-Radiative Decay of Ions and Molecules in Solids, (North Holland, Amsterdam, 1979).

2. Francis K. Fong, Theory of Molecular Relaxation, (Wiley, New York, 1975), see esp. chapters 3 and 5.

3. Bryan R. Henry and Michael Kasha, Radiationless molecular electronic transitions, Ann. Rev. Phys. Chem. 19, 161-192 (1968).

4. J. Jortner and S. Mukamel, Radiationless transitions, in International Review of Science, Physical Chemistry Series Two, Vol. 1, Theoretical Chemistry, edited by A. D. Buckingham and C. A. Coulson, (Butterworths, London, 1975), chapter 10, pp. 327-388.

5. P. T. Landsberg, Non-radiative transitions in semiconductors, Phys. Stat. Sol. 41, 457-489 (1970).

6. N. F. Mott, Recombination; A survey, Sol. State Elect. 21, 1275-1280 (1978).

7. R. Orbach and H. J. Stapleton, Electron spin-lattice relaxation, in Electron Paramagnetic Resonance, edited by S. Geschwind, (Plenum, 1972), chapter 2, pp. 121-216.

8. Yu. E. Perlin, Modern methods in the theory of many-phonon processes, Sov. Phys. Uspekhi 6, 542-565 (1964) [Usp. Fiz. Nauk 80, 553-595 (1963)].

9. A. M. Stoneham, Theory of Defects in Solids, (Oxford, 1975), chapter 14, Non-radiative processes and the interaction of free carriers with defects, pp. 477-551.

10. A. M. Stoneham, Non-radiative processes in insulators and semiconductors, Phil. Mag. 36, 983-997 (1977).

B. Early Papers on Radiationless Processes and Related Theoretical Methods

1. D. L. Dexter, C. C. Klick and G. A. Russell, Criterion for the occurrence of luminescence, Phys. Rev. 100, 603-605 (1955).

2. J. Frenkel, On the transformation of light into heat in

solids. I, Phys. Rev. 37, 17-44 (1931).

3. J. Frenkel, On the transformation of light into heat in solids. II, Phys. Rev. 37, 1276-1294 (1931).

4. Kun Huang and Avril Rhys, Theory of light absorption and non-radiative transitions in F-centers, Proc. Roy. Soc. (London) A 204, 406-423 (1950).

5. Ryogo Kubo, Thermal ionization of trapped electrons, Phys. Rev. 86, 929-937 (1952).

6. Ryogo Kubo and Yutaka Toyozawa, Application of the method of generating function to radiative and non-radiative transitions of a trapped electron in a crystal, Prog. Theor. Phys. 13, 160-182 (1955).

7. Melvin Lax, The Franck-Condon principle and its applications to crystals, J. Chem. Phys. 20, 1752-1760 (1952).

8. G. Rickayzen, In the theory of the thermal capture of electrons in semiconductors, Proc. Roy. Soc. A 241, 480-494 (1957).

9. H. D. Vasileff, Thermal ionization of impurities in polar crystals. I. Formal theory, Phys. Rev. 96, 603-609 (1954).

C. Papers on Which the Landau-Zener-Rosen Method is Developed or Applied

1. Anders Barany, Deflection functions for curve-crossing collisions, J. Phys. B 12, 2841-2855 (1979).

2. Gianni L. Bendazzoli, Mario Raimondi, Bruce A. Garetz, Thomas F. George and Keiji Morokuma, Semiclassical study of collision-induced predissociation: Comparison of the Landau-Zener model with the method of analytic continuation, Theoret. Chim. Acta. (Berl.) 44, 341-350 (1977).

3. John B. Delos, Walter R. Thorson and Seth A. Boorstein, Studies of the potential-curve-crossing problem. I. Analysis of Stuckelbergs method, Phys. Rev. A 4, 1052-1066 (1971).

4. John B. Delos and Walter R. Thorson, Studies of the potential-curve-crossing problem. II. General theory and a model for close crossings, Phys. Rev. A 6, 728-745 (1972).

5. John B. Delos, Walter R. Thorson and Stephen K. Knudson, Semiclassical theory of inelastic collisions, I. Classical

and semiclassical formulation, Phys. Rev. A 6, 709-720 (1972).

6. John B. Delos and Walter R. Thorson, Semiclassical theory of inelastic collisions. II. Momentum-space formulation, Phys. Rev. A 6, 720-727 (1972).

7. T. R. Dinterman and J. B. Delos, Generalization of the Rosen-Zener model of noncrossing interactions. I. Total cross sections, Phys. Rev. A 15, 463-474 (1977). II. Differential cross sections, Phys. Rev. A 15, 475-478 (1977).

8. Jean Heinrichs, Theory of electronic transitions in slow atomic collisions, Phys. Rev. 176, 141-176 (1968).

9. C. H. Henry and D. V. Lang, Nonradiative capture and recombination by multiphonon emission in GaAs and GaP, Phys. Rev. B 15, 989-1016 (1977).

10. E. C. Kemble and C. Zener, The two quantum excited states of the hydrogen moelcule, Phys. Rev. 33, 512-537 (1929).

11. John R. Laing, Thomas F. George, I. Harold Zimmerman and Ying-Wei Lin, The vibronic representation for collinear atom-diatom collisions: Two-state semiclassical model, J. Chem. Phys. 63, 842-851 (1975).

12. N. F. Mott, Multiphonon recombination processes, Phil. Mag. 36, 979-981 (1977).

D. Papers Dealing with the Correlation Function Approach to Radiationless Transitions, and Background for the Method

1. Sighart Fischer, Correlation function approach to radiationless transitions, J. Chem. Phys. 53, 3195-3207 (1970).

2. Francis K. Fong, Steven L. Naberhuis and Marvin M. Miller, Theory of radiationless relaxation of rare-earth ions in crystals, J. Chem. Phys. 56, 4020-4027 (1972).

3. E. W. Montroll, Nonequilibrium statistical mechanics, Lectures Theor. Phys. 3, 221-325 (1960).

4. Elliott W. Montroll and Kurt E. Shuler, Studies in nonequilibrium rate processes. I. The relaxation of a system of harmonic oscillators, J. Chem. Phys. 26, 454-464 (1957).

5. R. Zwanzig, Statistical Mechanics of irreversibility, Lectures Theor. Phys. 3, 106-141 (1960).

6. See also A.2 above.

E. Papers Applying a Perturbative Technique (e.g. The Golden Rule) to the Radiationless Transition Problem

1. A. D. Brailsford and Tai Yup Chang, Nonradiative decay of individual vibronic levels in large molecules, J. Chem. Phys. 53, 3108-3113 (1970).

2. S. H. Lin, Effect of temperature on radiationless transitions, J. Chem. Phys. 56, 2648-2653 (1972).

3. See also sections A and B, C.9 and Sections I and L.

F. Papers in which a Stationary State Approach to the Radiationless Transition Problem is Taken

1. Mordechai Bixon and Joshua Jortner, Intramolecular radiationless transitions, J. Chem. Phys. 48, 715-726 (1968).

2. Joshua Jortner and R. Stephen Berry, Radiationless transitions and molecular quantum beats, J. Chem. Phys. 48, 2757-2766 (1968).

3. William Rhodes, Radiationless transitions in isolated molecules. The effects of molecular size and radiation bandwidth, J. Chem. Phys. 50, 2885 (1969).

4. William Rhodes, Bryan R. Henry and Michael Kasha, A stationary state approach to radiationless transitions: Radiation bandwidth effect on excitation processes in polyatomic molecules, Proc. Nat. Acad. Sci. (USA) 63, 31-35 (1969).

5. See also A.3 above.

G. Papers Where Other Approaches Are Used

1. F. Auzel, Multiphonon interaction of excited luminescent centers in the weak coupling limit: Nonradiative decay and multiphonon side bands, in Luminescence of Inorganic Solids edited by Baldassare DiBartolo, (Plenum, New York, 1978), pp. 67-113.

2. Robert Englman, The status of discrete modes in multiphonon relaxation in solids, J. Chem. Phys. 66, 2212-2216 (1977).

3. V. L. Ermolaev and E. B. Sveshnikova, Nonradiative transitions as Forster's energy transfer to solvent vibrations, J. Luminescence, 20, 387-395 (1979).

4. T. H. Förster, Photochemische Primarprozesse bei Mehratomigen Molekulen, Festschrift Fur Electrochemie Berichte Der Bunsengesellschaft Fur Physikalische Chemie 56, 716 (1952).

5. Martin Gouterman, Radiationless transitions: A semiclassical model, J. Chem. Phys. 36, 2846-2853 (1962).

6. B. Halperin and R. Englman, Two-frequency description for the Jahn-Teller coupling of impurities in solids, Phys. Rev. B 9, 2264 (1974).

7. Melvin Lax, Cascade capture of electrons in solids, Phys. Rev. 119, 1502-1523 (1960).

8. S. H. Lin, Theory of vibrational relaxation and infrared absorption in condensed media, J. Chem. Phys. 65, 1053-1062 (1976).

9. S. H. Lin, H. P. Lin and D. Knittel, Effect of temperature and quencher concentration on vibrational relaxation in condensed media, J. Chem. Phys. 64, 441-451 (1976).

10. R. Pässler, Quantum-efficiency of multiphonon transitions according to the static coupling scheme, Phys. Stat. Sol. (b) 65, 561-569 (1974).

11. R. Pässler, Calculation of nonradiative multiphonon capture coefficients and ionization rates for neutral centers according to the static coupling scheme. I. Theory, Phys. Stat. Sol. (b) 68, 69-79 (1975).

12. R. Pässler, Calculation of nonradiative multiphonon capture coefficients and ionization rates for neutral centers according to the static coupling scheme. II. Alternative trap models, Phys. Stat. Sol. (b) 76, 647-659 (1976).

13. R. Pässler, Relationships between the nonradiative multiphonon carrier-capture properties of deep charged and neutral centers in semiconductors, Phys. Stat. Sol. (b) 78, 625-635 (1976).

14. N. Robertson and L. Friedman, Non-radiative transition probabilities, Phil. Mag. 36, 1013-1019 (1977).

15. Benjamin Sharf, Critical examination of the Herzberg-Teller theory, J. Chem. Phys. 55, 1379-1382 (1971).

16. M. D. Sturge, Temperature dependence of multiphonon non-

radiative decay at an isolated impurity center, Phys. Rev. B. 8, 6-14 (1973).

17. E. B. Sveshnikova and I. B. Neporent, Electronic energy radiationless transition as radiationless dipole-dipole energy transfer to vibrations of solvent molecules, in Luminescence of Crystals, Molecules and Solutions, edited by Ferd Williams, (Plenum, New York, 1973), pp. 188-195.

18. Yutaka Toyozawa, Multiphonon recombination processes, Sol. State Elect. 21, 1313-1318 (1978).

H. Papers Developing Theory and Applications Beyond the Condon Approximation

1. V. A. Kovarskii, Theory of nonradiative transitions in the "non-Condon" approximation. Low temperatures, Sov. Phys.- Solid State 4, 1200-1209 (1962) [Fiz. Tverd. Tela 4, 1636-1648 (1962)].

2. V. A. Kovarskii, I. A. Chaikovskii and E. P. Sinyavskii, Kinetic quantum equations for nonradiative recombinations, Sov. Phys.- Solid State 6, 1679-1989 (1965) [Fiz. Tverd. Tela 6, 2131-2145 (1964)].

3. V. A. Kovarskii and E. P. Sinyavskii, Theory of nonradiative transitions in crystals in the "non-Condon" approximation. High temperatures, Sov. Phys.-Solid State 4, 2345-2348 (1963) [Fiz. Tverd. Tela 4, 3202-3207 (1962)].

4. V. A. Kovarskii and E. P. Sinyavskii, Theory of radiationless transitions in the "non-Condon" approximation, Sov. Phys.- Solid State 6, 498-499 (1964) [Fiz. Tverd. Tela 6, 636-637 (1964)].

5. E. P. Sinyavskii and V. A. Kovarskii, A quasiclassical estimate of the cross section for multiphonon trapping in the non-Condon approximation for the deformation interaction, Sov. Phys.-Solid State 9, 1142-1149 (1967) [Fiz. Tverd. Tela 9, 1464-1472 (1967)].

I. Papers Dealing with Adiabatic and/or Diabatic Curves

1. M. R. Flannery, Ionic recombination, in Atomic Processes and Applications, edited by P. G. Burke and B. L. Moiseiwitsch, (North Holland, Amsterdam, 1976), chapter 12.

2. Helmut Gabriel and Knud Taulbjerg, Comment on electronic-state representations for atomic-collision problems, Phys.

Rev. A 10, 741-742 (1974).

3. William Lichten, Resonant charge exchange in atomic collisions, Phys. Rev. 131, 229-238 (1963).

4. William Lichten, Resonant charge exchange in atomic collisions. II. Further applications and extension to the quasi-resonant case, Phys. Rev. 139, A27-A34 (1965).

5. William Lichten, Molecular wave functions and inelastic atomic collisions, Phys. Rev. 164, 131-142 (1967).

6. Felix T. Smith, Diabatic and adiabatic representations for atomic collision problems, Phys. Rev. 179, 111-123 (1969).

7. J. von Neumann and E. Wigner, Uber das Verhalten von Eigenwerten bei adiabatischen Prozessen, Phys. Zschr. 30, 467-470 (1929), [reprinted in The Collected Works of John von Neumann, vol. 1, pp. 553-556].

8. Clarence Zener, Non-adiabatic crossing of energy levels, Proc. Roy. Soc. (London) A 137, 696-702 (1932).

J. Papers Containing Nuclear Configuration Space Evaluations of Nuclear Overlap Integrals and Matrix Elements of the Nuclear Position and Momentum Operators

1. Baurch Barnett and Robert Englman, Quantitative theory of luminescent centers in a configurational diagram model. II. Results and their interpretation, J. Luminescence 3, 55-73 (1970).

2. Robert Englman and Baruch Barnett, Quantitative theory of luminescent centers in a configurational diagram model. I. Description of the method, J. Luminescence 3, 37-54 (1970).

3. Robert Englman and Joshua Jortner, The Energy gap law for radiationless transitions in large molecules, Molec. Phys. 18, 145-164 (1970).

4. Karl F. Freed and Joshua Jortner, Multiphonon processes in the nonradiative decay of large molecules, J. Chem. Phys. 52, 6272 (1970).

5. Sheng Hsien Lin, Rate of interconversion of electronic and vibrational energy, J. Chem. Phys. 44, 3759-3767 (1966).

6. W. Siebrand, Radiationless transitions in polyatomic molecules. I. Calculation of Franck-Condon factors, J. Chem.

Phys. 46, 440-447 (1967).

7. Yitzhak Weissman and Joshua Jortner, Phonon dispersion effects on non-radiative transitions in solids, Phil. Mag. B 37, 21-34 (1978).

K. Papers in Which Particular Systems or Classes of Systems are Treated

1. I. S. Andriesh, V. Ya. Gamurar', D. N. Vylegzhanin, A. A. Kaminskii, S. I. Klokishner and Yu. E. Perlin, Electron-phonon interaction in $Y_3Al_5O_{12}$:Nd^{3+}, Sov. Phys.-Solid State 14, 2550-2558 (1973) [Fiz. Tverd. Tela 14, 2967-2979 (1972)].

2. Dianne Fletcher, Y. Fujimura and S. H. Lin, Calculation of rates of radiationless transitions in matrix-isolated CN, Chem. Phys. Lett. 57, 400-404 (1978).

3. V. Ya. Gamurar', Multiphonon radiationless relaxation in rare earth dopant ions, Opt. and Spect. 27, 524-525 (1969) [Opt. i Spekt. 27, 965-967 (1969)].

4. V. Ya. Gamurar', Yu. E. Perlin and B. S. Tsukerblat, Radiationless energy transfer in activated crystals, Sov. Phys.-Solid State 11, 970-974 (1969) [Fiz. Tverd. Tela 11, 1193-1199 (1969)].

5. V. Ya. Gamurar', Yu. E. Perlin and B. S. Tsukerblat, Multi-phonon radiationless transitions in rare-earth ion impurities, Bull. Acad. Sci. USSR, Phys. ser. 35, 1306-1309 (1971) [Izv. Akad. Nauk SSSR, Ser. fiz. 35, 1429 (1971)].

6. R. M. Gibb, G. J. Rees, B. W. Thomas, B. L. H. Wilson, B. Hamilton, D. R. Wight and N. F. Mott, A two stage model for deep level capture, Phil. Mag. 36, 1021-1034 (1977).

7. D. W. Howgate, Calculation of nonradiative electron transition rates in a lattice-localized-electron system. Phys. Rev. 177, 1358 (1969).

8. D. Knittel and S. H. Lin, Participation of rotational motion in vibrational relaxation of molecules in condensed media, Molec. Phys. 36, 893-906 (1978).

9. C. B. Layne, W. H. Lowdermilk and M. J. Weber, Multiphonon relaxation of rare-earth ions in oxide glasses, Phys. Rev. B 16, 10 (1977).

10. S. T. Lee, Y. H. Yoon, H. Eyring and S. H. Lin, A theoretical

investigation of symmetry-forbidden transitions in magnetic circular dichroism and electronic spectra of benzene, J. Chem. Phys. 66, 4349-4355 (1977).

11. S. H. Lin, Theory of electric field effect on electronic spectra and electronic relaxation with applications to F centers, J. Chem. Phys. 62, 4500-4524 (1975).

12. S. H. Lin, Study of vibronic, spin-orbit and vibronic-spin-orbit couplings of formaldehyde with applications to radiative and non-radiative processes, Proc. Roy. Soc. (London) A 352, 57-71 (1976).

13. S. H. Lin and R. Bersohn, Effect of partial deuteration and temperature on triplet-state lifetimes, J. Chem. Phys. 48, 2732 (1968).

14. R. Pässler, Temperature dependences of the nonradiative multiphonon carrier capture and ejection properties of deep traps in semiconductors, Phys. Stat. Sol. (b) 85, 203-215 (1978).

15. Yu. E. Perlin, Intra and inter-ionic multiphonon transitions in rare earth doped crystals, J. Luminescence, in press.

16. Yu. E. Perlin, A. A. Kaminskii, V. N. Enakii and D. N. Vylegzhanin, Nonradiative multiphonon relaxation in $Y_3Al_5O_{12}:Nd^{3+}$, Phys. Stat. Sol. (b) 92, 403-410 (1979).

17. M. H. Pilkuhn, Non-radiative recombination and luminescence in silicon, J. Luminescence 18/19, 81-87 (1979).

18. D. J. Robbins and A. J. Thomson, Non-radiative relaxation at metal-ion point deffects in solids, Phil. Mag. 36, 999-1012 (1977).

19. J. M. F. van Dijk, M. J. H. Kemper, J. H. M. Kerp, and H. M. Buck, *Ab initio* CI calculation of radiationless transition of formaldehyde, J. Chem. Phys. 69, 2462-2473 (1978).

L. Papers Dealing with Auger Processes

1. Albert Haug, Carrier density dependence of Auger recombination, Sol. State Elect. 21, 1281-1284 (1978).

2. P. T. Landsberg and M. J. Adams, Radiative and Auger processes in semiconductors, J. Luminescence 7, 3-34 (1973).

3. P. T. Landsberg and A. R. Beattie, Auger effect in semiconductors, J. Phys. Chem. Solids 8, 73-75 (1959).

4. P. T. Landsberg and D. J. Robbins, The first 70 semiconductor Auger processes, Sol. State Elect. 21, 1289-1294 (1978).

5. J. T. Rebsch, A combination of Auger and many-phonon processes in nonradiative recombination, Sol. State Commun. 31, 377-381 (1979).

6. Wolfgang Schmid, Experimental comparison of localized and free carrier Auger recombination in silicon, Sol. State Elect. 21, 1285-1287 (1978).

7. See also A.5 above.

M. Papers Using an Effective Hamiltonian Approach to Treat Dissipative Quantum Systems

1. V. V. Dodonov and V. I. Manko, Coherent states and the resonance of a quantum damped oscillator, Phys. Rev. A 20, 550-560 (1979).

2. Daniel M. Greenberger, A critique of the major approaches to damping in quantum theory, J. Math. Phys. 20, 762-770 (1979).

3. Daniel M. Greenberger, A new approach to the problem of dissipation in quantum mechanics, J. Math. Phys. 20, 771-780 (1979).

4. I. R. Sentzky, Dissipation in quantum mechanics. The harmonic oscillator, Phys. Rev. 119, 670-679 (1960).

N. Papers Using the Master Equation Approach

1. A. Blumen, S. H. Lin and J. Manz, Theory of vibrational energy transfer among diatomic molecules in inert matrices, J. Chem. Phys. 69, 881-896 (1978).

2. S. H. Lin, On the master equation approach of vibrational relaxation in condensed media, J. Chem. Phys. 61, 3810-3820 (1974).

O. Recent Condon Approximation Papers

1. W. H. Fonger and C. W. Struck, Temperature dependence of Cr^{+3} radiative and nonradiative transitions in ruby and emerald, Phys. Rev. B 11, 3251-3260 (1975).

2. C. W. Struck and W. H. Fonger, Transition rates in single $\hbar\omega$ models, J. Luminescence 18/19, 101-104 (1979).

P. Recent Non-Condon Approximation Papers

1. K. F. Freed and S. H. Lin, Nuclear Coordinate Dependence of Electronic Matrix Elements for Radiationless Transitions, Chem. Phys. 11, 409-432 (1975).

2. Y. Fujimura, H. Kono and T. Nakajima, Theory of Nonradiative Decays in the Non-Condon Scheme, J. Chem. Phys. 66, 199-206 (1977).

3. J. M. F. van Dijk, M. J. H. Kemper, J. H. M. Kerp, H. M. Buck, *Ab initio* CI calculation of radiative and non-radiative decay of formaldehyde ($^1A_2$) with application to its photochemical decomposition, Chem. Phys. Lett. 54, 353-356 (1978).

4. M. J. H. Kemper, J. M. F. van Dijk, H. M. Buck, *Ab initio* Calculation on the photochemistry of formaldehyde. The search for a hydroxycarbene intermediate, J. Am. Chem. Soc. (USA) 100, 7841-7846 (1978).

SPECTROSCOPY AND RADIATIONLESS DEACTIVATION

B. Di Bartolo

Boston College
Chestnut Hill, MA 02167, U.S.A.

ABSTRACT

This series of six lectures presents in a fundamental and comprehensive way some basic material necessary, in the view of the lecturer, to create a "common ground" for all the participants, in line with the purpose to give to the course the character of a didactical experience. The first lecture deals with the quantum theory of molecular systems. The adiabatic (Born-Oppenheimer) approximation is introduced and its implications for the use of the symmetry properties of the system are treated. The second lecture deals with the radiation field and the radiative processes it produces. The third lecture deals with the Franck-Condon principle and the basic processes of absorption, induced emission and spontaneous emission; vibrational-electronic processes are also treated. The fourth lecture presents simple models for molecular spectra leading to the Pekar formula and to the Huang and Rhys formula. The fifth lecture presents the basic mechanism for nonradiative relaxation, due to deviation from the adiabatic approximation. Finally, the last lecture deals with radiationless relaxation of ions in solids.

I. INTRODUCTION

In previous treatments, we have examined the problems of the interaction of radiation with atoms and molecules (1) and of the interaction of radiation with ions in solids(2). The task of this article is to extend such treatments to include other than radiative processes in the picture. The following steps are followed here: i) we treat first the most general molecular system consisting of an ensemble of nuclei and electrons, ii) we treat then briefly the radiative field and the radiative processes it produces, iii) we deal then with the interaction of the radiative field with a molecular system, iv) we consider simple models for molecular spectra, v) we deal, finally, with nonradiative processes and, vi) we examine these processes in solids.

II. QUANTUM THEORY OF A MOLECULAR SYSTEM

II.A. The Hamiltonian

The Hamiltonian of a molecular system, when spin orbit and other less important interactions are neglected, may be written as follows:

$$H = T_r + T_r + V(\vec{r}_i,\vec{R}_s) \quad , \tag{1}$$

where

$$T_r = \sum_{i=1}^{n} \frac{(\vec{p}_i)^2}{2m} = \text{kinetic energy of the electrons} \tag{2}$$

$$T_R = \sum_{s=1}^{N} \frac{(\vec{P}_s)^2}{2M} = \text{kinetic energy of the nuclei} \tag{3}$$

$$V(\vec{r}_i,\vec{R}_s) = V_{ee} + V_{nn} + V_{ne} \tag{4}$$

with

$$V_{ee} = \tfrac{1}{2} \sum_{\substack{i=1 \\ i\neq j}}^{n} \sum_{j=1}^{n} \frac{e^2}{|\vec{r}_i - \vec{r}_j|} \quad , \tag{5}$$

$$V_{nn} = \tfrac{1}{2} \sum_{\substack{s=1 \\ s\neq t}}^{N} \sum_{t=1}^{N} \frac{e^2 Z_s Z_t}{|\vec{R}_s - \vec{R}_t|} \quad , \tag{6}$$

$$V_{ne} = - \sum_{i=1}^{n} \sum_{s=1}^{N} \frac{e^2 Z_s}{|\vec{R}_s - \vec{r}_i|} \quad . \tag{7}$$

The Schroedinger equation is given by

$$H\Psi(\vec{r}_i,\vec{R}_s) = E\Psi(\vec{r}_i,\vec{R}_s) \tag{8}$$

or

$$T_r\Psi + T_R\Psi + V\Psi = E\Psi. \tag{9}$$

We assume at this point that the eigenfunctions Ψ of H have the following form, called a "Born-Oppenheimer product"

$$\Psi(\vec{r},\vec{R}) = \phi(\vec{R})\psi(\vec{r},\vec{R}), \tag{10}$$

where we have dropped the subscripts of the coordinates. Replacing (10) in (9) we obtain:

$$\phi(\vec{R})T_r\psi(\vec{r},\vec{R}) + T_R\phi(\vec{R})\psi(\vec{r},\vec{R}) + V(\vec{r},\vec{R})\phi(\vec{R})\psi(\vec{r},\vec{R}) = E\phi(\vec{R})\psi(\vec{r},\vec{R}). \tag{11}$$

But

$$T_r\phi(\vec{R})\psi(\vec{r},\vec{R}) = \phi(\vec{R})T_R\psi(\vec{r},\vec{R}) + \psi(\vec{r},\vec{R})T_R\phi(\vec{R}) - \\ - \sum_s \frac{\hbar^2}{M_s}\vec{\nabla}_s\phi(\vec{R}) \cdot \vec{\nabla}_s\psi(\vec{r},\vec{R}). \tag{12}$$

Therefore (11) becomes

$$\left\{- \sum_s \frac{\hbar^2}{M_s}\vec{\nabla}_s\phi(\vec{R})\cdot\vec{\nabla}_s\psi(\vec{r},\vec{R}) + \phi(\vec{R})T_R\psi(\vec{r},\vec{R})\right\} + \\ + \psi(\vec{r},\vec{R})T_R\phi(\vec{R}) + \phi(\vec{R})T_r\psi(\vec{r},\vec{R}) + V(\vec{r},\vec{R})\phi(\vec{R})\psi(\vec{r},\vec{R}) = E\phi(\vec{R})\psi(\vec{r},\vec{R}). \tag{13}$$

At this point, we shall neglect the terms in { } brackets; we shall say more about the validity of this (so-called "adiabatic") approximation later. Eq. (13) then becomes

$$\frac{\psi}{\phi}T_R\phi + \left[T_r + V\right]\psi = E\psi . \tag{14}$$

The terms in the [ ] brackets above represent the Hamiltonian of the system if the nuclei are kept fixed in space ($T_R = 0$); we call this Hamiltonian H_e

$$H_e = T_r + V(\vec{r},\vec{R}). \tag{15}$$

The functions $\psi(\vec{r},\vec{R})$ that enter the Born-Oppenheimer products are chosen to be eigenfunctions of H_e

$$T_r\psi(\vec{r},\vec{R}) + V(\vec{r},\vec{R})\psi(\vec{r},\vec{R}) = \varepsilon(\vec{R})\psi(\vec{r},\vec{R}). \tag{16}$$

We note that the energy eigenvalues depend parametrically on $\vec{R}$. Using (16), eq. (14) becomes:

$$T_R\phi(\vec{R}) + \varepsilon(\vec{R})\phi(\vec{R}) = E\phi(\vec{R}). \tag{17}$$

Eq. (17) determines the choice of the functions $\phi(\vec{R})$ that enter the Born-Oppenheimer products.

The possibility of using such products as representative of the eigenstates of the system is based on the fact that the terms in the {} brackets of eq. (13) are negligible; this implies that

$$\left|-\sum_s \frac{\hbar^2}{M_s}\left[\vec{\nabla}_s\phi\cdot\vec{\nabla}_s\psi + \phi T_R\psi\right]\right| \ll |\psi T_R\phi|. \tag{18}$$

We shall examine later the conditions under which this inequality holds.

For the moment, assuming the adiabatic approximation legitimate, we can say that the solution of the Schroedinger equation for the entire system reduces to the solution of the two equations.

$$T_r\psi_k(\vec{r},\vec{R}) + V(\vec{r},\vec{R})\psi_k(\vec{r},\vec{R}) = \varepsilon_k(\vec{R})\psi_k(\vec{r},\vec{R}) \tag{19}$$

$$T_R\phi_{k\ell}(\vec{R}) + \varepsilon_k(\vec{R})\phi_{k\ell}(\vec{R}) = E_{k\ell}\phi_{k\ell}(\vec{R}) \quad , \tag{20}$$

where we have attached proper subscripts to the eigenvalues and eigenfunctions. A stationary state of the system is then given by

$$\Psi_n(\vec{r},\vec{R}) = \psi_k(\vec{r},\vec{R})\phi_{k\ell}(\vec{R}), \tag{21}$$

where ψ_k and $\phi_{k\ell}$ are eigenfunctions of the "electronic" Hamiltonian

$$H_e = T_r + V(\vec{r},\vec{R}) \tag{22}$$

and of the "vibrational" Hamiltonian

$$H_v = T_R + \varepsilon_k(\vec{R}), \tag{23}$$

respectively.

II.B. The Meaning of the Adiabatic Approximation

The use of the adiabatic approximation has led us to the two eigenvalue equations (19) and (20) that we want now to consider again. First, eq. (19) can be written

$$H_e\psi_k(\vec{r},\vec{R}) = \varepsilon_k(\vec{R})\psi_k(\vec{r},\vec{R}). \tag{24}$$

The eigenfunctions of H_e represent the motion of the electrons when the nuclei are in fixed positions $\vec{R}$: the eigenfunctions and the energy eigenvalues depend parametrically on $\vec{R}$. This means that, as the positions $\vec{R}$ change in time, both ε_k and ψ_k follow that change "adiabatically", while the system continues to stay in the state represented by the quantum number k.

The validity of the approximation derives from the fact that, on account of their smaller mass, the electrons go through their orbits many times before the nuclei shift their positions by any considerable amount. This is expressed by the inequality (18) which is valid when

$$|\vec{\nabla}_s \psi(\vec{r}_i,\vec{R}_S)| << |\vec{\nabla}_s \phi(\vec{R}_s)|. \quad (25)$$

This inequality implies that the functions $\psi(\vec{r},\vec{R})$ which represents the motion of the electrons varies slowly with the nuclear coordinates $\vec{R}_s$.

From the physical point of view, if the conditions for the validity of the adiabatic approximation are present, the electrons in a certain state k move in a potential $V(\vec{r},\vec{R})$ and follow the lattice motion (i.e., change of $\vec{R}$) without making any transition to any other electronic state m ($m \neq k$).

The eigenvalue equation (20) can be written

$$H_v \phi_{k\ell}(\vec{R}) = E_{k\ell}\phi_{k\ell}(\vec{R}). \quad (26)$$

The eigenfunctions $\phi_{k\ell}$ represent the motion of the nuclei in the potential set by $\varepsilon_k(\vec{R})$. The eigenvalues of eq. (24) play here the role of a potential energy. We note that $\phi_{k\ell}$ depends on k, i.e., the electrons' quantum state; k is, however, only a subscript for ϕ and it does not represent a good quantum number:

$$\int \phi_{k\ell}(\vec{R})^* \phi_{k'\ell}(\vec{R}) d^3\vec{R} \neq \delta_{kk'} \quad . \quad (27)$$

II.C. The Adiabatic Approximation and the Role of Symmetry

The symmetry of a physical system is closely related to its quantum mechanical description. In particular, the degeneracy of the energy levels and the transformation properties of the eigenfunctions may be derived once the symmetry of the Hamiltonian is known. For this reason, it is important to examine the effect that the adiabatic approximation has on the symmetry properties of the Hamiltonian(s) considered.

Let us begin with some elementary considerations. The most general coordinate transformation consists of a rotation Q, followed by a translation $\vec{t}$; we shall use for it the symbol $\{Q|\vec{t}\}$. When operating on a position vector $\vec{x}$, the effect of this transformation is the following:

$$\{Q|\vec{t}\}\vec{x} = \vec{x}' = Q\vec{x}+\vec{t} \quad (28)$$

or

$$\begin{cases} x_1' = Q_{11}x_1 + Q_{12}x_2 + Q_{13}x_3 + t_1 \\ x_2' = Q_{21}x_1 + Q_{22}x_2 + Q_{23}x_3 + t_2 \\ x_3' = Q_{31}x_1 + Q_{32}x_2 + Q_{33}x_3 + t_3. \end{cases} \quad (29)$$

The coefficients Q_{ij} form a matrix Q which is real orthogonal:

$$\underset{\sim}{Q}\,\underset{\sim}{\tilde{Q}} = \underset{\sim}{1} \,, \qquad (30)$$

where $\underset{\sim}{1}$ is the "unit" matrix. From eq. (30) we find that the determinant of $\underset{\sim}{Q}$ can have only the values +1 or -1; the former value corresponds to a "proper" rotation and the latter value to an "improper" rotation.

The operation which consists of a pure rotation Q is represented by $\{Q|\vec{0}\}$; a pure translation is represented by $\{E|\vec{t}\}$. The operation "identify" is represented by $\{E|\vec{0}\}$.

Let us turn now to the consideration of the symmetry properties of the Hamiltonian H, the "exact" Hamiltonian of relation (1) we started with,

$$\begin{aligned} H = &- \sum_i \frac{\hbar^2}{2m} \nabla_i^{\,2} - \sum_s \frac{\hbar^2}{2M_s} \nabla_s^{\,2} + \\ &+ \frac{1}{2} \sum_{\substack{i \\ i\neq j}} \sum_j \frac{e^2}{|\vec{r}_i - \vec{r}_j|} + \frac{1}{2} \sum_{\substack{s \\ s\neq t}} \sum_t \frac{e^2 Z_s Z_t}{|\vec{R}_s - \vec{R}_t|} - \\ &- \sum_i \sum_s \frac{e^2 Z_s}{|\vec{R}_s - \vec{r}_i|} \;. \end{aligned} \qquad (31)$$

If we transform the coordinate system by performing a pure translation by a vector $\vec{d}$, the new position vectors are given by

$$\begin{aligned} \vec{R}'_s &= \vec{R}_s + \vec{d} \\ \vec{r}'_i &= \vec{r}_i + \vec{d} \;. \end{aligned} \qquad (32)$$

Once the transformation (32) is made, the Hamiltonian H has the same form in terms of the primed coordinates as it had in terms of the unprimed coordinates: this means that H has "translational invariance."

In order to investigate the rotational symmetry of H, let us now replace the 3n+3N coordinates of the system (n = number of electrons and N = number of nuclei) with the 3 coordinates of the center of mass and the (3n+3N-3) coordinates representing the relative vector distances; in the new coordinate system the Hamiltonian will be expressed as follows:

$$H = -\frac{\hbar^2}{2M} \nabla^2 + H', \qquad (33)$$

where M is the total mass, ∇^2 operates only on the coordinates of the center of mass, and H' is the Hamiltonian related to the "internal" energy of the system. Since the rotational symmetry properties of H are the same as those of H', we can restrict ourselves to the consideration of the rotational symmetry of H'.

The effects of a rotation Q on H' are the following:

a) The distances are unchanged, therefore the terms representing Coulomb interactions are invariant.

b) The Laplacians ∇^2 are also invariant. Therefore the terms representing the kinetic energy are also invariant.

The conclusion can now be made that the "exact" Hamiltonian H is invariant under all spatial transformations; these include translations, rotations (proper and improper) and any combinations of translations and rotations. Because of these non-restrictive conditions, no relevant information regarding the eigenvalues and the eigenfunctions of the system can be gained by symmetry considerations.

The situation is completely different when we consider the Hamiltonian

$$H_e = -\sum_i \frac{\hbar^2}{2m} \nabla_i^2 + V(\vec{r},\vec{R}). \tag{34}$$

This Hamiltonian represents the motion of the n electrons in the field of the nuclei which are assumed to be at <u>fixed</u> positions. In (34) the nuclear coordinates $\vec{R}$ represent parameters and not operators; as such, for the symmetry considerations of H_e, they are <u>not</u> to be operated upon. The result of a generic operation will be to change the coordinates of the electrons:

$$\vec{r}_i' = \{Q|\vec{t}\}\vec{r}_i. \tag{35}$$

Only the distances between electrons and nuclei will change. In the new coordinate system, the Hamiltonian H_e will appear as follows:

$$-\frac{\hbar^2}{2m}\sum_i \nabla_i^2 + \frac{1}{2}\sum_i\sum_{\substack{j \\ i\neq j}} \frac{e^2}{|\vec{r}_i-\vec{r}_j|} + \frac{1}{2}\sum_s\sum_{\substack{t \\ s\neq t}} \frac{e^2 Z_s Z_t}{|\vec{R}_s-\vec{R}_t|} - \sum_i\sum_s \frac{e^2 Z_s}{|\vec{R}_s-\vec{r}_i'|}. \tag{36}$$

Only the last sum has changed here. However, a particular operation

$\{Q|\vec{t}\}$ may be such as to make

$$|\vec{R}_s - \vec{r}_i'| = |\vec{R}_t - \vec{r}_i| , \tag{37}$$

where $\vec{R}_t$ is the coordinate of a nucleus identical with the nucleus in position R_s; in other words, the electron, because of a particular operation, may find itself in the same "environment" as before. The effect of operations of this type is to reshuffle the terms in the last sum in eq. (36), leaving the total sum invariant; therefore, the Hamiltonian H_e is invariant under such operations. On the other hand, as exemplified in Fig. 1, an operation $\{Q|\vec{t}\}$ that is consistent with eq. (37) is such that

$$\{Q|\vec{t}\}^{-1}\vec{R}_s = \vec{R}_t$$

or

$$\{Q|\vec{t}\}\vec{R}_t = \vec{R}_s. \tag{38}$$

We can then say that the electronic Hamiltonian H_e is invariant under all those operations on the electronic coordinates that, when applied to the nuclear coordinates, send identical nuclei into one another.

The Hamiltonian H_v represents the motion of the nuclei in the potential provided by the energy eigenvalues of H_e. H_v depends only on the nuclear coordinates and is invariant under all the operations that send identical nuclei into one another.

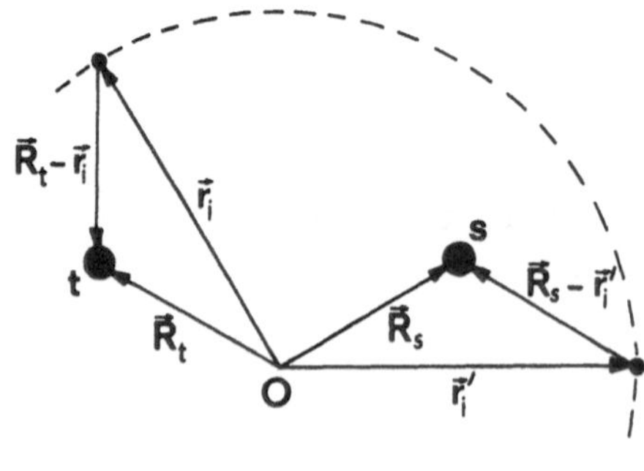

Fig. 1. Effect of a rotation on the electron coordinates.

III. RADIATION FIELD AND RADIATIVE PROCESSES

III.A. The Radiation Field. Classical Description

In a region free of currents and charges, a radiation field is defined by a vector potential $\vec{A}(\vec{r},t)$ (1,2)

$$\nabla^2\vec{A}(\vec{r},t) - \frac{1}{c^2}\frac{\partial^2\vec{A}(\vec{r},t)}{\partial t^2} = 0 \tag{39}$$

$$\vec{\nabla}\cdot\vec{A}(\vec{r},t) = 0 \tag{40}$$

$$\vec{E}(\vec{r},t) = -\frac{1}{c}\frac{\partial A(\vec{r},t)}{\partial r} \tag{41}$$

$$\vec{B}(\vec{r},t) = \vec{\nabla} \times \vec{A}(\vec{r},t) . \tag{42}$$

(39) is the so-called "field equation," (40) indicates that we have adopted the "Coulomb gauge," and (41) and (42) give the electric field and the magnetic field in terms of $\vec{A}(\vec{r},t)$, respectively.

A typical solution of the field equation is given by

$$A(\vec{r})q(t) = \vec{\pi}\left[\left(\frac{4\pi c^2}{V}\right)^{\frac{1}{2}} e^{i\vec{k}\cdot\vec{r}}\right]\left[|q|e^{-i\omega t}\right] . \tag{43}$$

We note that (40) implies

$$\vec{\pi}\cdot\vec{k} = 0 , \tag{44}$$

namely, that the polarization of the wave is perpendicular to the direction of the wave vector. The allowed values of $\vec{k}$ are determined by the boundary conditions of the problem.

Summing over all $\vec{k}$ and all polarizations (σ), we obtain the general solution of the field equation

$$\vec{A}(\vec{r},t) = \sum_\alpha\sum_\sigma\left[q_\alpha^\sigma(t)\vec{A}_\alpha^\sigma(\vec{r}) + q_\alpha^{\sigma*}(t)\vec{A}_\alpha^{\sigma*}(\vec{r})\right] , \tag{45}$$

where

$$q_\alpha^\sigma(t) = |q_\alpha^\sigma|e^{-i\omega_\alpha t} \tag{46}$$

$$\vec{A}_\alpha^\sigma(\vec{r}) = \vec{\pi}_\alpha^\sigma\left(\frac{4\pi c^2}{V}\right)^{\frac{1}{2}} e^{i\vec{k}_\alpha\cdot\vec{r}} . \tag{47}$$

The Hamiltonian of the radiation field can be derived from the expression for the energy of the field

$$\frac{1}{8\pi}\left[(\vec{E})^2 + (\vec{B})^2\right] d^3\vec{r} \tag{48}$$

and the relations (41) and (42) for $\vec{E}$ and $\vec{B}$, respectively. Because of the ortho onality of the various Fourier components of the field, cross terms with subscripts $\alpha\alpha'$ ($\alpha \neq \alpha'$) drop and we are left with (see ref. 1):

$$H = \sum_{\alpha} \sum_{\sigma} \omega_{\alpha}^{2} (q_{\alpha}^{\sigma} q_{\alpha}^{\sigma *} + q_{\alpha}^{\sigma *} q_{\alpha}^{\sigma}) \quad . \tag{49}$$

We note the following:

1) The Hamiltonian H of the radiation field is the sum of independent terms

$$H_{\alpha}^{\sigma} = \omega_{\alpha}^{2} (q_{\alpha}^{\sigma *} q_{\alpha}^{\sigma *} + q_{\alpha}^{\sigma *} q_{\alpha}^{\sigma}) \quad . \tag{50}$$

2) The coordinates q_{α}^{σ} represent the "normal coordinates" of the field.

3) No approximation has been made.

4) We use the form $\omega_{\alpha}^{2}(q_{\alpha} q_{\alpha}^{*} + q_{\alpha} q_{\alpha})$ rather than the form $2\omega_{\alpha}^{2} q_{\alpha} q_{\alpha}^{*}$ in preparation of our move into quantum mechanics: in the quantum mechanical treatment q and q* become non-commuting operators.

The values of $\vec{k}$, as we said before, are determined by the boundary conditions. When, as it is the case here, the wavelength of the radiation is much smaller than the dimension of the spatial region under consideration, any sum over $\vec{k}$ is in effect an integral and the relevant information is the density of states, i.e., the number of states with $\vec{k}$ in ($\vec{k}$, $\vec{k}+d\vec{k}$). In order to find this quantity, periodic boundary conditions can be used; these conditions give for the possible values of the three components of $\vec{k}$

$$\begin{aligned} k_x &= n_x \frac{2\pi}{L_x} \\ k_y &= n_y \frac{2\pi}{L_y} \\ k_z &= n_z \frac{2\pi}{L_z} \end{aligned} \quad , \tag{51}$$

where the volume has been taken as a cube of sides L_x, L_y and L_z and where

$$n_x, n_y, n_z = 0, \pm 1, \pm 2, \pm 3, \ldots \tag{52}$$

The subscript α used before stands for a particular choice of n_x, n_y and n_z.

The number of modes with $\vec{k}$ in ($\vec{k}$, $\vec{k}+d\vec{k}$) is given by

$$\frac{L_x dk_x}{2\pi} \frac{L_y dk_y}{2\pi} \frac{L_z dk_z}{2\pi} = \frac{L_x L_y L_z}{8\pi^3} dk_x dk_y dk_z =$$

$$= \frac{V}{8\pi^3} k^2 dk \sin\theta d\theta d\phi = \frac{V}{8\pi^3} \frac{\omega^3}{c^3} d\omega d\Omega \ , \quad (53)$$

where

$$d\Omega = \sin\theta d\theta d\phi \ .$$

III.B. The Quantum Radiation Field

Consider one term of the Hamiltonian (49)

$$H_\alpha = \omega_\alpha^2 (q_\alpha q_\alpha^* + q_\alpha q_\alpha) \quad (54)$$

where we have dropped for convenience the superscript σ. We introduce two new real variables for each α

$$Q_\alpha = q_\alpha + q_\alpha^*$$

$$P_\alpha = -i\omega_\alpha (q_\alpha - q_\alpha^*) = \dot{Q}_\alpha . \quad (55)$$

The Hamiltonian H_α, when written in terms of Q_α and P_α, takes the form

$$H_\alpha = \tfrac{1}{2}\omega_\alpha^2 Q_\alpha^2 + \tfrac{1}{2}P_\alpha^2 . \quad (56)$$

Q_α and P_α are real variables that satisfy Hamilton's equations:

$$\{Q_\alpha, P_{\alpha'}\} = \sum_i \left(\frac{\partial Q_\alpha}{\partial Q_i} \frac{\partial P_{\alpha'}}{\partial P_i} - \frac{\partial Q_\alpha}{\partial P_i} \frac{\partial P_{\alpha'}}{\partial Q_i} \right) = \delta_{\alpha\alpha'}$$

$$\{Q_\alpha, Q_{\alpha'}\} = \{P_\alpha, P_{\alpha'}\} = 0 . \quad (57)$$

The prescription for moving over from a classical to a quantum mechanical treatment is simple. Q_α and P_α become Hermitian operators and their commutator is obtained by replacing the Poisson brackets as follows:

$$\{Q_\alpha, P_{\alpha'}\} \rightarrow \frac{1}{i\hbar} \left[Q_\alpha, P_{\alpha'} \right] . \quad (58)$$

Then

$$\left[Q_\alpha, P_{\alpha'} \right] = i\hbar \, \delta_{\alpha\alpha'}$$

$$\left[Q_\alpha, Q_{\alpha'} \right] = \left[P_\alpha, P_{\alpha'} \right] = 0 . \quad (59)$$

q_α and q_α^*, which are related to Q_α and P_α by the relations (55), become two (non-Hermitian) operators which we shall call q_α and q_α^+, respectively. The commutation relation of these two operators are easily derived:

$$\left[q_\alpha, q_{\alpha'}^+\right] = \frac{\hbar}{2\omega_\alpha}\delta_{\alpha\alpha'}$$

$$\left[q_\alpha, q_{\alpha'}\right] = \left[q_\alpha^+, q_{\alpha'}^+\right] = 0. \tag{60}$$

We may replace q_α and q_α^+ by the dimensionless operators

$$a_\alpha = \left(\frac{2\omega_\alpha}{\hbar}\right)^{\frac{1}{2}} q_\alpha \quad , \quad a_\alpha^+ = \left(\frac{2\omega_\alpha}{\hbar}\right)^{\frac{1}{2}} q_\alpha^+. \tag{61}$$

It is

$$\left[a_\alpha, a_{\alpha'}^+\right] = \delta_{\alpha\alpha'}$$

$$\left[a_\alpha, a_{\alpha'}\right] = \left[a_\alpha^+, a_{\alpha'}^+\right] = 0. \tag{62}$$

The Hamiltonian of the radiation field can now be written

$$H = \sum_\alpha \omega_\alpha{}^2(q_\alpha q_\alpha^+ + q_\alpha^+ q_\alpha) = \sum_\alpha \hbar\omega_\alpha(a_\alpha^+ a_\alpha + \tfrac{1}{2}). \tag{63}$$

Reintroducing the polarization index σ

$$H = \sum_\alpha \sum_\sigma \hbar\omega_\alpha(a_\alpha^{\sigma+} a_\alpha^\sigma + \tfrac{1}{2}). \tag{64}$$

The Hamiltonian

$$H_\alpha^\sigma = \hbar\omega_\alpha(a_\alpha^{\sigma+} a_\alpha^\sigma + \tfrac{1}{2}) \tag{65}$$

has the energy eigenvalues

$$E_\alpha^\sigma = \hbar\omega_\alpha(n_\alpha^\sigma + \tfrac{1}{2}), \tag{66}$$

where $n_\alpha^\sigma = 0, 1, 2, \ldots.$ The eigenfunctions of H_α^σ are simply given by the kets $|n_\alpha^\sigma>$.

The Hamiltonian, eigenvalues and eigenfunctions of the radiation field are now listed:

$$H = \sum_\alpha \sum_\sigma H_\alpha^\sigma = \sum_\alpha \sum_\sigma \hbar\omega_\alpha(a_\alpha^{\sigma+} a_\alpha^\sigma + \tfrac{1}{2}) \tag{67}$$

$$E_{n_1^{\sigma_1} n_1^{\sigma_2} n_2^{\sigma_1} \ldots} = \sum_\alpha \sum_\sigma \hbar\omega_\alpha(n_\alpha^\sigma + \tfrac{1}{2}) \tag{68}$$

$$\psi_{n_1^{\sigma 1} n_1^{\sigma 2} n_2^{\sigma 1}} \ldots\ldots = \prod_\alpha \prod_\sigma \mid n_\sigma^\alpha > . \qquad (69)$$

One can see from the above relations that the radiation field may be thought of as a collection of an infinite number of harmonic oscillators, one for each (α,σ) component, with different degrees of excitation n_α^σ. Alternatively the radiation field may be thought of as an ensemble of photons: n_α^σ is the number of photons present for each wave vector $\vec{k}_\alpha$ and polarization σ.

In the quantum-mechanical treatment the vector potential represents an operator which can be expressed as follows:

$$\begin{aligned}\vec{A} &= \sum_\alpha \sum_\sigma \left[\vec{A}_\alpha^\sigma q_\alpha^\sigma + \vec{A}_\alpha^{\sigma *} q_\alpha^{\sigma +} \right] = \\ &= \sum_\alpha \sum_\sigma \left(\frac{4\pi c^2}{V}\right)^{\frac{1}{2}} \frac{\hbar}{2\omega_\alpha}^{\frac{1}{2}} \vec{\pi}_\alpha^\sigma \left(e^{i\vec{k}_\alpha \cdot \vec{r}} a_\alpha^\sigma + e^{-i\vec{k}_\alpha \cdot \vec{r}} a_\alpha^{\sigma +}\right) = \\ &= \sum_\alpha \sum_\sigma \left(\frac{hc^2}{\omega_\alpha V}\right)^{\frac{1}{2}} \vec{\pi}_\alpha^\sigma \left(e^{i\vec{k}_\alpha \cdot \vec{r}} a_\alpha^\sigma + e^{-i\vec{k}_\alpha \cdot \vec{r}} a_\alpha^{\sigma +}\right).\end{aligned} \qquad (70)$$

We note that the operators a_α^σ and $a_\alpha^{\sigma +}$ operate as follows:

$$\begin{aligned} a_\alpha^\sigma |n_\alpha^\sigma > &= \sqrt{n_\alpha^\sigma}\, |n_\alpha^\sigma - 1 > \\ a_\alpha^{\sigma +} |n_\alpha^\sigma > &= \sqrt{n_\alpha^\sigma + 1}\, |n_\alpha^\sigma + 1 > . \end{aligned} \qquad (71)$$

III.C. Interaction of a Radiation Field with Charged Particles

Assume that a particle of mass m and charge q is under the action of a radiation field $\vec{A}(\vec{r},t)$ and of a potential field $\phi(\vec{r},t)$.

The equation of motion of such a particle is given by the Hamiltonian

$$H = \frac{(\vec{p} - \frac{q}{c}\vec{A})^2}{2m} + q\phi , \qquad (72)$$

where $\vec{p}$ = linear momentum of the particle. This can be easily justified by considering the Hamilton's equations, which give the correct expression for the (Lorentz) force acting on the particle.

The Hamiltonian H can be written as follows:

$$H = \frac{(\vec{p})^2}{2m} - \frac{q}{2mc}(\vec{p}\cdot\vec{A}+\vec{A}\cdot\vec{p}) + \frac{q^2}{2mc^2}(\vec{A})^2 + q\phi$$

$$= \frac{(\vec{p})^2}{2m} - \frac{q}{mc}(\vec{p}\cdot\vec{A}) + \frac{q^2}{2mc^2}(\vec{A})^2 + q\psi \qquad (73)$$

since $[\vec{p},\vec{A}] = 0$, because of the Coulomb gauge. The interaction term, which is linear in the field, is relevant here. This term is

$$H_1 = -\frac{q}{mc}\vec{p}\cdot\vec{A}$$

$$= -\frac{q}{m}\sum_{\alpha\sigma}\left(\frac{h}{\omega_\alpha V}\right)^{\frac{1}{2}}\left(a_\alpha^\sigma e^{i\vec{k}_\alpha\cdot\vec{r}} + a_\alpha^{\sigma+} e^{-i\vec{k}_\alpha\cdot\vec{r}}\right)\vec{\pi}_\alpha^\sigma\cdot\vec{p} \quad . \qquad (74)$$

In case of several particles

$$H_1 = -\left\{\sum_{\alpha\sigma}\left(\frac{h}{\omega_\alpha V}\right)^{\frac{1}{2}}\sum_i\left[\frac{q_i}{m_i}\left(a_\alpha^\sigma e^{i\vec{k}_\alpha\cdot\vec{r}_i} + a_\alpha^{\sigma+} e^{-i\vec{k}_\alpha\cdot\vec{r}_i}\right)(\vec{\pi}_\alpha^\sigma\cdot\vec{p}_i)\right]\right\}. \qquad (75)$$

III.D. Absorption and Emission Processes

Let us consider the situation in which a charged particle is under the action of a potential ϕ and of a radiation field $\vec{A}$. The Hamiltonian of the system which consists of the particle <u>and</u> the radiation field is given by

$$H = \frac{1}{2m}\left(\vec{p} - \frac{q}{c}\vec{A}\right)^2 + q\phi + \frac{1}{8\pi}\int[(\vec{E})^2 + (\vec{B})^2]\,d^3\vec{r}$$

$$= \left[\frac{(\vec{p})^2}{2m} + q\phi\right] + \frac{1}{8\pi}\int[(\vec{E})^2 + (\vec{B})^2]\,d^3\vec{r} - \frac{q}{2mc}(\vec{p}\cdot\vec{A}) + \frac{q^2}{2mc^2}(\vec{A})^2. \qquad (76)$$

We can express H as follows:

$$H = H_o + H_1 + H_2 \qquad (77)$$

where

$$H_o = \frac{(\vec{p})^2}{2m} + q\phi + \frac{1}{8\pi}\int[(\vec{E})^2 + (\vec{B})^2]\,d^3\vec{r}$$

$$= -\frac{\hbar^2}{2m}\nabla^2 + q\phi + \sum_{\alpha\sigma}\hbar\omega_\alpha\left(a_\alpha^{\sigma+}a_\alpha^\sigma + \tfrac{1}{2}\right) \qquad (78)$$

and H_1 and H_2 are the terms linear and quadratic in the field, respectively.

The method to be applied here consists in considering H_o as the Hamiltonian of the "unperturbed" system, given simply by the sum of the Hamiltonian of the particle and the Hamiltonian of the radiation field and taking H_1 and H_2 as time dependent perturbations of the system which may induce transitions between the different eigenstates of H_o. These eigenstates are given by

$$\psi_{e;\, n_1^{\sigma_1},\, n_1^{\sigma_2},\, \ldots} = \psi^e \prod_\alpha \prod_\sigma \mid n_\alpha^\sigma > , \quad (79)$$

where ψ^e = eigenfunction of the particle and $|n_\alpha^\sigma >$ = eigenfunction of the (α,σ) radiation oscillator. The energies of the states are given by

$$E_{e;\, n_1^{\sigma_1},\, n_2^{\sigma_2},\, \ldots} = E^e + \sum_{\alpha\sigma}\sum \hbar\omega_\alpha (n_\alpha^\sigma + \tfrac{1}{2}) , \quad (80)$$

where E^e = energy of the particle and the sum over α and σ gives the energy of the radiation field.

In the case of one photon absorption, the initial and the final states are given by

$$\psi_i = |\psi_i^e >|n_\alpha^\sigma >|n_{\alpha'}^{\sigma'} > \ldots \quad (81)$$

$$\psi_f = |\psi_f^e >|n_\alpha^\sigma - 1 >|n_{\alpha'}^{\sigma'} > \ldots , \quad (82)$$

respectively. It is

$$< n_\alpha^\sigma - 1|a_\alpha^\sigma|n_\alpha^\sigma > = \sqrt{n_\alpha^\sigma} \;; \quad (83)$$

therefore the relevant matrix element for the process of absorption of one photon is

$$< \psi_f^e;n_\alpha^\sigma - 1|H_1|\psi_i^e;n_\alpha^\sigma >$$

$$= - \frac{q}{m} \left(\frac{h}{\omega_\alpha V} \right)^{\frac{1}{2}} < \psi_f^e|e^{i\vec{k}_\alpha \cdot \vec{r}} \vec{\pi}_\alpha^\sigma \cdot \vec{p}|\psi_i^e > \sqrt{n_\alpha^\sigma}. \quad (84)$$

In the case of one photon emission, the initial and the final states are given by

$$\Psi_i = |\psi_i^e>|n_\alpha^\sigma>|n_{\alpha'}^{\sigma'}> \ldots \quad (85)$$

$$\Psi_f = |\psi_f^e>|n_\alpha^\sigma + 1>|\, n_{\alpha'}^{\sigma'}> \ldots , \quad (86)$$

respectively. It is

$$< n_\alpha^\sigma + 1|a_\alpha^{\sigma+}|n_\alpha^\sigma > = \sqrt{n_\alpha^\sigma + 1} \;; \quad (87)$$

therefore the relevant matrix element for the process of emission of one photon is

$$<\psi_f^e;n_\alpha^\sigma + 1|H_1|\psi_i^e;n_\alpha^\sigma>$$

$$= -\frac{q}{m}\left(\frac{h}{\omega_\alpha V}\right)^{\frac{1}{2}} <\psi_f^e|e^{-i\vec{k}_\alpha \cdot \vec{r}}\vec{\pi}_\alpha^\sigma \cdot \vec{p}|\psi_i^e>\sqrt{n_\alpha^\sigma + 1}. \tag{88}$$

Since the radiative field has a continuous density of states (see for this section III.A), both absorption and emission processes are associated with a probability per unit time. By applying the Fermi Golden Rule, we derive that the probability per unit time of finding the system (particle + radiation field) with one less or one more photon of energy $\hbar\omega_\alpha$ and polarization $\vec{\pi}_\alpha^\sigma$ in the solid angle $(\Omega_\alpha,\Omega_\alpha + d\Omega_\alpha)$ is given by

$$p_\alpha^\sigma\, d\Omega_\alpha = \frac{2\pi}{\hbar^2}|M_\alpha^\sigma|^2\, g(\omega_\alpha) \tag{89}$$

where

$$g(\omega_\alpha) = \frac{V^2\omega_\alpha}{8\pi^3 c^3}\, d\Omega_\alpha \tag{90}$$

and

$$|M_\alpha^\sigma|^2 = \begin{cases} \dfrac{q^2}{m^2}\dfrac{h}{\omega_\alpha V}\,|<\psi_f^e|e^{i\vec{k}_\alpha\cdot\vec{r}}\,\vec{\pi}_\alpha^\sigma \cdot \vec{p}|\psi_i^e>|^2\, n_\alpha^\sigma \\[2ex] \dfrac{q^2}{m^2}\dfrac{h}{\omega_\alpha V}\,|<\psi_f^e|e^{-i\vec{k}_\alpha\cdot\vec{r}}\,\vec{\pi}_\alpha^\sigma \cdot \vec{p}|\psi_i^e>|^2\, (n_\alpha^\sigma + 1), \end{cases} \tag{91}$$

where the upper (lower) row corresponds to the process of absorption (emission) of one photon. Replacing (91) in (89) and taking (90) into account, we find:

$$p_\alpha^\sigma d\Omega_\alpha = \frac{\omega_\alpha q^2}{hc^3m^2}\,|\,\psi_f^e \begin{vmatrix} e^{i\vec{k}_\alpha\cdot\vec{r}}\,\vec{\pi}_\alpha^\sigma \cdot \vec{p} \\[2ex] e^{-i\vec{k}_\alpha\cdot\vec{r}}\,\vec{\pi}_\alpha^\sigma \cdot \vec{p} \end{vmatrix} \psi_i^e>|^2 \begin{pmatrix} n_\alpha^\sigma \\[2ex] n_\alpha^\sigma + 1 \end{pmatrix} d\Omega_\alpha. \tag{92}$$

Let us consider two quantum states of the particle ψ_ℓ^e (ℓ stands for lower) and ψ_u^e (u stands for upper) with energies E_ℓ

and E_u, respectively and $E_\ell < E_\mu$. It is possible to show that

$$|\langle\psi_\ell^e|d^{-i\vec{k}\cdot\vec{r}}(\vec{\pi}\cdot\vec{p})|\psi_u^e\rangle|^2 = |\langle\psi_u^e|e^{i\vec{k}\cdot\vec{r}}(\vec{\pi}\cdot\vec{p})|\psi_\ell^e\rangle|^2. \quad (93)$$

The last squared matrix element is the one that would enter the transition probability for an $\ell\to u$ (absorption) process.

On the basis of the above result, (92) becomes

$$p_\alpha^\sigma d\Omega_\alpha = \frac{\omega_\alpha q^2}{hc^3m^2}|\langle\psi_u^e|e^{i\vec{k}\cdot\vec{r}}\vec{\pi}_\alpha^\sigma\cdot\vec{p}|\psi_\ell^e\rangle|^2\begin{pmatrix} n_\alpha^\sigma \\ \\ n_\alpha^\sigma + 1\end{pmatrix}d\Omega_\alpha. \quad (94)$$

The transition probability for absorption is always proportional to the number of photons n_α^σ present; the transition probability for emission consists of one part, called "induced emission," which is proportional to n_α^σ and of another part, called "spontaneous emission," which is present even when $n_\alpha^\sigma = 0$. We note here that the transition probability of absorption and the transition probability for induced emission between two states are equal. If two or more charged particles are present

$$p_\alpha^\sigma d\Omega_\alpha = \frac{\omega_\alpha}{hc^3}|\langle\sum_t\frac{q_t}{m_t}(\vec{\pi}_\alpha^\sigma\cdot\vec{p}_t)e^{i\vec{k}_\alpha\cdot\vec{r}_t}\rangle u\ell|^2\begin{pmatrix} n_\alpha^\sigma \\ \\ n_\alpha^\sigma + 1\end{pmatrix}d\Omega_\alpha. \quad (95)$$

III.E. Absorption and Emission in the Electric Dipole Approximation

The dimension d of free atoms and molecules is in general such that

$$|kd| \ll 1 \quad (96)$$

or

d << λ (= wavelength of absorbed or emitted light); in this case, it is possible to use the approximation

$$e^{i\vec{k}\cdot\vec{r}} \simeq 1 \quad (97)$$

and the electric dipole transitions are predominant. On the other hand, impurity ions in non-centrosymmetric solids may sometimes present electric dipole and magnetic dipole transitions with probabilities of the same order of magnitude (see for example ref. 3).

In the electric dipole approximation, the expression (95) becomes

$$p_\alpha^\sigma \, d\Omega_\alpha = \frac{\omega_\alpha}{hc^3} \left| \langle \sum_t \frac{q_t}{m_t} (\vec{\pi}_\alpha^\sigma \cdot \vec{p}_t) \rangle_{fi} \right|^2 \begin{pmatrix} n_\alpha^\sigma \\ \\ n_\alpha^\sigma + 1 \end{pmatrix} . \tag{98}$$

Given a function $F(x_i,p_i,t)$, we have classically

$$\begin{aligned} \frac{dF(x_i,p_i,t)}{dt} &= \frac{\partial F}{\partial t} + \sum_i \left(\frac{\partial F}{\partial x_i} \frac{\partial x_i}{\partial t} + \frac{\partial F}{\partial p_i} \frac{\partial p_i}{\partial t} \right) \\ &= \frac{\partial F}{\partial t} + \sum_i \left(\frac{\partial F}{\partial x_i} \frac{\partial H}{\partial p_i} - \frac{\partial F}{\partial p_i} \frac{\partial H}{\partial x_i} \right) \\ &= \frac{\partial F}{\partial f} + \{F,H\} \end{aligned} \tag{99}$$

where the {} brackets indicate a Poisson bracket. If we want to express the above relation quantum mechanically, we replace the Poisson bracket by a commutator as follows:

$$\{F,H\} \xrightarrow[\text{Q.M.}]{} \frac{1}{i\hbar} [F,H] = \frac{i}{\hbar} [H,F] . \tag{100}$$

When this is done, (99) becomes

$$\frac{dF}{dt} = [H,F] + \frac{\partial F}{\partial t} \tag{101}$$

If we set $F = x$,

$$\dot{x} = \frac{i}{\hbar} [H,x] \tag{102}$$

and

$$p_x = m\dot{x} = i \frac{m}{\hbar} [H,x] . \tag{103}$$

Taking a matrix element of $[H,x]$ between two electronic eigenfunctions, we find

$$\begin{aligned} \langle \psi_f^e | [H,x] | \psi_i^e \rangle &= \langle \psi_f^e | Hx-xH | \psi_i^e \rangle \\ &= (E_f^e - E_i^e) \langle \psi_f^e | x | \psi_i^e \rangle \end{aligned} \tag{104}$$

and

$$\langle\psi_f^e|\vec{p}|\psi_i^e\rangle = \frac{im}{\hbar}(E_f^e - E_i^e)\langle\psi_f^e|\vec{r}|\psi_i^e\rangle$$

$$= im\omega_{fi}\langle\psi_f^e|\vec{r}|\psi_i^e\rangle , \tag{105}$$

where

$$\omega_{fi} = \frac{E_f^e - E_i^e}{\hbar} . \tag{106}$$

Taking the result (105) into account, (98) becomes

$$P_\alpha^\sigma d\Omega_\alpha = \frac{\omega_\alpha^3}{hc^3}\left|\langle\vec{\pi}_\alpha^\sigma \cdot \vec{M}\rangle_{fi}\right|^2 \begin{pmatrix} n_\alpha^\sigma \\ n_\alpha^\sigma + 1 \end{pmatrix} d\Omega_\alpha , \tag{107}$$

where

$$\vec{M} = \sum_t q_t \vec{r}_t = \text{electric dipole operator.} \tag{108}$$

The radiation energy density per unit frequency range at $\omega = \omega_\alpha$, unit solid angle and polarization $\vec{\pi}_\alpha^\sigma$ is given by

$$\frac{g(\omega_\alpha)\hbar\omega_\alpha n_\alpha^\sigma}{Vd\Omega_\alpha} = \frac{\omega_\alpha^2}{8\pi^3c^3}\hbar\omega_\alpha n_\alpha^\sigma = \frac{\hbar\omega_\alpha^3}{8\pi^3c^3} n_\alpha^\sigma , \tag{109}$$

and the radiation energy intensity per unit frequency range at $\omega=\omega_\alpha$, unit solid angle and polarization $\vec{\pi}_\alpha^\sigma$ if given by

$$I(\omega_\alpha,\sigma) = \frac{\hbar\omega_\alpha^3}{8\pi^3c^2} n_\alpha^\sigma . \tag{110}$$

It is

$$\frac{4\pi^2 I}{\hbar^2 c} = \frac{4\pi^2}{\hbar^2 c}\frac{\hbar\omega_\alpha^3}{8\pi^3c^2} n_\alpha^\sigma = \frac{\omega_\alpha^3}{hc^3} n_\alpha^\sigma . \tag{111}$$

Therefore we can write the formula (107) as follows:

$$P_\alpha^\sigma = \begin{cases} \dfrac{4\pi^2 I(\omega_\alpha;\sigma)}{\hbar^2 c}|M_{fi}|^2 \\ \left\{\dfrac{4\pi^2 I(\omega_\alpha;\sigma)}{\hbar^2 c} + \dfrac{\omega_\alpha^3}{hc^3}\right\}|M_{fi}|^2 , \end{cases} \tag{112}$$

where the upper (lower) row corresponds to absorption (emission) and where

$$|M_{fi}|^2 = |< \vec{\pi}_\alpha^\sigma \cdot \vec{M} >_{fi}|^2 . \tag{113}$$

IV. INTERACTION OF A RADIATION FIELD WITH A MOLECULAR SYSTEM

IV.A. The Franck-Condon Principle

The process of absorption or emission of ultraviolet light by molecules is accompanied by a change of their electronic state. Stated very simply, the "Franck-Condon Principle" (4,5) recognizes the fact that the change in the electronic structure occurring in an electronic transition is much more rapid than the possible changes in the internuclear distances occurring during the same transition.

By using purely classical arguments, the Franck-Condon principle is stated as follows:

> "During an electronic transition, the electronic state changes so fast that 1) the nuclei do not move and 2) the nuclei do not change their momenta."

This is best illustrated considering a diatomic molecule. Potential curve diagrams for such a system are given in Fig. 2; in it R is the distance between the two nuclei. The condition 1) above means that during an electronic transition, the internuclear distance R must remain constant; the condition 2) implies that the kinetic energy must also remain constant and this means that if the molecule's initial state is given by the point A' the molecule after the electronic transition will be found at B' with AA' = BB'. A third condition may be added considering that the harmonic oscillator which represents the vibrational motion of the molecule spends most of its time at its turning points at which the kinetic energy is zero. This condition 3) states that at the instant at which the electronic transition takes place, the molecule is found at these turning points. Therefore, according to this condition, the following transitions are allowed:

AB, CD

and the following transitions are forbidden

A'B', C'D'.

In its semiclassical formulation the Franck-Condon principle consists of three conditions:

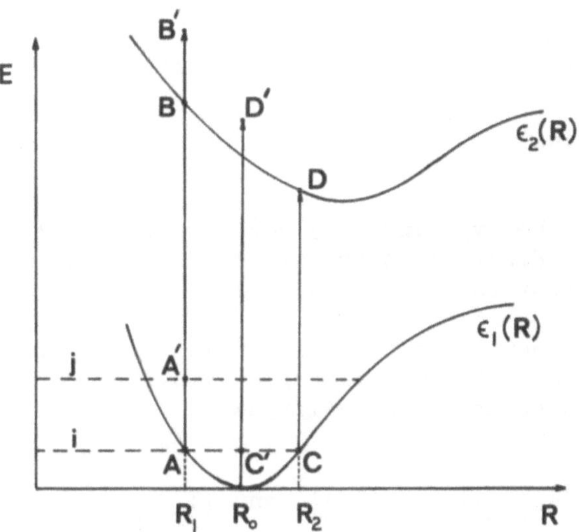

Fig. 2. Diagram illustrating the Franck-Condon principle.

1) R = const, 2) p = const and 3) an electronic transition can take place for any value of R with a probability W(R), as given by quantum mechanics.

If the system is in a particular vibrational quantum state, say the ith vibrational state, then

$$W(R) = |\phi_i(R)|^2. \tag{114}$$

If the system is in thermal equilibrium at a temperature T

$$W_T(R) = \frac{\sum_i e^{-E_i/kT} |\phi_i(R)|^2}{\sum_i e^{-E_i/kT}} . \tag{115}$$

For most diatomic milecules at ordinary temperature $\hbar\omega$ is so large (of the order 0.1 eV) that

$$\hbar\omega >> kT \tag{116}$$

and the vibrational states other than the ground state are practically unoccupied; under these conditions, the relevant probability W(R) may well be given by eq. (114) with i = 0.

The above considerations can be generalized by treating a molecule with many vibrational degrees of freedom. In these more complex cases, the relevant probability is given by

$$W_T = \prod_K W_{KT}(q_K), \tag{117}$$

where W_{KT} is of the type in (115) with $\omega = \omega_K$ and $R = q_K$ (vibrational coordinate); in the limit of low temperatures or high ω, W_{KT} is simply of the type (114).

Considering now again the diagram in Fig. 2, we note that the semiclassical Franck-Condon principle allows the transition AB, CD, A'B' and C'D' which have probabilities

$$|\phi_i(R_i)|^2, |\phi_i(R_2)|^2, |\phi_j(R_1)|^2 \text{ and } |\phi_i(R_o)|^2 ,$$

respectively; in particular, if $i = 0$, the probability will have its maximum in correspondence to R_o and transitions such as C'D' have the greatest importance.

Finally, the quantum mechanical formulation of the Franck-Condon principle is perhaps the most illuminating. The transition probability for a radiative transition in the dipole approximation, is proportional to the square of the matrix element

$$M_{fi} = \langle \vec{\pi}^{\sigma}_{\alpha} \cdot \vec{M} \rangle_{fi} = \langle \vec{\pi}^{\sigma}_{\alpha} \cdot \sum_t q_t \vec{r}_t \rangle_{fi} , \tag{118}$$

where q_t and $\vec{r}_t$ are the charge and the position of the tth charged particle, respectively. The initial state of the system is given in the adiabatic approximation by

$$\Psi_i(\vec{r},\vec{R}) = \psi_k(\vec{r},\vec{R})\phi_{k\ell}(\vec{R}) \tag{119}$$

and the final state by

$$\Psi_f(\vec{r},\vec{R}) = \psi_m(\vec{r},\vec{R})\phi_{mn}(\vec{R}). \tag{120}$$

The electric dipole operator can be written as follows:

$$\vec{M} = - e \sum_i \vec{r}_i + e \sum_s Z_s \vec{R}_s , \tag{121}$$

where the sums over i and s extend to the electrons and the nuclei, respectively. We can write

$$\vec{\pi} \cdot \vec{M} = - e \sum_i \vec{r}_i \cdot \vec{\pi} + e \sum_s Z_s \vec{R}_s \cdot \vec{\pi}$$

$$= D_e(\vec{r}) + D_n(\vec{R}) \ . \qquad (122)$$

The relevant matrix element is now given by

$$M_{fi} = M_{mn,k\ell}$$

$$= \int\int d^3\vec{r} d^3\vec{R} \ \Psi_f(\vec{r},\vec{R})^* \left[D_e(\vec{r}) + D_n(\vec{R}) \right] \Psi_i(\vec{r},\vec{R}) =$$

$$= \int\int d^3\vec{r} d^3\vec{R} \ \psi_m(\vec{r},\vec{R})^* \ \phi_{mn}(\vec{R})^* \left[D_e(\vec{r}) + D_n(\vec{R}) \right] \psi_k(\vec{r},\vec{R}) \phi_{k\ell}(\vec{R}) =$$

$$= \int d^3\vec{R} \ \phi_{mn}(\vec{R})^* \left[\int d^3\vec{r} \ \psi_m(\vec{r},\vec{R})^* D_e(\vec{r}) \psi_k(\vec{r},\vec{R}) \right] \phi_{k\ell}(\vec{R}) \ +$$

$$+ \int d^3\vec{R} \ \phi_{mn}(\vec{R})^* D_n(\vec{R}) \phi_{k\ell}(\vec{R}) \left[\int d^3\vec{r} \ \psi_m(\vec{r},\vec{R})^* \ \psi_k(\vec{r},\vec{R}) \right] =$$

$$= \int d^3\vec{R} \ \phi_{mn}(\vec{R})^* D_{mk}(\vec{R}) \phi_{k\ell}(\vec{R}) + \int d^3\vec{R} \ \phi_{mn}(\vec{R}) D_n(\vec{R}) \phi_{k\ell}(\vec{R}) \delta_{mk} , \qquad (123)$$

where

$$D_{mk}(\vec{R}) = \int d^3\vec{r} \ \psi_m(\vec{r},\vec{R})^* D_e(\vec{r}) \psi_k(\vec{r},\vec{R}) \ . \qquad (124)$$

The following observations can be made:

1) The matrix element M_{fi} consists of two terms. The second term contributes to the matrix element only when m = k, namely when the electronic state does not change in the transition and gives rise to <u>infrared absorption</u> by the vibrations.

2) The first term contains the dipole moment of the electrons and corresponds to transitions between different electronic states.

3) If m = k the first term becomes

$$\int d^3\vec{R} \ \phi_{mn}(\vec{R})^* D_{mm}(\vec{R}) \phi_{m\ell}(\vec{R}) \quad , \qquad (125)$$

where

$$D_{mm}(\vec{R}) = \int d^3\vec{r} |\psi_m(\vec{r},\vec{R})|^2 \ D_e(\vec{r}) \ . \qquad (126)$$

Since $D_e(\vec{r})$ is an odd operator, $D_{mm}(\vec{R})$ is different from zero only if the molecule lacks inversion symmetry.

If this is the case, the first term in M_{fi} also contributes to the infrared transition.

For transitions between different electronic states, the relevant matrix element is given by

$$M_{fi} = M_{mn;k\ell}$$

$$= \int \phi_{mn}(\vec{R})^* D_{mk}(\vec{R}) \phi_{k\ell}(\vec{R}) R^2 \sin\theta dR d\theta d\phi \ , \quad (127)$$

where $D_{mk}(\vec{R})$ is given by (124).

We shall consider now for simplicity the case of a diatomic molecule; for such a system, the eigenfunctions appearing in the integral (127) are of the type:

$$\phi(R,\theta,\phi) = \mathcal{R}_{NK}(R)\theta_{KM}(\theta)\phi_M(\phi). \quad (128)$$

The equation for $\mathcal{R}(R)$ is

$$\frac{1}{R^2}\frac{d}{dR}\left(R^2 \frac{d\mathcal{R}}{dR}\right) + \left\{ -\frac{K(K+1)}{R^2} + \frac{2\mu}{\hbar^2}\left[E - \varepsilon(R)\right]\right\}\mathcal{R} = 0, \quad (129)$$

where μ = reduced mass of the molecule (6). This equation may be simplified by setting

$$\mathcal{R}(R) = \frac{S(R)}{R} \ ; \quad (130)$$

with this (129) reduces to

$$\frac{d^2S}{dR^2} + \left\{ -\frac{K(K+1)}{R^2} + \frac{8\pi^2\mu}{h^2}\left[E - \varepsilon(R)\right]\right\} S = 0 \ . \quad (131)$$

In order to find S(R) and (R), we need to know $\varepsilon(R)$, the form of the adiabatic potential. If it is assumed that the force between the atoms is proportional to the displacement of the internuclear distance from its equilibrium value, then the functions S(R) take the form of harmonic oscillator eigenfunctions. In any case, regardless of the form of the adiabatic potential $\varepsilon(R)$, the matrix element will be proportional to the radial integral

$$\int S_{mn}(R)^* D_{mk}(R) S_{k\ell}(R) dR. \quad (132)$$

If we assume that $D_{mk}(R)$ does not depend strongly on the internuclear distance, we may expand it in terms of the displacement from its equilibrium value. According to the Franck-Condon principle, we retain only the first (constant) term in this expansion:

$$D_{mk}(R) \simeq D^o_{mk} \ . \quad (133)$$

Then for electronic transitions

$$|M_{fi}|^2 \sim |D^o_{mk}|^2 | \int S_{mn}(R)^* S_{k\ell}(R)dR|^2 \quad . \tag{134}$$

The transition probability is then proportional to the square of the overlap integral of the vibrational eigenfunctions of the initial and final states. It is this overlap that controls (apart from D^o) the strength of the transition probabilities. It is to be noted, here that $S_{mn}(R)$ and $S_{k\ell}(R)$ are solution of different Schroedinger equations; they actually belong to different sets of orthonormal eigenfunctions and the subscripts m and k do not play the role of good quantum numbers. Therefore if $m \neq k$, the overlap integral in general may not go to zero even if $n \neq \ell$.

The diagram in Fig. 3 illustrates the role played by the overlap integral in determining the strength of a transition. It is evident from this diagram that the "vertical" transition AA' is

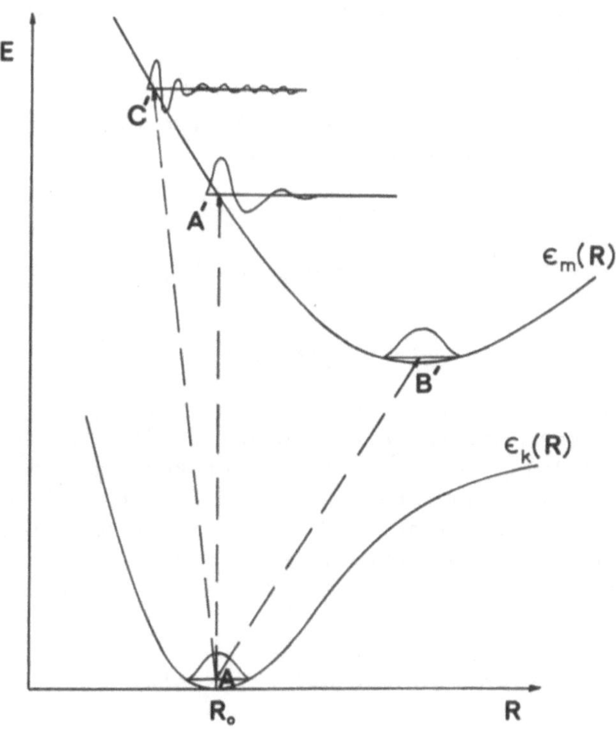

Fig. 3. Diagram illustrating the quantum mechanical Franck-Condon principle.

the one of highest probability and that the "sloped" transitions AB' and AC' have much lower probabilities; in particular in the former case (AB'), there is practically no overlap, and in the latter case (AC'), the overlap is small because of the oscillating nature of the upper vibrational eigenfunction.

The following considerations can be made:

1) The quantum mechanical Franck-Condon principle replaces the requirements R = const., p = const. of its semi-classical formulation with the notion that transitions for which such conditions are not fulfilled are very unlikely.

2) The Franck-Condon principle in its quantum mechanical formulation is not a selection rule in the conventional sense. It does not say that certain transitions cannot occur but, rather, that they are highly improbable. Selection rules are generally derived from the matrix element of an operator taken between two eigenfunctions of the <u>same</u> orthonormal set, contrary to the present situation where the relevant entity is an overlap integral of two wavefunctions belonging to different orthonormal sets.

3) If the potential curves of the two electronic states have the same shape, i.e., if

$$\varepsilon_k(R) = \varepsilon_m(R) + \text{const.}, \qquad (135)$$

then the two functions S_{mn} and $S_{k\ell}$ belong to the same set and

$$\int S_{mn}(R)^* S_{k\ell}(R)dR = \delta_{n\ell} \quad . \qquad (136)$$

In this case, the Franck-Condon principle can be expressed as the selection rule: "If the adiabatic potentials of two electronic states are the same, no transition can take place in which the molecule changes its vibrational state." This selection rule is independent from the nature (electric dipole, magnetic dipole, or electric quadrupole) of the transition.

IV.B. Probability of Radiative Transitions

We have already derived the following formula (see eq. 112) for the transition probability per unit time in the electric dipole approximation for one-photon processes ($\vec{k} = \vec{k}_\alpha$, polarization $\vec{\pi}_\alpha^\sigma$):

$$p^{\sigma}_{\alpha} d\Omega_{\alpha} = \begin{cases} \dfrac{4\pi^2 I(\omega_\alpha;\sigma)}{\hbar^2 c} |M_{fi}|^2 \, d\Omega_\alpha \\ \left\{ \dfrac{4\pi^2 I(\omega_\alpha;\sigma)}{\hbar^2 c} + \dfrac{\omega_\alpha^3}{hc^3} \right\} |M_{fi}|^2 \, d\Omega_\alpha \end{cases} , \tag{137}$$

where the upper (lower) row corresponds to absorption (emission) and where

$$|M_{fi}|^2 = |\langle \vec{\pi}^{\sigma}_{\alpha} \cdot \vec{M} \rangle|^2 = |\langle \vec{\pi}^{\sigma}_{\alpha} \cdot \sum_t q_t \vec{r}_t \rangle|^2 . \tag{138}$$

The probability of spontaneous emission is given by

$$P^{\sigma}_{\alpha}(sp)\, d\Omega_\alpha = \frac{\omega_\alpha^3}{hc^3} |M_{fi}|^2 d\Omega_\alpha \tag{139}$$

and the radiated power due to spontaneous emission by

$$P^{\sigma}_{\alpha}(sp)\, d\Omega_\alpha \, \hbar\omega_\alpha = \frac{\omega_\alpha^4}{2\pi c^3} |M_{fi}|^2 \, d\Omega_\alpha \quad . \tag{140}$$

If several pairs of initial and final states are connected by transitions corresponding to the same frequency, wave vector, and polarization, the power radiated per unit energy range, is

$$\frac{\omega_\alpha^4}{2\pi c^3} \sum_i N_i \sum_f |M_{fi}|^2 \, \delta(E_f - E_i + \hbar\omega_\alpha) d\Omega_\alpha =$$

$$= A_+ \sum_i N_i \sum_f |M_{fi}|^2 \, \delta(E_f - E_i + \hbar\omega_\alpha) d\Omega_\alpha \quad , \tag{141}$$

where N_i = number of molecules occupying the state i and

$$A_+ = \frac{\omega_\alpha^4}{2\pi c^3} \quad . \tag{142}$$

The expression for the absorption and emission transition probability is given by

$$p^{\lambda}_{\alpha}(abs;emi)\, d\Omega_\alpha = \frac{4\pi^2 I(\omega_\alpha;\sigma)}{\hbar^2 c} |M_{fi}|^2 \, d\Omega_\alpha \, . \tag{143}$$

Correspondingly, the power absorbed is

$$P^{\sigma}_{\alpha} d\Omega_\alpha \hbar\omega_\alpha = \frac{4\pi^2 \omega_\alpha I(\omega_\alpha;\sigma)}{\hbar c} |M_{fi}|^2 \, d\Omega_\alpha \quad , \tag{144}$$

and if several pairs of levels may be connected by transitions of the same frequency ω_α, wave vector, and polarization, the power absorbed per unit energy range is given by

$$\frac{4\pi^2\omega_\alpha I(\omega_\alpha;\sigma)}{\hbar c} \sum_i N_i \sum_f |M_{fi}|^2 \delta(E_f-E_i-\hbar\omega_\alpha)d\Omega_\alpha =$$

$$= A_- \sum_i N_i \sum_f |M_{fi}|^2 \delta(E_f-E_i-\hbar\omega_\alpha)d\Omega_\alpha \quad , \qquad (145)$$

where

$$A_- = \frac{4\pi^2\omega_\alpha I(\omega_\alpha;\sigma)}{\hbar c} \quad . \qquad (146)$$

We can write (141) and (145) concisely by using the expression

$$A_\pm \sum_i N_i \sum_f |M_{fi}|^2 \delta(E_f-E_i \pm h\omega_\alpha)d\Omega_\alpha \qquad (147)$$

where the plus and minus signs stand for spontaneous emission and absorption, respectively, and A_+ and A_- are given by (142) and (146), respectively.

IV.C. Integrated Intensity of a Vibronic Band

In anticipation of our treatment of solids, we shall call "Zero-phonon" line a spectral line corresponding to a transition in which the vibrational state of the system does not change.

An absorption or emission zero-phonon line is accompanied by other electronic transitions in which the vibrational state of the system undergoes a change; these transitions are called "vibrational-electronic" or "vibronic" transitions. The zero-phonon transition and the accompanying vibronic transitions produce a spectral "band."

In order to calculate the integrated intensity of such a band, we have to use the formula

$$A \sum_i N_i \sum_f |M_{fi}|^2 \quad , \qquad (148)$$

where i is the initial state, f is the final state, and M_{fi} is given by (138). With the notation

i : 0(electronic), n_i(vibrational)

f : 1(electronic), n_f(vibrational)

the integrated intensity of the band is given by

$$A \sum_i N_i \sum_f \left| \int d^3\vec{R} \phi_{1n_f}(\vec{R})^* D_{10}(\vec{R}) \phi_{0n_i}(\vec{R}) \right|^2 \tag{149}$$

where

$$D_{10}(\vec{R}) = \int d^3\vec{r}\, \psi_1(\vec{r},\vec{R})^* \left[-e\sum_t \vec{r}_t \cdot \vec{\pi} \right] \psi_0(\vec{r},\vec{R}). \tag{150}$$

Then

$$A \sum_i N_i \sum_f |M_{fi}|^2 =$$

$$= A \sum_i N_i \sum_f \int d^3\vec{R} \phi_{1n_f}(\vec{R})^* D_{10}(\vec{R}) \phi_{0n_i}(\vec{R}) \int d^3\vec{R}' \phi_{1n_f}(\vec{R}') D_{10}(\vec{R}')^* \phi_{0n_f}(\vec{R}')^* =$$

$$= A \sum_i N_i \iint d^3\vec{R} d^3\vec{R}' \left[\sum_f \phi_{1n_f}(\vec{R})^* \phi_{1n_f}(\vec{R}') \right] \left[D_{10}(\vec{R}) D_{10}(\vec{R}')^* \right] \phi_{0n_i}(\vec{R}) \phi_{0n_i}(\vec{R}')^* =$$

$$= A \sum_i N_i \iint d^3\vec{R} d^3\vec{R}' \delta(\vec{R}-\vec{R}') D_{10}(\vec{R}) D_{10}(\vec{R}')^* \phi_{0n_i}(\vec{R}) \phi_{0n_i}(\vec{R}')^* =$$

$$= A \sum_i N_i \int d^3\vec{R} |D_{10}(\vec{R})|^2 |\phi_{0n_i}(\vec{R})|^2 \tag{151}$$

This result was derived with the assumption of the validity of the adiabatic approximation, but with no assumption regarding the Franck-Condon approximation. If the latter approximation is considered valid, then

$$A \sum_i N_i \sum_f |M_{fi}|^2 = A \sum_i N_i |D_{10}^0|^2 \int d^3\vec{R} |\phi_{0n_i}(\vec{R})|^2 =$$

$$= A |D_{10}^0|^2 \sum_i N_i = A |D_{10}^0|^2 N \quad , \tag{152}$$

where N = total number of molecules.

The above result indicates that the overall integrated intensity of the <u>entire</u> vibronic band, consisting of the zero-phonon line and the accompanying vibronic lines, is independent of temperature within the limits of the present model:

a. the coefficient A is constant, independent of frequency,

b. the adiabatic approximation is valid, and

c. the Franck-Condon approximation is valid.

V. SIMPLE MODELS FOR MOLECULAR SPECTRA

V.A. Radiative Transitions in Diatomic Molecules

We shall examine this problem under the following assumptions:

a) ω, the vibrational frequency of the diatomic molecule is such that $\hbar\omega >> kT$, which implies that only the ground vibrational level is populated and

b) the frequency ω is the same in the ground and excited electronic states of the molecule.

Under these conditions, the intensity of a vibronic (absorption) line of frequency ω corresponding to a transition by which the molecule undergoes a change of both its electronic and its vibrational states from 0 to n is given by

$$A\sum_i N_i \sum_f |M_{if}|^2 \delta(E_i - E_f + \hbar\omega) = AN|M_{fi}|^2 = AN|D_{10}^0|^2 |\int \phi_{1n}(q)\phi_{00}(q)dq|^2. \tag{153}$$

This transition is represented in Fig. 4.

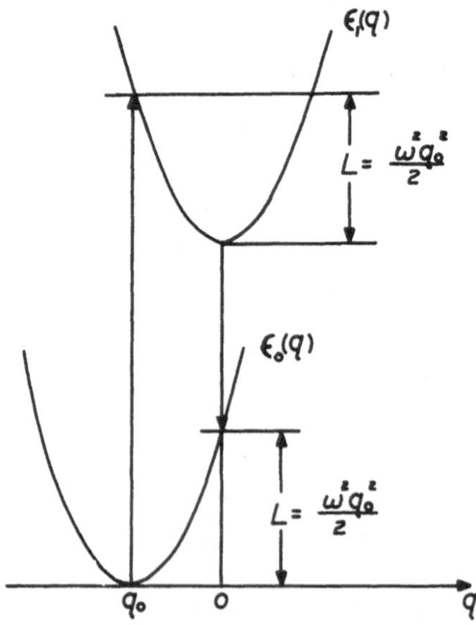

Fig. 4. Adiabatic potential curves for a diatomic molecule.

The quantity

$$L = \frac{\omega^2 q_o{}^2}{2} \quad , \tag{154}$$

where q_o = displacement of the two adiabatic potential curves, is the "Stokes loss" and the quantity

$$\ell = \frac{\omega^2 q_o{}^2}{2} / \hbar\omega \tag{155}$$

is the Stokes loss in number of vibrational quanta.

The overlap integral of interest is now

$$\int \phi_{1n}(q)\,\phi_{00}(q)\,dq. \tag{156}$$

The functions ψ above are harmonic oscillator eigenfunctions. These functions are given by the formula

$$\phi_n(q) = \frac{1}{\sqrt{\bar{q}}}\,\frac{(-1)^n}{\sqrt{2^n n!\sqrt{\pi}}}\; e^{y^2/2}\,\frac{d^n}{dy^n}\,e^{-y^2} \tag{157}$$

where

$$y = \frac{q}{\bar{q}} = \frac{q}{\sqrt{\hbar/\omega}} = \text{pure number} \tag{158}$$

$$\bar{q} = \sqrt{\frac{\hbar}{\omega}} = \text{zero point energy amplitude.} \tag{159}$$

In particular

$$\phi_0(q) = \frac{1}{\sqrt{\bar{q}}}\,\frac{1}{\sqrt[4]{\pi}}\,e^{-\frac{1}{2}y^2} \quad . \tag{160}$$

Therefore

$$\phi_{00}(q) = \frac{1}{\sqrt{\bar{q}}}\,\frac{1}{\sqrt[4]{\pi}}\,e^{-\frac{1}{2}(y+y_o)^2} = \frac{1}{\sqrt{\bar{q}}}\,\frac{1}{\sqrt[4]{\pi}}\,e^{-\frac{y^2}{2} - \frac{y_o{}^2}{2} - yy_o} \tag{161}$$

where

$$y_o = \frac{q_o}{\sqrt{\hbar/\omega}} = \frac{q_o}{\bar{q}} \tag{162}$$

Also

$$\phi_{1n}(q) = \frac{1}{\sqrt{\bar{q}}}\,\frac{(-1)^n}{\sqrt{2^n n!\sqrt{\pi}}}\; e^{\frac{y^2}{y}}\,\frac{d^n}{dy^n}\,e^{-y^2} \tag{163}$$

and

$$\int_{-\infty}^{+\infty} \phi_{1n}(q)\phi_{00}(q)dq =$$

$$= \frac{1}{q} \frac{(-1)^n}{\sqrt{2^n n! \pi}} e^{-\frac{y_o^2}{2}} \int_{-\infty}^{+\infty} e^{-yy_o} \frac{d^n}{dy^n} e^{-y^2} dq =$$

$$= \frac{(-1)^n}{\sqrt{2^n n! \pi}} e^{-\frac{y_o^2}{2}} \int_{-\infty}^{+\infty} e^{-yy_o} \frac{d^n}{dy^n} e^{-y^2} dy. \tag{164}$$

By successive integrations by part

$$\int_{-\infty}^{+\infty} e^{-yy_o} \frac{d^n}{dy^n} e^{-y^2} dy = \int_{-\infty}^{+\infty} e^{-yy_o} \frac{d}{dy}\left(\frac{d^{n-1}}{dy^{n-1}} e^{-y^2}\right) dy =$$

$$= \left[e^{-yy_o} \frac{d^{n-1}}{dy^{n-1}} e^{-y^2}\right]_{-\infty}^{+\infty} + y_o \int_{-\infty}^{+\infty} e^{-yy_o} \left(\frac{d^{n-1}}{dy^{n-1}} e^{-y^2}\right) dy =$$

$$= y_o \int_{-\infty}^{+\infty} \left(\frac{d^{n-1}}{dy^{n-1}} e^{-y^2}\right) e^{-yy_o} dy = y_o^n \int_{-\infty}^{+\infty} e^{-y^2 - yy_o} dy . \tag{165}$$

We solve this integral by completing the square in the exponent

$$\int_{-\infty}^{+\infty} e^{-y^2 - yy_o} dy = e^{\frac{y_o^2}{4}} \int_{-\infty}^{+\infty} e^{-\left(\frac{y_o}{2} + y\right)^2} dy . \tag{166}$$

We set

$$\frac{y_o}{2} + y = z. \tag{167}$$

Then

$$\int_{-\infty}^{+\infty} e^{-y^2 - yy_o} dy = e^{\frac{y_o^2}{4}} \int_{-\infty}^{+\infty} e^{-z^2} dz = e^{\frac{y_o^2}{4}} \sqrt{\pi} \tag{168}$$

and

$$\int_{-\infty}^{+\infty} \phi_{1n}(q)\phi_{00}(q)dq = \frac{(-1)^n}{\sqrt{2^n n! \pi}} e^{-\frac{y_o^2}{2}} y_o^n e^{\frac{y_o^2}{4}} \sqrt{\pi} =$$

$$= \frac{(-y_o)^n}{\sqrt{2^n n!}} e^{-\frac{y_o^2}{4}} . \tag{169}$$

Also

$$\left|\int_{-\infty}^{+\infty} \phi_{1n}(q)\phi_{00}(q)dq\right|^2 = e^{-\frac{y_o^2}{2}} \left(\frac{y_o^2}{2}\right)^n \frac{1}{n!} . \tag{170}$$

But

$$\frac{y_o^2}{2} = \frac{q_o^2}{2q} = \frac{q_o^2}{2\hbar} = \frac{q_o^2\omega}{2\hbar} = \frac{q_o^2\omega^2}{2} \Big/ \hbar\omega =$$

$$= \ell = \text{Stokes loss in number of vibrational quanta.} \tag{171}$$

Therefore

$$\left|\int\phi_{1n}(q)\phi_{00}(q)dq\right|^2 = e^{-\ell}\frac{\ell^n}{n!} . \tag{172}$$

This formula is known as the "Pekarian Formula." The function (172) is represented in Fig. 5 for different values of the Stokes loss ℓ. We note here that similar patterns correspond to those observed in the electronic spectra of diatomic molecules, apart from the effects caused by the rotation of the molecules, which produces additional fine structure to each vibronic line.

We can now make the following observations:

1) The Pekarian formula expresses the Poisson distribution

$$P(n) = e^{-\bar{n}}\frac{(\bar{n})^n}{n!} , \tag{173}$$

if we replace ℓ with $\bar{n}$.

2) For large $\ell(\bar{n})$, the Pekarian (Poisson) distribution goes over to a Gaussian distribution.

3) If n = 0, we obtain

$$|\int\phi_{10}(q)\phi_{00}(q)dq|^2 = e^{-\ell} . \tag{174}$$

The intensity of the n = 0 line is the smaller, the larger is ℓ.

4) If n = 1, we get

$$|\int\phi_{11}(q)\phi_{00}(q)dq|^2 = e^{-\ell}\ell \approx \ell , \tag{175}$$

for small ℓ .

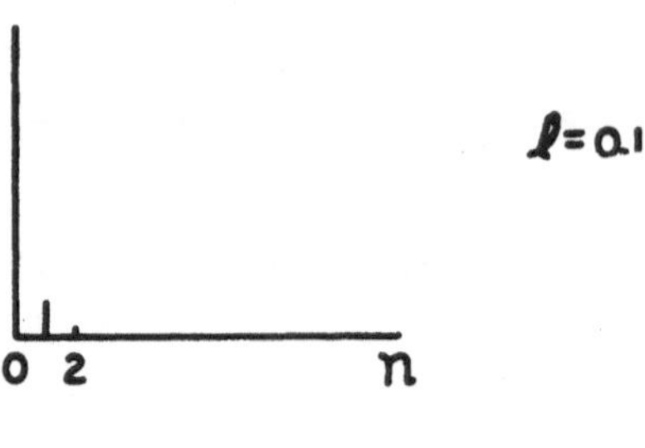

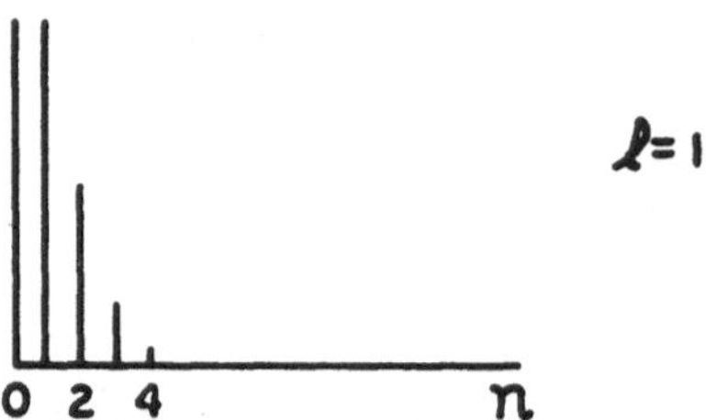

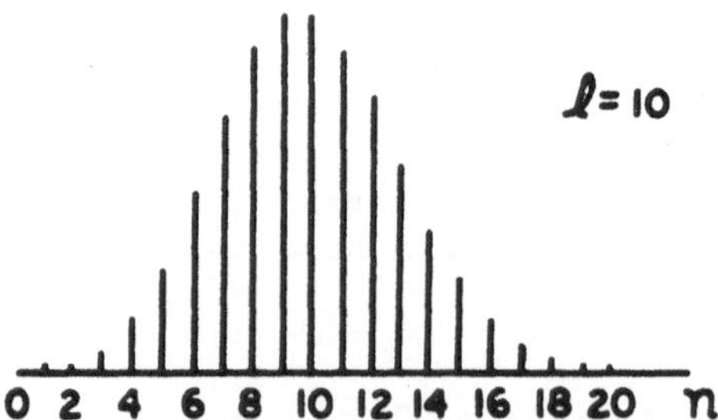

Fig. 5. Typical vibronic transitions with creation of vibrational quanta.

5) If $\ell = 0$, the expression (172) becomes zero for $n \neq 0$. This means: no vibronic lines without Stokes shift.

V.B. The Formula of Huang and Rhys

We shall treat now a case similar to the one just examined. As previously, the vibrational frequency ω will be considered to be the same in the ground and in the excited electronic states of the molecule. However, the restriction on the temperature ($\hbar\omega >> kT$) is removed.

The most general (absorption) transition initiates in the nth vibrational level of the ground electronic state and ends up in the mth vibrational level of an excited electronic state.

In this process, n vibrational quanta are destroyed and m vibrational quanta are created. The net number of vibrational quanta created is

$$p = m - n. \qquad (176)$$

We shall now obtain the formula of Huang and Rhys by following the derivation made by Curie (7).

The probability of creating a quantum of vibration may be taken to be

$$\pi_e \sim \ell(N + 1) \qquad (177)$$

where ℓ = Stokes loss in number of vibrational quanta, measure of the strength of the electron-vibration coupling and

$$N = \frac{1}{e^{\hbar\omega/kT} - 1} \quad . \qquad (178)$$

N represents the average degree of excitation of the quantum vibrational oscillator at temperature T.

Similarly, the probability of destroying a quantum of vibration may be taken to be

$$\pi_a \sim \ell N \quad . \qquad (179)$$

Note that if $T = 0$, $N = 0$, $\pi_a = 0$ and $\pi_e \sim \ell$.

The probability of a process in which m vibrational quanta are created and n vibrational quanta are destroyed is then

$$\sim \frac{\pi_e^m}{m!} \; \frac{\pi_a^n}{n!} \qquad (180)$$

Normalizing

$$\left[e^{-\pi_e}\frac{\pi_e^{\ m}}{m!}\right]\left[e^{-\pi_a}\frac{\pi_a^{\ n}}{n!}\right] = e^{-(\pi_e+\pi_a)}\frac{\pi_e^{\ m}}{m!}\frac{\pi_a^{\ n}}{n!} =$$

$$= e^{-\ell(2N+1)}\frac{[\ell(N+1)]^m}{m!}\frac{(\ell N)^n}{n!} = e^{-\ell(2N+1)}\frac{\ell^{n+m}}{n!m!}N^n(N+1)^m . \tag{181}$$

The probability that a net number of p = m - n vibrational quanta are created is given by

$$W(p,\ell) = \sum_{n=0}^{\infty} e^{-\ell(2N+1)}\frac{\ell^{2n+p}}{n!(n+p)!}N^n(N+1)^{n+p} =$$

$$= e^{-\ell(2N+1)}\sum_{n=0}^{\infty}\frac{\ell^{2n+p}}{n!(n+p)!}\frac{N^{n+p/2}}{N^{p/2}}(N+1)^{n+p/2}(N+1)^{p/2} =$$

$$= e^{-\ell(2N+1)}\left(\frac{N+1}{N}\right)^{p/2}\sum_{n=0}^{\infty}\frac{\left[\ell\sqrt{N(N+1)}\right]^{2n+p}}{n!(n+p)!} =$$

$$= e^{-\ell(2N+1)}\left(\frac{N+1}{N}\right)^{p/2} I_p\left[2\ell\sqrt{N(N+1)}\right] \tag{182}$$

where (8)

$$I_p(x) = \sum_{n=0}^{\infty}\frac{\left(\frac{x}{2}\right)^{2n+p}}{n!(n+p)!} = (i)^{-p}J_p(ix)$$

= modified Bessel function of the first kind, of order p. (183)

Therefore

$$A\sum_i N_i \sum_f |M_{fi}|^2\ \delta(E_f - E_i - \hbar\omega)$$

$$= A\ N|D_{10}^{o}|^2\ W(p,\ell) , \tag{184}$$

where

$$W(p,\ell) = e^{-\ell(2N+1)}\left(\frac{N+1}{N}\right)^{p/2} I_p\left[2\ell\sqrt{N(N+1)}\right] \tag{185}$$

and N = total number of molecules.

We can now make the following observations:

1) Since

$$\sum_{p=-\infty}^{+\infty} W(p,\ell) = 1 \tag{186}$$

we find, in agreement with (152)

$$A \sum_i N_i \sum_f |M_{fi}|^2 = A\, N|D^o_{10}|^2 \quad . \tag{187}$$

2) If $\ell = 0$ all I_p are zero except I_o; namely, only zero-phonon lines are allowed.

3) At low T only the n = 0 term is relevant. In this case, m = p and

$$W(p,\ell) \rightarrow e^{-\ell}\left(\frac{1}{N}\right)^{p/2} \frac{\ell\sqrt{N}}{p!}^{p} = e^{-\ell}\frac{\ell}{p!} \quad . \tag{188}$$

We reobtain in this case the Pekarian formula.

4) For n = 0, p > 0 and small ℓ, we get

$$W(p,\ell) \rightarrow \left(\frac{N+1}{N}\right)^{p/2} \frac{\left[\ell\sqrt{N(N+1)}\right]^{p}}{p!} = \frac{\ell^p}{p!}\,(1+N)^p. \tag{189}$$

VI. NONRADIATIVE PROCESSES

VI.A. Deviations from the Adiabatic Approximation

The "exact" Hamiltonian of a molecular system is given as in Section II.A. by

$$H = T_r + T_R + V(\vec{r},\vec{R}) \tag{190}$$

where

T_r = kinetic energy of the electrons

T_R = kinetic energy of the nuclei

$V(\vec{r},\vec{R}) = V_{ee} + V_{nn} + V_{ne}$.

In the adiabatic approximation, the eigenstates of the system are given by

$$\Psi_i(\vec{r},\vec{R}) = \psi_k(\vec{r},\vec{R})\phi_{k\ell}(\vec{R}) \; , \tag{191}$$

where

$$\left[T_r + V(\vec{r},\vec{R})\right]\psi_k(\vec{r},\vec{R}) = \varepsilon_k(\vec{R})\psi_k(\vec{r},\vec{R}) \quad (192)$$

$$\left[T_R + \varepsilon_k(\vec{R})\right]\phi_{k\ell}(\vec{R}) = E_{k\ell}\,\phi_{k\ell}(\vec{R}) \; . \quad (193)$$

If we operate with H on Ψ_i we find

$$H\Psi_i(\vec{r},\vec{R}) = \left[T_r + T_R + V(\vec{r},\vec{R})\right]\;\psi_k(\vec{r},\vec{R})\phi_{k\ell}(\vec{R}) =$$

$$= \left[T_r + V(\vec{r},\vec{R})\right]\psi_k(\vec{r},\vec{R})\,\phi_{k\ell}(\vec{R}) + T_R\psi_k(\vec{r},\vec{R})\phi_{k\ell}(\vec{R}) =$$

$$= \varepsilon_k(\vec{R})\psi_k(\vec{r},\vec{R})\phi_{k\ell}(\vec{R}) + T_R\psi_k(\vec{r},\vec{R})\phi_{k\ell}(\vec{R}) \; , \quad (194)$$

where we have taken advantage of (192). From (193), we obtain

$$\psi_k(\vec{r},\vec{R})T_R\phi_{k\ell}(\vec{R}) + \varepsilon_k(\vec{R})\psi_k(\vec{r},\vec{R})\phi_{k\ell}(\vec{R}) = E_{k\ell}\psi_k(\vec{r},\vec{R})\phi_{k\ell}(\vec{R}) \quad (195)$$

and

$$\varepsilon_k(\vec{R})\psi_k(\vec{r},\vec{R})\phi_{k\ell}(\vec{R}) = E_{k\ell}\psi_k(\vec{r},\vec{R})\phi_{k\ell}(\vec{R}) - \psi_k(\vec{r},\vec{R})T_R\phi_{k\ell}(\vec{R}) \, . \quad (196)$$

Therefore (194) becomes

$$H\Psi_i(\vec{r},\vec{R}) = E_{k\ell}\psi_k(\vec{r},\vec{R})\phi_{k\ell}(\vec{R}) - \psi_k(\vec{r},\vec{R})T_R\phi_{k\ell}(\vec{R}) + T_R\psi_k(\vec{r},\vec{R})\phi_{k\ell}(\vec{R}) =$$

$$= E_{k\ell}\psi_k(\vec{r},\vec{R})\phi_{k\ell}(\vec{R}) + A\psi_k(\vec{r},\vec{R})\phi_{k\ell}(\vec{R}) \; , \quad (197)$$

where A, the "nonadiabatic" operator, is defined by

$$A\psi_k(\vec{r},\vec{R})\phi_{k\ell}(\vec{R}) = T_R\psi_k(\vec{r},\vec{R})\phi_{k\ell}(\vec{R}) - \psi_k(\vec{r},\vec{R})T_R\phi_{k\ell}(\vec{R}) \; . \quad (198)$$

Let us follow up

$$A\psi_k\phi_{k\ell} = T_R\psi_k\phi_{k\ell} - \psi_k T_R\phi_{k\ell} =$$

$$= \psi_k T_R\phi_{k\ell} + \phi_{k\ell}T_R\psi_k - \sum_s \frac{\hbar^2}{M_s}\vec{\nabla}_s\psi_k\cdot\vec{\nabla}_s\phi_{k\ell} - \psi_k T_r\phi_{k\ell} =$$

$$= \phi_{k\ell}T_R\psi_k - \sum_s \frac{\hbar^2}{M_s}\vec{\nabla}_s\psi_k\cdot\vec{\nabla}_s\phi_{k\ell} \; . \quad (199)$$

The first term of this last expression is negligible in respect to the second term in the assumption that

$$|\vec{\nabla}_s\psi_k(\vec{r},\vec{R})| << |\vec{\nabla}_s\phi_{k\ell}(\vec{R})| \, . \quad (200)$$

Therefore

$$A\;\psi_k(\vec{r},\vec{R})\phi_{k\ell}(\vec{R}) = -\sum_s \frac{\hbar^2}{M_s}\vec{\nabla}_s\psi_k(\vec{r},\vec{R})\cdot\vec{\nabla}_s\phi_{k\ell}(\vec{R}) \; . \quad (201)$$

VI.B. Mechanism for Radiationless Transitions

The work we did in the previous section has led us to the following result. When we operate on a Born-Oppenheimer product state function of the type (191), we obtain

$$H\Psi_i(\vec{r},\vec{R}) = H\psi_k(\vec{r},\vec{R})\phi_{k\ell}(\vec{R}) = E_i\Psi_i(\vec{r},\vec{R}) + A\Psi_i(\vec{r},\vec{R}), \tag{202}$$

where $E_i = E_{k\ell}$.

The general solution of the time dependent Schroedinger equation

$$H\Psi(t) = i\hbar \frac{\partial\Psi(t)}{\partial t} \tag{203}$$

can be expanded in terms of the eigenfunctions $\Psi_i(\vec{r},\vec{R})$ as follows:

$$\Psi(t) = \sum_i c_i(t)\ \Psi_i(\vec{r},\vec{R})e^{-\frac{i}{\hbar}E_i t} = \sum_i c_i(t)\Psi_i(t)\ , \tag{204}$$

where the wavefunctions

$$\Psi_i(t) = \psi_k(\vec{r},\vec{R})\phi_{k\ell}(\vec{R})e^{-i\frac{E_{k\ell}}{\hbar}t}$$

represent states of definite energy $E_{k\ell} = E_i$. Replacing the expression (204) of $\Psi(t)$ in (203)

$$H\sum_i c_i(t)\Psi_i(t) = i\hbar\{\sum_i c_i(t)\frac{\partial\Psi_i(t)}{\partial t} + \sum_i \frac{\partial c_i(t)}{\partial t}\Psi_i(t)\}\ . \tag{205}$$

Then

$$\sum_i c_i(t)H\Psi_i(t) = \sum_i c_i(t)E_i\Psi_i(t) + i\hbar\frac{\partial c_i(t)}{\partial t}\Psi_i(t). \tag{206}$$

Taking (202) into account, (206) becomes

$$\sum_i c_i(t)E_i\Psi_i(t) + \sum_i c_i(t)A\Psi_i(t) = \sum_i c_i(t)E_i\Psi_i(t) + ih\sum_i\frac{\partial c_i(t)}{\partial t}\Psi_i(t)$$

or

$$\sum_i c_i(t)A\Psi_i(t) = i\hbar\sum_i \frac{\partial c_i(t)}{\partial t}\quad . \tag{207}$$

Multiplying by $\Psi_j^*(t)$ and integrating over spatial coordinates, we obtain

$$i\hbar\frac{\partial c_j(t)}{\partial t} = \sum_i c_i(t) < \Psi_j(t)|A|\Psi_i(t) > =$$

$$= \sum_i c_i(t)\, e^{i\frac{E_j-E_i}{\hbar}t} < \Psi_j(0)|A|\Psi_i(0) > \quad . \tag{208}$$

The matrix element of importance is

$$<\Psi_j(0)|A|\Psi_i(0)> = <\psi_m(\vec{r},\vec{R})\phi_{mn}(\vec{R})|A|\psi_k(\vec{r},\vec{R})\phi_{k\ell}(\vec{R})> \quad . \tag{209}$$

VI.C. Radiationless Transition Probability per Unit Time

We get

$$\frac{E_j - E_i}{\hbar} = \omega_{ji} \tag{210}$$

$$<\Psi_j(0)|A|\Psi_i(0)> = A_{ji} \quad . \tag{211}$$

Then (208) becomes

$$i\hbar\, \dot{c}_j(t) = \sum_i c_i(t) A_{ji}\, e^{i\omega_{ji}t} . \tag{212}$$

Assume that only two states are involved

i : k(electronic), ℓ(vibrational)

j : m(electronic), n(vibrational) .

Then (212) gives

$$\begin{cases} i\hbar\, \dot{c}_j(t) = c_i(t) A_{ji}\, e^{i\omega_{ji}t} \\ i\hbar\, \dot{c}_i(t) = c_j(t) A_{ij}\, e^{i\omega_{ij}t} \quad . \end{cases} \tag{213}$$

These two equations can be solved if we know the initial conditions. We assume these to be

$$\begin{cases} c_i(0) = 1 \\ c_j(0) = 0 \quad . \end{cases} \tag{214}$$

We find

$$\begin{cases} c_j(t) = \dfrac{A_{ji}}{i\hbar a}\, e^{i\frac{\omega_{ji}}{2}} \sin a\, t \\ c_i(t) = e^{-i\frac{\omega_{ji}}{2}t}\left(\cos a\, t + i\, \dfrac{\omega_{ji}}{2a} \sin at\right), \end{cases} \tag{215}$$

where

$$a = \tfrac{1}{2}\left[\omega_{ji}^2 + \frac{4|A_{ji}|^2}{h^2}\right]^{\frac{1}{2}} \quad . \tag{216}$$

Therefore

$$|c_j(t)|^2 = \frac{|A_{ji}|^2}{\hbar^2 a^2} \sin^2 at \tag{217}$$

$$|c_i(t)|^2 = \cos^2 at + \frac{\omega_{ji}^2}{4a^2} \sin^2 at \quad . \tag{218}$$

$|c_j(t)|^2$ is the probability that, if the system is in the state Ψ_i at time t = 0, it can be found in state Ψ_j at time t. We can see that this probability oscillates in time. Accordingly, the probability of finding the system in the state Ψ_i which is 1 at time t = 0, becomes again 1 at times t = π/a, $2\pi/a$, $3\pi/a$, etc. This situation is clearly at variance with the irreversibility we expect, following a radiationless transition.

But now let us assume that j is in a continuum. Then the equations (213) can be written as follows:

$$\begin{cases} i\hbar\, \dot{c}_j(t) = c_i(t) A_{ji} e^{i\omega_{ji}t} \\ \\ i\hbar\, \dot{c}_i(t) = \int c_j(t) A_{ij} e^{-i\omega_{ji}t} dj \quad . \end{cases} \tag{219}$$

Let us set

$$c_i(t) = e^{-(\gamma/2)t} \quad . \tag{220}$$

Then the first equation of the set (219) integrates as follows

$$c_j(t) = \frac{1}{i\hbar} A_{ji} \frac{e^{(i\omega_{ji} - \gamma/2)t} - 1}{i\omega_{ji} - \gamma/2} \quad . \tag{221}$$

We replace this expression in the second equation of the set (219) and take (220) into account. We obtain then

$$\gamma = \frac{2i}{\hbar^2} \int |A_{ji}|^2 \frac{1 - e^{(i\omega_{ij} + \gamma/2)t}}{\omega_{ij} - i\,\gamma/2} dj \quad . \tag{222}$$

But

$$dj = \frac{dj}{d\omega_{ij}} d\omega_{ij} = \rho(\omega_{ij}) d\omega_{ij} \quad ,$$

where $\rho(\omega_{ij})$ = density of states in the continuum. Then

$$\gamma = \frac{2i}{\hbar^2} \int |A_{ji}|^2 \frac{1 - e^{(i\omega_{ij} + \gamma/2)t}}{\omega_{ij} - i\,\gamma/2} \rho(\omega_{ij}) d\omega_{ij} \, . \tag{223}$$

In evaluating the integral above, we shall neglect γ in the integrals and write

$$\gamma = \frac{2i}{\hbar^2} \int |A_{ji}|^2 \; \frac{1 - e^{i\omega_{ij}t}}{\omega_{ij}} \rho(\omega_{ij}) d\omega_{ij} \; . \tag{224}$$

Consider now the function

$$\frac{1 - e^{i\omega t}}{\omega} = \frac{1 - \cos\omega t}{\omega} - i \, \frac{\sin\omega t}{\omega} \quad . \tag{225}$$

For very large t, the first term may be replaced by $1/\omega$ for $\omega \neq 0$, because the rapidly oscillating $\cos \omega t$ does not give any contribution to the integral; at $\omega = 0$ this term goes to zero. When it is multiplied by the rest of the integrand and integrated over ω, the result is the principal value of the integrand.

For very large t, the second term of (225) can be expressed as $i\pi\delta/\omega$. In fact, the function $\delta(x)$ can be written as follows:

$$\delta(x) = \frac{1}{\pi} \lim_{\alpha\to\infty} \frac{\sin\alpha x}{x} \quad . \tag{226}$$

Therefore, we may write

$$\lim_{t\to\infty} \frac{1 - e^{i\omega t}}{\omega} = \frac{P}{\omega} - i\pi\delta(\omega) \quad , \tag{227}$$

where P indicates the principal value.

Now γ can be expressed as follows for very large t

$$\gamma = \frac{2i}{\hbar^2} \int |A_{ji}|^2 \rho(\omega_{ij}) \left\{ \frac{P}{\omega_{ij}} - i\pi\delta \; \omega_{ij} \right\} d\omega_{ij} =$$

$$= \frac{2\pi}{\hbar^2} \; |A_{ji}|^2 \rho(\omega_i = \omega_j) + i \; \frac{2}{\hbar^2} P \int \frac{|A_{ji}|^2 \rho(\omega_{ij})}{\omega_{ij}} \, d\omega_{ij} \tag{228}$$

or

$$\gamma = W + I \tag{229}$$

where

$$W = \mathrm{Re}(\gamma) = \frac{2\pi}{\hbar^2} \; |A_{ji}|^2 \rho(\omega_i = \omega_j) \tag{230}$$

$$I = \mathrm{Im}(\gamma) = \frac{2}{\hbar} P \int \frac{|A_{ji}|^2 \rho(\omega_{ij})}{\hbar\omega_{ij}} \, d\omega_{ij} \; . \tag{231}$$

The quantity W is the transition probability per unit time from state i to state j.

It is

$$c_i(t) = e^{-(\gamma/2)t} = \exp(- \frac{W}{2} \, t - \frac{i}{\hbar} \, \Delta E \; t) \tag{232}$$

where

$$\Delta E = P\int \frac{|A_{ji}|^2 \rho(\omega_{ij})}{\hbar\omega_{ij}} d\omega_{ij} \tag{233}$$

represents a correction to the energy to second order in the perturbation. Also from (221)

$$c_j(t) = -\frac{i}{\hbar} A_{ji} \frac{e^{(i\omega_{ji} - \gamma/2)t} - 1}{i\omega_{ji} - \gamma/2} \xrightarrow[t>>1/W]{}$$

$$\longrightarrow \frac{i}{\hbar} A_{ji} \frac{1}{i(\omega_j - \omega_i) - \frac{1}{2}(W + iI)} = \frac{i}{\hbar} A_{ji} \frac{1}{i[\omega_j - (\omega_i + \frac{1}{2}I)] - \frac{1}{2}W} \tag{234}$$

and

$$|c_j(t)|^2 = \frac{1}{\hbar^2} |A_{ji}|^2 \frac{1}{[(\omega_i + \frac{1}{2}I) - \omega_j]^2 + W^2/4} \ . \tag{235}$$

The relations (232) and (233) indicate that the $i \rightarrow j$ transition process is now irreversible! We can then state that in order to obtain irreversible transitions and a progressive depletion of the initial state, it is necessary that the discrete initial state be coupled to a large number of states of similar energy.

VII MULTIPHONON RELAXATION IN SOLIDS

VII.A. Introduction

We shall now focus our attention on the spectra of impurity ions in solids. We shall assume the following:

a. The concentration of optically-active ions is very low: no ion-ion interaction.
b. The electronic energy levels of the impurity ions do not overlap with the electronic energy bands of the solid.

In dealing with the radiative processes of ions in solids, we have to consider the fact that these ions are part of a structure which undergoes vibrational (thermal) motion. We may at this point make the following observations:

1. The "electronic" energy eigenvalue $\varepsilon_k(\vec{R})$ plays the role of a potential in which the atoms of the solid perform their vibrational motions.

2. These motions can be thought of as a superposition of normal modes of vibration which are represented

by normal coordinates q_i (i = 1,2,......, 3N), where N=number of atoms in the solid. (More precisely, since the solid has also three translational and three rotational degrees of freedom, the number of normal modes of vibration is 3N-6).

3. A normal mode is a pattern of motion in which, in general, all the atoms of the solid participate. There may be, however, normal modes, called "localized," which are related to the motion of a relatively small number of atoms.

4. In treating the vibrations of solids, the "harmonic approximation" is used. According to this approximation, no exchange of energy takes place between the normal modes and the "potential" energy $\varepsilon_k(\vec{R})$ and the kinetic energy are both sums of independent quadratic ($\sim q^2, \sim \dot{q}^2$) terms, one for each normal mode.

5. Each normal mode is equivalent to a harmonic oscillator. The vibrations of a solid are then equivalent to a collection of 3N harmonic oscillators. If an oscillator of frequency ω is in its nth excited state, this fact is also expressed by saying that n "phonons" of energy $\hbar\omega$ are present in the solid.

6. The different normal modes are not completely isolated, but are, rather, <u>in speaking terms</u> due to their anharmonicity: this provides the mechanism for reaching thermal equilibrium.

7. When considering the thermal vibrations of a solid, it may be interesting to have an idea of how many phonons may be present in a solid at, say, room temperature. The number of phonons in a frequency interval (ω, $\omega + d\omega$) is given by $\bar{n}(\omega)\rho(\omega)\,d\omega$ where: $\bar{n}(\omega) = (e^{\hbar\omega/kT}-1)^{-1}$ and $\rho(\omega)$ = density of phonon states $\propto \omega^2/c_s{}^3$. It is the value of c_s = velocity of sound in solids $\sim 5 \times 10^5$ cm/sec (versus c = velocity of light = 3×10^{10} cm/sec) that makes the number of phonons extremely large. For the sake of comparison, the total number of photons/cm^3 in black body radiation at T=300°K is $\sim 6.4 \times 10^8$, the number of photons/cm^3 in a typical laser medium (λ = 6300 Å, 1W/cm^2) is 10^8, the number of phonons/cm^3 in a solid (with a Debye temperature T_D = 1000K) at T = 300°K is $\sim 3.5 \times 10^{23}$! The sheer number of phonons may give us an idea of their importance in affecting the spectral characteristics of ions in solids.

8. The electronic energy $\varepsilon_k(\vec{R})$ can be represented as a surface in a (3N+1)-fold space. If the system of normal

coordinates q_i is used, a potential "curve" will correspond to each coordinate q_i as in Fig. 2. This figure represents the potential curve diagrams of a diatomic molecule: for such a system, only one vibrational mode is present ($q_i = R$).

9. The question now arises: is it legitimate, when considering the potential curves in correspondence to a certain normal coordinate q_i, to put the potential curves for the ground state and for an excited electronic state in the same diagram as in Fig. 2? Electronic excitation corresponds always to a rearrangement of the electron charges in the internuclear spaces: this may in turn produce changes in the molecular architecture, i.e., the actual positions of the nuclei. In a diatomic molecule, the electronic excitation produces a change in the relative distance of the two nuclei, and in the force constant that controls the vibrational motion; however, since we have only one vibrational coordinate, it is legitimate to represent all the potential curves in the same diagram. This is not the case for a complex molecule, where, in general, angular distortions and changes in the force constants may result from an electronic excitation. This means that, to be precise, a change of electronic state produces also a change in the system of normal coordinates: this change has been called pictorally by Prof. Rebane (9) the "scrambling of normal coordinates."

10. According to the adiabatic approximation that we introduced in section II of this article, the quantum states of a solid are represented by the Born-Oppenheimer products

$$\Psi_i(\vec{r},\vec{R}) = \psi_k(\vec{r},\vec{R})\phi_{k\ell}(\vec{R}) \tag{236}$$

We note here that, due to the presence of non-adiabatic terms in the Hamiltonian (these terms are in {} in eq. 13)), the states exemplified in (236) are not truly stationary. The non-adiabaticity may furnish the mechanism by which a system moves non-radiatively from a state ψ_i, given by (236) above, to a state

$$\Psi_f(\vec{r},\vec{R}) = \psi_m(\vec{r},\vec{R})\phi_{mn}(\vec{R}). \tag{237}$$

11. Finally, we need to mention the so-called "Jahn-Teller effect" (10): complexes with degenerate electronic states are unstable to symmetry - lowering distortions that take place in the way of vibrational modes. This fact was established by Jahn and Teller on the basis of symmetry considerations which do not furnish any

quantitative information.

VII.B. Generalization of the Franck-Condon Approximation

In the general case of a solid, the Franck-Condon approximation consists of the following:

1) We neglect the dependence of the matrix element D_{mk} on the position of the nuclei.

2) We set D_{mk} equal to the value it takes in correspondence to the equilibrium configuration of the nuclei.

The transition probability between two quantum states Ψ_i and Ψ_f given in (236) and (237), respectively, is proportional to the squared matrix element

$$|M_{fi}| = |D^o_{mk}|^2 |\int \phi_{mn}(\vec{R})^* \phi_{k\ell}(\vec{R}) d^3\vec{R}| \tag{238}$$

Neglecting the scrambling of normal coordinates, we can write

$$\phi_{k\ell}(\vec{R}) = \phi_{kn_1''}(q_1)\phi_{kn_2''}(q_2) \quad \ldots\ldots\ldots \tag{239}$$

$$\phi_{mn}(\vec{R}) = \phi_{mn_1'}(q_1)\phi_{mn_2'}(q_2) \quad \ldots\ldots \tag{240}$$

Taking the last two relations into account the square of the overlap integral in (238) becomes

$$|\int \phi_{mn}(\vec{R})^* \phi_{k\ell}(\vec{R}) d^3\vec{R}|^2 = \prod_i |\int \phi_{mn_i'}(q_i)^* \phi_{kn_i''}(q_i) dq_i|^2 \quad . \tag{241}$$

We note that the integral which appears in eq. (238) exists even if the scrambling of the normal coordinates is not negligible; in this case, however, the integral does not appear in the form $\prod_i \ldots$

The radiative and non-radiative processes of ions in solids are summarized in Fig. 6 in which we have established a similarity of a sort between the atom-photon and the atom-phonon interaction. The arrows indicate the direction in which the energy moves. Vibronic processes are a result of the simultaneous interaction of the atomic system with the radiation field and the phonon field.

Other effects like "natural width" and "Lamb shift" with their equivalent "thermal broadening" and "thermal shift" of sharp lines are also reported to call the attention of the reader on these important processes (11).

The direct interaction of the radiation field with the thermal

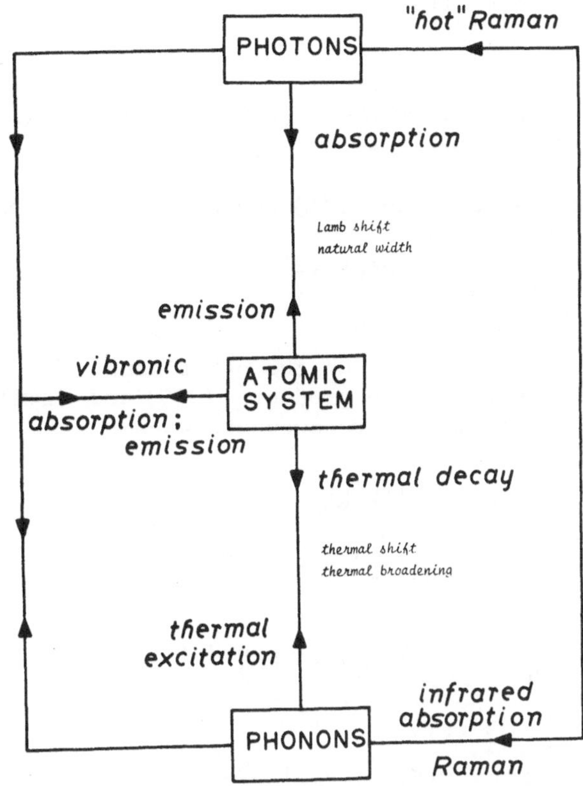

Fig. 6. Diagram illustrating various processes which are present in solids.

vibrations produces the infrared and Raman spectra of solids. Other, more extensive treatments should be consulted (12) for an account of these effects.

VII.C. Radiationless Transitions

We consider now radiationless transitions between two quantum states defined by (236) and (237). We apply to these states the formula (230) already found.

$$W(k\ell \to mn) = \frac{2\pi}{\hbar} \left|\langle mn|A|k\ell\rangle\right|^2 \rho(E_{mn}=E_{k\ell}) \qquad (242)$$

where

$$|k\ell\rangle = \psi_k(\vec{r},\vec{R})\,\phi_{k\ell}(\vec{R}) \qquad (243)$$

$$|mn\rangle = \psi_m(\vec{r},\vec{R})\phi_{mn}(\vec{R}) \tag{244}$$

and

$$A\psi_k(\vec{r},\vec{R})\phi_{k\ell}(\vec{R}) = -\sum_s \frac{\hbar^2}{M_s} \vec{\nabla}_s\psi_k(\vec{r},\vec{R})\cdot\vec{\nabla}_s\phi_{k\ell}(\vec{R}) \ . \tag{245}$$

We can express the relation (245) above in terms of the vibrational normal coordinates

$$A\psi_k(\vec{r},q)\phi_{k\ell}(q) = -\sum_i \frac{\partial\psi_k(\vec{r},q)}{\partial q_i} \frac{\partial\phi(q)}{\partial q_i} \ , \tag{246}$$

where q stands for $q_1,q_2\ldots$ On the other hand, the functions $\phi_{k\ell}$ and ϕ_{mn} are expressed as the products (239) and (240) respectively. By taking these relations into account, we may write

$$\langle mn|A|k\ell\rangle = \langle\psi_m\phi_{mn}|A|\psi_k\phi_{k\ell}\rangle$$

$$= \langle\psi_m(\vec{r},q)\phi_{mn_1'}(q_1)\phi_{mn_2'}(q_2)\ldots\phi_{mn_i'}(q_i)\ldots|$$

$$|-\sum_i\hbar^2 \frac{\partial\psi_k(\vec{r},q)}{\partial q_i} \phi_{kn_1''}(q_1)\phi_{kn_2''}(q_2)\ldots\ldots \frac{\partial\phi_{kn_i''}(q_i)}{\partial q_i} \ldots\ldots\rangle =$$

$$= -\sum_i\hbar^2\langle\psi_m(\vec{r},q)|\frac{\partial\psi_k(\vec{r},q)}{\partial q_i}\rangle\langle\phi_{mn_i'}(q_i)|\frac{\partial\phi_{kn_i''}(q_i)}{\partial q_i}\rangle\prod_{s\neq i}\langle\phi_{mn_s'}(q_s)|\phi_{kn_s''}(q_s)\rangle =$$

$$= -\hbar^2\sum_i R_i(mk)\langle\phi_{mn_i'}(q_i)|\frac{\partial\phi_{kn_i''}(q_i)}{\partial q_i}\rangle\prod_{s\neq i}\langle\phi_{mn_s'}(q_s)|\phi_{kn_s''}(q_s)\rangle, \tag{247}$$

where

$$R_i(mk) = \langle\psi_m(\vec{r},q)|\frac{\partial\psi_k(\vec{r},q)}{\partial q_i}\rangle \ . \tag{248}$$

The modes q_i responsible for the electronic transition $m\rightarrow k$ are called "promoting modes," the other modes q_s are called "accepting modes."

Finally

$$W(k\rightarrow m) = \frac{2\pi}{\hbar}\sum_{\ell n} P_\ell\,|\langle\psi_m\phi_{mn}|A|\psi_k\phi_{k\ell}\rangle|^2\delta(E_{mn}-E_{k\ell}) \ , \tag{249}$$

where

$$P_\ell = \frac{e^{-E_{k\ell}/kT}}{\sum_\ell e^{-E_{k\ell}/kT}} \tag{250}$$

is the probability of occupancy of the vibrational state ℓ.

This result is the starting point of more elaborate treatments (13-16).

REFERENCES

1. B. Di Bartolo in Spectroscopy of the Excited State, B. Di Bartolo, ed., Plenum Press, New York and London, 1976, p. 17.

2. B. Di Bartolo in Luminescence of Inorganic Solids, B. Di Bartolo, ed., Plenum Press, New York and London, 1978, p. 1.

3. E. V. Sayre and S. Freed, J. Chem. Phys. 24, 1213 (1956).

4. J. Franck, Trans. Faraday Soc. 21, 536 (1925).

5. E. U. Condon, Phys. Rev. 32, 858 (1928).

6. L. Pauling and E. B. Wilson, Introduction to Quantum Mechanics, McGraw-Hill, New York and London, 1935, p. 265.

7. D. Curie in Optical Properties of Ions in Solids, B. Di Bartolo, ed., Plenum Press, New York and London, 1975, p. 84.

8. F. B. Hildebrand, Advanced Calculus for Engineers, Prentice Hall, Inc., Englewood Cliffs, N.J., 1956, p. 160.

9. K. K. Rebane, Impurity Spectra of Solids, Plenum Press, New York and London, 1970, p. 13.

10. M. D. Sturge, in Solid State Physics, Vol. 20, F. Seitz, D. Turnbull and H. Ehrenreich, eds., Academic Press, New York and London, 1967, p. 91.

11. B. Di Bartolo, Optical Interactions in Solids, Wiley, New York, 1968.

12. B. Di Bartolo and R. C. Powell, Phonons and Resonances in Solids, Wiley, New York, 1976.

13. K. Huang and A. Rhys, Proc. Roy. Soc. A204, 406 (1950).

14. M. Lax, J. Chem. Phys. 20, 1752 (1952).

15. R. Kubo and Y. Toyozawa, Progr. Theoret. Phys. 13, 160 (1955).

16. T. Miyakawa and D. L. Dexter, Phys. Rev. B1, 2961 (1970).

SPECTROSCOPY AND RADIATIONLESS PROCESSES IN SIMPLE MOLECULES

D.A. Ramsay

Herzberg Institute of Astrophysics
National Research Council of Canada
100 Sussex Drive, Ottawa, Ontario, Canada K1A 0R6

ABSTRACT

Radiationless processes depend on the mixing of two or more levels in a molecule by various terms in the molecular Hamiltonian. The first two lectures deal with the theory of perturbations in molecular spectra and the various selection rules involved. Examples are given of the application of the theory to the spectra of diatomic and simple polyatomic molecules. Predissociation is discussed as a special case of a perturbation leading to dissociation of a molecule.

The third lecture deals with singlet-triplet perturbations in polyatomic molecules. Magnetic rotation spectroscopy has recently been shown to provide a sensitive method for revealing such perturbations. The application of this technique to the study of singlet-triplet perturbations in the near ultraviolet spectrum of formaldehyde is discussed.

I. GENERAL CONSIDERATIONS

Suppose that the solution of the complete Schrödinger wave equation

$$H\psi = E\psi \,, \tag{1}$$

gives rise to a set of non-degenerate eigenvalues and eigenfunctions E_1,-----E_N and ψ_1,-----ψ_N, respectively. In general it is not possible to solve rigorously the complete equation though it may be possible to give exact solutions for a modified (zero-order) equation

$$H^0\psi = E^0\psi \,, \tag{2}$$

where

$$H = H^0 + H', \tag{3}$$

and $H' \ll H^0$. (4)

If $E_1{}^0$,-----$E_N{}^0$ and $\psi_1{}^0$,-----$\psi_N{}^0$ are the eigenvalues and eigenfunctions for the zero-order equation, then by perturbation theory

$$E_n = E_n{}^0 + W_{nn} + \sum_{\substack{i=1 \\ i\neq n}}^{N} \frac{|W_{ni}|^2}{E_n{}^0 - E_i{}^0} + \cdots \,, \tag{5}$$

and

$$\psi_n = \psi_n{}^0 + \sum_{\substack{i=1 \\ i\neq n}}^{N} \frac{W_{in}}{E_n{}^0 - E_i{}^0}\psi_i{}^0 + \cdots \,, \tag{6}$$

where

$$W_{ni} = \langle\psi_n{}^0|H'|\psi_i{}^0\rangle \;. \tag{7}$$

The quantities W_{ni} are the matrix elements of the perturbation function H'. The eigenvalues E_n in eq. (5) are correct to the second order of approximation; the eigenfunctions ψ_n in eq. (6) are correct to the first order.

Consider now a system for which two energy levels lie close to each other and for which the interactions with other energy levels of the system can be neglected. Suppose that the "unperturbed" energies are considered to include the effects of the terms W_{nn},

i.e., $E_1 = E_1{}^0 + W_{11}$, (8)

and $E_2 = E_2{}^0 + W_{22}$. (9)

Then the perturbed energies E_a and E_b are the roots of the secular equation

$$\begin{vmatrix} E_1-E & W_{12} \\ W_{21} & E_2-E \end{vmatrix} = 0, \tag{10}$$

with $W_{21} = W_{12}$. The values of E are given by

$$E = \tfrac{1}{2}(E_1+E_2) \pm \tfrac{1}{2}\sqrt{4|W_{12}|^2+\delta^2} \quad , \tag{11}$$

where $\delta = E_1 - E_2$ is the separation of the "unperturbed" levels. The shifts of the two levels produced by the interaction matrix element W_{12} are thus equal in magnitude but opposite in sign, i.e., the two levels "repel" each other. The magnitudes of the shifts increase with larger W_{12} and smaller δ. When $\delta = 0$ the shifts are $\pm W_{12}$. For large separations between the "unperturbed" levels, i.e., $\delta >> 2W_{12}$, the shifts are $\pm|W_{12}|^2/\delta$. A table of typical shifts as functions of W_{12} and δ is given in Table 1.

TABLE 1

MAGNITUDES OF THE SHIFTS AS FUNCTIONS OF W_{12} AND δ (IN cm^{-1})

$\delta \backslash W$	10	5	2	1
0	10.0	5.0	2.0	1.0
1	9.5	4.5	1.6	0.62
2	9.0	4.1	1.2	0.41
5	7.8	3.1	0.7	0.19
10	6.2	2.1	0.4	0.10

The eigenfunctions of the two resulting states are given by

$$\psi_a = a\psi_1 - b\psi_2 , \tag{12}$$

$$\psi_b = b\psi_1 + a\psi_2 , \tag{13}$$

where ψ_1 and ψ_2 are the "unperturbed" eigenfunctions corresponding to the eigenvalues E_1 and E_2, and

$$a = \left(\frac{\sqrt{4|W_{12}|^2+\delta^2}+\delta}{2\sqrt{4|W_{12}|^2+\delta^2}}\right)^{\frac{1}{2}} , \tag{14}$$

$$b = \left(\frac{\sqrt{4|W_{12}|^2+\delta^2}-\delta}{2\sqrt{4|W_{12}|^2+\delta^2}}\right)^{\frac{1}{2}} . \tag{15}$$

When $\delta = 0$ the two eigenfunctions are fifty-fifty mixtures of the unperturbed eigenfunctions. As $\delta \to \infty$, $\psi_a \to \psi_1$ and $\psi_b \to \psi_2$. A table showing the mixing of the eigenfunctions for a given value of W_{12} and different values of δ is given in Table 2.

TABLE 2

VALUES OF THE MIXING COEFFICIENTS a^2 AND b^2 FOR A GIVEN VALUE OF W_{12} (=1 cm^{-1}) AND DIFFERENT VALUES OF δ

δ (cm^{-1})	a^2	b^2
0	0.500	0.500
1	0.724	0.276
2	0.854	0.146
5	0.964	0.036
10	0.990	0.010

II. SELECTION RULES

The perturbation matrix element W_{12} is different from zero only for pairs of levels which have the same value of the resultant angular momentum quantum number and the same overall symmetry species. More detailed selection rules for diatomic and polyatomic molecules will now be considered.

II.A. Diatomic Molecules

For diatomic molecules the selection rules derived by Kronig may be summarized as follows (Herzberg, 1950):

(1) Both states must have the same value of the resultant angular momentum quantum number, which is usually J, i.e., $\Delta J = 0$.

(2) Both states must have the same multiplicity, i.e., $\Delta S = 0$.

(3) The Λ values of the two states can differ by only 0 or ± 1, i.e., $\Delta\Lambda = 0, \pm 1$.

(4) Both states must have the same parity, i.e., $+ \leftrightarrow +$, $- \leftrightarrow -$, $+ \not\leftrightarrow -$.

(5) For homonuclear diatomic molecules, the nuclear spin symmetry must be the same, i.e., $s \leftrightarrow s$, $a \leftrightarrow a$, $s \not\leftrightarrow a$.

The first, fourth and fifth rules are perfectly rigorous. The second rule can be violated by spin-orbit interaction and the third rule holds only if Λ is defined, i.e., for Hund's cases (a) and (b).

For Hund's case (c), rule (3) must be replaced by $\Delta\Omega = 0, \pm 1$. For Hund's case (b) the quantum number N for the total angular momentum apart from spin is defined and the selection rule $\Delta N = 0$ also applies.

Perturbations with $\Delta\Lambda = 0$ are sometimes called *homogeneous* perturbations and perturbations with $\Delta\Lambda = \pm 1$ are called *heterogeneous* perturbations.

In addition to the above selection rules for the electronic and rotational quantum numbers, the vibrational quantum numbers are subject to the considerations of the Franck-Condon principle. If the interaction matrix element W is split into three components depending on the electronic, vibrational and rotational coordinates, respectively, then the magnitude of the vibrational contribution to W_{12} is given by

$$W_{12}{}^{v} = \langle\psi_1{}^{v}|W^{v}|\psi_2{}^{v}\rangle , \qquad (16)$$

where W^v is the vibrational contribution to the perturbation function. A strong perturbation can occur only if the vibrational eigenfunctions show a favourable overlap.

II.B. Polyatomic Molecules

Four different types of perturbation are possible for a molecule with C_{2v} symmetry and are characterized by the product of the symmetry species ($\Gamma_1 \times \Gamma_2$) of the two interacting levels. The various selection rules for singlet states are (Mills, 1965):

(i) Fermi:

$$\Gamma_1 \times \Gamma_2 = A_1,\ \Delta J = 0,\ \Delta K_a = 0,\ \Delta K_c = 0 ,$$

(ii) Type A Coriolis:

$$\Gamma_1 \times \Gamma_2 = A_2,\ \Delta J = 0,\ \Delta K_a = 0,\ \Delta K_c = \pm 1 ,$$

(iii) Type B Coriolis:

$$\Gamma_1 \times \Gamma_2 = B_1,\ \Delta J = 0,\ \Delta K_a = \pm 1,\ \Delta K_c = \mp 1 ,$$

(iv) Type C Coriolis:

$$\Gamma_1 \times \Gamma_2 = B_2,\ \Delta J = 0,\ \Delta K_a = \pm 1,\ \Delta K_c = 0.$$

For multiplet states in which the electron spin is weakly coupled, the above rules can be supplemented by the rule $\Delta N = 0$. If the coupling of the electron spin is stronger then additional rules, e.g., $\Delta N = \pm 1$ are also possible.

The above rules are applicable to perturbations between two states of the same multiplicity. As an example of the selection rules applicable when the two states have different multiplicities we shall consider the mutual perturbations which are possible between the $^1A_2$ and $^3A_2$ (π*-n) excited states of formaldehyde. No direct spin-orbit interaction is allowed between these two states but Stevens and Brand (1973) have discussed two mechanisms by which perturbations can occur. The mechanisms and the corresponding selection rules are:

(i) spin-orbit orbital-rotation mechanism

$$^1\Gamma_e = {}^3\Gamma_e,\ {}^1\Gamma_{ev} = {}^3\Gamma_{ev},\ \Delta J = 0,\ \Delta N = 0, \pm 1,\ \Delta K_a = 0, \pm 2\ ,$$

(ii) vibronic spin-orbit mechanism

$$^1\Gamma_{ev} \times {}^3\Gamma_{ev} \supset R_z;\ \Delta J = 0,\ \Delta N = 0, \pm 1,\ \Delta K_a = 0$$

$$^1\Gamma_{ev} \times {}^3\Gamma_{ev} \supset R_x, R_y;\ \Delta J = 0,\ \Delta N = 0, \pm 1,\ \Delta K_a = \pm 1.$$

III. EXAMPLES

Four examples of perturbations, two involving diatomic molecules and two involving polyatomic molecules will now be given to illustrate the points discussed above.

III.A. CN

One of the classic examples of a perturbation in the spectrum of a diatomic molecule is the perturbation found in the v' = 11 - v" = 11 band of the violet system of CN ($B^2\Sigma^+ - X^2\Sigma^+$). The perturbation involves the v = 11 level of the $X^2\Sigma^+$ state and is produced by a crossing with the v = 7 level of the $^2\Pi_{3/2}$ component of the $A^2\Pi$ state near N = 14. A reproduction of the band is given in Fig. 1a and a graphical representation of the perturbation is shown in Fig. 1b. It is seen that only one component of the $^2\Sigma^+$ state is affected by the perturbation since in addition to the requirement that the two perturbing levels should have similar energies, only levels with the same J (and N) and the same parity can interact. Additional lines are observed in the spectrum in the region of the perturbation and are caused by the mixing of the eigenfunctions of the $X^2\Sigma^+$ and $A^2\Pi$ states.

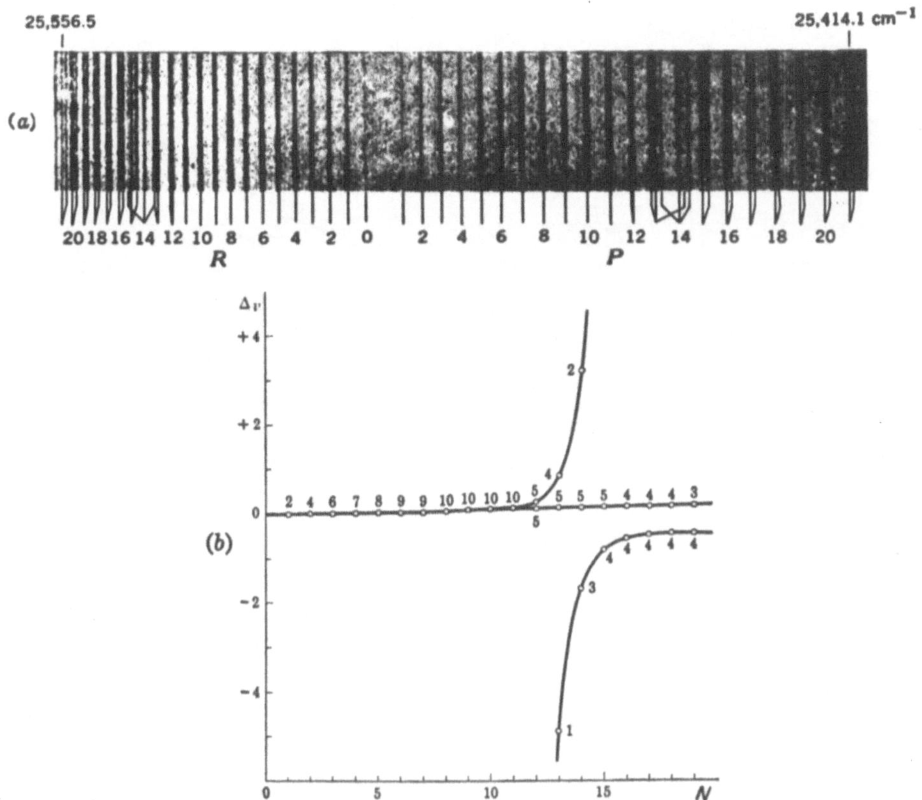

Fig. 1. (a) Spectrum of the 11-11 band of the violet system of CN. (b) Graphical representation of the perturbation in the P branch. The numbers on the curves are estimates of the relative intensities of the lines (from Herzberg, 1950).

III.B. $\underline{C_2}$

Several perturbations have been found in the spectrum of the C_2 molecule. One of the most interesting involves the small perturbations (up to 0.3 cm^{-1}) which have been found between the levels of the lowest singlet state $X^1\Sigma_g^+$ and the $b^3\Sigma_g^-$ state. It was on the basis of these perturbations that Ballik and Ramsay (1963a) showed that the lowest $^1\Sigma_g^+$ state is the ground state of the molecule and not the lowest $^3\Pi_u$ state as was formerly thought. A diagram showing the rotational energy levels for various vibrational levels of the $X^1\Sigma_g^+$ and $b^3\Sigma_g^-$ states is reproduced in Fig. 2. Perturbations are found in the regions of the crossing points. The displacements are found to be equal in magnitude and opposite in sign in the two interacting states and occur at the same J values. For example, the analysis of the 0-0 band of the $b^3\Sigma_g^-$ - $a^3\Pi_u$

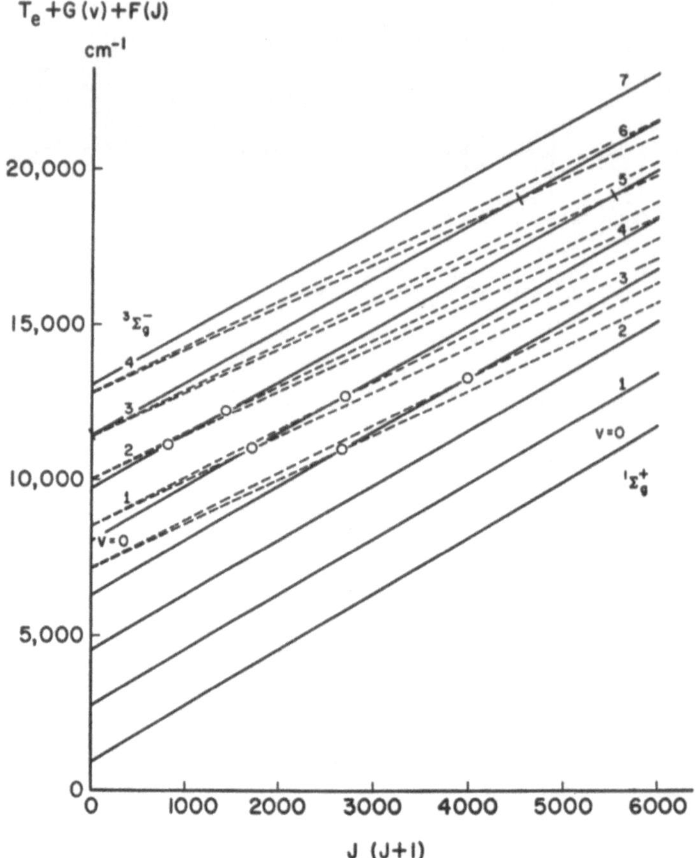

Fig. 2. Rotational energy levels and mutual perturbations for the $X^1\Sigma_g^+$ and $b^3\Sigma_g^-$ states of C_2. Full lines refer to the $X^1\Sigma_g^+$ state and dashed lines to the $b^3\Sigma_g^-$ state. The lower and upper dashed lines for each value of v refer to the F_1 and F_3 levels, respectively. Observed perturbations are indicated by circles, and perturbations which are predicted but not yet observed are indicated by short oblique lines (after Ballik and Ramsay, 1963b).

system reveals that the F_1 levels of the $b^3\Sigma_g^-$ state with J = 50 and 52 are displaced by +0.14 and -0.15 cm^{-1}, respectively from their unperturbed positions. Similarly, the analysis of the 7-3 band of the $A^1\Pi_u - X^1\Sigma_g^+$ system shows that the J = 50 and 52 levels for the v = 3 level of the $X^1\Sigma_g^+$ state are perturbed by -0.13 and +0.16 cm^{-1}, respectively. The mutual perturbation between the v = 0 level of the $b^3\Sigma_g^-$ state and the v = 3 level of the $X^1\Sigma_g^+$ state is thus clearly established. Singlet-triplet perturbations are forbidden according to the Kronig selection rules but are weakly allowed owing to second order effects. Small perturbations can indeed be very informative.

III.C. CO_2

The classic example of Fermi resonance involves the 100 and 020 levels of the CO_2 molecule. When the Raman spectrum was first investigated, two strong polarized bands were found at 1285.5 and 1388.3 cm^{-1} whereas only one strong band, viz., ν_1 might have been expected (Herzberg, 1945). However, the unperturbed frequency for the ν_1 vibration lies very close to twice the frequency for the bending vibration ν_2 and a strong interaction results ($W \sim 50$ cm^{-1}). It is interesting to note that the $2\nu_2$ vibration has two components with vibrational angular momentum quantum numbers $\ell = 0$ and 2. The vibrational symmetries are, respectively, σ_g^+ and δ_g. Since the vibrational symmetry for the ν_1 vibration is σ_g^+, the strong interaction takes place only between ν_1 and the $\ell = 0$ component of ν_2. The $\ell = 2$ component of ν_2 is unaffected and has been found from difference bands to lie approximately half-way between ν_1 and $2\nu_2$. The wavefunctions for ν_1 and $2\nu_2$ are nearly 50-50 mixtures, hence it is realistic to consider the two levels together as a Fermi diad rather than as individual levels.

III.D. H_2CO

Several examples of Coriolis interactions are found in the spectrum of formaldehyde. One example taken from current research involves the $2^1 4^1$ level of the $\tilde{A}^1A_2$ state of $H_2{}^{13}CO$ (Birss, Gordon, Ramsay and Till, 1979). The $J_{1,J-1}$ levels show a crossover between J = 7 and J = 8 (Fig. 3) but no perturbation is found in the $J_{1,J}$ levels. The perturbation is produced by the $J_{0,J}$ levels of the 3^1 vibrational level. The selection rules are $\Delta J = 0$, $\Delta K_a = \pm 1$,

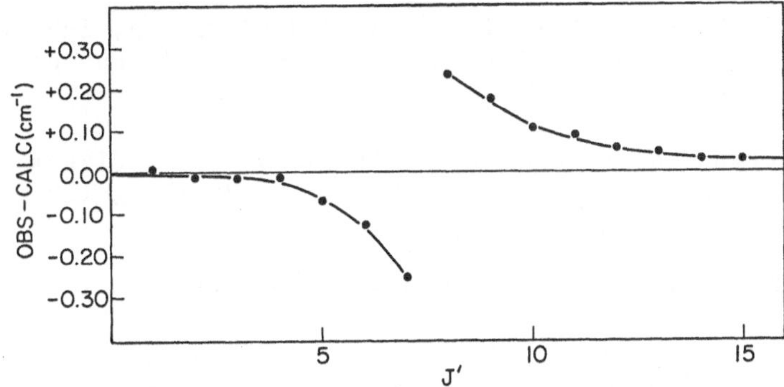

Fig. 3. Coriolis perturbations involving the $J_{1,J-1}$ rotational levels in the $2^1 4^1$ vibrational level of the $\tilde{A}^1A_2$ state of $H_2{}^{13}CO$ (after Birss, Gordon, Ramsay and Till, 1979).

$\Delta K_c = \mp 1$. The perturbation is therefore caused by a type B Coriolis interaction in agreement with the product of the symmetry species of the vibrational levels ($\Gamma_1 \times \Gamma_2 = b_1 \times a_1 = b_1$).

The Coriolis matrix element is $\sim 0.04\sqrt{J(J+1)}$ and increases with J; similarly the separation between the unperturbed levels increases roughly with J(J+1). The displacement of the levels caused by the perturbation therefore tends towards a constant value at high J.

IV. PREDISSOCIATION

Predissociation is a special case of a perturbation involving the interaction between a stable molecular state and a continuum. Since the angular momentum and symmetry properties of the energy levels are still defined for a continuum, the selection rules for predissociation are the same as those discussed above for perturbations. The selection rule $\Delta J = 0$ can always be satisfied since all J values are possible for each energy in the continuum. The selection rules based on symmetry properties, however, may provide important restrictions on the possible predissociation processes.

A stable Σ^+ state cannot be predissociated by a continuous Σ^- state and vice versa, on account of the selection rule $+ \not\leftrightarrow -$. An example is found in the excited state of the $^2A''\Pi - {}^2A'$ system of the HCO free radical (Herzberg and Ramsay, 1955). The excited state has a linear configuration and all the vibronic levels are predissociated except for the Σ levels which have Σ^- symmetry. The continuum formed by the combination of $H(^2S) + CO(^1\Sigma^+)$ has Σ^+ symmetry and no predissociation is possible.

If a stable Π state is overlapped by the continuum of a Σ state, only one Λ component of the Π state is predissociated since for only one component are the selection rules $\Delta J = 0$ (or $\Delta N = 0$) and $+ \not\leftrightarrow -$ simultaneously satisfied. An example is found in the 0-0 band of the $B^2\Pi - X^2\Sigma^+$ system of MgH (Herzberg, 1950). The P and R branches break off in emission at $N' = 11$ but no breaking-off is observed in the Q branch.

For further discussion on predissociation in diatomic and polyatomic molecules the reader is referred to Herzberg (1950, 1966).

V. SINGLET-TRIPLET PERTURBATIONS IN FORMALDEHYDE

Magnetic optical activity in the 3260 Å band of formaldehyde was first reported by Kusch and Loomis (1939). A detailed Zeeman study of this band was carried out by Brand and coworkers and reported in a series of three papers. A further study of this band and of other bands of the $\tilde{A}^1A_2 - \tilde{X}^1A_1$ system by magnetic rotation spectroscopy has been reported by Ramsay and his colleagues in a further series of three papers.

The theory of the mutual perturbations between the $\tilde{A}^1A_2$ and $\tilde{a}^3A_2$ states has been treated by Stevens and Brand (1973) and is discussed above. The 3260 Å band which is the $2_0^2 4_0^1$ band of the $\tilde{A}^1A_2$ - $\tilde{X}^1A_1$ system shows many perturbations (up to 0.5 cm^{-1}) in the various K manifolds of the excited state. By using pulsed magnetic fields up to 13 kG, Brand and Stevens (1973) found that the levels which showed the largest perturbations also showed Zeeman effects. They therefore concluded that the perturbations are produced by crossings with the levels of the $\tilde{a}^3A_2$ state. In this paper and a subsequent paper by Brand and Liu (1974), evidence is presented that the levels responsible for the perturbations are the $1^1 2^1 4^1$ and $1^1 2^2$ vibrational levels of the triplet state.

The magnetic rotation spectrum of the 3260 Å band was studied later under high resolution by Birss, Ramsay and Till (1978). The levels which were found to be active in the Zeeman spectrum were also found to be active in the magnetic rotation spectrum. In addition many new magnetically sensitive levels were found. Magnetic rotation spectra were obtained with much smaller flux densities (100-1000 G) than were needed in the Zeeman studies. The greater sensitivity of the magnetic rotation technique is connected with the fact that it operates with crossed polarizers, thus giving rise to a higher signal to noise ratio. Furthermore the technique has the added advantage that it selects only those lines that are magnetically sensitive.

The magnetic rotation lines are readily identified since they lie close to corresponding absorption lines. Furthermore it is found that ^rR and ^pP lines are usually the strongest, ^pR and ^rP are weaker, while ^rQ and ^pQ lines are very weak and rarely observed. The magnetically sensitive levels which were identified using a flux density of only 160 G are given in Table 3. A further 22 levels were identified using higher flux densities. The number of magnetically sensitive levels is too large to be explained on the basis of only one or two mechanisms. It is likely that several vibrational levels of the triplet state are involved.

A much simpler situation exists for the 4_0^1 and 4_0^3 bands (Ramsay and Till, 1979). In the 4_0^1 band only the $12_{0,12}$ level of the excited state is found to be involved in a singlet-triplet perturbation. It is probable that a $\Delta K_a = \pm 1$ mechanism is involved but no definite vibrational assignment for the triplet level can be given at this time. By contrast a complete interpretation can be given for the magnetic activity found in the 4_0^3 band. The $19_{0\ 19}$, $19_{1,19}$, $17_{1,16}$, $17_{3,14}$, $17_{3,15}$, $18_{3,15}$ and $18_{3,16}$ rotational levels of the excited state are found to be involved in singlet-triplet perturbations; weaker activity is also found for the $16_{2,14}$, $17_{2,15}$,

TABLE 3

MAGNETICALLY SENSITIVE LEVELS ASSOCIATED WITH THE 3260 Å BAND OF FORMALDEHYDE[a)]

Singlet level	Displacement (cm^{-1})	Singlet level	Displacement (cm^{-1})
$7_{0,7}$	-0.03	$12_{4,9}$ [b)]	+0.038
$11_{0,11}$ [b)]	+0.230	$15_{4,11}$	-0.031
$18_{0,18}$	-0.02_5	$15_{4,12}$ [b)]	-0.029
$20_{0,20}$	-0.02	$17_{4,13}$	+0.010
$11_{2,9}$ [b)]	+0.061	$17_{4,14}$	+0.019
$12_{2,10}$ [b)]	-0.061	13_5 [b)]	+0.078
$12_{2,11}$ [b)]	-0.200	16_5	-0.018
$15_{2,14}$	+0.003	12_6	-0.005
$18_{2,16}$	+0.000	15_6	+0.045
$11_{3,8}$ [b)]	-0.045	16_6 [b)]	-0.016
$13_{3,10}$ [b)]	-0.110	12_7	+0.047
$13_{3,11}$ [b)]	-0.074	13_8	-0.06

a) Identified using magnetic rotation spectra at 160 G.
b) Denotes magnetically sensitive level reported by Brand and Stevens (1973).

$17_{2,16}$ and $18_{2,17}$ levels. All these perturbations can be explained by $\Delta J = \Delta N = \Delta K_a = 0$ interactions with a b_1 vibrational level of the $\tilde{a}^3A_2$ state. From the vibrational frequencies for the triplet state the only candidate is the level $4^1 5^1 6^1$. Further perturbations affecting the $10_{2,8}$, $10_{2,9}$ and $11_{2,9}$ levels of the singlet state are then readily accounted for by $\Delta J = \Delta N = 0$, $\Delta K_a = \pm 1$ interaction with the $5^1 6^1$ level of the triplet state, lying $\sim 31._5$ cm^{-1} below the $4^1 5^1 6^1$ level. It is interesting to note that the perturbations involving these levels are too small to be detected by the rotational analysis of the 4^3_0 band. Once again this emphasizes the sensitivity of magnetic rotation spectroscopy as a technique for revealing singlet-triplet perturbations. Similar studies have also been carried out for the 4^1_0 and 4^3_0 bands of D_2CO (Ramsay and Till, 1979).

I wish to thank Mrs. F.L. Chester for kindly typing this manuscript.

REFERENCES

Ballik, E.A., and Ramsay, D.A., 1963a, Astrophys. J. 137, 61.
Ballik, E.A., and Ramsay, D.A., 1963b, Astrophys. J. 137, 84.
Birss, F.W., Ramsay, D.A., and Till, S.M., 1978, Chem. Phys. Lett. 53, 14.
Birss, F.W., Gordon, R.M., Ramsay, D.A., and Till, S.M., 1979, Can. J. Phys. In press.
Brand, J.C.D., and Stevens, C.G., 1973, J. Chem. Phys. 58, 3331.
Brand, J.C.D., and Liu, D.S., 1974, J. Phys. Chem. 78, 2270.
Herzberg, G., 1945, Infrared and Raman Spectra of Polyatomic Molecules, D. Van Nostrand and Co., Inc., Princeton, J.J., U.S.A.
Herzberg, G., 1950, Spectra of Diatomic Molecules, D. Van Nostrand Co., Inc., Princeton, N.J., U.S.A.
Herzberg, G., and Ramsay, D.A., 1955, Proc. Roy. Soc. A233, 34.
Herzberg, G., 1966, Electronic Spectra of Polyatomic Molecules, D. Van Nostrand and Co., Inc., Princeton, N.J., U.S.A.
Kusch, P., and Loomis, F.W., 1939, Phys. Rev. 55, 850.
Mills, I.M., 1965, Pure Appl. Chem. 11, 325.
Ramsay, D.A., and Till, S.M., 1979, Can. J. Phys. 57, 1224.
Ramsay, D.A., and Till, S.M., 1979, Chem. Phys. Lett. In press.
Stevens, C.G., and Brand, J.C.D., 1973, J. Chem. Phys. 58, 3324.

ELECTRONIC RELAXATION IN LARGE MOLECULES

J. Jortner

Department of Chemistry
Tel-Aviv University
Tel-Aviv, Israel

ABSTRACT

We shall be concerned with intramolecular, nonreactive, interstate electronic relaxation processes in electronically excited states of collision-free isolated large molecules. These radiationless electronic relaxation processes involve the intramolecular conversion of electronic energy into vibrational energy. It is appropriate to start with a

I PROLOGUE, where we discuss some of the general features of the acquisition, storage and disposal of energy in molecular systems, as explored from the microscopic point of view. This will be followed by another introductory section, where we dwell on the

II HISTORY of the development of the experimental background, which provided conclusive evidence for the occurrence of electronic relaxation within a bound level structure of isolated large molecules. As this experimental work opened up some challenging theoretical issues, we proceed to outline the

III METHODOLOGY underlying the theoretical description of time evolution within a bound level structure. This will be followed by the introduction of the

IV BASIC MOLECULAR MODEL for interstate coupling and electronic relaxation, which provides the basis for outlining the theory of

V TIME EVOLUTION involving the decay of a metastable state into a continuum or a quasicontinuum, which results in the exponential decay law and the golden rule decay rate. Next, deviations from the exponential decay law are considered, going

VI BEYOND THE GOLDEN RULE, introducing the effective Hamiltonian formalism to handle the decay of a manifold of discrete states into a common continuum and bringing up the notion of quantum beats in the radiative decay of some complex molecular level structures. While up to this point only time-resolved experimental observables were considered, we proceed to describe the diverse sources of experimental information discussing

VII OBSERVABLES, which pertain to determination of populations and to the interrogation of retention of phase relationships, involving time-resolved observables, energy-resolved observables and observables pertaining to coherent optical effects. As an example for energy-resolved observables, we consider

VIII ABSORPTION LINESHAPES, focusing attention on the information regarding interstate coupling which stems from optical spectroscopy. The last stage of the exposition will discuss

IX THE RISE AND FALL OF EXCITED STATES of isolated polyatomic molecules, where we distinguish between implications of interstate coupling and relaxation, briefly discuss excitation modes and digress on criteria for practical irreversibility in a bound level structure. We conclude with a discussion of

X FUTURE TRENDS, where some novel and intriguing problems in the area of intramolecular dynamics are exposed.

I. PROLOGUE

Excited-state intramolecular and intermolecular dynamics is the study of the basic photophysical and chemical mechanisms of the acquisition, storage and disposal of energy in molecules and in condensed phases, as explored from the microscopic point of view (1-13). The aspect of energy acquisition pertains to photoselective excitation, i.e., the "preparation" of well-defined excited states by optical excitation. The advent of laser sources (14-18) surpassed and eclipsed previous experimental work, taking advantage of many of the unique features of these optical excitation sources, such as broad spectral range, tunability, high power, high energy, ultrashort duration and coherence. Photoselectively excited systems involve collision-free "isolated" molecules in the bulb (1-12), in thermal beams (19-21) and in supersonic beams (22-36), as well as medium-perturbed molecules suffering collisions in the gas phase (37,38), molecules embedded in low-temperature matrices (39) or mixed crystals (40) and molecules in solutions. The problem of energy storage addresses the basic aspects of intramolecular and intermolecular time evolution of excited states of "isolated" and of medium-perturbed molecules, considering the mechanisms of energy exchange and the time scales for such processes. Finally, the problem of energy disposal is concerned with the basic microscopic mechanism for the dumping of the excitation energy. It should be recognized that the radiative decay channel just provides one possible energy decay route and that the exploration of energy disposal requires the characterization and interrogation of a variety of decay channels. An outstanding goal of research in the area of molecular dynamics is the elucidation of a wide class of radiationless processes, which occur without the emission of radiation and which are responsible for a variety of intramolecular and intermolecular phenomena, such as electronic-electronic energy exchange, electronic-vibrational exchange, vibrational-vibrational exchange, dissociation and ionization as direct processes, and also as indirect processes in the form of autoionization and predissociation, as well as more complex chemical phenomena occurring in excited states.

The simplest radiationless phenomena involve basic molecular relaxation processes (Table 1), which fall into two distinct categories:

(a) Reactive processes, e.g., molecular autoionization (41) and predissociation (42,43) (rotational (42), vibrational (43) and electronic (42,43)), which result in ionization or dissociation.

(b) Nonreactive processes, e.g., intramolecular electronic relaxation (1-13) and intramolecular vibrational redistribution (9,13) in large molecules.

Another useful classification of the basic molecular processes separates those (intrastate) processes occurring on a single electronic potential surface from the (interstate) processes involving at least two electronically excited configurations:

TABLE 1. CLASSIFICATION OF BASIC INTRAMOLECULAR RELAXATION PROCESSES

NATURE OF DECAY CHANNEL	NATURE OF COUPLING	
	INTRASTATE (1 Electronic Configuration)	INTERSTATE (2 Electronic Configurations)
REACTIVE	ROTATIONAL PREDISSOCIATION VIBRATIONAL PREDISSOCIATION	AUTOIONIZATION ELECTRONIC PREDISSOCIATION
NONREACTIVE	INTRAMOLECULAR VIBRATIONAL ENERGY REDISTRIBUTION IN LARGE MOLECULES	ELECTRONIC RELAXATION (INTERNAL CONVERSION AND INTERSYSTEM CROSSING) IN LARGE MOLECULES

(A) Intrastate dynamics involving some forms of rotational predissociation, as well as intramolecular vibrational energy redistribution, which occurs between bound vibrational levels of large molecules.

(B) Interstate dynamics incorporating the processes of autoionization, electronic predissociation, as well as electronic relaxation in large molecules.

The phenomena of reactive autoionization and predissociation have been understood since the early days of quantum mechanics. As early as 1927 Wentzel (44) advanced the modern theory of electronic predissociation utilizing quantum mechanical time-dependent perturbation theory to derive the decay rate of a metastable discrete vibronic level to another, repulsive, electronic configuration. The Wentzel formula is nowadays usually referred to as the Fermi golden rule (45). The theory required to understand rotational predissociation was provided in 1928 by the tunnelling formalism of Gamow (46) and of Condon and Gurney (47). The conceptual framework for the understanding of vibrational predissociation in polyatomic molecules was provided by the 1933 work of Rosen (48). At about the same time Fano (49) developed the basic theory of autoionization. On the other hand, the experimental and theoretical exploration of nonreactive excited-state dynamics in polyatomic molecules is of much more recent vintage. The field of electronic relaxation (ER) in large molecules was explored most extensively during the last fifteen years, with particular emphasis on intramolecular, interstate, nonreactive dynamics in a bound level structure of "isolated" large molecules (1-13). These ER processes fall into two broad categories: (i) internal conversion, spin-multiplicity conserving transitions and (ii) intersystem crossing between states of different spin multiplicity. These relaxation phenomena, essentially involving the intramolecular interconversion of electronic energy into vibrational energy, are of interest because of several reasons. First, from the point of view of general methodology, these nonreactive processes provide interesting examples for dynamic relaxation in a bound spectrum of large, but finite, molecular systems. Second, the ubiquity of these processes in excited electronic states of large molecules require the understanding of these phenomena, both on the formal level and on the "working hypothesis" level. Third, intramolecular nonreactive processes play an important role in determining excited-state reactivity in large molecules, so that the elucidation of nonreactive relaxation will provide the basis for the understanding of a variety of photochemical phenomena on the molecular level. Fourth, the theoretical concepts developed for the description of interstate coupling and ER in large molecules can also be adopted for the study of intrastate anharmonic coupling between nuclear states and vibrational energy flow in vibrationally-electronically excited states of large molecules (9,13), providing a unified conceptual framework for the description of interstate and intrastate nonreactive intramolecular dynamics (13). The present review will dwell on some problems in the broad and diverse area of interstate ER in isolated large molecules in an attempt to explore the perspectives and future of intramolecular dynamics in excited states.

II. HISTORY

The theoretical understanding of ER in polyatomic molecules embedded in a dense medium (e.g., solutions, glasses or mixed crystals) originated with the phenomenological three-level model of Jablonski (50) advanced in 1933. The modern focus on the importance and generality of these processes was strongly emphasized by the work of Lewis and collaborators (51) and of Kasha (52) in the forties. Since then a vast amount of information on intramolecular relaxation in medium-perturbed molecules has accumulated (2,3), resulting in several important generalizations which involve the Kasha rules (3), the deuterium isotope effect on ER of aromatics (53), as well as the energy gap law for ER (54).

About fifteen years ago it became apparent that ER processes can be exhibited in collision-free, isolated molecules. The notion of an isolated molecule deserves some elucidation. In a bulb room-temperature experiment at pressure p, the time, t_c, between gas-kinetic collisions, which are characterized by a gas kinetic cross section of 75 $Å^2$, is $t_c \simeq [1.5 \times 10^{-7}/p(\mathrm{Torr})]$ sec, so that $t_c \sim 10^{-4}$ sec at $p = 10^{-3}$ Torr. Considerably higher collisional cross sections were reported for rotational-vibrational relaxation in electronically-vibrationally molecular excited states. For example, the cross section for collisional-rotational relaxation of electronically excited benzene in the $^1B_{2u}$ state by a ground-state benzene molecule is 500 $Å^2$ (55). Thus, in a bulb experiment at $p = 10^{-3}$ Torr, an excited molecule can be considered as "isolated" on a time scale which is shorter than a μsec, or so. Such low pressure bulb experiments led to the first conclusive evidence regarding ER in isolated large molecules. In considering this information, we shall view the molecular level structure of large molecules, adopting the point of view of the spectroscopist (Fig. 1) by focusing attention on Born-Oppenheimer pure-spin states. The pertinent evidence regarding ER in isolated molecules, which was accumulated around 1966, was:

(a) Lack of resonance fluorescence from the S_2 state of some large molecules, such as naphthalene and anthracene (56, 57).

(b) Short radiative decay times. The radiative decay lifetime of the S_1 state of anthracene $\tau = 4.5$ nsec (58) is considerably shorter than the pure radiative lifetime $\tau_r \simeq 14$ nsec deduced from the integrated oscillator strength, implying an emission quantum yield of $Y = 0.28$.

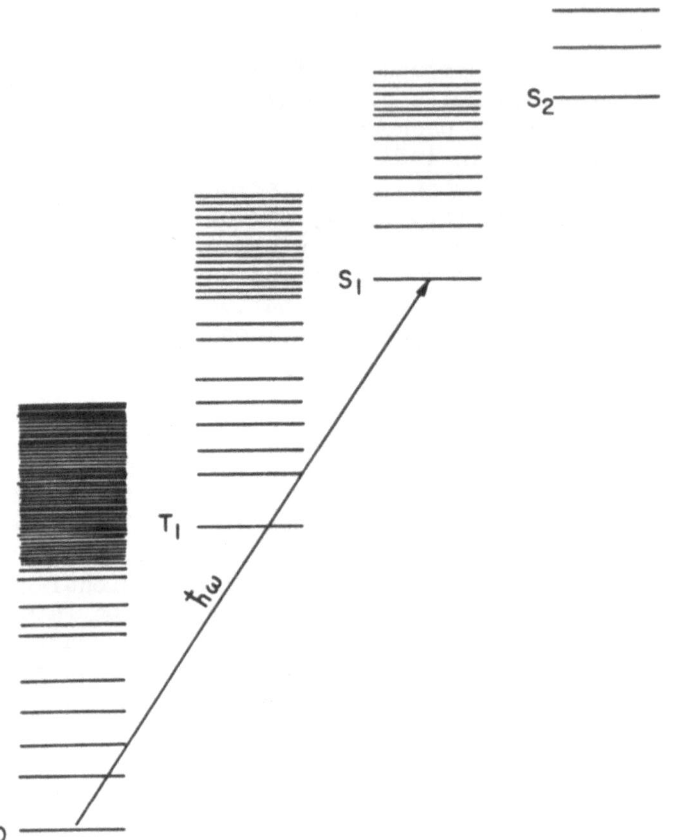

Fig. 1. The spectroscopist view of the bound level structure of a large polyatomic molecule. Horizontal lines describe Born-Oppenheimer, pure-spin states. The Born-Oppenheimer (approximate) separability enables us to consider vibronic levels corresponding to distinct electronic configurations. The ground electronic configuration is taken to be a singlet state.

(c) Reduction of emission quantum yields. The fluorescence quantum yield from the S_1 state of benzene is $\gamma = 0.34$, being pressure independent at low pressures (59,60).

This last observation led Kistiakowsky and Parmenter (59) to point out that their observation of S_1- T_1 intersystem crossing in the isolated benzene molecule may be "incompatible with the laws of quantum mechanics", as it implies the occurrence of a relaxation process in an energy range where all reactive channels are closed, so that intramolecular relaxation occurs within a bound level structure. This experimental work triggered hectic theoretical activity, which led to the elucidation of the nature of irreversibility in a nonreactive relaxation process and to the understanding of the gross features of ER in isolated molecules (61,62,1-12).

Subsequently, low-pressure bulb experiments provided detailed information regarding ER in isolated molecules. Notable are the intensive and exhaustive studies of Schlag (63), Parmenter (64) and Rice (65) and their colleagues on the dependence of the ER rate on the excess vibrational energy within the excited-state manifold. Such optical selection studies can be conducted only in collision-free molecular systems. Another class of fancy experimental methods used to interrogate intramolecular dynamics involve the techniques of picosecond spectroscopy, using mode-locked lasers developed by Rentzepis (15,66), which were successfully utilized to probe a variety of ER phenomena. The ultrafast relaxation dynamics of the S_1 state of azulene was probed (Fig. 2) in the gas phase (67), when the time between collisions is considerably shorter than the relaxation rate. The lifetime of the S_1 state of azulene excited at 625 nm was found to be 4±3 psec in the collision-free sample (67), providing conclusive evidence for efficient ER in this system. The corresponding lifetime of the S_1 state of azulene excited again at 625 nm in solution was found to be 3±2 psec, so that ER in a large molecule essentially involves an intramolecular process, being insensitive to perturbations exerted by an inert medium.

The experimentalist concerned with bulb experiments on isolated molecules is often haunted by the distinct possibility that the pressure is not sufficiently low to eliminate all intermolecular collision-induced processes, which may be characterized by very large cross sections. To overcome these inherent difficulties electronic relaxation studies of large molecules in effusive thermal molecular beams were conducted by Zare and colleagues (19,20). In their first experiment a laser beam was crossed with a thermal molecular beam and the $S_1 \rightarrow S_0$ ER process in pentacene was explored by interrogating the population of the vibrationally excited S_0 manifold (19) (Fig. 3). The only difficulty inherent in the interpretation of these results involves the possibility of excitation by the probe laser of those molecules which decayed radiatively to low-lying vibrationally excited states of S_0. In another experiment

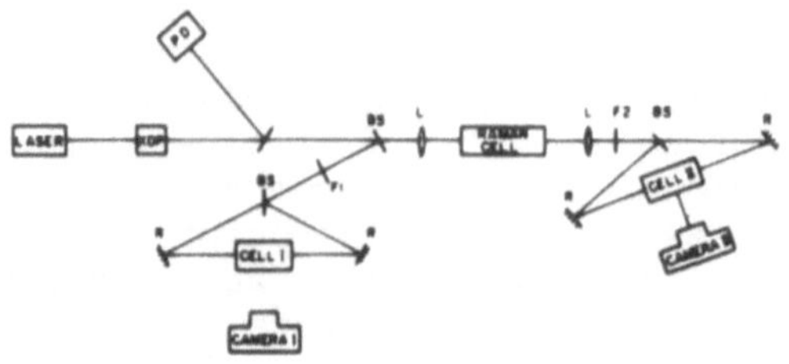

(A) Experimental setup

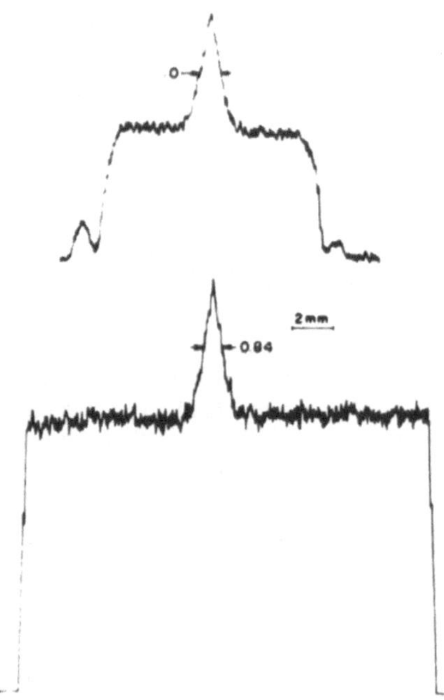

(B) Microdensitometer Traces

Fig. 2. Picosecond decay of the S_1 state of the azulene molecule in the gas phase, studied by Huppert, Jortner and Rentzepis [Ref. (67)]. The experimental setup consists of a pair of two-photon-fluorescence (TPF) triangles. One triangle is used for the measurement of the pulse width by simultaneous TPF from an organic dye (BBOT in solution), while the second triangle contains the azulene cell. The two microdensitometer traces correspond to the pulse width pattern (upper trace) and to the azulene pattern (lower trace).

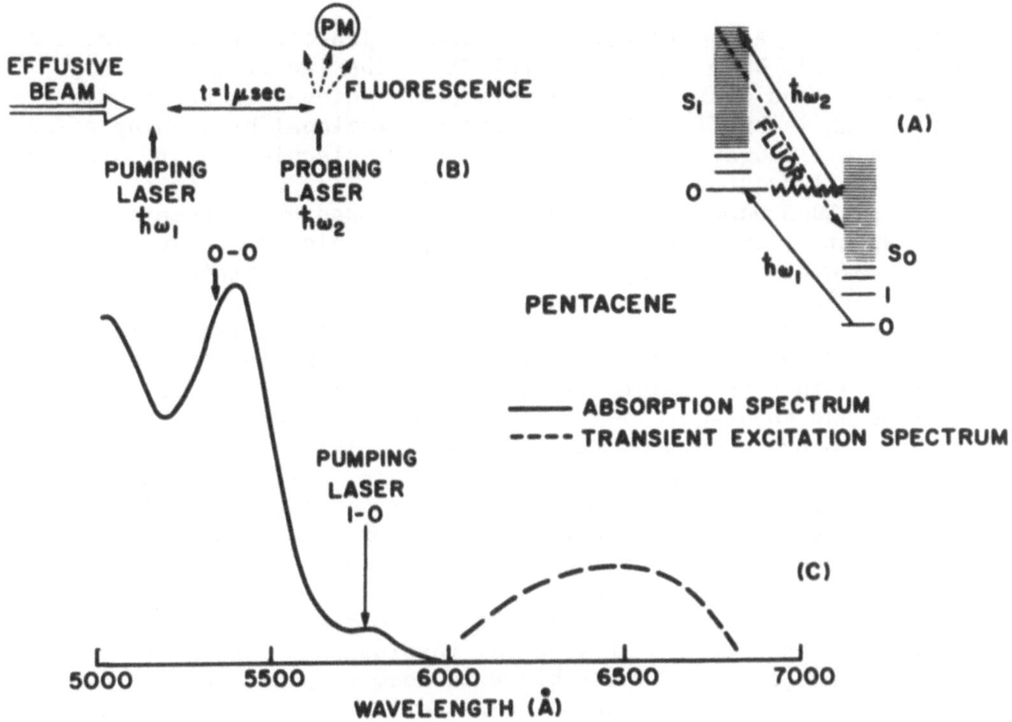

Fig. 3. Internal conversion from S_1 to S_0 of the pentacene molecule in a molecular beam, studied by Sander, Soep and Zare [Ref. (19)]. $S_0 \rightarrow S_1$ excitation of the molecule is conducted by the pumping laser at the frequency $\hbar\omega_1$, corresponding to the $1 \rightarrow 0$ hot band. After 1 μsec the molecule is hit by the probing laser of frequency $\hbar\omega_2$ and fluorescence is monitored. The transient excitation spectrum corresponds to the dependence of this fluorescence intensity on the frequency $\hbar\omega_2$. The transient excitation spectrum is red-shifted by ~3000 cm^{-1} relative to the first broad peak in the absorption spectrum. The boardening of the absorption spectrum originates from TIB effects.

the decay dynamics of the S_1 state of benzophenone was investigated by monitoring its radiative decay (20). The intrinsic limitations of probing the consequences of ER, resulting from laser excitation of large molecules in effusive molecular beams, is due to thermal inhomogeneous effects (TIB), i.e., thermal rotational broadening and vibrational sequence congestion, which preclude truly photoselective excitation of an ensemble of large molecules at room temperature. A powerful way to overcome TIB effects rests on the use of nozzle beam expansions (22-38). The low translational, rotational and vibrational temperature achieved in supersonic free expansions is sufficient to eliminate rotational broadening effects and achieve vibrational cooling in small molecules and to avoid all vibrational sequence congestion effects in large molecules. As early as 1963 Douglas and Huber (22) proposed and attempted to perform spectroscopic studies on internally cooled diatomic and triatomic molecules in a supersonic free expansion, obtaining spectroscopic evidence for rotational cooling of NO, NO_2 and SO_2. Subsequently in 1973 Zare and colleagues (23) utilized laser spectroscopy to demonstrate rotational-vibrational cooling of diatomic molecules in a supersonic free expansion. The merger between laser sources and supersonic beams provided a multitude of powerful methods to probe excited state dynamics. The recent use of supersonic beams of rare gases seeded with large molecules provide a novel experimental approach for internal cooling of large molecules. Photoselective laser excitation can be performed and the excited-state energetics (24-31) and dynamics (32-36) of isolated, ultracold large molecules can be explored. As is apparent from Fig. 4, laser spectroscopy of a large molecule seeded in supersonic beams allows for an increase of spectral resolution by about three orders of magnitude over that possible with room-temperature bulb experiments. Excited-state intramolecular dynamics in the first singlet state of several large isolated molecules, such as naphthalene (31,32), phtalocyanine (28,33), tetracene (29), pentacene (35) and ovalene (36), was recently experimentally explored. Some of these studies provided novel information on the features of ER from well-defined, genuinely-photoselected, vibrational electronic excited states of isolated large molecules.

From the foregoing historical review, it is apparent that there exists compelling experimental evidence for the occurrence of intramolecular nonradiative electronic relaxation in a bound level structure of isolated large molecules. We shall now proceed to discuss the theoretical foundations underlying the interpretation of these phenomena.

III. METHODOLOGY

The phenomenon of ER in an excited vibronic manifold of a large isolated molecule involves time evolution with a dense, congested, bound level structure. To gain some insight into the general

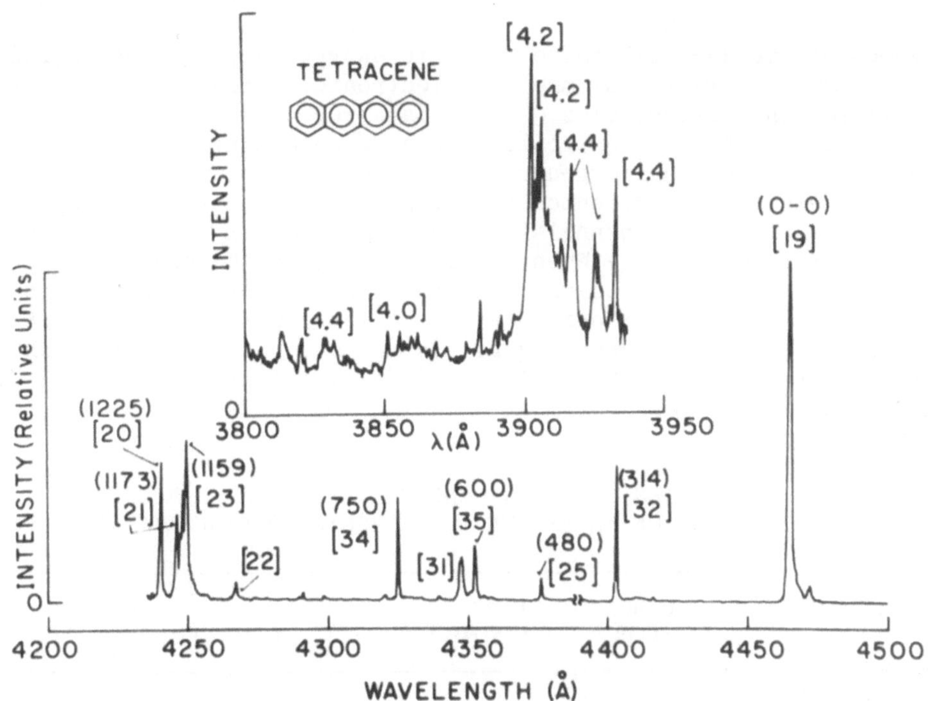

Fig. 4. Fluorescence excitation spectrum of gas phase isolated tetracene molecule cooled in a supersonic expansion. This spectrum is reproduced from the work of Amirav, Even and Jortner [Ref. (29)]. Tetracene was seeded into Ar and expanded from a pressure of 150 Torr through a 150 μ nozzle. The exciting dye laser had a bandwidth of 0.3 cm^{-1}. All the fluorescence excitation spectra are normalized to the laser intensity. The fluorescence intensity in the region 4235-4385 Å should be scaled by (¼) relative to the range 4400-4500 Å. The origin is labelled as (0-0). Numbers in round brackets ( ) denote frequencies of several fundamental vibrations (in cm^{-1}). Numbers in square brackets [ ] denote fluorescence decay lifetimes [in nsec].

features of this dynamic process, it will be useful to consider rough estimates of densities of states in polyatomic molecules. In Table 2 we present a compilation of densities of states for a lower electronic configuration, which are quasidegenerate with the electronic origin of a higher electronic state. The density of states, ρ_ℓ, calculated within the framework of the harmonic approximation, is determined by three factors: (1) The number of vibrational degrees of freedom, (2) the electronic energy gap, ΔE, separating the electronic origins of the two electronic configurations and (3) the vibrational frequencies. These data reveal several interesting features. First, an electronic origin of an excited electronic configuration can be quasidegenerate with several lower vibronic manifolds of distinct electronic states, providing interstate coupling to several different channels. Second, ρ_ℓ can vary widely for different large molecules. When ΔE is large ρ_ℓ can be huge, as expected. However, quite small background densities of states can be encountered in an electronic state of a large molecule characterized by a small electronic energy gap. Third, one has to distinguish carefully between the implications of interstate coupling and relax-

TABLE 2. DENSITIES OF BACKGROUND STATES FOR AN ELECTRONIC ORIGIN OF AN EXCITED ELECTRONIC CONFIGURATION

MOLECULE	UPPER STATE	LOWER MANIFOLD	ΔE (cm^{-1})	ρ_ℓ (cm)
ANTHRACENE	S_1	T_1	12000	$5x10^{10}$
NAPHTHALENE	T_1	S_0	20000	$8x10^{15}$
PENTACENE	S_1	S_0	19000	10^{14}
AZULENE	S_1	S_0	14000	10^{10}
BENZENE	S_1	T_1	84000	10^5
NAPHTHALENE	S_2	S_1	3400	$2x10^3$
OVALENE	S_2	S_1	1800	$\sim 10^4$
BENZOPHENONE	S_1	T_1	2800	10^3

ation. As will become apparent from the subsequent discussion, the existence of a high density of background states constitutes a necessary condition for the occurrence of ER in an isolated molecule. On the other hand, in excited states of some large molecules which reveal interstate coupling with a relatively sparse manifold, e.g., S_2-S_1 coupling in naphthalene, genuine ER will not be exhibited. Fourth, the number of effectively coupled levels constitutes a subset of the total states, so that in this sense the ρ_ℓ data of Table 2 provide an overestimate of the level density of effectively coupled states. However, anharmonicity corrections will lead to an increase of ρ_ℓ and of the density of effectively coupled levels. Fifth, the densities of states considered herein do not provide any direct information concerning the features of relaxation and neither do they yield any quantitative information regarding relaxation rates. We shall now proceed to examine these problems.

We shall now address the central problem of nonreactive relaxation and inquire under what conditions do transitions occur in a bound level structure? It has been realized since the early days of quantum mechanics that a stationary state of the molecular Hamiltonian exhibits only radiative decay, and that only nonstationary states of the molecular Hamiltonian are metastable with respect to intramolecular dynamics. To demonstrate this cardinal point, we consider a molecule specified by the molecular Hamiltonian H_M and molecular eigenstates $|m\rangle$, so that

$$H_M \; |m\rangle = \varepsilon_m \; |m\rangle \; , \tag{1}$$

where ε_m are the corresponding energies in the discrete spectrum. The initial state

$$\psi(0) = |m\rangle \; , \tag{2}$$

corresponding to a stationary state of H_M will exhibit the trivial time evolution

$$\psi(t) = |m\rangle \exp(-i\varepsilon_m t/\hbar) \, , \tag{3}$$

so that the population probability of the initial state at time t

$$P_0(t) \equiv |\langle\psi(0)|\psi(t)\rangle|^2 = 1 \; , \tag{4}$$

is, of course, time independent. On the other hand, when the initial state corresponds to a superposition of molecular eigenstates

$$\psi(0) = \sum_m A_m \; |m\rangle \; , \tag{5}$$

where the constant coefficients A_m are preparation amplitudes, the initial state is metastable and the system will exhibit a meaningful time evolution. The state of the system at time t is

$$\psi(t) = \sum_m A_m \, |m\rangle \exp(-i\varepsilon_m t/\hbar) , \tag{6}$$

while the population probability of the initial state is

$$P_0(t) = |\langle\psi(0)|\psi(t)\rangle|^2 = |\sum_m |A_m|^2 \exp(-i\varepsilon_m t/\hbar)|^2 . \tag{7}$$

The discrete system, starting in a nonstationary state, exhibits dynamic time evolution, which is governed by the Fourier sum of the initial population probabilities. The simple result, eq. (7), which disregards radiative decay, incorporates many of the basic features of intramolecular dynamics in a bound level structure. A systematic description of these dynamic processes has to provide a proper specification of the following three ingredients which appear in eq. (7):

1. Characterization of the Molecular Level Structure. This pertains to the specification of the eigenstates of the total molecular Hamiltonian in the relevant energy range. Excited-state level structure in bound electronically-vibrationally excited molecular states may include (1.1) sparse level structure in diatomics, in small polyatomics, as well as in low-lying vibrational excitations of large molecules; (1.2) an intramolecular quasicontinuum consisting of a dense manifold of bound vibronic states.

2. Characterization of the Excitation Amplitudes. The energetic spread of the transition moments connecting the ground state with the excited-state level structure determines the accessibility of these excited states to optical excitation.

3. Specification of the Initial Conditions. A variety of excitation methods can be applied to a molecular system, e.g., short-time excitation, energy-resolved excitation, coherent excitation, just to mention a few examples. The metastable molecular state "prepared" by optical excitation is specified in terms of the preparation amplitudes $\{A_m\}$, which appear in eqs. (5)-(7), and which are determined by the excited-state level structure, by the energetic spread of the excitation amplitudes, as well as by the energetic, temporal and coherence properties of the excitation pulse. We can conclude that the intramolecular dynamics of a given system provide the signature of the initial conditions.

The dynamic problem of ER in a bound level structure essentially involves the dynamics of initially "prepared" wavepackets of bound states. Information concerning the basic input data for specification of ER, which involve the excited-state level structure as well as on the energetic spread of the transition moments, can be obtained from spectroscopic data. The distribution of the transition moments among the molecular eigenstates is expected to exhibit sharp structure in range (1.1), while in range (1.2) quasicontinuous structure characterized by resonances will be exhibited. This resonance structure can conventionally be described in terms of zero-order states. Such zero-order states are also extremely useful for the specification of initial nonstationary states of the molecular system. We shall now consider the important technical problem of the choice of zero-order states for the proper description of intramolecular dynamics. The traditional segregation of the molecular Hamiltonian

$$H_M = H_{MO} + V , \tag{8}$$

into a zero-order part H_{MO} and a perturbation term should accomplish two goals. First, we should be able to get away with a relatively small basis set, rather than considering the entire spectrum of H_M. Second, it will be useful, but not imperative, to obtain a physically transparent "spectroscopic-type" description of the initially prepared state. The guidelines for the separation of H_M, eq. (8), are

A. Similarity criteria

A.1 The energies of the zero-order states should be close to those of the molecular eigenstates.

A.2 Small off-resonance coupling. The zero-order basis set is chosen to minimize off-resonance interactions. The effect of off-resonance interactions can either be entirely disregarded or incorporated as perturbative correction terms.

The similarity criteria imply that the spectrum of H_{MO} is close to that of H_M. From the point of view of general methodology, these criteria make it possible to use a truncated basis set, i.e., a two-electronic level structure for interstate coupling. Next, we have to invoke an additional criterion for the choice of a zero-order basis which pertains to the "preparation" of the initial metastable state.

B. Accessibility Criterion

Only a single (or a small number of) doorway state(s) is (are) accessible to optical excitation. In the case of one-photon excitation only the doorway state carries oscillator strength from the ground state, while for multiphoton excitation a small number of zero-order states are radiatively coupled in each energy region.

The accessibility criterion is not necessary for the choice of the zero-order molecular basis. Nevertheless, the concept of a

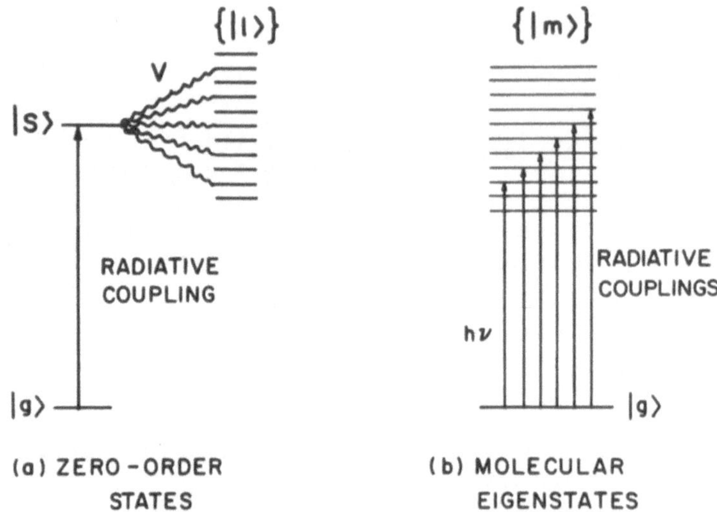

Fig. 5. A molecular energy level model used to discuss interstate coupling and nonreactive relaxation in a bound level structure in excited states of large molecules, originally introduced by Bixon and Jortner [Ref. (62)]. The zero-order molecular levels $|g\rangle$, $|s\rangle$ and $\{|\ell\rangle\}$ are Born-Oppenheimer states. They correspond, respectively, to the ground state $|g\rangle$, the optically accessible doorway state $|s\rangle$ and the background manifold $\{|\ell\rangle\}$. Radiative coupling prevails only between $|g\rangle$ and $|s\rangle$. The wiggly arrows represent interstate coupling. The molecular eigenstates $|m\rangle$ diagonalize the molecular Hamiltonian and are radiatively coupled to the ground state.

doorway state is essential for a meaningful characterization and specification of time-resolved observables.

The general criteria for the practical dissection of the molecular Hamiltonian will be utilized to advance the basic molecular model for ER.

IV. BASIC MOLECULAR MODEL

ER has been explained on the basis of the molceular level scheme presented in Fig. 5. Such a ladder diagram was first advanced about ten years ago (62) and is expected to be common for low-lying electronic-vibrational excitations of polyatomic molecules. The zero-order states correspond to the Born-Oppenheimer energy levels. The Born-Oppenheimer (BO) approximation separability conditions satisfy the similarity criteria of section III for the choice of appropriate zero-order states. For large molecules built of moderately light atoms, e.g., aromatic hydrocarbons, the zero-order states are chosen as pure-spin states.

We shall not dwell here on the details of the BO separability, which is discussed in other chapters of this volume (68). What is important for our purpose is that we can get away with a small zero-order BO basis set, consisting of the ground state $|g\rangle$, a single state $|s\rangle$ corresponding to a single vibronic level of an electronic configuration and a manifold $\{|\ell\rangle\}$ of levels belonging to lower electronic configurations, including also the ground electronic state which is isoenergetic with $|s\rangle$. The ground state $|g\rangle$ is so far detached in energy that it can be taken as a genuine eigenstate of the molecular Hamiltonian. The excited zero-order states are coupled by the interstate interaction V which involves:

1. Nonadiabatic coupling due to nuclear momentum, which connects states of the same spin multiplicity. This involves a near-resonant interstate coupling.

2. Spin-orbit coupling between states of different spin multiplicity, which may involve either near-resonant first-order spin-orbit coupling or weak off-resonant

second-order coupling, incorporating both spin-orbit and nuclear momentum coupling with another distant electronic configuration. In this case, $V_{s\ell}$ is an effective interaction subsuming the effects of other electronic configurations.

It is also important to emphasize an additional important technical aspect of the perturbation $V_{s\ell}$, in that it does not couple zero-order states within the same electronic manifold. Thus, $V_{ss} = 0$ and $V_{\ell\ell'} = 0$, for all ℓ and ℓ', are appropriate for non-adiabatic as well as for spin-orbit coupling, and only interstate coupling prevails between zero-order states. Interstate coupling effects essentially originate from the breakdown of the BO separability conditions.

At this stage we should inquire whether the notion of the failure of the BO separability is compatible with the traditional concepts of molecular spectroscopy, which imply that in many cases the BO approximation provides a faithful representation of the genuine molecular energy levels. An answer to this question may readily be obtained by a cursory examination of Table 2.

When the density of background states is tremendously high, the level spacing between the $|s\rangle$ state and the background levels is very small. Then a moderately weak interstate coupling can go a long way as far as dynamic implications are concerned. The basic conditions of effective strong interstate coupling is

$$V_{s\ell}\rho_{\ell} >> 1 \tag{9}$$

which is easily satisfied.

Next, we have to consider the aspect of accessibility of zero-order states. The excited state $|s\rangle$ is radiatively coupled to the ground state by a dipole transition. The transition moment μ_{gs}, connecting $|g\rangle$ and $|s\rangle$, may be intrinsic or may be induced by vibronic coupling with other electronic manifolds. In the latter case, μ_{gs} is an effective term. On the other hand, radiative coupling between $|g\rangle$ and the background manifold $\{|\ell\rangle\}$ is negligibly small. When the $\{|\ell\rangle\}$ manifold corresponds to highly vibrationally excited levels of the ground state electronic configurations, the transition moments from $|g\rangle$ to very high vibrational overtones are exceedingly small. When the $\{|\ell\rangle\}$ states correspond to high vibrational excitations of electronically excited configurations, radiative coupling from $|g\rangle$ is prohibited by very small Franck-Condon vibrational overlap factors, and/or spin selection rules. The $|s\rangle$ state, which carries oscillator strength from the ground state, constitutes the doorway state which is accessible to optical excitation from the ground state in the relevant energy range. The specification of a doorway state (8,10,11) is straightforward for the simple level structure of Fig. 5. In more complex molecular level structures, a general definition of a doorway state, $|N\rangle$, can be obtained (8,10,11) in terms of the projection of the radiative interaction on the ground state, $|g\rangle$, which for the case of dipole radiative coupling is (8,10,11)

$$|N\rangle = \hat{\mu}\,|g\rangle \ , \tag{10}$$

which can subsequently be expanded in terms of an appropriate zero-order or in the molecular eigenstates basis.

Up to this point we have been concerned with a useful zero-order representation of the bound molecular level structure. For this level structure of Fig. 5 the molecular eigenstates are given in terms of a superposition of the doorwary state and the background manifold

$$|m\rangle = a_s^{\,m}|s\rangle + \sum_{\ell} b_{\ell}^{\,m}|\ell\rangle \quad , \tag{11}$$

with energies ε_m, which satisfy eq. (1). The time evolution of the system should be envisioned in terms of the dynamics of wavepackets, eq. (5)-(7), of the $|m\rangle$ states. We shall now proceed to discuss the intramolecular dynamics in a simple bound level structure.

V. TIME EVOLUTION

The time evolution of the simple bound molecular level structure of Fig. 5 will provide insight into the implications of interstate coupling and dynamics in isolated large molecules. Let us consider an optical excitation of the system by a weak light pulse described in terms of a photon wavepacket, where the energetic spread spans the entire energy range of those $|\ell\rangle$ states which are effectively coupled to $|s\rangle$. Under these circumstances, the doorway state is photoselected, the initial state of the system being

$$\psi(t=0) = |s> . \tag{12}$$

For the sake of simplicity, we shall consider only intramolecular time evolution, disregarding for the moment the radiative decay. The initial state (eq. (12)) is obviously nonstationary and will exhibit time evolution. The time evolution of the system in the interaction representation is

$$\psi(t) = A(t)\exp(-iE_s t/\hbar)|s> + \sum_\ell B_\ell(t)\exp(-iE_\ell t/\hbar)|\ell> , \tag{13}$$

the initial conditions for the problem being

$$A(0) = 1 \; ; \; B_\ell(0) = 0 \; ; \quad \text{all } \ell . \tag{14}$$

The time-dependent coefficients A(t) and $\{B_\ell(t)\}$ are obtained from the time-dependent Schrödinger equation. The amplitude A(t) of the doorway state satisfies the integral equation

$$\dot{A}(t) = -\int_0^t dt' \, F(t-t')A(t') , \tag{15}$$

where $\dot{A}(t) \equiv dA(t)/dt$ and $F(t-t')$ is the memory function for the problem

$$F(t-t') = \hbar^{-2}\sum_\ell |V_{s\ell}|^2 \exp[-i(E_s-E_\ell)(t-t')/\hbar] . \tag{16}$$

It is important to emphasize that this result is general for the bound level structure of Fig. 5 subjected to the initial condition in eq. (14). The time dependence of the amplitude of the initial state and of the initial state occupation probability

$$P_0(t) = |A(t)|^2 , \tag{17}$$

are determined by the energetic spread of the energy levels $\{|E_\ell-E_s|\}$ and the coupling $\{V_{s\ell}\}$ terms. Two extreme limiting cases of time evolution are an oscillatory pattern and an exponential decay over a long time domain. The latter case is pertinent for the elucidation of genuine intramolecular relaxation. This brings us to the derivation of the Wigner-Weisskopf decay law. This derivation is of pedagogical value as it will result in the celebrated golden rule for the dynamics of the system of Fig. 5 specified by the initial conditions in eq. (14), which is obtained as a general, "infinite-order" result, rather than a perturbative solution. To proceed, we introduce the definition of the strength function,

$\Gamma(E)$, sometimes referred to as the imaginary part of the (complex) level shift, for our problem

$$\Gamma(E) = 2\pi \sum_{\ell} |V_{s\ell}|^2 \delta(E-E_{\ell}) , \tag{18}$$

which is used to recast F the memory function (eq. (16)) in the form

$$F(t-t') = \hbar^{-2} \int_0^t dE\ \Gamma(E) \exp[-i(E_s-E)(t-t')/\hbar] . \tag{19}$$

We now invoke the basic assumptions:

1. The strength function $\Gamma(E)$ is smooth, varying slowly with energy around $E \sim E_s$, so that we can take $\Gamma(E) = \Gamma(E_s)$ in the integral.

2. The level structure is unbound, so that edge effects can be ignored.

3. Energetic level shifts are neglected.

When these conditions hold, the memory function takes the simple form

$$F(t-t') = (\Gamma/2\hbar)\ \delta(t-t') , \tag{20}$$

where $\Gamma \equiv \Gamma(E_s)$. It is apparent that the system does not exhibit any memory effects. Eq. (20), together with eqs. (15) and (17), immediately results in the exponential decay law

$$P_0(t) = \exp(-\Gamma t) , \tag{21}$$

for the population of the doorway state $|s\rangle$. The nonradiative decay rate is

$$\Gamma \equiv \Gamma(E_s) = 2\pi \sum_{\ell} |V_{s\ell}|^2 \delta(E_s-E_{\ell}) , \tag{22}$$

which together with the mathematical definition of the density of states

$$\rho_{\ell}(E) = \sum_{\ell} \delta(E-E_{\ell}) , \tag{23}$$

can be expressed in terms of the celebrated golden rule

$$\Gamma = 2\pi\ |V_{s\ell}|^2 \rho_{\ell}|_{E_{\ell} \simeq E_s} , \tag{24}$$

where both the coupling and the density of states are taken in the vicinity of $E_\ell \sim E_s$. We note in passing that when the discrete level structure in the $\{|\ell>\}$ manifold and the interstate coupling terms vary in an irregular way, adequate coarse graining and averaging procedures have to be applied to specify the decay rate. Finally, it is worthwhile to point out that the golden rule can be expressed in terms of a correlation function. Defining the interstate coupling in the interaction representation

$$V(t) = \exp(-iH_{MO}t/\hbar)\, V\exp(iH_{MO}t/\hbar) \quad , \tag{25}$$

it can readily be shown that the decay rate (eq. (22)) can be expressed in terms of a zero-frequency Fourier transform of a correlation function

$$\Gamma = \int_{-\infty}^{\infty} dt \langle s|V(0)V(t)|s\rangle \quad , \tag{26}$$

establishing contact with Kubo's formalism (69) of relaxation processes.

These results provide the basis for the understanding of relaxation phenomena, establishing a set of general conditions for the occurrence of relaxation in a bound level structure. The central condition (1) for the occurrence of ER implies qualitatively that the level structure in the $\{|\ell>\}$ manifold is dense. Under these circumstances, one can inquire on what time scale can the strength function, which appears in the memory function of eq. (19) be considered to be smooth. The answer to this question for some model systems was provided about ten years ago, when it was shown that practical ER occurs on a time scale $t \ll h\rho_\ell$, defining a recurrence time (61), which for dense level structures is exceedingly long. This brings up the central issue of practical irreversibility in a bound level structure, which will be considered further in section IX.

The golden rule exponential decay law was derived for the simplest case of intramolecular dynamics, neglecting the radiative decay channel. When radiative decay is also incorporated one should consider the parallel decay of the doorway state $|s\rangle$ to the intramolecular channel governed by the rate Γ, eqs. (22) and (24) and into the radiative channel, which is determined by the golden rule radiative decay rate

$$\Gamma^{(r)} = 2\pi|V_{sg}^{(r)}|^2\rho_r \quad , \tag{27}$$

where $V_{sg}^{(r)}$ is the radiative coupling between $|s\rangle$ and the ground state $|g\rangle$, while ρ_r represents the density of states in the radiation field. The total decay rate is the sum of the partial rates

$(\Gamma+\Gamma^{(r)})$, while the decay law of the initially excited $|s\rangle$ state now takes the form

$$P_0(t) = \exp[-(\Gamma+\Gamma^{(r)})t/h] . \quad (28)$$

Finally, it should be emphasized that the golden rule exponential decay law is of wide applicability. The case of ER on the time scale $t \ll h\rho_\ell$ is just one example which pertains to the interesting case of a bound level system. The golden rule is applicable for a broad spectrum of decay processes of a single metastable state into a genuine continuum. Simple examples are the radiative decay of excited states into a photon continuum and reactive interstate predissociation processes involving a dissociative continuum. The golden rule also provides the ideological basis for the understanding of a variety of nonradiative relaxation phenomena in condensed phases, which are of considerable interest in physical chemistry, solid-state physics and biophysics.

VI. BEYOND THE GOLDEN RULE

We have been concerned with the basic problem in the area of relaxation theory, where a single zero-order metastable state decays into a genuine continuum, e.g., a radiative or dissociative channel, or into a bound intramolecular quasicontinuum. The exponential decay law is expected to hold under the circumstances specified in section V. An additional hidden assumption underlying the validity of the exponential decay law is the existence of a single metastable zero-order state decaying into a background continuum or quasicontinuum. The exponential decay pattern exhibited in this case signals the absence of interference effects. When a set of close-lying zero-order discrete states are coupled to the same common continuum, the system will reveal interference effects manifested by nonexponential decay. Interference effects in the decay of closely coupled levels are well known in the field of level-crossing spectroscopy (70), where close-lying Zeeman components of an atom or a large molecule reveal nonexponential oscillatory radiative decay. What is of interest to us is the radiative decay of a complicated excited-state level structure of a manifold of molecular eigenstates in a large molecule. An interesting question is, under what circumstances will the radiative decay from an excited-state congested level structure exhibit nonexponentional time evolution? This problem is of considerable interest as probing of radiative decay can provide central information concerning intramolecular excited-state dynamics in isolated large molecules. We have already commented in section V on the golden-rule treatment of the radiative decay of a single optically active doorway state. In what follows we shall consider the problem of interference effects in radiative decay of a complex molecular level structure. This problem is of considerable methodological interest and will provide a set of predictions which have not yet been subjected to an experimental test.

To go beyond the golden rule (71) we consider the problem of a few discrete levels, say $|s1\rangle$ and $|s2\rangle$ coupled to a common continuum $\{|\ell\rangle\}$ (Fig. 6). We can define the level widths, Γ_{s1} and Γ_{s2}, of these two states in a manner analogous to eq. (23),

$$\Gamma_{sj} \equiv 2\pi|V_{sj,\ell}|^2\rho_\ell \; ; \; j = 1,2 \; , \tag{29}$$

where $V_{sj,\ell}$ (j = 1,2) is the coupling between the state $|s_j\rangle$ and the continuum, which is characterized by the density of states ρ_ℓ. The basic condition for interference in the decay is

$$\Gamma_{s1}, \Gamma_{s2} \gtrless |E_{s1} - E_{s2}| \; , \tag{30}$$

implying that the level widths exceed the level spacing. When eq. (30) is obeyed, interference effects will lead to nonexponential oscillatory decay.

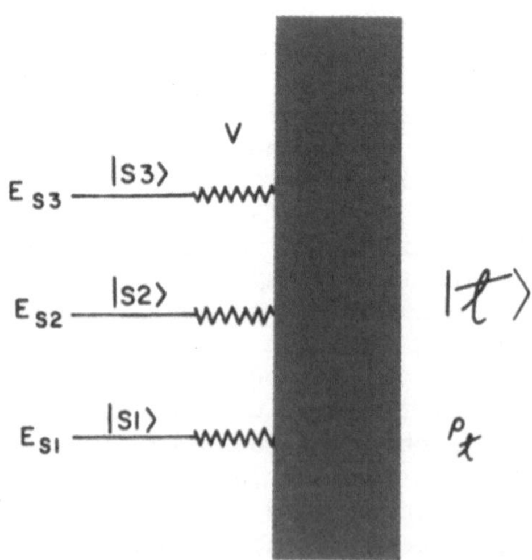

Fig. 6. The coupling of several discrete levels to a common continuum.

To provide a description of the time evolution of such overlapping resonances, we start with the initial state

$$\psi(t=0) = A_{s1}|s1\rangle + A_{s2}|s2\rangle \quad , \tag{31}$$

where A_{s1} and A_{s2} are the preparation (excitation) amplitudes of the discrete zero-order states. The initial condition (eq. (31)) implies that the continuum is inactive with respect to excitation. The dynamic time evolution of the system is

$$\psi(t) = A(t)\exp(-iE_{s1}t/\hbar)|s1\rangle + B(t)\exp(-iE_{s2}t/\hbar)|s2\rangle + \\ + \sum_{\ell} C_{\ell}(t)\exp(-iE_{\ell}t/\hbar)|\ell\rangle \quad , \tag{32}$$

with the initial condition determined by eq. (31). Without alluding to any detailed calculations, it is apparent that the initial state population probability

$$\begin{aligned} P_0(t) &= |\langle\psi(0)|\psi(t)\rangle|^2 = \\ &= |A(t)\exp(-iE_{s1}t/\hbar) + B(t)\exp(-iE_{s2}t/\hbar)|^2 = \\ &= |A(t)|^2 + |B(t)|^2 + 2\mathrm{Re}\, A^*(t)B(t)\exp\left[-i(E_{s2}-E_{s1})t/\hbar\right], \end{aligned} \tag{33}$$

contains an interference term which is determined by the phases of the amplitudes. This is the essence of the interference effects exhibited in the decay of close-lying levels.

The decay law for the two-discrete-level system, eq. (32), is obtained (71) by the use of the time dependent Schrödinger equation invoking assumptions similar to those introduced in section V for the decay of a single doorway state. The time evolution of the amplitudes of eq. (32) is governed by the matrix equation (71)

$$i\hbar \begin{pmatrix} \dot{A}(t) \\ \dot{B}(t) \end{pmatrix} = \begin{pmatrix} (E_{s1} - \frac{i}{2}\Gamma_{11}) & -\frac{i}{2}\Gamma_{12} \\ -\frac{i}{2}\Gamma_{21} & (E_{s2} - \frac{i}{2}\Gamma_{22}) \end{pmatrix} \begin{pmatrix} A(t) \\ B(t) \end{pmatrix}, \tag{34}$$

where the decay matrix, $\underset{\approx}{\Gamma}$, is defined by its matrix elements,

$$\underset{\approx}{\Gamma}_{ij} = 2\pi \langle si|V|\ell\rangle \langle\ell|V|sj\rangle \rho_\ell \quad , \tag{35}$$

and it is assumed that Γ_{ij} varies weakly with energy in the relevant region.

It is possible to provide an immediate generalization of this result for the decay of a manifold of n discrete zero-order states $\{|sj\rangle\}$; $j = 1,2,\ldots n$ into a common continuum. The time evolution

$$\psi(t) = \sum_{j=1}^{n} A_j(t)\exp(-iE_{sj}t/\hbar)\ |sj\rangle \ , \tag{36}$$

can be expressed in terms of the effective Hamiltonian formalism. The vector of the amplitudes appearing in eq. (36)

$$\underset{\sim}{a}(t) = \begin{vmatrix} A_1(t) \\ A_2(t) \end{vmatrix} \quad , \tag{37}$$

obeys the equation

$$i\hbar \frac{\partial}{\partial t}\ \underset{\sim}{a}(t) = \underset{\approx}{H}_{eff}\ \underset{\sim}{a}(t) \quad , \tag{38}$$

where the effective Hamiltonian matrix is defined by

$$(\underset{\approx}{H}_{eff})_{jj} = E_j - \frac{i}{2}(\underset{\approx}{\Gamma})_{jj} \ ; \ (\underset{\approx}{H}_{eff})_{ij} = -(\underset{\approx}{\Gamma})_{ij},\ i\neq j \ , \tag{39}$$

and where the decay matrix $\underset{\approx}{\Gamma}$ is defined in terms of eq. (35).

Eqs. (38) and (39) provide the generalization of the golden rule for the dynamics of the decay of a manifold of states into a joint continuum. The following general features of this result should be noted (8,10,11):

(1) It exhibits the time evolution of the discrete manifold, while the entire Hilbert space contains both this discrete manifold and the continuum. Thus, we have obtained the time evolution within a subspace of the Hilbert space.

(2) The time evolution within this subspace is determined by the effective Hamiltonian matrix

$$\underset{\approx}{H}_{eff} = \underset{\approx}{H}_{M} - \frac{i}{2}\underset{\approx}{\Gamma} \quad , \tag{40}$$

which is invariant with respect to representation in any adequate basis. The contribution of the continuum states is subsumed in the decay matrix.

(3) $\underset{\approx}{H}_{eff}$ is a nonhermitian matrix. This is the price we pay for subsuming the contribution of the continuum states.

(4) Provided that the $\{|sj\rangle\}$ states can be expressed in a real form, then $\underset{\approx}{H}_{eff}$ is a complex symmetric matrix.

(5) $\underset{\approx}{H}_{eff}$ can then be diagonalized by a complex orthogonal matrix $\underset{\approx}{S}$

$$\underset{\approx}{S}\,\underset{\approx}{H}_{eff}\,\underset{\approx}{S}^{-1} = \underset{\approx}{\Lambda} \,, \tag{41}$$

where $\underset{\approx}{\Lambda}$ is a (complex) diagonal matrix

$$\Lambda_{ij} = (\varepsilon_j - \frac{i}{2}\gamma_j)\,\delta_{ij} \quad . \tag{42}$$

Here, the real part ε_j constitutes the "new" energies, while the imaginary part γ_j represents the widths of the "new" states of the system.

(6) The eigenvalues of eq. (42) obey the diagonal sum rules

$$\sum_{j=1}^{n} \varepsilon_j = \sum_{j=1}^{n} E_j \quad ; \tag{43a}$$

$$\sum_{j=1}^{n} \gamma_j = \sum_{j=1}^{n} \Gamma_{jj} \quad . \tag{43b}$$

(7) The "new" (complex) eigenvectors $|m\rangle$, which diagonalize $\underset{\approx}{H}_{eff}$ according to eq. (41), correspond to the independently decaying levels of the system, providing a generalization of the concept of molecular eigenstates to incorporate the effect of genuine decay channels.

The time evolution of the system can now be expressed in terms of the eigenvectors matrix $\underset{\approx}{S}$ and the eigenvalues matrix $\underset{\approx}{\Lambda}$, eq. (42). The general result is

$$\underset{\sim}{a}(t) = \underset{\approx}{S}^{-1} \exp[-\frac{i}{\hbar}\underset{\approx}{\Lambda}\, t]\, \underset{\approx}{S}\, \underset{\sim}{a}(0) \,, \tag{44}$$

which, together with the specification of the initial conditions $\underset{\sim}{a}(0)$, provide an adequate general solution to the dynamic problem. To gain some insight into the characteristics of this solution we specialize to a two-level system with the initial condition of eq. (31).

The initial state population probability, eq. (33), takes the form

$$P_0(t) = \eta_1 \exp(-\gamma_1 t/\hbar) + \eta_2 \exp(-\gamma 2\ t/\hbar) +$$

$$+ \mathrm{Re}\left\{ \eta_{12} \exp\left[-\frac{\gamma_1+\gamma_2}{2\hbar} t \right] \exp[i(\varepsilon_1-\varepsilon_2)t/\hbar] \right\}, \tag{45}$$

where the coefficients η_1, η_2 and η_{12} are determined by the matrix elements of $\underset{\approx}{S}$ and by the preparation amplitudes and will not be specified. The first two terms on the RHS of eq. (45) correspond to direct exponential decay terms. On the other hand, the third term includes an oscillatory time evolution with a period $\hbar/|\varepsilon_1-\varepsilon_2|$ as well as an exponential decay contribution. The oscillatory time dependence of the third, interference-type term corresponds to quantum beats in the decay of this simple system. Generalization to the decay of a multilevel system is conceptually straightforward. For a n level system one expects, in general, a set of pair contributions of the form $\exp[(i/\hbar)(\varepsilon_j-\varepsilon_{j'})t]$, with $jj' = 1...n$. When the number of such oscillatory contributions is not too large, quantum beats will be exhibited.

Let us now consider the conditions for the observation of interference effects manifested by quantum beats in the radiative decay of a many-level system, which can be spelled out as follows:

(a) The zero-order states $\{|sj\rangle\}$ are coupled to a common continuum; a sufficient condition is that these states are of the same symmetry.

(b) A coherent excitation of the manifold is practically possible.

(c) The zero-order levels are not too widely spaced, i.e., we require that

$$|(E_{si} - E_{sj}) - \frac{i}{2}(\Gamma_{ii} + \Gamma_{jj})| \sim \Gamma_{ij} , \tag{46}$$

for all i and j. This condition is essentially equivalent to condition (30) for the interference effects.

(d) The independently decaying levels are not too close:

$$\hbar/|\varepsilon_i-\varepsilon_j| \sim \gamma_i^{-1} , \gamma_j^{-1} , \tag{47}$$

otherwise, when the levels are too closely spaced, the oscillators will be too fast to be amenable to experimental observation.

(e) When the number of $\{|sj\rangle\}$ levels is large, there will be a large number of interference contributions to the temporal evolution of the system. As the zero-levels are expected to be irregularly spaced and irregularly coupled to the radiation field, a practically random energetic distribution of the ε_j levels and of the widths γ_j is expected to occur. Under these circumstances, there will be mutual cancellation between individual pair contributions, resulting in the erosion of the quantum beats.

An interesting candidate for the observation of quantum beats in the radiative decay of close-spaced levels in large isolated molecules involves a scrambled singlet-triplet manifold, where a low-lying vibronic singlet state, which acts as a doorway state and which is active in emission, is quasidegenerate with a moderately sparse triplet manifold. This state of affairs corresponds to the intermediate level structure, which will be discussed in section IX. As the density of states in the triplet manifold is not too high, the spacing of the mixed molecular eigenstates (Fig. 7) can be comparable to their widths. Furthermore, as the spin-orbit coupling can be reasonably weak, the energetic spread of the molecular eigenstates is not excessive to prohibit their coherent excitation. Accordingly, conditions (a)-(d) for the occurrence of quantum beats can be satisfied for such a level structure. The difficulties inherent in the observation of quantum beats from such a system are of three types.

(i) Practical limitations. Thermal rotational and vibrational sequence congestion prohibit coherent excitations of a molecular level structure at room temperature. To overcome this difficulty the molecule has to be cooled in a supersonic beam, where effective vibrational and rotational cooling can be accomplished.

(ii) Possibility of coherent excitation (condition (a)). The energetic spread of the contaminated molecular eigenstates should not exceed $\sim 1\ \text{cm}^{-1}$. This condition can be met.

(iii) Level structure. The density of states has to be moderate, obeying conditions (b)-(d). Such a system can be found.

(iv) Irregularity of level structure. The wide and wild variation of the interstate coupling and the energetic spread may result (according to condition (e)), to the erosion of the quantum beats. It is not clear whether this inherent difficulty can be overcome in real life.

A recent experimental study by McDonald and Chaiken (72) reported the observation of molecular quantum beats in the decay of the scrambled singlet-triplet manifold of biacetyl, cooled in a supersonic beam. The level structure of this system conforms to conditions (a)-(d). It is intriguing to inquire how the observable quantum beats survive the effects of irregularity in the level structure.

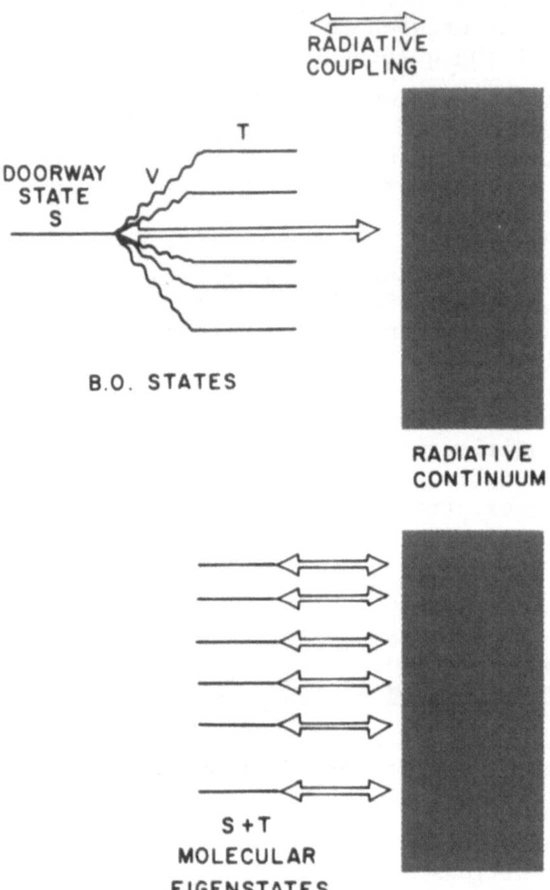

Fig. 7. A molecular candidate for the study of quantum beats in the radiative decay of a manifold of closely spaced levels of the same symmetry. The doorway singlet state (S) is coupled by weak spin-orbit interaction V to a moderately sparse triplet (T) manifold. The intrinsic decay widths of the zero-order triplet states are small relative to their spacing. Only the zero-order S state decays radiatively. The mixed S + T molecular eigenstates are coupled to a common radiative channel. The decay matrix $\underset{\approx}{\Gamma}$ due to radiative decay is nondiagonal.

VII. OBSERVABLES

Up to this point we have been concerned with time evolution of metastable states. This information emerges from "short-time" excitation experiments, where the temporal duration of the exciting light pulse is considerably shorter than all the relevant decay times of the molecular systems. Such time-resolved experimental data correspond to one class of observables, which provide information on intramolecular dynamics. The basic physical information emerging from the experimental observables falls into two classes:

1. Populations of individual excited states and/or populations of discrete (quasicontinuum) or continuous decay channels. Temporal and energy-resolved information is obtained from some of the time-resolved and energy-resolved observables. This information, with regard to the population of excited electronic-vibrational-rotational states of the parent molecule, is of central importance.

2. Phase relationships. Information on phase relationships between excited states is obtained from the analysis of interference effects exhibited in the time-resolved observables and in the energy-resolved variables. Phase relationships between the ground and excited doorway states can be monitored by studies of optical coherence effects (73,74).

To be more specific, let us now consider these observables in some detail.

(A) Time-resolved observables. Direct information regarding the temporal decay of excited states is obtained from "short-time" excitation followed by:

(i) monitoring radiative decay by time-resolved photon counting;

(ii) monitoring populations of reactive and nonreactive decay channels by time-resolved absorption methods.

The different patterns of time-resolved decay modes can be classified as follows:

(1) Exponential decay. The simplest example involves the parallel decay of a single doorway state into intramolecular and radiative channels, according to eq. (44).

(2) Nonexponential decay consisting of a superposition of exponentials. This interesting decay mode will be discussed in section IX.

(3) Quantum beats in the decay, which has already been dealt with in section VI.

(B) Energy-resolved observables. A variety of cross sections for optical absorption, photon scattering, etc., yields basic data regarding reactive and nonreactive molecular processes. These studies sacrifice the time resolution for the sake of energy resolution, providing a blending of spectroscopic data with basic information concerning intramolecular dynamics. These observables are:

(a) absorption cross sections;
(b) photon scattering cross sections;
(c) cross sections for population of decay channels;
(d) angular distribution of products;

(C) Observables pertaining to coherent optical effects. Optical excitation by coherent light pulses can establish definite phase relationships between the ground state and excited doorway states (73,74). The temporal persistence of the phase relationships can be interrogated by studies of optical nutation, free-induction decay and photon echo experiments. These phase relationships are destroyed by dephasing phenomena, which are of three distinct types. First, intramolecular dephasing in reactive processes is equivalent to T_1 level depletion processes. Second, intramolecular dephasing in interstate and intrastate nonreactive relaxation in large isolated molecules can be regarded as intramolecular T_2 processes (75,76). Third, erosion of phase coherence by medium perturbations provide important chemical and physical information on intermolecular T_2 processes, which pertain to the consequences of the coupling of excited molecular states with the host medium (73, 74). Most of the coherent effects in the optical region have been limited to two-level systems (73). The studies of retention of phase relationships in multilevel systems (77-79) are of considerable interest for the elucidation of multiphoton excitation of large molecules.

In sections V and VI, we have already discussed a few features of time-resolved observables. We shall now proceed to outline some of the information content emerging from the study of energy-resolved variables, focusing attention on absorption cross sections.

VIII. ABSORPTION LINESHAPES

The effect of line broadening in the optical absorption spectra of molecular metastable states undergoing reactive intermolecular relaxation, e.g., predissociation, was discovered fifty years ago (80). The linewidth, $\Delta\nu$, of the Lorentzian lineshape is related to the lifetime, τ, of the metastable state by the uncertainty relation $\tau\Delta\nu = \hbar$. The intensity distribution in one-photon absorption from the ground state to the excited congested bound level structure of a large molecule (Fig. 6) can provide useful information regarding the energetic spread of the molecular eigenstates. This spectroscopic information is useful only provided that all sources of inhomogeneous broadening are eliminated. In what follows we shall outline the theory of absorption lineshapes, focusing attention on the information regarding interstate coupling which emerges from spectroscopic studies.

We consider a bound molecular level structure consisting of the ground state $|g\rangle$ and a set of molecular eigenstates $\{|m\rangle\}$, invoking the following simplifying assumptions:

(1) The intensity of the electromagnetic field used to probe the absorption spectrum is weak.

(2) Linear response theory is utilized, treating the radiative interaction to first order.

(3) Zero temperature limit is considered, justifying the use of a single ground state.

(4) The optical $|g\rangle$-$|m\rangle$ transitions are dipole allowed, being specified by the transition moments $\mu_{gm} = \langle g|\mu|m\rangle$, where μ is the transition moment operator.

The absorption lineshape L(E) at the photon energy E is now given, apart from irrelevant numerical factors, in the form

$$L(E) = E\sum_m |\mu_{gm}|^2 \, \delta(E-E_m) \tag{48}$$

which can be expressed in the equivalent form (3,81)

$$L(E) = -(E/\pi) \, \mathrm{Im} \langle g|\mu G(E)\mu|g\rangle \tag{49}$$

where G(E) is the Green's operator

$$G(E) = (E - H_M + i\eta)^{-1} \; ; \quad \eta \rightarrow 0^+ \quad . \tag{50}$$

Here, E is a real variable and η is an infinitesimally small quantity providing an upper limit for the exceedingly long time scale of the experiment.

It is important to emphasize at this point that eq. (49) is general within the framework of assumptions (1)-(4), being valid for any level structure. This formalism is extremely useful as L(E) can be recast in terms of any acceptable basis set. To provide an instant application of this powerful formalism, we consider the lineshape for the basic molecular model of Fig. 5, where a single doorway state $|s\rangle$ carries oscillator strength from the ground state, i.e., $\mu_{gs} \neq 0$, while $\mu_{g\ell} = 0$ for all $|\ell\rangle$. For this system we get

get

$$L(E) = \left(\frac{|\mu_{gs}|^2 E}{\pi} \right) \mathrm{Im} \langle s|G(E)|s\rangle \ . \tag{51}$$

The diagonal matrix element of the Green's function appearing in eq. (51) takes the explicit form (8,11)

$$\langle s|G(E)|s\rangle = \frac{1}{E-E_s - \sum_\ell \frac{|V_{s\ell}|^2}{E-E_\ell+i\eta} + i\eta} \ . \tag{52}$$

The (complex) sum appearing in the denominator of the Green's function in eq. (52) is usually referred to as the level shift operator (82). Defining the diagonal matrix element of the level shift operator

$$R_{ss}(E) = \sum_\ell \frac{|V_{s\ell}|^2}{E - E_\ell + i\eta} \ , \tag{53}$$

we rewrite the lineshape

$$L(E) = \left(\frac{|\mu_{gs}|^2 E}{\pi} \right) \frac{\mathrm{Im}[R_{ss}(E)]}{(E-E_s-\mathrm{Re}[R_{ss}(E)])^2 + (\mathrm{Im}[R_{ss}(E)])^2} \ . \tag{54}$$

The level shift operator incorporates all the pertinent information on interstate coupling. The imaginary part of the level shift is

$$\mathrm{Im}[R_{ss}(E)] = -\pi \sum_\ell |V_{s\ell}|^2 \ \delta(E-E_\ell) \ , \tag{55}$$

so that the strength function, eq. (18), is

$$\Gamma(E) = -2\mathrm{Im}[R_{ss}(E)] \ . \tag{56}$$

The corresponding real part, sometimes known as the energetic level shift, can be expressed in terms of the dispersion relation (82)

$$\mathrm{Re}[R_{ss}(E)] = \frac{1}{\pi} \ \mathrm{pp} \int \frac{\Gamma(E')dE'}{E-E'} \ , \tag{57}$$

where pp stands for the principal part of the integral. It should be noted that for the problem at hand the diagonal matrix element of The level shift operator, which determines the absorption lineshape, depends on coupling terms of the form $|V_{s\ell}|^2$, thus being independent of the phases of the wavefunctions.

Eqs. (54)-(57) constitute a general theory of absorption line-shape. It is important to realize that, in general, one cannot invoke the assumption of section V concerning the smoothness of the strength function. In many cases, the density of states is not sufficiently high to warrant the validity of the smoothness assumption and the energy dependence of $\Gamma(E)$ is of intrinsic value.

To demonstrate the usefulness of spectroscopic data for providing information concerning interstate coupling and intramolecular dynamics, we have reproduced in Fig. 8 the recent experimental data

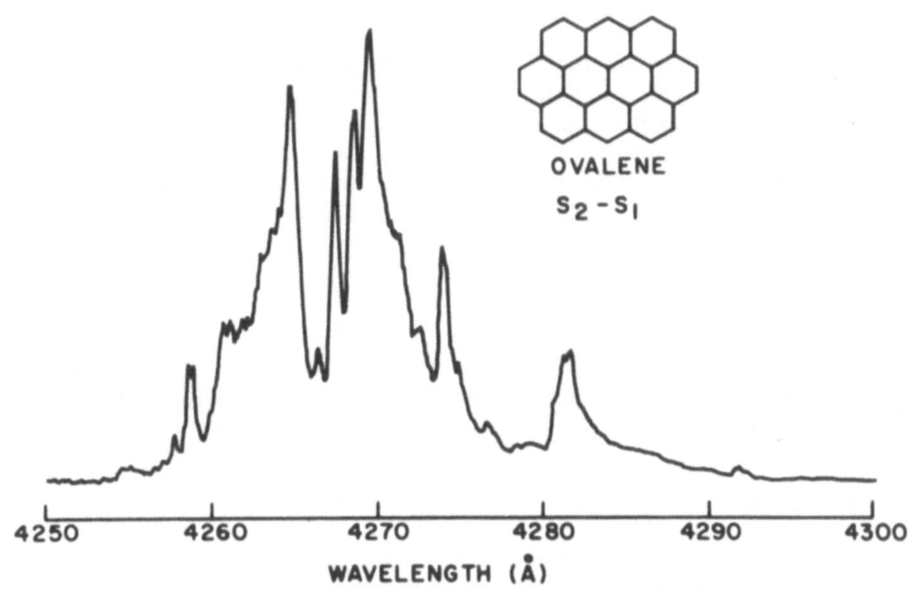

Fig. 8. Fluorescence excitation spectrum in the region 4300-4250 Å of ovalene cooled in a supersonic expansion. This spectrum is reproduced from the work of Amirav, Even and Jortner [Ref. (36)]. Ovalene (vapor pressure $\sim 10^{-2}$ Torr) was seeded into Ar at 210 Torr and was expanded through a 200 μ nozzle. The exciting dye laser has a bandwidth of 0.3 cm^{-1}. All fluorescence spectra are normalized to the laser intensity. This entire spectrum corresponds to the transition $S_0(0) \rightarrow S_2(0)$, i.e., to the electronic origin of the S_2 state. The irregular closely-spaced structure reflects the distribution of the excitation amplitudes of the molecular eigenstates originating from $S_2(0) - S_1$ scrambling.

of Amirav et al. (36) on the $S_0 \rightarrow S_2$ absorption spectrum of the isolated ovalene molecule ($C_{32}H_{14}$) in a supersonic beam. In this huge molecule, characterized by s = 134 vibrational degrees of freedom, the energy gap between the electronic origins of the S_2 and S_1 states is $\Delta E \simeq 1800$ cm^{-1}. This small electronic energy gap implies that the density of S_1 vibronic states, which are quasidegenerate with the electronic origin of S_2, is relatively low, $\rho \sim 10^3$ cm or so, while the density of effectively coupled levels is even lower. For such moderately low values of ρ_ℓ, coarse-graining procedures for the interstate coupling are inapplicable. It is remarkable that statistical concepts cannot be applied to such a large molecule as ovalene in view of the small energy gap. The $S_0 \rightarrow S_2$ absorption spectrum of Fig. 8 reveals an irregular, closely spaced structure consisting of many lines whose spacing is 5-15 cm^{-1}. The entire spectrum in the range 4250-4300 Å is due to the electronic origin of S_2. The intensity of this single doorway state is spread among a large number of molecular eigenstates, due to $S_2 - S_1$ scrambling, all of which are active in absorption. The strength function for the scrambling of the S_2 origin with the background S_1 manifold was extracted from the experimental data. As the lineshape just monitors the energy dependence of the imaginary part of the Green's function, it makes contact between formal theory and experiment. The strength function can be related to this observable by the relations (8,36,83) which follow from eqs. (52) and (53)

$$\Gamma(E) = \frac{-\mathrm{Im}\, G_{ss}(E)}{[\mathrm{Re}(G_{ss}(E))]^2 + (\mathrm{Im}[G_{ss}(E)])^2} \quad . \tag{58}$$

The real part of $G_{ss}(E)$ appearing in eq. (58) can be related to the corresponding real part by a dispersion relation (82)

$$\mathrm{Re}\, G_{ss}(E) = -\frac{1}{\pi}\, \mathrm{pp} \int \frac{dE'\ \mathrm{Im}\, G_{ss}(E')}{E - E'} \quad . \tag{59}$$

From eqs. (58) and (59), together with Eq. (49), one can extract the strength function. $\Gamma(E)$ for the interstate coupling of S_2(origin) $- S_1$ was calculated by Amirav et al. (36) and is reproduced in Fig. 9. This procedure is similar to solid-state techniques, which extract a weighted density of states from experimental data. The important information, which emerges from the strength function of Fig. 9, pertains to the energy dependence of the interstate coupling. The following conclusions emerge from the analysis of the specific system. First, we are encountering the consequences of effective interstate $S_2 - S_1$ scrambling, the coupling obeying relation (9). Second, the strong and irregular energy dependence of $\Gamma(E)$ is characteristic for an intermediate level structure, (to be discussed in the next section), where the density of background $\{|\ell\rangle\}$ states is modest. Third, the energetic spread of the molecular eigenstates originating from the S_2 (origin) - S_1 scrambling is $\sim$100 cm^{-1}. Fourth, we would like to inquire whether it is possible

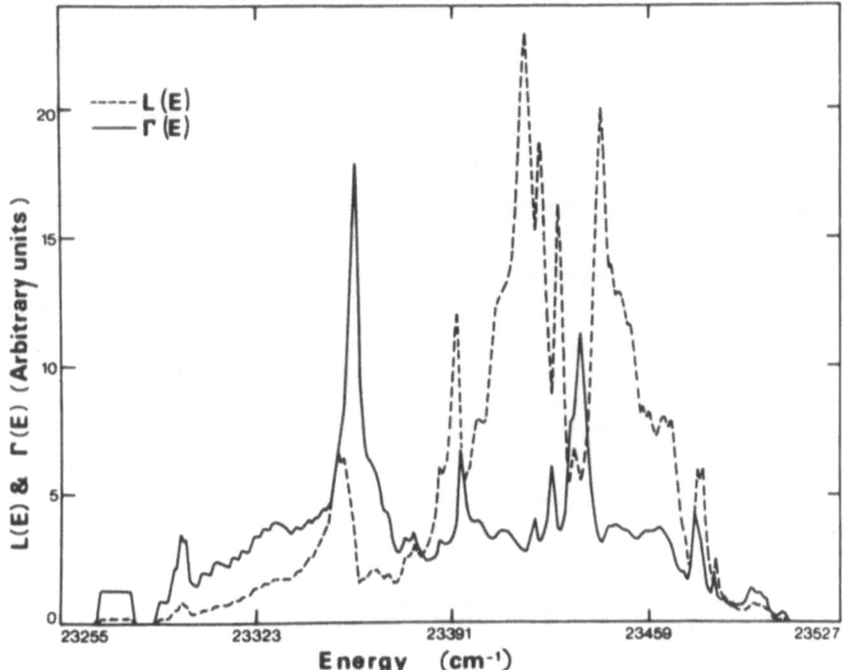

Fig. 9. The absorption spectrum, L(E), for the scrambled electronic origin S(0) of the S_2 electronic state and the strength function, Γ(E), for the $S_2(0)$ - S_1 coupling of ovalene, which corresponds to the intermediate level structure. The strength function was calculated by Amirav, Even and Jortner [Ref. (36)] from their experimental lineshape data.

to excite coherently the S_2 (origin) doorway state, using an ultrashort light pulse. The answer to this cardinal question is negative on practical grounds, as such time-resolved excitation process would require a light pulse whose coherent temporal spread spans an energy range of $\sim$100 cm^{-1}, so that its temporal duration should be shorter than $\sim 10^{-14}$ sec. Such light sources are not available at present.

This discussion of the spectroscopic and dynamic implications of interstate coupling in a huge molecule characterized by a small electronic energy gap brings up an important distinction between the implications of interstate coupling and electronic relaxation.

While interstate coupling prevails in general in electronically excited states of large molecules, we argue that intramolecular relaxation in a bound level system is determined by the level structure and the initial conditions. This brings us to consider the general classification of level structure, coupling and dynamics in bound electronically excited states of polyatomics.

IX. THE RISE AND FALL OF EXCITED STATES OF LARGE MOLECULES

Nonreactive relaxation processes, such as electronic relaxation in electronically-vibrationally excited states of large molecules, are characterized by the following universal features:

(1) The intramolecular level structure is bound.

(2) The bound intramolecular level structure is characterized by a high density of states.

(3) The level spacing between adjacent states varies in an irregular way.

(4) The interstate coupling terms vary widely and wildly with respect to the quantum number which specify the zero-order states.

We can thus assert that interstate electronic relaxation processes involve time evolution within a bound "bumpy" quasicontinuum. To provide an overview of the dynamics of electronically-vibrationally excited states of polyatomic molecules, it is imperative to consider not only the level structure of Fig. 5 and the initial excitation conditions, but also to discuss the fate of the states in the background manifold. Up to now it was implicitly assumed that the $\{|\ell\rangle\}$ manifold is inactive in absorption and in emission. In other words, when the system evolves in time into the $\{|\ell\rangle\}$ manifold, it will remain there forever, at least on the relevant experimental time scale. However, sequential decay process of the background $\{|\ell\rangle\}$ states may be characterized by decay probabilities, i.e., decay widths, which exceed the extremely small level spacing. Under these circumstances, the sequential decay problem, which is of intrinsic theoretical interest, should be considered.

No time evolution process in a bound level structure terminates in the $\{|\ell\rangle\}$ manifold. Even under "astrophysical" conditions, when the molecule is isolated on an exceedingly long time scale, sequential dumping processes of the $\{|\ell\rangle\}$ states occur. These sequential decay processes involve:

(a) Infrared photon emission from the $\{|\ell\rangle\}$ states, which are quasidegenerate with the doorway state $|s\rangle$ to lower states in the ℓ manifold (61). This sequential decay process is important when the $\{|\ell\rangle\}$ manifold corresponds to high vibrational levels of the ground electronic state, or to a triplet state. The decay rates for such infrared decay are $\sim 10^3$ sec^{-1}, corresponding to sequential decay widths of $\gamma_\ell \simeq 10^{-8}$ cm^{-1}.

(b) Optical emission of states in the $\{|\ell>\}$ manifold may sometimes occur to high vibrational levels of the ground state electronic configuration. Such a sequential decay process will be exhibited when the $\{|\ell>\}$ manifold and the ground electronic configuration correspond to the same spin states. The decay rates for such optical decay are $\sim 10^8$-10^6sec^{-1}, corresponding to decay widths of $\gamma_\ell \simeq 10^{-3}$-$10^{-5}$cm^{-1}.

For both sequential decay processes of types (a) and (b), we can introduce a privacy rule, asserting that each $|\ell>$ state is coupled to its own private continuum. Accordingly, we can now assign an independent decay width γ_ℓ to each of the background states. Thus, the decay matrix (section VI) for this problem is diagonal. At this stage it is imperative to extend the concept of molecular eigenstates introduced in section IV to incorporate the role of sequential decay of the $\{|\ell>\}$ states and the radiative decay of the doorway state. This is done by assigning the decay widths γ_s and $\{\gamma_\ell\}$ to the doorway states (Fig. 10).

The effective Hamiltonian for the problem is

$$\underset{\approx}{H}_{eff} = \begin{pmatrix} (E_s - \frac{i}{2}\gamma_s) & V_{s,\ell=1} & V_{s,\ell=2} & \cdots \\ V_{\ell=1,s} & (E_{\ell=1} - \frac{i}{2}\gamma_{\ell=1}) & 0 & \cdots \\ V_{\ell=2,s} & 0 & (E_{\ell=2} - \frac{i}{2}\gamma_{\ell=2}) & \cdots \\ \cdot & \cdot & \cdot & \\ \cdot & \cdot & \cdot & \\ \cdot & \cdot & \cdot & \end{pmatrix} . \quad (60)$$

The molecular input parameters involve the energy levels $\{|E_\ell - E_s|\}$ (or the energy E_s together with density of states ρ_ℓ), the interstate coupling terms $\{V_{s\ell}\}$ and the decay widths γ_s and $\{\gamma_\ell\}$. Application of the effective Hamiltonian formalism outlined in section VI, which rests on the diagonalization of $\underset{\approx}{H}_{eff}$, eq. (60), results in a set $\{|m>\}$ of (complex) molecular eigenstates. These $\{|m>\}$ states are characterized by the complex energies (Fig. 10),

$$\varepsilon_m = E_m - \frac{i}{2}\gamma_m \quad (61)$$

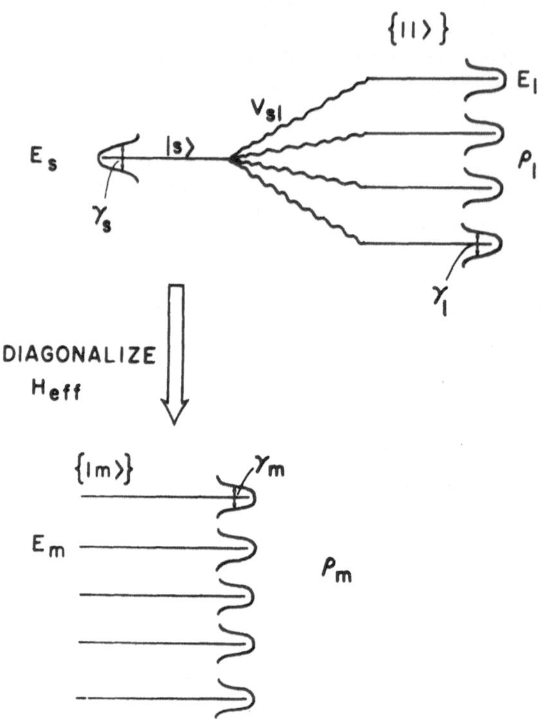

Fig. 10. The effective Hamiltonian formalism. The zero-order states $|s\rangle$ and $\{|\ell\rangle\}$ are characterized by the energies E_s and $\{E_\ell\}$, respectively, and by the decay widths γ_s and γ_ℓ. $V_{s\ell}$ represents the intramolecular (interstate or intrastate) coupling between the doorway state $|s\rangle$ and the $\{|\ell\rangle\}$ manifold, which is characterized by the density of states ρ_ℓ. Diagonalization of the effective Hamiltonian results in a set of independently decaying levels $\{|m\rangle\}$, i.e., generalized molecular eignestates, characterized by energies $\{E_m\}$, decay widths $\{\gamma_m\}$ and density of states ρ_m.

where E_m are the energies and γ_m are the decay widths of the independently decaying levels of the molecular system.

A complete description of the dynamics of the molecular system driven by a weak radiation field requires the specification of the following ingredients:

(A) The excited-state molecular level structure determined by the energies E_m, eq. (61).

(B) The genuine decay rates γ_m, eq. (61), which provide a self-consistent generalization of the golden rule decay law.

(C) The independently decaying levels $\{|m>\}$ which determine the excitation amplitudes from the ground state, as well as the decay amplitudes of the independently decaying levels.

Accordingly, we can consider a general initial state

$$\psi(0) = \sum_m A_m |m> , \tag{62}$$

where A_m are the excitation amplitudes. The initial state occupation probability at time t is

$$P_0(t) = \sum_m |A_m|^2 \exp[-iE_m t/\hbar - \gamma_m t/2\hbar] , \tag{63}$$

providing a generalization of eq. (7). The time evolution (63) is determined by Fourier sums damped by real decay components. To provide a physically meaningful description of the preparation process, we also have to specify

(D) The light pulse. This weak excitation source will be described in terms of a one-photon wavepacket (10,11). Let us take, for the sake of simplicity, a Lorentzian wavepacket in energy peaking at energy E_p and characterized by a width γ_p. The complementary temporal description involves exponential decay $\exp(-\gamma_p t)$ in time. For time-resolved experiments we require that $\gamma_p >> \gamma_m$ for all m. The range of molecular energies spanned by the pulse determines the initial "preparation" conditions (10,11).

Significant molecular information stems from the classification of the bound level structure, which rests on the following guidelines.

(i) It always corresponds to effective interstate scrambling, i.e., $V_{s\ell}\rho_\ell >> 1$.

(ii) To distinguish between sparse and dense level structures, we inquire how the level spacings compare to the sequential decay widths, considering the dimensionless parameter $\gamma_\ell \rho_\ell$.

(iii) To characterize initial excitation conditions, we inquire what energetic range of molecular eigenstates is spanned by the one-photon wavepacket, which represents the radiation source. Two limiting excitation modes are distinguished

(iii.a) Coherent excitation

$$\gamma_p >> \gamma_m \tag{64a}$$

$$\gamma_p >> |E_m - \bar{E}_p| , \tag{64b}$$

for all m. The pulse width exceeds the energy spread of the relevant excited-state distribution.

(iii.b) Energy-weighted excitation

$$\gamma_p >> \gamma_m , \tag{65a}$$

for all m, but for many of these states

$$\gamma_p << |E_m - \bar{E}_p| . \tag{65b}$$

Now, only a small subset, or even a single $|m\rangle$ state, can be photoselected.

We can now proceed to advance an extremely useful classification of interstate coupling and relaxation, when effective interstate scrambling, condition (i), is satisfied. The three limiting situations of intramolecular dynamics are (3,8-12,84)

(I) The Statistical Limit. The density of states is exceedingly high, so that

$$\gamma_\ell \rho_\ell >> 1 . \tag{66}$$

The density of states can be very high, as is evident from Table 2, to insure the validity of condition (66), which implies the efficiency of sequential coupling. When relation (i) for strong interstate coupling and condition (66) are satisfied, a huge number of levels are involved in interstate mixing. The energetic spread of the mixed molecular eigenstates in the statistical limit can be specified by the width

$$\Gamma_s = 2\pi \langle V_{s\ell}^2 \rho_\ell \rangle , \tag{67}$$

with the average being taken over the $\{|\ell\rangle\}$ manifold. In the statistical limit Γ_s is the only intramolecular nonradiative experimental observable. Useful information can be obtained from short-time experiments. The validity conditions, eqs. (64), for coherent excitation of a level structure, which corresponds to the statistical limit, are essentially

$$\gamma_p >> \Gamma_s \quad . \tag{64c}$$

Provided that a wavepacket of all the molecular eigenstates contaminated by the doorway state can be excited by a coherent excitation mode in the short-time domain $\gamma_m t << 1$, the decay mode of the wavepacket assumes the familiar form of a Fourier sum of the absorption strengths. On the time scale shorter than the characteristic recurrence time (61), $t << h\rho_\ell$, the exponential decay law is obtained

$$P_0(t) \propto \exp[-(\Gamma_s+\gamma_s)t/\hbar] \quad , \tag{68}$$

where the nonradiative rate is given by eq. (67). One is lead to the familiar results concerning (nearly) Lorentzian broadening of the doorway state in the absorption spectrum, (nearly) exponential decay mode, shortening of the experimental lifetime relative to the pure radiative lifetime and decrease of the quantum yield below the unity. The width Γ_s, eq. (67), can be considered to constitute a genuine intramolecular decay rate, as the condition (66) for overlapping resonances together with the long recurrence time $h\rho_\ell$, insure irreversible decay in the bound level structure which corresponds to the statistical limit. When the energetic spread Γ_s exceeds the coherent width γ_p of the available light sources, coherent excitation cannot be conducted and the pertinent information on the statistical limit will stem from energy-resolved variables, such as absorption lineshapes.

(II) The Small Molecule Limit. Here the density of states in the $\{|\ell\rangle\}$ manifold is sparse, i.e., $\gamma_\ell\rho_\ell << 1$ and the molecular eigenstates are well separated, their energetic spread being large relative to the coherent energetic width γ_p of any available light sources. In general, coherent excitation is not feasible and only energy-weighted excitation is possible. Each molecular eigenstate acts on its own in absorption and in emission so that only photoselective excitation of individual molecular eigenstates can be accomplished. The resulting decay is expected to be exponential, being determined by the decay rate (8,10,11,85)

$$\gamma_m \simeq \gamma_\ell + \gamma_s/N \quad , \tag{69}$$

where $N = \pi^2|V_{s\ell}|^2\rho_\ell^2$ is the dilution factor. As $N >> 1$ resulting in $\gamma_m << \gamma_s$, a dilution of the radiative and nonradiative decay probabilities of the doorway state will be exhibited. This general pattern of a "diluted" exponential decay can be violated by thermal rotational congestion and vibrational congestion effects, or by more subtle effects arising from accidental near-degeneracy of a small number of levels which can then be coherently excited. In the latter case, oscillatory quantum beats (discussed in section V) are expected to be exhibited in the decay, although such effect has not yet been experimentally documented. The small molecule limit exhibits the consequences of intramolecular scrambling but no relaxation effects are expected to be exhibited in the isolated molecule.

(III) Intermediate Level Structure. This situation, intermediate between the small molecule and the statistical limit, is characterized by a moderately large density of states. However, these states are still well separated relative to their decay widths, i.e., $\gamma_\ell \rho_\ell < 1$. Under these circumstances, the coarse-graining procedures inherent in the description of the statistical limit are inapplicable. When a coherent excitation of the wavepacket is possible, two decay components are expected to be exhibited,

$$P_0(t) \propto \exp[-(\gamma_s + \Gamma_s)t] \; ; \; t < \rho_\ell \, , \quad \gamma_\ell^{-1} \tag{70}$$

$$P_0(t) \propto \sum_m |A_m|^4 \exp(-\gamma_m t) \; ; \; t > \rho_\ell , \quad \gamma_\ell^{-1} \, . \tag{71}$$

The short-time component (eq. (70)) exhibits the dephasing of the initially excited wavepacket and, as in the statistical limit, can be interrogated by photon counting or by probing of coherent optical effects. The long-time decay component (eq. (71)) reveals a sum of exponentials when all phase relationships between the molecular eigenstates were eroded and each molecular eigenstate decays on its own. The long-time decay rates γ_m exhibit the dilution effect relative to the radiative and nonradiative decay widths of the doorway state.

To strike the last chord in our symphony, we have presented in Table 3 a schematic outline of the many possible modes of behavior of electronically excited states of large molecules, which elucidates the many faceted features of the rise and fall of excited states.

X. FUTURE TRENDS

We have been concerned with some aspects of the basic physical and chemical phenomena of the acquisition, storage and disposal of energy in excited electronic states of large, isolated molecules. We have attempted to provide an overview of the problems, concepts, experiments and ideas underlying the diverse and interesting processes of intramolecular interstate coupling and electronic relaxation. The prominence of the microscopic point of view in the description of excited-state dynamics reflects the blending between the development of theoretical concepts and of sophisticated experimental

TABLE 3. FLOW CHART DESCRIBING INTERSTATE COUPLING AND NONREACTIVE ELECTRONIC RELAXATION IN ELECTRONICALLY EXCITED BOUND MOLECULAR STATES

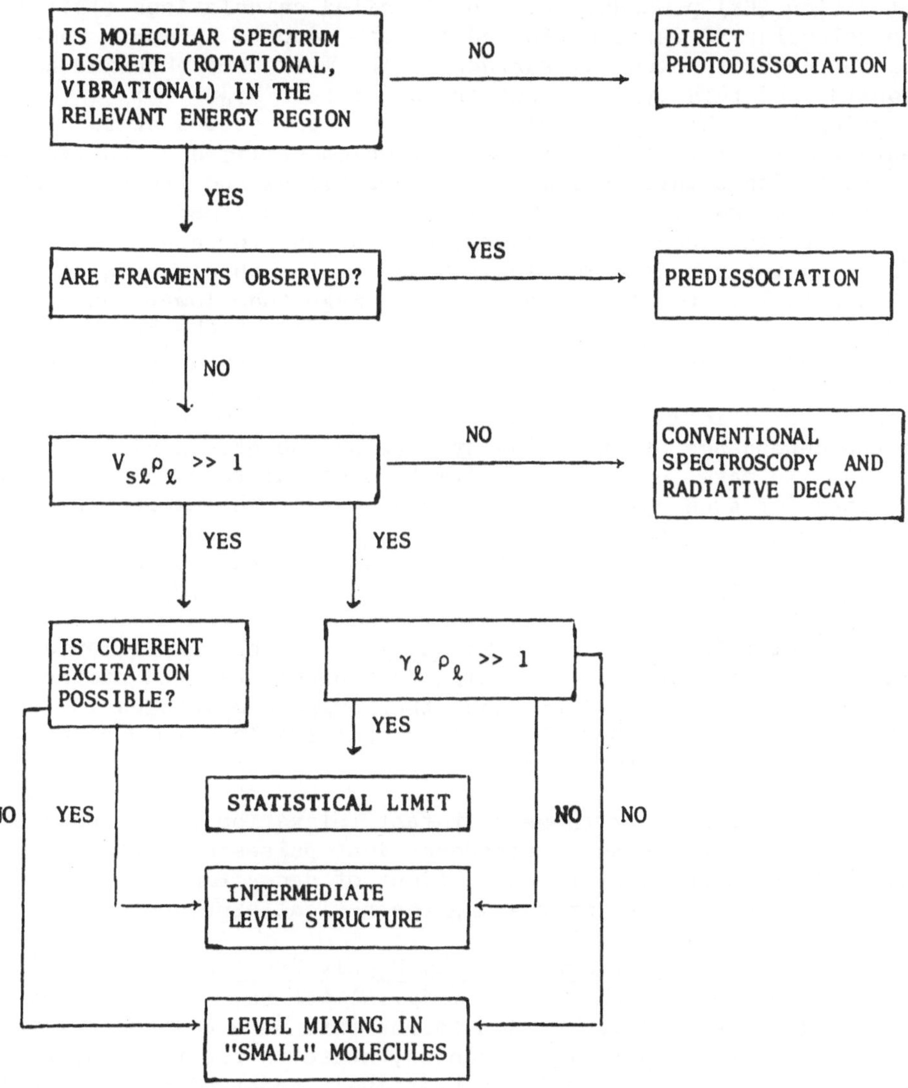

methods. We have reviewed the present state of the art in the field of intramolecular electronic relaxation. We shall now proceed to explore some of the perspectives and future problems of intramolecular dynamics in isolated molecules.

1. Details of Electronic Relaxation Processes. To the best of our knowledge there is not a single calculation available which provides a quantitative result for the rate of a molecular electronic relaxation (ER) process. All the detailed calculations (18,19) of the molecular ER rates performed to date adopt a harmonic model for the nuclear potential surfaces. The most comprehensive calculation (86) of this type was performed for the $^3B_{1u} \rightarrow {}^1A_{1g}$ intersystem crossing from the lowest triplet state to the ground state of the benzene molecule. This calculation, which utilized extensive spectroscopic input information on configurational and frequency changes between the two electronic configurations, confirmed some quantitative features, such as the energy gap law and the deuterium isotope effect on intersystem crossing. However, the absolute value of the ER rate (86) is two to three orders of magnitude lower than the experimental results (2). This failure reflects the breakdown of the harmonic approximation, which is not expected to hold for large ($\sim$3.9 eV) energy gaps. Another central issue pertaining to consequences of optical excitation of isolated large molecules pertains to the dependence of the ER decay rate on the excess vibrational energy. To date, only qualitative theoretical results have been obtained for such optical selection studies. The experimental exploration of excited-state dynamics of large molecules in supersonic beams (29) indicate that the details of the current theories of optical selection studies require a gross modification.

2. Ultrafast Electronic Relaxation. Recent experimental studies of emission quantum yields from highly excited states, e.g., S_2 and S_3 states of aromatic and other organic molecules (87,88), provide strong experimental evidence that the lifetimes of these states are $\sim 10^{-13} - 10^{-14}$ sec.

Direct studies of these ultrafast relaxation phenomena are presently impossible as fantosecond light pulses are not yet available. These experimental observations of decay times in the 10-100 fsec range poses some interesting theoretical problems.

3. Electronic Relaxation from Highly Excited Vibrational States. The theoretical models (Fig. 5) for ER, which we have considered in some detail, pertain to the fate of a molecular level structure containing a single doorway state in the relevent energy region. When we climb up the vibrational ladder of the "optically active" electronic configuration of an isolated molecule, the

density of doorway states increases. In very large molecules at moderately excess vibrational energies a single doorway state cannot be selected. One has then to consider simultaneously the implications of intrastate anharmonic coupling and interstate ER. This problem is of considerable interest regarding general aspects of intramolecular energy flow in large molecules.

4. Intramolecular Vibrational Energy Redistribution. A second class of intramolecular nonreactive relaxation processes, occurring within a bound level structure of large isolated molecules, involve intrastate anharmonic coupling and intramolecular vibrational energy redistribution, which occur within the vibronic manifold of a single electronic configuration (9,13). Experimental information concerning these phenomena stems from several sources.

(a) Absorption spectroscopy of high vibrational overtones (89,90).

(b) Visible emission spectra of mixed crystals (91).

(c) Infrared luminescence in molecular beams (92).

(d) Picosecond optical spectroscopy (93).

(e) Multiphoton optical double resonance (94).

(f) Vibrational predissociation of van der Waals molecules (26,27,95).

(g) Observation of Fermi resonances in electronically excited states of large molecules (29).

(h) High-order multiphoton molecular excitation and isomerization (96).

The outstanding experimental and theoretical questions in this area pertain to the energetics and dynamics of high vibrational excitations and the photophysical and chemical consequences of such excitation. Regarding the intrastate level structure, one can separate three bound energy regions in the order of increasing energy: (i) sparse vibrational level structure; (ii) intermediate level structure, where interstate anharmonic scrambling prevails, but the density of states is low; and (iii) a vibrational quasicontinuum corresponding to a dense, congested, bound level structure. We note in passing that this classification of level structure for intrastate dynamics is analogous to that provided in section IX for interstate coupling and ER. Region (i) is analogous to the small molecule limit, region (ii) corresponds to the intermediate level structure, while region (iii) is isomorphous with the statistical limit. Turning now to comment briefly on intrastate dynamics in a bound level structure (13), we would just like to point out that these processes should and can be adequately described in terms of the time evolution of wavepackets of bound states, in complete analogy with the interstate ER which we have discussed. Thus, both intrastate intramolecular vibrational redistribution and interstate ER can be described within a unified conceptual framework.

5. Complex Chemical Processes. Such processes in isolated large molecules involve a sequence of nonreactive processes, which may be followed by a reactive process. For example, cis-trans isomerization within the isolated stilbene molecule (97) involves just a sequence of nonreactive ER and intramolecular vibrational energy redistribution, while photofragmentation of a large molecule involves nonreactive processes followed by reactive predissociation. The understanding of basic molecular relaxation phenomena will provide the basis for the exploration of photochemistry on the microscopic level.

6. Astrophysical Applications. The nature of intramolecular relaxation processes of isolated molecules in the interstellar space will provide useful information regarding the formation, decomposition and stabilization mechanisms of interstellar molecular species. An example that comes to mind is internal conversion in long polyacetylene chains C_n (n = 5-15), which may be responsible for some, but as yet unidentified, broadened interstellar lines (98). The processes involved incorporate possibly internal (nonreactive) conversion of electronically excited states to a vibrationally hot ground state which is stable in respect to decomposition on the time scale of infrared emission.

7. Multiphoton Processes in Intense Radiation Fields. The nature and characteristics of the photophysical and chemical processes induced by high-order infrared multiphoton excitation (MPE) of collision-free, isolated molecules (96) triggered hectic experimental and theoretical activity and generated a considerable amount of interest and excitement because of several reasons. Firstly, MPE processes provide a novel technique for high-order multiphoton photoselective excitation on the ground state potential surface of a large molecule. Secondly, this excitation process may provide an avenue for specific excitation of some nuclear modes, resulting in unconventional photochemical consequences. Thirdly, diverse and interesting chemical phenomena were induced already by MPE and deserve further study. Fourthly, these MPE processes open up a new research area in photophysics and photochemistry, which is concerned with the interaction of isolated polyatomic molecules with intense radiation fields.

The prominence of the microscopic point of view, underlying the description of these novel and exciting excited-state phenomena, reflects the blending between the development of theoretical concepts and sophisticated theoretical methods in providing a firm conceptual framework for the understanding of intramolecular dynamics.

REFERENCES

1. Molecular Energy Transfer (R.D. Levine and J. Jortner, eds.), John Wiley and Sons, New York (1975).

2. B.R. Henry and M. Kasha, Ann. Rev. Phys. Chem. 19, 161 (1968).

3. J. Jortner, S.A. Rice and R.M. Hochstrasser, Adv. Photochem. 7, 149 (1969).

4. E.W. Schlag, S. Schneider and S.F. Fischer, Ann. Rev. Phys. Chem. 22, 465 (1971).

5. K.F. Freed, Topics Current Chem. 31, 105 (1972).

6. B.R. Henry and W. Siebrand, Org. Mol. Photophys. 1, 153 (1973).

7. G.W. Robinson, in Excited States, Vol. I (E.C. Lim, ed.), Academic Press, New York, 1 (1974).

8. J. Jortner and S. Mukamel, in The World of Quantum Chemistry, Vol. II (R. Daudel and B. Pullman, eds.), Neidel, Boston, 225 (1974).

9. S.A. Rice, in Excited States, Vol. II (E.C. Lim, ed.), Academic Press, New York (1975).

10. J. Jortner and S. Mukamel, in Molecular Energy Transfer (R.D. Levine and J. Jortner eds.), Wiley, New York, 178 (1975).

11. J. Jortner and S. Mukamel, in MTP Series in Science, Vol. 13, Theoretical Chemistry (C.A. Coulson and A.D. Buckingham, eds.), Butterworth, London (1976).

12. K.F. Freed, in Radiationless Transitions in Molecules and Condensed Phases (F.K. Fong, ed.), Springer-Verlag, Berlin, 23 (1976).

13. J. Jortner and R.D. Levine, in Photoselective Chemistry (J. Jortner, R.D. Levine and S.A. Rice eds.), Advances in Chemical Physics, Academic Press, New York (1980).

14. Chemical and Biochemical Applications of Lasers, Vols. 1, 2 and 3 (C.B. Moore, ed.), Academic Press, New York, (1976, 1977, 1978).

15. Picosecond Phenomena (C.V. Chank, E.I. Ippen and S.L. Shapiro, eds.), Springer-Verlag, Berlin (1979).

16. Laser Induced Processes in Molecules (K.L. Kompa and S.D.

Smith, eds.), Springer-Verlag, Berlin (1979).

17. Laser Handbook, Vol. 1, (F.T. Arecchi and E.O. Schultz, eds.), Dubois, North Holland, Amsterdam, (1972).

18. Handbook of Chemical Lasers, (R.W.F. Gross and J.F. Bott, eds.), Wiley, New York, (1976).

19. R.K. Snader, B. Soep and R.N. Zare, J. Chem. Phys. 64, 1242 (1976).

20. R. Naaman, D.M. Luban and R.N. Zare, Chem. Phys. 32, 17 (1978).

21. M.J. Coggiola, P. Schultz, Y.T. Lee and Y. Shen, Phys. Rev. Letters 38, 17 (1977).

22. K.P. Huber and A.E. Douglas, Quarterly Progress Report, National Research Council of Canada, Ottawa Division of Pure Physics, Report 161 (1963), (1964) and (1966).

23. M.P. Sinha, A. Schultz and R.N. Zare, J. Chem. Phys. 58, 549 (1973).

24. R.E. Smalley, B.L. Ramakrishna, D.H. Levy and L. Wharton, J. Chem. Phys. 61, 4363 (1974).

25. R.E. Smalley, L. Wharton and D.H. Levy, J. Chem. Phys. 63, 4977 (1975).

26. D.H. Levy, L. Wharton and R.E. Smalley, in Chemical and Biochemical Applications of Lasers, Academic Press, New York, Vol. 2, 1 (1977).

27. D.H. Levy, R. Wharton and R.E. Smalley, Accounts Chem. Res. 10, 139 (1977).

28. P.S.H. Fitch, L. Wharton and D.H. Levy, J. Chem. Phys. 69, 3424 (1978).

29. A. Amirav, U. Even and J. Jortner, J. Chem. Phys. 71, 2319 (1979).

30. S.M. Beck, M.G. Liverman, D.L. Monts and R.E. Smalley, J. Chem. Phys. 70, 232 (1979).

31. S.M. Beck, D.L. Morris, M.G. Liverman and R.E. Smalley, J. Chem. Phys. 70, 1062 (1979).

32. F.M. Behlen, N. Mikami and S.A. Rice, Chem. Phys. Letters 60, 364 (1979).

33. P.S.H. Fitch, L. Wharton and D.H. Levy, J. Chem. Phys. 70, 2019 (1979).

34. A. Amirav, U. Even and J. Jortner, Chem. Phys. Letters 67, 9 (1979).

35. A. Amirav, U. Even and J. Jortner, Optics Comm. (in press).

36. A. Amirav, U. Even and J. Jortner, Chem. Phys. Letters (in press); J. Chem. Phys. (submitted).

37. J. Tusa, M. Sulkes and S.A. Rice, J. Chem. Phys. 70, 3136 (1979).

38. G.M. McClelland, K.L. Saenger, J.J. Valentini, and D.R. Herschbach, J. Phys. Chem. 83, 947 (1979).

39. E.V. Shposkii, Sov. Phys. Uspekhi, 6, 411 (1963).

40. R.M. Hochstrasser and P. Prasad, in Excited States, Vol. 1, (E.C. Lim, ed.), Academic Press, New York, 79 (1974).

41. G. Gerzberg, Commentarii Pontifica, Academia Scientiarum, Vol. II-49, p. 1 (1972).

42. G. Herzberg, in Spectra of Diatomic Molecules, D. Van Nostrand Company, Toronto (1950).

43. G. Herzberg, in Electronic Spectra of Polyatomic Molecules, D. Van Nostrand Company, Toronto (1966).

44. G. Wenzel, Z. Phys. 43, 524 (1927).

45. E. Fermi, in Nuclear Physics, University of Chicago Press (1949).

46. G. Gamow, Zeit f. Phys. 51, 204 (1928).

47. R.W. Gurney and E.V. Condon, Nature 122, 439 (1928).

48. N. Rosen, J. Chem. Phys. 1, 319 (1933).

49. U. Fano, Novo Cimento 12, 156 (1935).

50. A. Jablonski, Nature 131, 839 (1933).

51. G.N. Lewis and M. Kasha, J. Am. Chem. Soc. 66, 2100 (1944).

52. M. Kasha, Discuss. Farad. Soc. 9, 14 (1950).

53. (a) C.A. Hutchison Jr. and B.W. Mangum, J. Chem. Phys. 32, 1261 (1960).
(b) M.R. Wright, R.P. Frosch and G.W. Robinson, J. Chem. Phys. 33, 934 (1960).

54. G.W. Robinson and R.P. Frosch, J. Chem. Phys. 37, 1962 (1962); 38, 1187 (1963).

55. J.H. Callamon, V.M. Anderson, J.R. Christie and A.R. Lacey, paper 25 at General Discussion Meeting on Radiationless Processes, Schliersee, September (1974).

56. R.J. Watts and S.J. Strickler, J. Chem. Phys. 44, 2423 (1966).

57. R. Williams and G.J. Goldsmith, J. Chem. Phys. 44, 4364 (1966).

58. W.R. Ware and P.T. Cunningham, J. Chem. Phys. 44, 4364 (1966).

59. G.B. Kistiakowsky and C.S. Parmenter, J. Chem. Phys. 42, 2942 (1965).

60. E.M. Anderson and G.B. Kistiakowsky, J. Chem. Phys. 48, 4787 (1968).

61. M. Bixon and J. Jortner, J. Chem. Phys. 48, 715 (1968); 50, 4061 (1969); 50, 3284 (1969).

62. M. Bixon and J. Jortner, Mol. Crystals 213, 237 (1969).

63. E.W. Schlag and H. von Weyssenhoff, J. Chem. Phys. 53, 3108 (1970).

64. C.S. Parmenter and M.W. Schuyler, Chem. Phys. Letters 6, 339 (1970).

65. K.G. Spears and S.A. Rice, J. Chem. Phys. 55, 5561 (1971).

66. P.M. Rentzepis, Science 169, 239 (1970).

67. D. Huppert, J. Jortner and P.M. Rentzepis, Israel J. Chem. 16, 277 (1978).

68. B. Di Bartolo (This Volume).

69. R. Kubo, J. Phys. Soc. Japan 12, 570 (1957).

70. P.A. Franken, Phys. Rev. 121, 508 (1961).

71. M. Bixon, J. Jortner and Y. Dothan, Mol. Phys. 17, 109 (1969).

72. J. Chaiken and J.D. McDonald, Chem. Phys. Letters 61, 195 (1979).

73. L. Allen and J.H. Eberly, in Optical Resonance and Two-Level Atoms, Wiley, New York (1975).

74. R.G. Brewer, in Frontiers in Laser Spectroscopy (R. Balian, S. Haroche and S. Liberman, eds.), North Holland, 341 (1977).

75. R. Lefebvre and J. Savolamen, Chem. Phys. Letters 3, 449 (1969).

76. J. Jortner and J. Kommandeur, Chem. Phys. 28, 273 (1978).

77. R.G. Brewer and E.L. Hahn, Phys. Rev. A11, 1641 (1975).

78. A. Amirav, U. Even, J. Jortner and L. Kleinman, J. Chem. Phys. (in press).

79. C.D. Cantrell, W.H. Louisell and J.F. Lam, in Laser Induced Processes in Molecules (K.L. Kompa and S.D. Smith, eds.), Springer-Verlag, Berlin, 138 (1979).

80. K.F. Bonhoeffer and L. Farkas, Z. Phys. Chem. A134, 337 (1927).

81. D. Chock, J. Jortner and S.A. Rice, J. Chem. Phys. 49, 610 (1968).

82. M.L. Goldberger and K.M. Watson, in Collision Theory, John Wiley, New York (1964).

83. J.O. Berg, Chem. Phys. Letters 41, 547 (1976).

84. J. Jortner, in J. Chim. Phys. Special Issue "Transitions Non-Radiative dans le Molecules", 1 (1970).

85. A. Nitzan, J. Jortner and P.M. Rentzepis, Proc. Roy. Soc. A327, 367 (1972).

86. A. Nitzan and J. Jortner, Theoret Chimica Acta. 30, 217 (1973).

87. I. Kaplan and J. Jortner, Chem. Phys. 32, 381 (1978).

88. H.B. Lin and M.R. Topp, Chem. Phys. Letters 47, 442 (1977); 48, 251 (1977).

89. R.L. Swofford, M.S. Burberry, J.A. Morrell and A.C. Albrecht, J. Chem. Phys. 66, 5245 (1977).

90. K.V. Reddy, R.G. Bray and M.J. Berry in Advances in Laser Chemistry, (A.H. Zewail, ed.), Springer-Verlag, Berlin, 48 (1978).

91. D.D. Smith and A.H. Zewail, J. Chem. Phys. 71, 540 (1979).

92. J.G. Moehlam, J.T. Gleaves, J.W. Hudgens and J.D. McDonald, J. Chem. Phys., 60, 4790 (1974).

93. J.P. Maier, S. Selmeier, A. Laubereau and W. Kaiser, Chem. Phys. Letters 46, 527 (1977).

94. R.V. Ambartzumian, Yu A. Gorokhov, V.S. Letokhov, G.N. Makarov, Zh ETF Pis. Red. (Sov.), 22, 96 (1975).

95. J.A. Beswick and J. Jortner, J. Chem. Phys. 68, 2277 (1968).

96. See for example (a) R.V. Ambartzumian and V.S. Letokhov, Accts. Chem. Res. 10, 61 (1977); (b) C.D. Cantrell, S.M. Freund and J.L. Lyman, Laser Handbook, Vol. III, North Holland, Amsterdam (1978); (c) N. Blumbergen and E. Yablonovitch, Phys. Today, 31 (5) 23, 1978.

97. B.I. Greene, R.M. Hoshstrasser and R.B. Weisman, J. Chem. Phys. 71, 544 (1979).

98. A.E. Douglas, Nature, 269, 130 (1977).

INTERSYSTEM CROSSING IN COMPLEX MOLECULES

R. G. Pappalardo

Member of Technical Staff
GTE Laboratories Incorporated
Waltham, Massachusetts 02154

ABSTRACT

The general question of singlet-triplet intersystem crossing is addressed in the context of large organic molecules, i.e., "complex" molecules capable of self-relaxation in the absence of collisions. Examples of spectral properties of such molecules in the vapor phase are discussed, relying on extensive Russian literature in this area. Formal expressions for the relaxation rate in electronic excited states are derived on the basis of the formalism of collision theory, and are applied to the specific case of intersystem crossing. The derivation of the "energy-gap" law for triplet-singlet conversion in aromatic hydrocarbons is briefly outlined. The steep rise of internal conversion rates as a function of excess excitation energy, and its competition with the intersystem crossing process, are reviewed for the case of naphthalene vapor.

A general expression for the spin-orbit interaction Hamiltonian in molecular systems is outlined. Experimental observations on singlet-triplet conversion rates and the factors that can drastically affect such rates are discussed, with emphasis on the "internal" and "external" heavy-atom effects. Basic relations of ESR spectroscopy and magnetophotoselection are reviewed.

Technological implications of the singlet-triplet crossing in complex molecules are discussed in the context of chelate lasers, dye lasers and luminescent displays.

Effects related to singlet-triplet crossing, and generally to excited-state energy-transfer in biological systems, are exemplified by the role of aromatic amino-acids in the phosphorescence of proteins, by some recent studies of energy-transfer in models of biomembranes, and by the clustering of triplet-energy donor-acceptor pairs in micelles.

I. INTRODUCTION

Electronic excitation of molecular systems generally involves a transition from ground singlet state(s) to excited singlet states via absorption of photons, or inelastic collisions with energetic electrons. The resulting nonequilibrium situation is followed by relaxation to equilibrium conditions by radiative and/or nonradiative decay. The latter proceeds either by internal conversion to the ground singlet state, or by conversion to a state of different spin-multiplicity (generally a triplet state); in other words, "intersystem crossing" takes place. At a later stage the triplet state in turn will relax by a combination of radiative and nonradiative decay.

Intersystem crossing is therefore a very general phenomenon in the relaxation of excited molecular states. In order to restrict the range of processes to be considered, we shall limit our treatment to "complex" molecules, namely to molecular entities wherein the excitation energy is redistributed even in the absence of collisions with surrounding molecules in a gas or in a liquid, or in the absence of coupling with the vibrational modes of a solid. According to Robinson and Frosch (1) these molecules can act as their own heat bath even in a rarefied gas.

In Section II the spectral behavior of aromatic molecules in the excited state is considered in solutions and in the vapor phase. Molecules are classified according to the resulting spectral pattern for emission and absorption in the gaseous phase, and the effect on these spectra of temperature, pressure and excitation wavelength are noted.

Formal expressions for the decay rates from excited electronic states are derived in Section III on the basis of collision theory. In the process extensive use is made of the Green function operator for the system Hamiltonian. The final step in the derivation relies on Kubo and Toyozawa's formalism of generating functions. The specific form of the rate for intersystem crossing is finally obtained. In Section IV the phonon-dependent part of the rate is calculated for the low-temperature case, following Englman and Jortner's approach. The result is used to rationalize the observed triplet-energy dependence of the $T_1 \rightarrow S_0$ conversion in aromatic hydrocarbons. The influence of excess excitation energy in the excited state, already discussed in general terms in Section II, is reexamined in detail in Section V in the specific case of naphthalene vapor. Excess excitation energy leads to a steep, exponential increase in the internal conversion rate $S_1 \rightarrow S_0$, and this decay channel dominates the radiative decay from S_1 and intersystem crossing. In Section VI the mechanism responsible for the electronic coupling of singlet and triplet levels is considered in the framework of molecular systems. The main components of the spin-orbit Hamiltonian are listed in terms of one-electron and two-electron operators. The question of suitable molecular wavefunctions to be acted upon by the spin-orbit Hamiltonian operator is also outlined. Section VII gives background notes on the identification of triplet states as responsible for the phosphorescence emission, and on techniques to detect and enhance intersystem crossing, namely ESR spectroscopy, magnetophotoselection, optically-detected magnetic resonance (ODMR) on the one hand, and the internal and external heavy-atom effects as conducive to increased intersystem rates. In Section VIII we look at a specific aspect of a very general effect, the role played by the long-lived triplet state in transferring excitation energy to other molecules. In this particular instance we consider the sensitization of emission from lanthanide ions, and some of its technological spin-offs, such as chelate lasers and luminescent displays. The generally harmful effect associated with triplet-state population is analyzed in Section IX in a computer simulation of dye-laser operation. Structural guidelines for reducing singlet-triplet rates in laser dyes are also discussed. In Section X the important function of triplet states as energy-storage centers for biological processes is examined in the case of proteins. In this wide class of substances only two aromatic amino-acids, tyrosine and tryptophan, are mainly responsible for the absorption and transfer of photon energy. An interesting example of how biological energy-transfer processes can be aided by segregation and selective proximity of reactants is offered by recent studies of sensitized rare-earth luminescence, whereby the excited-triplet species is shielded from quenching impurities by its encapsulation in micelles.

II. RELAXATION OF COMPLEX MOLECULES IN THE GASEOUS PHASE

In the context of the processes of electronic relaxation in the excited state, molecules are grouped into "simple," "intermediate," and "complex" types. Since intersystem crossing, together with internal conversion, constitutes the dominant nonradiative path of relaxation of excited singlets in molecules, its detailed course will be dictated by the nature of the molecule under consideration (simple, intermediate, or complex type).

Electronic relaxation, between "pure" electronic levels of the type occurring in isolated atomic systems, requires the levels to overlap in energy. In view of the wide energy separation between singlet and triplet states in molecules (electronic correlation energy), electronic relaxation between "pure" electronic levels will not take place. The bridging of the energy gap between the electronic levels involved in the relaxation process will be provided by the molecular roto-vibronic levels. Energy overlap of the pertinent levels will be further aided by any level-broadening mechanisms of existing roto-vibronic levels, such as coupling to lattice phonons in the condensed phase, and molecular collisions in the gaseous and liquid phase.

In order to study the molecular nature of the electronic relaxation, the influence of the surroundings on the molecular system under consideration is best eliminated or minimized by considering the behavior of the molecules, whenever feasible, as noninteracting species in a rarefied gas, ideally at pressures as low as $\sim 5 \times 10^{-3}$ torr. Under these conditions the electronic relaxation will be mainly intramolecular, except for occasional collisions with the container walls.

In the gaseous phase "simple" molecules exhibit line spectra or finely-banded spectra, in both absorption and emission. Occasionally, as for SO_2 and NO_2, the emission lifetimes (2) are longer than calculated from the absorption strengths. The sparseness of the existing levels is a serious obstacle to efficient relaxation, and these simple molecules should therefore exhibit both resonant emission and near-unity quantum efficiency in the low-pressure limit.

Benzene has been investigated in great detail since the classical paper by Kistiakowsky and Parmenter (3). It is considered typical of "intermediate" systems: the emission spectrum, the quantum yield for fluorescence, and the triplet yield from the intersystem crossing in the excited state are all sensitive and complicated functions of gas pressure and excitation wavelengths.* In the case of benzene, vapor collisions are apparently critical in changing the relaxation process from a "no-go" to a "go" situation (5).

The transition from "simple" to "complex" molecules can be characterized as the gradual setting in of vibrational relaxation, concomitant to electronic relaxation, or as a progressive spreading of the energy-level densities with the increased size of the molecular system. Early Russian literature (6) contains extensive reports on the properties of complex molecules in the gaseous phase. In this area, Russian authors tend to follow a nomenclature introduced by Neporent (7), based on the shape of emission and absorption spectra of the corresponding molecules. Broad, structured bands are considered diagnostic of semicomplex molecules, while complex molecules will be characterized by broad, structureless bands. The latter spectra are further subdivided into "modulation spectra" if Levshin's mirror-symmetry rule (8) is obeyed, and "damping" spectra if differences in the shape of emission and excitation bands are indicative

*Similar results were reported much earlier by Vartanian (4) for the case of aniline.

of different strengths of electron-vibration coupling in the ground and excited states. In the latter case the bandwidths are found to be proportional to the square of the average transition frequency.

We shall now proceed with some typical examples of spectra of complex and semicomplex molecules. Most of the molecular systems to be discussed in what follows pertain to conjugated molecules with extensive delocalization of the valence electrons into π-electron orbitals. Transitions between ground and excited states will involve the π-electrons in $\pi \rightarrow \pi^*$ or $n \rightarrow \pi^*$ processes, the latter transitions taking place in conjugated systems with heteroatoms (N, O or S) in the conjugated system, or with substituents containing the N, O or S atoms. Fig. 1 illustrates the differences and similarities in the emission and absorption spectra of low-pressure vapors and of solutions, for the case of phthalimide and some of its derivatives. From the gaseous to the liquid phase no substantial changes are noticeable in the <u>shape</u> of the bands, only a long-wavelength shift of the bands in going to the solutions. The spectra clearly demonstrate that in the gaseous phase the relaxation of complex molecules is an intramolecular phenomenon and that the molecule can act as its own heat reservoir.

In Fig. 1 and some of the following figures the absorption and emission spectra are expressed via the normalized absorption coefficient χ_υ and the normalized luminescence power W_υ, mutually related by Stepanov "universal relation" (9):

$$\frac{W_\upsilon}{\chi_\upsilon} = D(T)\ \upsilon^3\ e^{-h\upsilon/kT} \tag{1}$$

with D(T) a function of the absolute temperature. Gas-phase spectra should specify the vapor pressure or the temperature at which measurements were carried out, and also the excitation wavelength used to obtain the emission spectra.

In the next two figures we present examples of <u>semicomplex</u> spectra. In Fig. 2 and Table 2 anthraquinone and some of its derivatives are considered. The electronic processes involved in absorption and emission are of low probability and are interpreted as arising from $n \rightarrow \pi^*$ transitions in the carbonyl group. Anthraquinone shows a structured emission spectrum, with peaks spaced by ~1600 cm^{-1}, corresponding to stretching vibrations of the carbonyl group in the ground state. Addition of substituents produces a smearing out of the spectra toward a situation typical of complex spectra. The observed change is considered to be diagnostic of vibronic coupling of the substituent(s) with the carbonyl group.

Typical semicomplex spectra (Fig. 3 and Table 3) are exhibited by compounds containing fused benzene rings. With the exception of

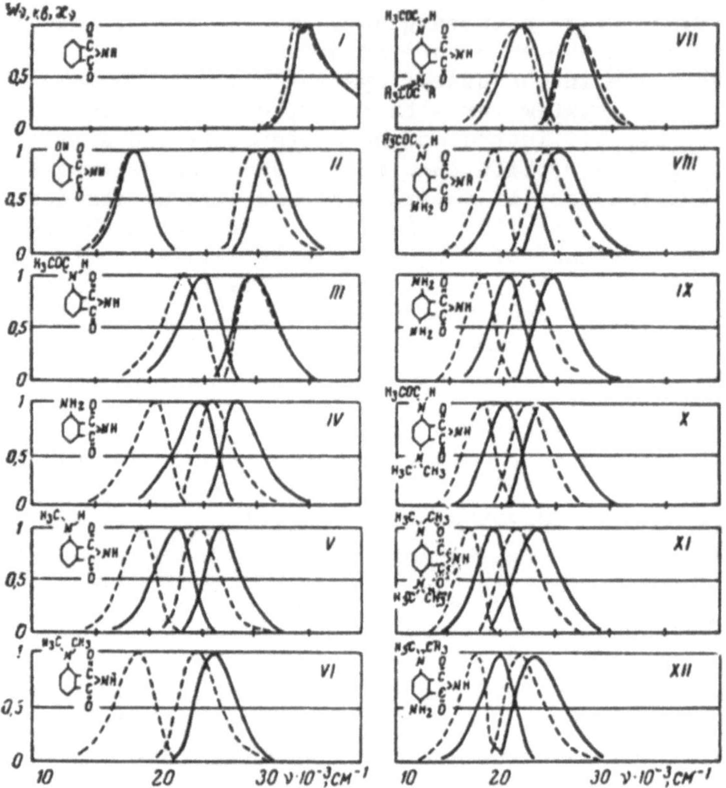

Fig. 1. Comparison of solution spectra and vapor-phase spectra in phthalimide and its derivatives. Solid lines: vapor spectra; dashed lines: solution spectra. Normalized emission spectra (in units of luminescence power W_υ) and absorption spectra (in units of the absorption coefficient χ_υ). The substances are identified in Table 1. Data from Ref. (6), p. 80.

TABLE 1. SUBSTANCES WHOSE ABSORPTION AND EMISSION SPECTRA IN SOLUTIONS AND IN THE GASEOUS PHASE ARE REPORTED IN FIG. 1

I	phthalimide
II	hydroxy-phthalimide
III	acetamido-phthalimide
IV	3-aminophthalimide
V	3-methylamino-phthalimide
VI	3-dimethylamino-phthalimide
VII	3,6-diacetamido-phthalimide
VIII	3-acetamido-6-amino-phthalimide
IX	3,6-diamino-phthalimide
X	3-acetamido-6-dimethylamino-phthalimide
XI	3,6-tetramethyldiamino-phthalimide
XII	3-dimethylamino-6-amino-phthalimide

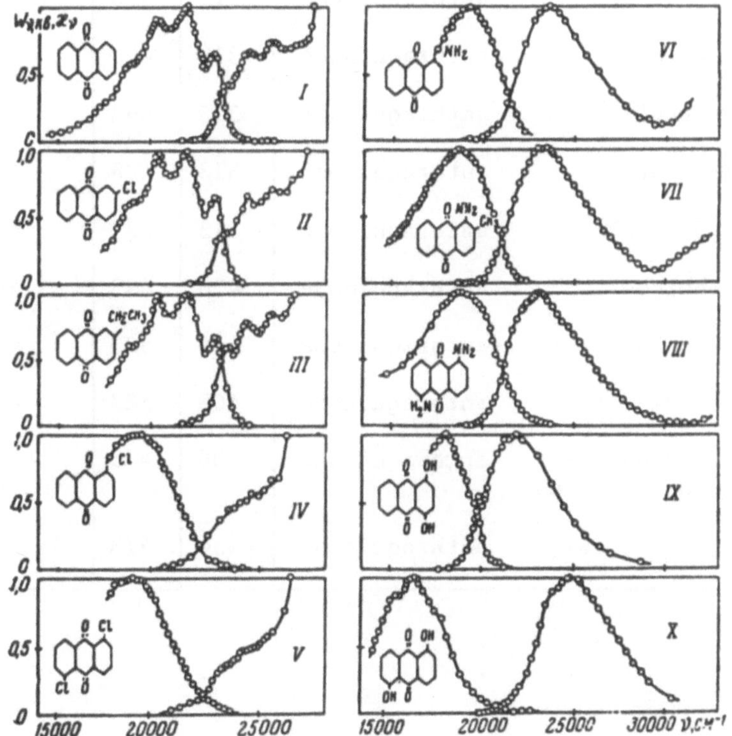

Fig. 2. Vapor-phase emission and absorption spectra of anthraquinone and its derivatives. Listings of the compounds and of measurement parameters are given in Table 2 [Ref. (6), p. 81].

TABLE 2. EXPERIMENTAL CONDITIONS FOR THE LUMINESCENCE MEASUREMENTS ON VAPOR-PHASE ANTHRAQUINONE AND ITS DERIVATIVES [FROM REF. (6), P. 81]

	Compound	λ_{exc} (nm)	T (°K)	T_v or Press. (torr)
I	anthraquinone	313	523	6.5×10^{-2}
II	2-chloro- anthraquinone	313	493	428K
III	2-ethyl- anthraquinone	313	493	428K
IV	1-chloro- anthraquinone	313	488	470K
V	1,5-dichloro- anthraquinone	313	529	513K
VI	1-amino- anthraquinone	436	493	453K
VII	1-amino,2-methyl-anthraquinone	436	523	493K
VIII	1,5-diamino- anthraquinone	405	523	493K
IX	1,4-dihydroxy- anthraquinone (quinizarin)	436	473	1.4×10^{-1}
X	1,5-dihydroxy- anthraquinone	436	473	3.3×10^{-2}

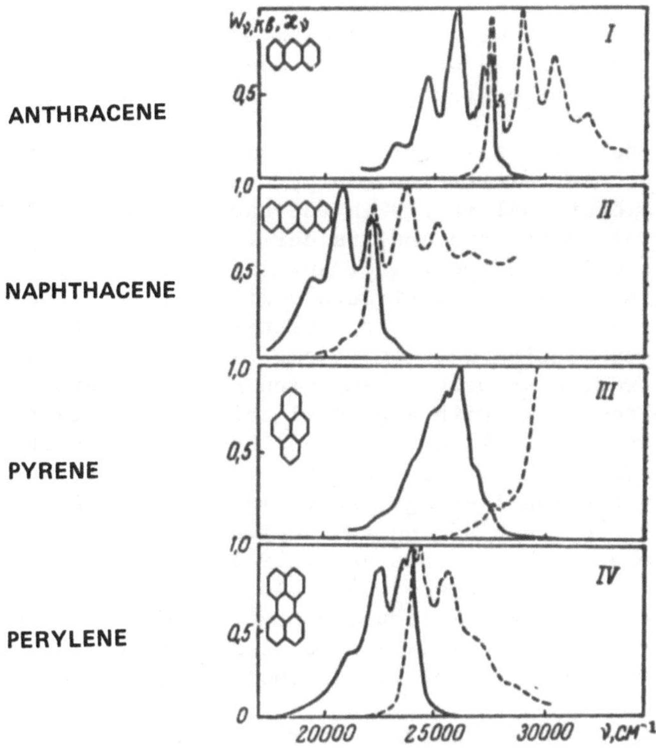

Figure 3. Vapor-phase absorption and emission spectra of cata-condensed aromatic hydrocarbons. Experimental details in Table 3.

TABLE 3. EXPERIMENTAL CONDITIONS FOR THE MEASUREMENTS OF THE VAPOR-PHASE SPECTRA OF FIG. 3 [FROM REF. (6), P. 85]

Compound	λ_{exc} (nm)	T (°K)	Vapor Pressure (torr)
Anthracene	365	413	5×10^{-2}
Naphthacene	365	emis. 493 abs. 593	em. 3.3×10^{-2}
Pyrene	365	473	6.5×10^{-2}
Perylene	405	513	6.3×10^{-3}

pyrene, the spectra have a well-defined structure, common to absorption and emission, and with peak separation of ~1400 cm^{-1}. This is indicative of potential hypersurfaces undisplaced and unchanged in shape in going from the ground to the first excited singlet.

If the model of a self-relaxing isolated molecule is correct, both temperature and excitation energy could be used equally to increase the vibrational energy in the excited state. A case in point is that of anthracene and its derivatives. At a pressure of 0.05 torr, the emission spectra of anthracene become progressively broader and finally lose any structure with the decrease in the excitation wavelength (Fig. 4). As a matter of fact, by a suitable choice of the excitation wavelength one can obtain a quasi-linear emission spectrum, a structured emission spectrum, and a structureless emission spectrum, with a gradual shift of the center of gravity of the emission band to longer wavelengths, as the excitation wavelength decreases. As for a perfect correspondence between the distribution in vibrational energy, resulting from an increase in temperature on the one hand and the decrease in excitation wavelengths on the other hand, this equivalence is expected to hold more strictly in the case of complex molecules where all of the vibrational modes seem to be well-coupled together, than in semicomplex molecules where the optical excitation tends to be distributed preferentially among a set of vibrational levels.

The previous considerations apply to fast-decaying emitting states. The emission of benzophenone and anthraquinone, for instance, does not change in shape on decreasing the excitation wavelength. The reason is that during the long lifetime of the emitting state, enough collisions with like molecules and the container walls effectively bring the excited molecules down to system temperature. Conversely, in the case of fast emitters, a wavelength-independent emission could be obtained by the addition of an inert gas capable of providing enough thermalizing collisions during the emitting-state lifetime. This effect due to the addition of inert gases (Fig. 5) has been experimentally observed (10).

Probe times in the picosecond range have now been made available by laser technology for the study of molecules in the gaseous phase. Processes occurring in the 10^{-11}s time scale are effectively collision-free for pressures of less than a torr. Greene and coworkers (11) have measured transient excited-state absorption of trans-stilbene vapor and found that the transient absorption has decay times of 17 ps for excitation at 267 nm, namely when the excited state has excess vibrational energy, to be contrasted with the 55 ps decay for excitation at longer wavelengths, at 287 nm.

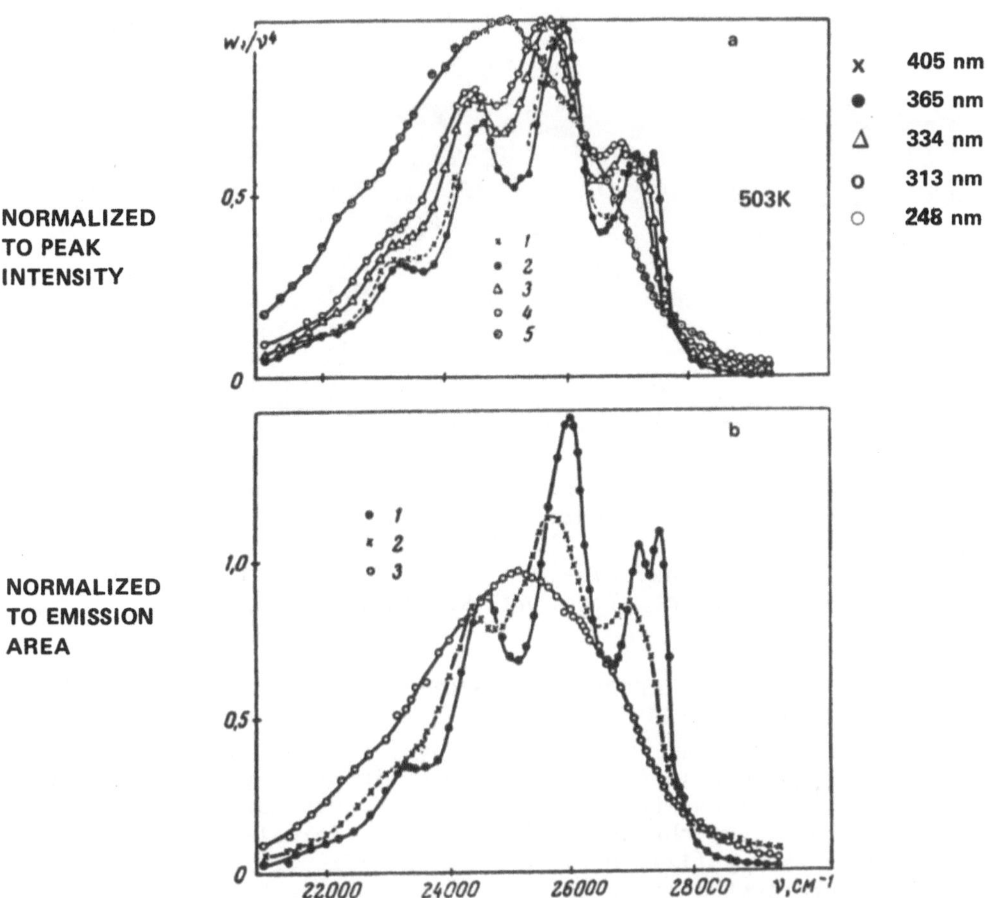

Fig. 4. (a) Spectral changes in the emission spectra of anthracene vapor on decreasing the excitation wavelength. Temp. = 503°K; 0.05 torr. Emission spectra normalized to peak intensity. Excitation wavelength: 1 - 405 nm; 2 - 365 nm; 3 - 334 nm; 4 - 313 nm; 5 - 248 nm. (b) Emission intensity normalized to emission area. Excitation wavelength: 1 - 365 nm; 2 - 313 nm; 3 - 248 nm. The quantum yield for fluorescence in anthracene vapor does not depend on the excitation wavelengths used. [Ref. (6), pp. 88 and 89]

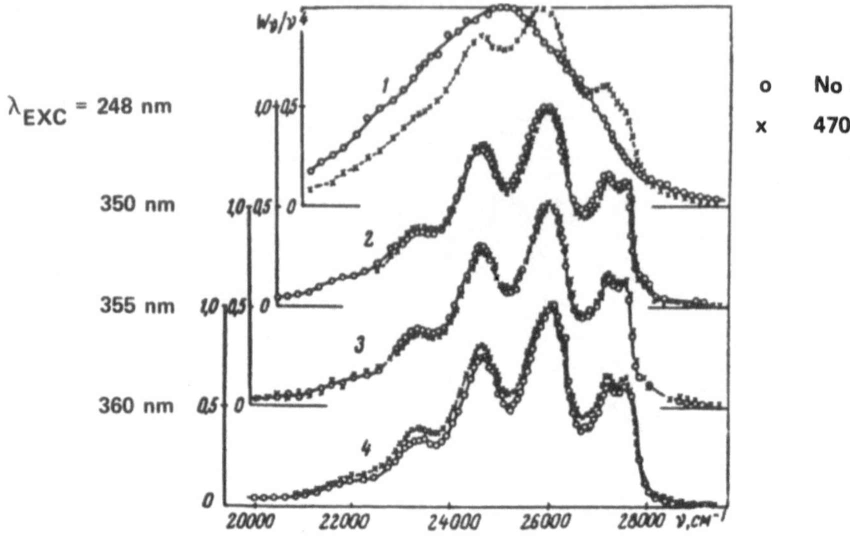

Fig. 5. Effect of added gases on the fluorescence spectra of vapor-phase anthracene. Circles (o): no added gases; Crosses (×): 470 torr pentane. Excitation wavelengths: 1 - 248 nm; 2 - 350 nm; 3 - 355 nm; 4 - 360 nm. Temp. 473°K. [Ref. (6), p. 235]

III. DERIVATION OF FORMAL EXPRESSIONS FOR EXCITED-STATE RELAXATION RATES

A formalism will now be introduced in order to describe the kinetics of excited-state relaxation into both the radiative and nonradiative decay channels. This general formalism, borrowed from scattering theory (12), will then be specifically applied to the case of intersystem crossing. The derivation makes extensive use of the properties of the Green function (and corresponding operator) for the energy of the relaxing system. This Green function can be conveniently introduced in stationary-state perturbation theory. The starting point is, as usual, a total Hamiltonian

$$H = K + V \ , \tag{2}$$

comprised of an unperturbed component K and a perturbation V. The wavefunction corresponding to H is:

$$|\Psi\rangle = |\Psi_o\rangle + |\Phi\rangle \ , \tag{3}$$

where it is assumed that:

$$\langle\Psi_o|\Phi\rangle = 0 \ . \tag{4}$$

Next the projection operator M and its complementary operator P are introduced:

$$P = 1 - M = 1 - |\Psi_o\rangle\langle\Psi_o| \ . \tag{5}$$

The perturbed eigenvalues and eigenfunctions are expressed as:

$$|\Psi\rangle = |\Psi_o\rangle + (E - K)^{-1} PV|\Psi\rangle \ , \tag{6}$$

and

$$E = E_o + \langle\Psi_o|V|\Psi\rangle \ . \tag{7}$$

The recursive nature of eq. (6), whereby $|\Psi\rangle$ appears on both sides of the relation, leads to the so-called Brillouin-Wigner series (Appendix 1). The operator in eq. (6)

$$G_o(E) = \frac{1}{E - K} \ , \tag{8}$$

is the Green function for the Hamiltonian of the unperturbed system, while

$$G(E) = \frac{1}{E - H} \ , \tag{9}$$

is the corresponding Green function for H. The eigenvalues of H, say E_λ, are obtained by finding the poles of G(E) in a given matrix representation of H:

$$\langle a'|G(E)|a\rangle = \sum_\lambda \frac{\langle a'|\lambda\rangle\langle\lambda|a\rangle}{E - E_\lambda} \ . \tag{10}$$

Of special interest in the process of finding the eigenvalues of the perturbed system is the diagonal operator

$$G_a(E) = \langle a|G(E)|a\rangle \ . \tag{11}$$

In deriving $G_a(E)$ it is convenient to introduce the operator product Fg whereby F is an operator with unit diagonal matrix elements, and g is the diagonal matrix

$$\langle a'|g|a\rangle = \delta_{a'a} \langle a'|G(E)|a\rangle = \delta_{a'a}\, G_a(E) \ . \tag{12}$$

Furthermore, the product of F and the perturbation operator V is labelled as R, and in the case of discrete eigenvalues for H, it plays the role of "level-shift" operator by shifting the energy (E_a^o) of the unperturbed system. Finally, $G_a(E)$ is expressed as:

$$G_a = \frac{1}{E - E_a^o - R_a(E)} \quad . \tag{13}$$

The eigenvalues of the perturbed system are therefore the solutions of the algebraic equation:

$$E_\lambda = E_a^o + R_a(E_\lambda) \quad . \tag{14}$$

Assuming now that we have prepared the molecule in an excited state, we want to follow its evolution in time, under the effect of the perturbation V, which includes both the coupling to the radiation field and the coupling to another manifold of levels.

The time evolution of the system can be followed by means of either the Goldstone development operator

$$U(t) = \exp\,(-iHt/\hbar) \quad , \tag{15}$$

or by means of the Hugenholtz resolvent (13)

$$Q = (H - z)^{-1} \quad . \tag{16}$$

The two are connected as follows (assuming $\hbar = 1$):

$$Q(z) = -i \int_{-\infty}^{0} U(t)\exp(izt)dt \qquad \text{for Im } z < 0 \tag{17a}$$

$$Q(z) = i \int_{0}^{\infty} U(t)\exp(izt)dt \qquad \text{for Im } z > 0 \; . \tag{17b}$$

The resolvent is seen to be related to the Green function for the Hamiltonian, and Q is simply the (imaginary) Laplace transform of the development (or time displacement) operator. Conversely, we can start from the relation:

$$\Psi(t) = e^{-iHt/\hbar}\, X_a \quad , \tag{18}$$

where X_a is the initial state of the system. After performing two successive Laplace transformations, we can express the time development of the system via the Green function for the Hamiltonian. We have, in general,

$$\Psi(t) = \frac{1}{2\pi i} \int_{C_2} dE\, e^{-iEt/\hbar}\, G(E)X_a \quad . \tag{19}$$

The contour C_2 in the E plane is from $\infty + ic$ to $-\infty + ic$ and does not include singularities of G(E). Now if the Hamiltonian has discrete, real eigenvalues, the value of c can be made infinitesimally small and so we can write:

$$\Psi(t) = \frac{1}{2\pi i} \int_{-\infty}^{\infty} dE\ e^{-iEt/\hbar}\ G^{+}(E)X_a \tag{20}$$

The probability of finding the system at time t in a state X_b is found by evaluating first the probability amplitude

$$\Upsilon_b = \langle X_b|\Psi(t)\rangle = \frac{1}{2\pi i} \int_C dE\ e^{-iEt/\hbar}\langle X_b,\ G^{+}(E)X_a\rangle\ , \tag{21}$$

and is proportional [see eq. (10)] to:

$$\sum_\lambda e^{-iE_\lambda t/\hbar} \langle X_b|\lambda\rangle\langle\lambda|X_a\rangle\quad .$$

As for the probability of decay out of the state a, it can be inferred from the corresponding expression:

$$\Upsilon_a = \sum_\lambda e^{-iE_\lambda t/\hbar} |\langle X_a|\lambda\rangle|^2\quad , \tag{22}$$

and it has an oscillatory behavior in time, but does not tend to zero for large values of t. If there are many terms in the sum, though, the resulting effect is to approximate a transition out of state a.

The situation is different when the Hamiltonian has a continuous range of eigenvalues. Both $G_a(E)$ and $R_a(E)$ are analytic throughout the complex plane E, except for a cut on the real axis corresponding to the range of continuous eigenvalues. When $R_a(E)$ approaches the cut from either side of the real axis, its imaginary component does not vanish; instead, we have:

$$R_a^{\pm}(E) = D(E) \mp iI(E)\quad , \tag{23}$$

with

$$I(E) = \pi\langle a|R^{+\dagger}(E)\delta(E - K)PR^{+}(E)|a\rangle\quad , \tag{24}$$

P being again the complementary projection operator.

The physical consequence is that if the system was originally prepared in a state X_a, the probability amplitude that the system will be in the same state at time t is given by:

$$\Upsilon_a = \frac{1}{2\pi i} \int_{+\infty}^{-\infty} \frac{dE\ e^{-iEt/\hbar}}{E - E_a - D(E) + iI(E)} \quad . \tag{25}$$

After performing the integration

$$|\Upsilon_a| \propto e^{-I(E_a)t/\hbar} \quad , \tag{26}$$

and the state will decay irreversibly in time.

This can be viewed as the mathematical expression for the model of a molecule relaxing into a continuum of excited-state levels. If the density of states is sufficiently high, the initially-prepared state will decay, even in the absence of coupling with the radiation field, into various relaxation channels. The excitation energy will be redistributed, if the perturbing potential V is capable of coupling the initial state with a continuum of other states of the same or different multiplicity.

We are now interested in deriving expressions for the decay rates from the excited state. We start from the relation:

$$\Upsilon_b = \frac{1}{2\pi i} \int_{C_2} dE\ e^{-iEt/\hbar} \langle X_b, G(E)X_a \rangle \quad , \tag{27}$$

and the probability:

$$P_{ba}(t) \propto |\Upsilon_b|^2 \quad , \tag{28}$$

for discrete levels, or $dP_{b(a)}$ for a continuous distribution of levels b. Since

$$G_{ba}(E) = \frac{1}{E - E_b} R_{ba}(E) \frac{1}{E - E_a - R_a(E)} \quad , \tag{29}$$

we find on integration that we have to consider contributions from both poles, namely:

$$\Upsilon_b \sim e^{-iE_b t/\hbar} \frac{R^+_{ba}(E_b)}{E_b - E_a - R^+_a(E_b)} - \frac{e^{-i[E_a + D_a(E_a)]t/\hbar}\ e^{-\frac{\Gamma}{2}t}\ R^+_{ba}(E_a)}{E_b - E_a - R^+_a(E_a)} \quad , \tag{30}$$

after setting $I = \hbar\Gamma/2$. In the case of a distribution of b levels we introduce:

$$d^2P_b \rightarrow \lim_{t \rightarrow \infty} dP_b(t) \quad ,$$

and obtain for the yield:

$$d^2P_b = \frac{d\rho'(E_b)|R^+_{ba}(E_b)|^2 dE_b}{[E_b-E_a-D(E_a)]^2 + \frac{\hbar^2\Gamma^2}{4}} \quad .$$

Finally, the initial decay rate into b is given by:

$$d\dot{P}_b = \frac{2\pi}{\hbar} d\rho(E_a)|R^+_{ba}(E_a)|^2 \quad . \tag{31}$$

We now specify these relations for the case of radiative and non-radiative decay (14).

$$\Delta_s(E_s) = \frac{2\pi}{\hbar}\int \frac{dE_\ell \rho_\ell(E)|v_{\ell s}(E_\ell)|^2 \Gamma_\ell(E_s)\hbar/2}{(E_s-E_\ell)^2 + [\hbar\Gamma_\ell(E_s)/2]^2} \quad , \tag{32}$$

where s labels an originally prepared excited state, while ℓ is the label for the electronic state connected to s by the coupling operator $V_{s\ell}$; Γ_ℓ is the radiative width of the ℓ levels, and derives from the weighted integral for the coupling with the radiation continuum of the states ℓ, which would be expressed as:

$$\int d\rho(E_b)|R^+_{ba}(E_b)|^2 = \frac{1}{2\pi}\Gamma\hbar \ ,$$

in the previous formula. For the limit $\Gamma_\ell \to 0^+$, the familiar relation:

$$\Delta_s(E_s) \sim (2\pi/\hbar) \sum_\ell |v_{s\ell}|^2 \delta(E_\ell - E_s) \ , \tag{33}$$

is obtained, where for the sake of dimensionality $\delta(E_\ell - E_s)$ is to be viewed as a singular case of the density of states ρ.

Eq. (33) can be adapted to the "statistical" case of a complex molecule, wherein the vibrational relaxation proceeds very rapidly after excitation, and leads to a Boltzmann distribution over the i vibrational levels of s. Indicating by j the possible vibrational components of ℓ, we rewrite eq. (32) in a slightly modified form as:

$$\Delta_{si}(E_{si}) = \frac{2\pi}{\hbar}\int \frac{dE_{\ell j}\rho_\ell(E_{\ell j})|v_{si,\ell j}|^2 \hbar\Gamma_{\ell j}(E_{si})/2}{(E_{si}-E_{\ell j})^2 + [\hbar\Gamma_{\ell j}(E_{si})/2]^2} \ , \tag{34}$$

and if $\rho_\ell(E_{\ell j})$ and $v_{si,\ell j}$ vary slowly with energy, we arrive at the expression:

$$\Delta_{si}(E_{si}) = \frac{2\pi}{\hbar}\, \rho_\ell(E_{si})|v_{si,\ell j}|^2 \tag{35a}$$

The total rate is the sum of terms such as (35a) over the vibrational levels i and j, each term weighted by the Boltzmann probability of the vibrational level i:

$$k_{nr} = \sum_{ij} p(si;\beta)\Delta_{si} = \frac{2\pi}{\hbar}\sum_{ij} p(si)|\langle si|v_{si,\ell j}|\ell j\rangle|^2\delta(E_{si}-E_{\ell j}). \tag{35b}$$

The calculation of the last relation is made easier by the introduction of Kubo and Toyazawa's generating function (15a). One starts from the generalized line-shape function:

$$F(E) = Z^{-1} \sum_{\ell,j} |v_{si,\ell j}|^2 \exp(-\beta E_{si})\delta(\Delta E + E_{\ell j} - E_{si} - E), \tag{36}$$

with Z the vibrational partition function:

$$Z = \sum_i \exp(-\beta E_{si}) \;. \tag{37}$$

Its Fourier transform is:

$$f(t) = Z^{-1} \sum_{ij} v_{si,\ell j} \exp(iE_{\ell j}t/\hbar) v_{\ell j,si} \exp(-iE_{si}\tau/\hbar) \;, \tag{38}$$

with $\tau = t - i\hbar\beta$ and $\beta = 1/kT$.

Since

$$F(E) = \frac{1}{2\pi\hbar} \int_{-\infty}^{\infty} f(t) \exp[-i(E - \Delta E)t/\hbar]dt \;, \tag{39}$$

the expression for the nonradiative rate Δ_{si}, or $k_{nr}(s\to\ell,\beta)$ using the more common notation, is obtained, setting E=0 in eq. (39).

If the purpose of the approach being taken was to bypass the need to evaluate the double sum over the vibrational states, we have made no great progress, because the sums still appear in the expression for f(t). The sums can be eliminated by introducing the Green function for the nuclear motion, typically:

$$G_s(\underline{Q}^s,\tau;\overline{\underline{Q}}^s_s,0) = \sum_i \chi_{si}(\underline{Q}^s)\chi^*_{si}(\overline{\underline{Q}}^s)\exp(-iE_{si}\tau/\hbar) \;, \tag{40}$$

and the expression $v_{s\ell}(Q)$ for the coupling operator averaged over the electronic motion:

$$v_{s\ell}(\underline{Q}) = \int d\underline{q}\; \Phi_s(\underline{q},\underline{Q})\; v(\underline{q},\underline{Q})\; \Phi_\ell(\underline{q},\underline{Q}) \;. \tag{41}$$

The resulting expression for f(t) is:

$$f(t) = Z^{-1} \iint dQ\, d\bar{Q}\, [v_{s\ell}(Q) G_\ell(\bar{Q}^\ell, -t; Q^\ell, 0)][v_{s\ell}(\bar{Q}) G_s(Q^s, \tau; \bar{Q}^s, 0)]. \quad (42)$$

If the vibrational motion can be represented by individual oscillators, then:

$$G_s(Q^s, T; \bar{Q}^s, 0) = \prod_{j=1}^{N} g_j^s(Q_j^s, T; \bar{Q}_j^s, 0) \quad . \quad (43)$$

For a one-dimensional harmonic oscillator*

$$g_j^\alpha\ (Q_j^\alpha, T; \bar{Q}_j^\alpha, 0) = \left[\frac{M_j \omega_\alpha^j}{2\pi i\hbar\ \sin(\omega_j^\alpha T)}\right]^{\frac{1}{2}} \exp\{(iM_j\omega^\alpha/4\hbar)[(\bar{Q}_j^\alpha - Q_j^\alpha)^2 \cot(\omega_j^\alpha T/2)$$

$$- (\bar{Q}_j^\alpha + Q_j^\alpha)^2 \tan(\omega_j^\alpha T/2)]\} \quad . \quad (44)$$

The remaining quantity to be specified is the coupling operator $v_{s\ell}(Q)$ averaged over the electronic motion. The dominant coupling mechanism is the nonadiabatic component of the Hamiltonian, and can be expressed as:

$$v_{s\ell}(Q) = \sum_{k=1}^{p} C_{s\ell}^k i\hbar M_k^{-1/2} \frac{\partial}{\partial Q_k^\ell} \quad . \quad (45)$$

For nonradiative transitions between levels of the same multiplicity, we have:

$$C_{s\ell}^k \equiv J_{s\ell}^k = \hbar(M_k)^{-1/2} \langle \Phi_s(q, Q^s) | i \frac{\partial}{\partial Q_k^\ell} | \Phi_\ell(q, Q^\ell) \rangle \quad . \quad (46)$$

For intersystem crossing, the spin-orbit mechanism H_{SO} is relied upon to scramble the singlet-triplet characteristics prior to the action of the nonadiabatic coupling operator. The corresponding expression will therefore be:

$$C_{s\ell}^k = \sum_{m \neq \ell, s} \left[\frac{\langle \Phi_s | H_{SO} | \Phi_m \rangle J_{m\ell}^k}{E_s(Q) - E_m(Q)} + \frac{J_{sm}^k \langle \Phi_m | H_{SO} | \Phi_\ell \rangle}{E_m(Q) - E_\ell(Q)}\right] \quad . \quad (47)$$

*See also Ref. (15a) for the corresponding q-representation of the density matrix.

The evaluation of the electronic coupling constant from first principles is a very difficult undertaking, especially in the case of intersystem crossing, since it depends critically on the spin-orbit mixing of the initial and final states of the nonradiative transition. The strengths of the electronic coupling constant $|C^k_{s\ell}|^2$ are of the order of ~10^6 cm^2 for internal conversion and of ~10^{-4} cm^2 for intersystem crossing.

IV. THE ENERGY-GAP LAW IN INTERSYSTEM CROSSING

The study of phosphorescence of organic molecules provides information not only on the radiative decay rates from the lowest triplet to the ground singlet state, but also on the corresponding nonradiative rates. A remarkable success of the radiationless relaxation theory has been its ability to rationalize the observed nonradiative rates $^3T_1 \rightarrow {}^1S_0$ in a series of normal and deuterated aromatic hydrocarbons.

On plotting the log of these rates vs. the triplet energy, Siebrand (16) observed a clustering of the points into two plot regions, one pertaining to deuterated hydrocarbons and the other to the normal hydrocarbons (Fig. 6). In order to rationalize the experimental observations, Siebrand concentrated on the evaluation of the Franck-Condon factors F, as a function of the triplet energy, in the expression for the nonradiative decay rate of the type of eq. (35). He also noticed that the experimental points clustered around two well-defined straight lines after performing some "ad-hoc" modifications of the independent variable. Specifically, the abscissae were taken to be $(E - E_o)/\eta$, with $E^H_0 = 4000$ cm^{-1} and $E^D_0 = 5500$ cm^{-1} and $\eta = 0.4$ (Fig. 7).

The exponential dependence of the radiationless rate on the energy separation of initial and final electronic levels ("energy-gap law") was formally derived by Englman and Jortner (17a). We follow the highlights of their derivation, starting from the expression of the rate, of a form similar to eq. (35b), namely:

$$k_{nr} = \frac{2\pi}{\hbar} C^2 \sum_{ij} e^{-\beta E_{si}} |v_{si,\ell j}|^2 \delta(E_{si} - E_{\ell j}) / \Sigma_i \, e^{-\beta E_{si}} \; . \quad (48)$$

The expression is then cast in the form of a Fourier integral:

$$k_{nr} = \frac{C^2}{\hbar^2} e^{-G} \int_{-\infty}^{\infty} \exp[-i\Delta Et/\hbar + G_+(t) + G_-(t)]dt \; , \quad (49)$$

where the quantities $G_\pm$ are phonon-generating functions:

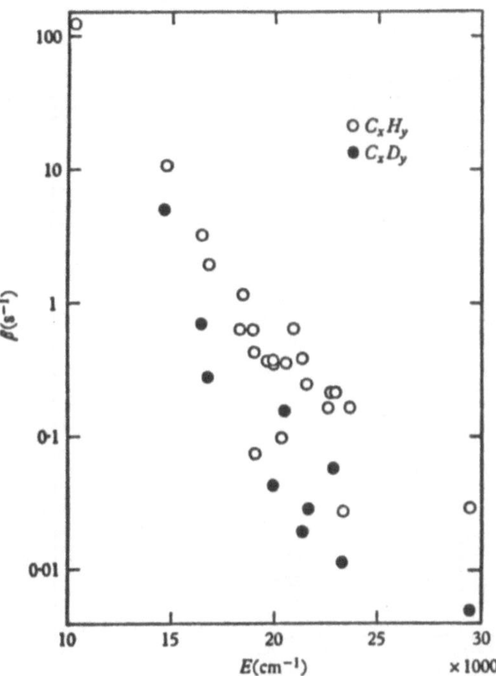

Fig. 6. Observed nonradiative rates β for intersystem crossing from $^3T_1$ to $^1S_0$, plotted vs. the energy of the first excited triplet, in a series of normal and deuterated aromatic hydrocarbons (16).

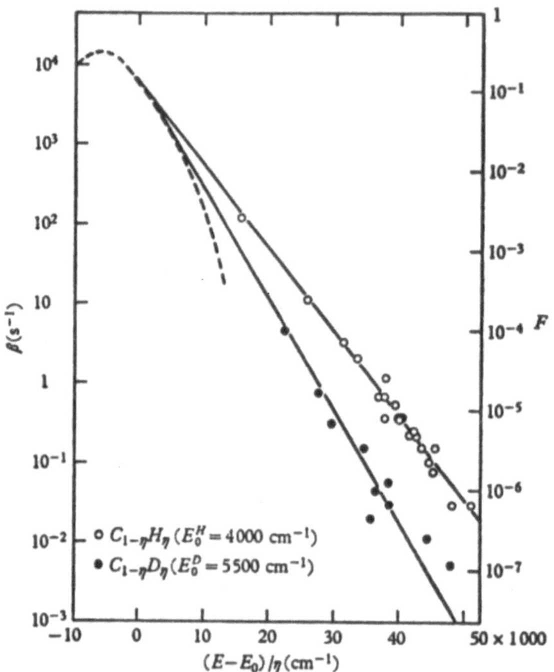

Fig. 7. Replotting of the data of Fig. 6, using as abscissae the modified triplet energy $(E - E_0)/\eta$ [see text and Ref. (16)].

$$G_+(t) = \frac{1}{2} \sum_j \Delta_j^2(\bar{n}_j + 1)\, e^{i\omega_j t} \quad , \tag{50}$$

$$G_-(t) = \frac{1}{2} \sum_j \Delta_j^2\, \bar{n}_j e^{-i\omega_j t} \quad , \tag{51}$$

$$\bar{n}_j = \left[e^{\beta\hbar\omega_j} - 1 \right]^{-1} \quad , \tag{52}$$

and

$$G = G_+(0) + G_-(0) = \frac{1}{2} \sum_j \Delta_j^2(2\bar{n}_j + 1) \quad . \tag{53}$$

Δ_j^2 is the <u>coupling constant</u> for mode j with

$$\Delta_j = \left(\frac{M_j\omega_j}{\hbar} \right)^{1/2} [Q_j^{0(\ell)} - Q_j^{0(s)}] \quad . \tag{54}$$

The expression for the rate is similar to the expression obtained by Fischer (17b) and by Fong (17c) using the correlation function approach. At this point, further elaboration of the integral of eq. (49) proceeds by a familiar route -- that is, that of the "saddle point" approximation, namely:

$$\int_{-\infty}^{\infty} dt\, \exp[f(t+i\hbar\beta)] = (2\pi)^{1/2}\, [-f''(t_s)]^{-1/2} \exp[f(t_s)], \tag{55}$$

where the saddle point t_s is determined by the condition:

$$f'(t+i\hbar\beta)_{t+i\hbar\beta=t_s} = 0 \quad . \tag{56}$$

The expression for the saddle point in this specific case is given by:

$$\frac{1}{2} \sum_j \hbar\omega_j\Delta_j^2 \operatorname{cosech}(\tfrac{1}{2}\beta\hbar\omega_j) \sinh(\tfrac{1}{2}\beta\hbar\omega_j + it\omega_j) = \Delta E \quad . \tag{57}$$

At low temperatures the condition reduces to:

$$\frac{1}{2} \sum_j \hbar\omega_j\Delta_j^2 [\cosh i\omega_j t + \coth \tfrac{1}{2}\beta\hbar\omega_j \sinh i\omega_j t] = \Delta E \quad , \tag{58}$$

$$\frac{1}{2} \sum_j \hbar\omega_j\Delta_j^2 \left[\frac{e^{i\omega_j t} + e^{-i\omega_j t}}{2} + \frac{e^{\beta\hbar\omega_j} + e^{-\beta\hbar\omega_j}}{e^{\beta\hbar\omega_j} - e^{-\beta\hbar\omega_j}} \, \frac{e^{i\omega_j t} - e^{-i\omega_j t}}{2} \right] = \Delta E, \tag{59}$$

$$\frac{1}{2}\sum_j \hbar\omega_j \Delta_j^2 e^{i\omega_j t} = \Delta E \quad . \tag{60}$$

Since normally (weak-coupling case)

$$\frac{1}{2}\sum_j \hbar\omega_j \Delta_j^2 << \Delta E \quad , \tag{61}$$

for the equality to be satisfied t must equal $i\tau$ with τ negative. Considering only the highest values of ω_j, we find:

$$e^{i\omega_M t} = \frac{2\Delta E}{\sum_M \hbar\omega_M \Delta_M^2} \quad , \tag{62}$$

hence:

$$t_s = \frac{1}{i\omega_M} \ln \frac{2\Delta E}{\sum_M \left(\hbar\omega_M \Delta_M^2\right)} \quad ; \tag{63}$$

the final expression for the rate is given, at low temperatures, by:

$$k_{nr} = \frac{C^2 \sqrt{2\pi}}{\hbar^2 \sqrt{\hbar\omega_M \Delta E}} \exp\left(-\frac{1}{2}\sum_j \Delta_j^2\right) \exp\left(-\frac{\Delta E}{\hbar\omega_M}\gamma\right) \quad , \tag{64}$$

with

$$\gamma = \ln\left(2\Delta E / \sum_M \hbar\omega_M \Delta_M^2\right) - 1 \quad . \tag{65}$$

Hence the energy dependence of the rate is mainly through the ratio $\Delta E/\hbar\omega_M$ in the exponential.

V. EXCITATION-ENERGY DEPENDENCE OF INTERSYSTEM CROSSING AND INTERNAL CONVERSION

Many of the qualitative conclusions reached by Russian workers (6),(18) on the interpretation of gas-phase molecular spectra are being validated by recent spectroscopic investigations and their detailed theoretical interpretation. One important effect occurring during excitation of gaseous species in the collisionless regime is the steep increase in the internal-conversion rate with increasing energy of the exciting radiation. In such cases internal conversion to the ground state competes very efficiently with both fluorescence and intersystem crossing.

A thorough study of these processes was reported for the case of naphthalene by Hsieh and Lim (19), while a detailed theoretical interpretation of the observed effects was discussed by Fischer and and coworkers (20). Hsieh and Lim (19) irradiated a mixture of naphthalene vapor (0.07 torr) and biacetyl (0.1 torr) into the excited singlets of naphthalene and measured, as a function of the energy of the excited photons, the quantum yield for sensitized fluorescence from biacetyl (ratio of quanta emitted by biacetyl to the number of quanta absorbed by naphthalene). In addition, they measured as a function of the excitation energy the intersystem crossing yield, defined as $(1 - \phi_F)k_{ISC}/k_{nr}$. From a comparison of the two functions (Fig. 8) it can be concluded that the production of biacetyl triplets follows faithfully the production of naphthalene triplets and that the latter drops drastically when the excess excitation energy (difference between zero energy of the excited electronic state and the exciting-photon energy) exceeds 10,000 cm^{-1}. The observed drop in naphthalene triplet formation was attributed to a steep rise in the internal conversion rate. In effect, the fluorescent lifetime of naphthalene in these experiments was found to decrease from 79 ns for excitation at 289 nm to 1.3 ns for excitation at 200 nm. The internal conversion rate increases almost exponentially with excitation energy, from an initial value of 1.2×10^5 s^{-1} to $\sim0.8\times10^9$ s^{-1} for an excess excitation energy of 18,000 cm^{-1}. Fischer, et al. (20) attempted to account for the observed rapid rise in internal-conversion rate by considering the mechanisms of redistribution of excess excitation into the available vibrational modes. Two models were utilized; one (the "communicating modes" model) assumes strong coupling amongst modes, and should lead to a statistical distribution of the excitation energy amongst available modes, and corresponding to an effective temperature T* of the excited molecules. A second set of calculations was based on the model of "retention modes," whereby the excitation energy remains mostly in the class of strongly accepting modes. The model that fitted the observations best was a modification of the "retention modes" with partial mixing. At high excess energy, the excitation is no longer restricted to the high-frequency C-H stretching modes, but is partially distributed to the remaining lower-frequency modes. Since this type of distribution corresponds to that occurring in the case of smaller energy gaps ΔE, the observed increase in the internal conversion rate is thereby rationalized.

In a sense, Fischer and coworkers' conclusions are a quantitative validation of Neporent's basic description of mode coupling in "complex" and "semicomplex" molecules (21), naphthalene being a typical semicomplex molecule.

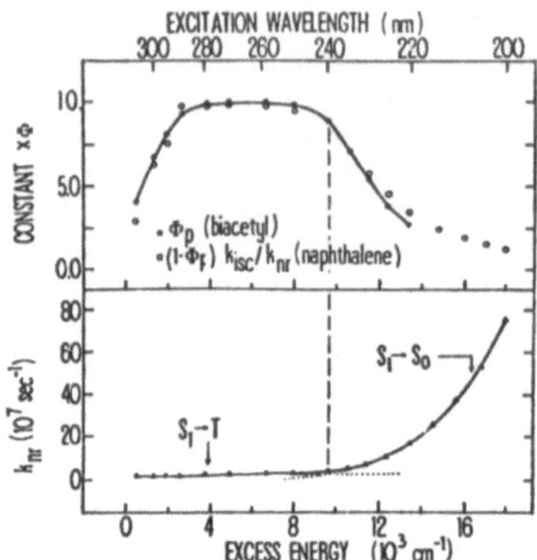

Fig. 8. Plot of the relative quantum yield of sensitized biacetyl phosphorescence vs. the excess energy (excitation energy minus zero-point energy of S_1) of the naphthalene molecule in the vapor phase. Also, a similar plot for the nonradiative decay rates from the excited singlet-state of naphthalene. The steep increase in nonradiative rates (19) with excess energy is interpreted as arising from the internal conversion $S_1 \rightarrow S_0$.

VI. THE SPIN-ORBIT INTERACTION IN MOLECULES

The spin-orbit coupling H_{SO} is responsible for the admixing of singlet and triplet states in organic molecules. Contrary to the case of transition-metal ions and lanthanides and actinides, higher spin-multiplicities than three are seldom considered in organic molecules.

In atomic systems the spin-orbit interaction can be expressed via a sum of one-electron factors involving a radial integral (central field approximation) times the scalar product of one-electron orbital-momentum and spin-momentum operators:

$$H_{SO} = \sum_j \xi(r_j)\, \underline{\ell}_j \cdot \underline{s}_j \,. \tag{66}$$

More rigorously, already in atomic system, three components of H_{SO} can be distinguished (22). The first is analogous to eq. (66),

$$H_{SO}^{(1)} = \frac{e^2}{2m^2c^2} \sum_j \left(Zr_j^{-3} - \sum_{k \neq j} r_{j,k}^{-3} \right) \left(\underline{\ell}_j \cdot s_j \right) , \tag{67}$$

since it involves one-electron operators. The second component has the self-explanatory label of spin-other-orbit operator:

$$H_{SO}^{(2)} = \frac{-e^2}{m^2c^2} \sum_j \sum_{k \neq j} r_{j,k}^{-3} \; \underline{\ell}_k \cdot \underline{s}_j , \tag{68}$$

while in the third term the $\underline{r}$ and $\underline{p}$ operators of different electrons are coupled:

$$H_{SO}^{(3)} = \frac{e^2}{2m^2c^2} \sum_j \sum_{k \neq j} r_{j,k}^{-3} \{ 2(\underline{r}_j \times \underline{p}_k) - (\underline{r}_k \times \underline{p}_j) \} \cdot \underline{s}_j \quad . \tag{69}$$

As for the radial dependence of the operators, the central field approximation of the atomic case is definitely inapplicable in the molecular case.

Having derived the formal expression for the spin-orbit operator, suitable expressions for the molecular eigenfunctions to be operated on by $H_{SO}^{(i)}$ are needed next. Such state functions are built up from properly antisymmetrized products of one-electron orbitals. For instance, for the ground singlet-state of a molecule containing 2N electrons, we can write:

$$\begin{aligned} {}^1\Phi_0 = \{(2N)!\}^{-\frac{1}{2}} \sum_P P\delta_P \, [&\eta_1(1)\eta_1(2)\alpha(1)\beta(2)\eta_2(3)\eta_2(4)\alpha(3)\beta(4) \cdots \\ &\eta_k(2k-1)\eta_k(2k)\alpha(2k-1)\beta(2k) \cdots \eta_N(2N-1)\eta_N(2N)\alpha(2N-1)\beta(2N)] , \end{aligned} \tag{70}$$

using the one-electron molecular orbitals $\eta(i)$ occupied by electron i and the corresponding spin functions $\alpha(i)$ and $\beta(i)$. Excited singlet states are obtained promoting an electron from an orbital η_ℓ to an excited η_n orbital. The spin component of the resulting wavefunction may contain only $\alpha\beta$ products (singlet states), or both $\alpha\beta$ and $\alpha\alpha$ and $\beta\beta$ products (triplet states).

$$\begin{aligned} {}^1\Psi_{\ell,n} = \{2(2N)!\}^{-\frac{1}{2}} \sum_P P\delta_P [&\Pi_{k \neq \ell} \{\eta_k(2k-1)\eta_k(2k)\alpha(2k-1)\beta(2k)\} \\ &\times \{\eta_\ell(2\ell-1)\eta_n(2\ell) + \eta_n(2\ell-1)\eta_\ell(2\ell)\}\alpha(2\ell-1)\beta(2\ell)] , \end{aligned} \tag{71}$$

$$^{3}\Psi_{\ell,n,i} = \{(2N)!\}^{-\frac{1}{2}} \sum_{P} P\delta_{P}[\Pi_{k\neq\ell}\{\eta_{k}(2k-1)\eta_{k}(2k)\alpha(2k-1)\beta(2k)\}$$

$$\times\ \eta_{\ell}(2\ell-1)\eta_{n}(2\ell)\ ^{3}\zeta_{i}(2\ell-1,2\ell)]\ , \qquad (72)$$

with

$$^{3}\zeta_{1}(\ell,n) = \alpha(\ell)\alpha(n)$$

$$^{3}\zeta_{0}(\ell,n) = 2^{-\frac{1}{2}}\{\alpha(\ell)\beta(n) + \beta(\ell)\alpha(n)\} \qquad (73)$$

$$^{3}\zeta_{-1}(\ell,n) = \beta(\ell)\beta(n)\quad .$$

In turn, the molecular orbitals are generally constructed from symmetry-determined linear combinations of atomic orbitals ϕ_j,

$$\eta_k = \sum_{n=1}^{N} a_{k,n}\phi_{j,n}\ , \qquad (74)$$

and since H_{SO} contains two-electron operators, the radial factors in the expression for the spin-orbit matrix elements may involve up to four-center integrals. In general one-center integrals dominate over multi-center integrals.

Since spin-orbit matrix elements admix singlet and triplet states, they will determine the magnitude of the electronic factor in the rate for both radiative and nonradiative processes connecting states of different multiplicity [eqs. (47) and (48)].

VII. EXPERIMENTAL OBSERVATIONS ON SINGLET-TRIPLET TRANSITIONS

In the previous sections we have summarized the derivation of formal expressions for the nonradiative, singlet-triplet rate (intersystem crossing). It would be over-optimistic to expect that the expressions in question will allow one to predict actual rates in real systems. Rather, the previous derivation is valuable in clarifying, correlating and rationalizing the body of available experimental information, and in evaluating relative rates. Applications of the previously derived formal rates have been reported in the literature, especially for benzene, a typical intermediate-case molecule, which has been thoroughly investigated.

In the subsections to follow we shall review the highlights of available experimental data bearing on the phenomenon of intersystem crossing in complex molecules.

VII.A. Existence and Detection of the Triplet State

Many organic, complex molecules capable of fluorescing, especially some of the dyes, have been known for a long time to exhibit, in addition to a prompt emission (fluorescence), a long-lived emission at longer wavelengths than fluorescence, especially in evidence at low temperatures (phosphorescence). Terenin (23), and later Lewis and Kasha (24), interpreted the phosphorescence as proceeding from a triplet level, populated from the excited singlet via the "intersystem-crossing" process. The existence of a paramagnetic species following photoexcitation was confirmed by Evans' observations (25). Remaining reservations that the observed magnetism might be due to photoproduced radicals were removed by Hutchison and Mangum's landmark work (26) on the ESR spectra of excited naphthalene in durene crystals.

Characterization of phosphorescence, together with ESR spectroscopy, is still the main source of available information on the triplet state. The limitations of both techniques are complemented by other optical and magnetic techniques. For instance, excited-state absorption, using steady-state, flashlamp, or laser radiation has been used to determine both the energy of higher triplet states and the decay kinetics of the lowest triplet state. The long lifetime of the latter has also been used to advantage in either exciting other light-emitting species (triplet sensitization) or in producing selective photochemical processes. In particular, triplet-state-sensitized isomeric transformation of butene (27) and dideutero-ethylene (28) are used as a means of detecting and measuring the yield of formation of triplet states in cases where the triplets formed are not conveniently detected by phosphorescent emission.

VII.B. Methods to Enhance Triplet-Singlet Transitions

Both the radiative and nonradiative coupling of singlet and triplet states are governed by the strength of the admixing caused by the spin-orbit coupling. It is known from atomic spectroscopy that the intensity ratio of the spin-allowed ${}^1S_0 \rightarrow {}^1P_1$ to spin-forbidden transitions ${}^1S_0 \rightarrow {}^3P_1$ decreases monotonically down the periodic table. The effect is a reflection of the dependence of the spin-orbit coupling constant on the atomic number Z, specifically a proportionality to Z^5. Therefore, the presence within, or in the vicinity of, an organic molecule of a heavy atom or ion is likely to introduce a more effective singlet-triplet mixing. The manifestation of the presence of such a center in an organic molecule is defined as "internal heavy-atom" effect (29). If the molecule comes into the proximity of heavy atoms, for instance as constituents in solvent molecules, the resulting enhancement in multiplet mixing is defined as the "external heavy-atom effect" (30).

Table 4 summarizes data on singlet-triplet mixing in halogen-substituted naphthalenes. On replacement of H^- with increasingly heavier halogen ions, the quantum yield for fluorescence drops from the 0.55 value by almost four orders of magnitude. The quantum yields for phosphorescence show a complementary behavior. The phosphorescence decay times are similarly decreased by almost three orders of magnitude in the series, evidence that not only the $^1S_1 \rightarrow {}^3T_1$ process is affected by the heavy atom substitution, but also the $^3T_1 \rightarrow {}^1S_0$ transition. The decrease in the phosphorescence decay time τ_p obeys the relations:

$$\frac{\tau_p(\text{Hal})}{\tau_p(\text{F})} = \left(\frac{\zeta_F}{\zeta_{Hal}}\right)^2 , \tag{75}$$

with ζ the heavy-atom spin-orbit coupling constant (31). The non-radiative $^3T_1 \rightarrow {}^1S_0$ rates in tetraphenyls are also a good example of the operation of the internal heavy-atom effect (Table 5).

TABLE 4. EFFECT OF INTERNAL SPIN-ORBIT PERTUBATION ON THE TRANSITION RATES FROM GROUND-SINGLET TO FIRST EXCITED TRIPLET (AND VICE VERSA)

Internal Spinorbit Perturbation in 1-Halonaphthalenes[a]

Halogen	τ_P[d] (sec)	f' (cm/mole)[b]	ζ[c] (cm^{-1})	$\frac{\tau_P(F)}{\tau_P(\text{halogen})}$	$\frac{f'(\text{halogen})}{f'_F}$	$\frac{\zeta^2(\text{halogen})}{\zeta^2_F}$
F	1.4	0.42	272	1	1	1
Cl	0.23	2.86	587	6.08	6.81	4.66
Br	0.014	42.11	2460	100.0	100.26	81.72
I	0.0023	386.60	5060	608	920.5	346

[a]Ref. 31, page 264

[b]Integrated molar decadic extinction

[c]One-electron radial spin-orbit coupling constant

[d]Is the observed phosphorescence decay time $1/\tau_p = (1/\tau_p^0) + k_{QP}$

TABLE 5. EFFECT OF INTERNAL SPIN-ORBIT PERTURBATION ON THE TRANSITION RATES EXCITED-SINGLET-TO-TRIPLET IN A SERIES OF TETRAPHENYLS [FROM REF. (31), P. 266]

Molecule	Fluorescence Band Maximum (Å)	Phosphorescence Band Maxima (Å)	Φ_P/Φ_F [a]	τ_P (sec) [b]	ζ^2_{np} [c] $(cm^{-1})^2$	k_{ISC} (sec^{-1}) [d]
$C(C_6H_5)_4$	3200	4500, 4700, 5100	$\leqslant$ 0.1	2.9	7.84×10^2	$10^6 - 10^7$
$Si(C_6H_5)_4$	3100	4300, 4600	0.1	1.1	2.02×10^4	10^8
$Ge(C_6H_5)_4$	3200	4500, 4700, 5100	1	0.055	7.74×10^5	10^9
$Sn(C_6H_5)_4$	3200	4500, 4700	10	0.003	4.39×10^6	10^{11}
$Pb(C_6H_5)_4$	—	4250, 4550	$\geqslant$ 10	(0.0008)	5.32×10^7	—

[a] ϕ_P, ϕ_F quantum yields
[b] τ_P is the observed phosphorescence decay
[c] radial, one electron spin-orbit constant
[d] intersystem crossing rate

In view of its relevance to a topic to be discussed in what follows, it should be mentioned that the heavy-atom effect is operative in metallo-organic chelate complexes, as shown by the data of Table 6 on dibenzoylmethane chelates of various trivalent and monovalent metal ions. The external heavy-atom effect is demonstrated in Fig. 9 for 1-chloronaphthalene mixed with halogenated solvents. The $^1S_0 \rightarrow {}^3T_1$ intersystem transition, barely discernible in the pure liquid on the tail of the singlet-singlet absorption, becomes increasingly stronger in the halogenated solvents (32).

Both molecular oxygen and nitric oxide are paramagnetic species capable of forming collision complexes in both the gaseous and liquid phases with a variety of organic molecules. These paramagnetic ions, by a process intermediate in character between the external and internal heavy-atom effect, can cause increased multiplet mixing, and therefore enhance both intersystem crossing and quenching of the triplet state, by favoring the $^3T_1 \rightarrow {}^1S_0$ nonradiative transition.

VII.C. Electron Paramagnetic Resonance (EPR) Spectroscopy

A very powerful and versatile technique for the detection and characterization of the triplet state of organic molecules is the electron paramagnetic resonance (EPR) spectrocopy. Conclusive evidence on the triplet nature of the phosphorescent state in molecules was obtained by Hutchinson and Mangum (22) in 1958 using this particular technique. The threefold degeneracy of the triplet state in condensed media can be lifted (even in the absence of an external magnetic field) by crystal-field effects, spin-orbit coupling and spin-spin dipolar perturbation. The latter mechanism appears to be the dominant one in the systems under consideration.

TABLE 6. SPECTROSCOPIC DATA ON DIBENZOYLMETHANE CHELATES[a,b]

Ion	Symmetry of Complex	$P_{0,0}$ (cm^{-1})	$F_{0,0}$ (cm^{-1})	Φ_F/Φ_P	τ_P (s)	Polarization of Phosphorescence[c]
Al^{3+}	D_3	20900	23950	6.5 ±1	0.50±0.05	Unpolarized
Sc^{3+}	D_3	20600	23500	-	0.30±0.05	-
Y^{3+}	D_3	20350	23350	2.3 ±0.06	0.24±0.05	-
Lu^{3+}	D_3	20350	-	0.86±0.02	0.12±0.03	-
La^{3+}	D_3	20250	-	0.43±0.10	0.09±0.01	Unpolarized
Gd^{3+}	D_3	20350	-	Zero	0.002	Strong Positive
Li^+	C_{2v}	-	-	-	-	Weak Positive
K^+	C_{2v}	-	-	-	-	Weak Positive

[a]From Ref. (31), p. 272.

[b]All data refer to glassy solutions at ~77°K.

[c]Relative to $S_1 \leftarrow S_0$ exciting light polarization.

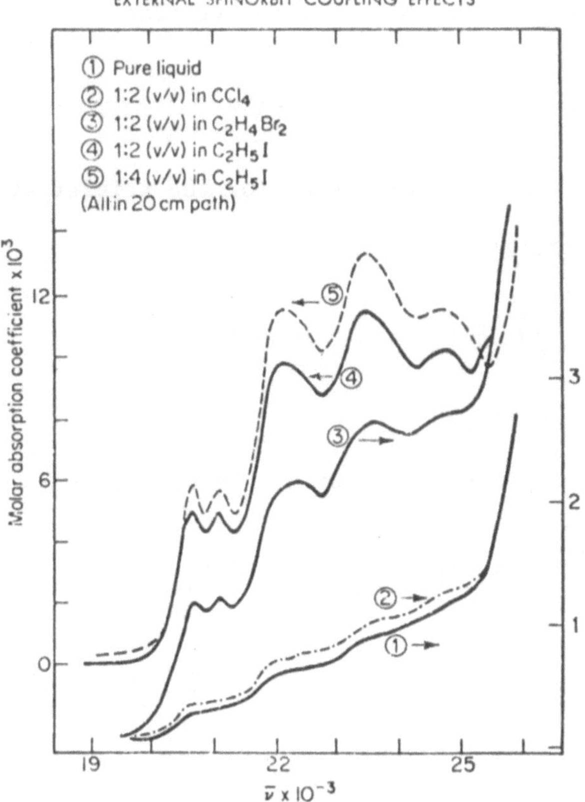

Fig. 9. Illustration of the external heavy-atom effect (31) on the strength of the intersystem transition $S_0 \rightarrow T_1$ in liquid chloronaphthalene, following its admixture with liquids containing heavy atoms. Note the change in left and right ordinate scales.

In order to calculate the expression for the zero-field splitting (ZFS) Hamiltonian, we can neglect the closed-shell electrons and write simply the pertinent wavefunctions of the two electrons with paired spins.

$$^3\Psi_k^{M_S}(1,2,3 \cdots n) \rightarrow \phi_k(1,2)\Theta_{SM_S}(1,2) \quad , \tag{76}$$

with S = 1 and $M_S = \pm 1;0$ where ϕ_k is the function of the spatial coordinates and Θ_{SM_S} the function for the spin coordinates.

Using molecular orbitals labeled as a,b, we can write the antisymmetrized space wavefunction for the triplet state as:

$$\Phi_k(1,2) = \left(\frac{1}{2}\right)^{\frac{1}{2}}[a(1)b(2) - b(1)a(2)] \quad , \tag{77}$$

and the symmetric spin wavefunction for the triplet state as:

$$\begin{aligned} \Theta_{11}(1,2) &= \alpha(1)\alpha(2) \\ \Theta_{10}(1,2) &= \left(\frac{1}{2}\right)^{\frac{1}{2}} [\alpha(1)\beta(2) + \beta(1)\alpha(2)] \\ \Theta_{1-1}(1,2) &= \beta(1)\beta(2) \quad . \end{aligned} \tag{78}$$

Taking $\hbar$=1, we write the operator of the spin-spin Hamiltonian (33) as:

$$\begin{aligned} \mathcal{H}_{ss} &= g^2\beta^2 \, [\underline{r}_{12}^2(\underline{s}_1\cdot\underline{s}_2) - 3(\underline{r}_{12}\cdot\underline{s}_1)(\underline{r}_{12}\cdot\underline{s}_2)] \, r_{12}^{-5} \\ &= \sum_{pq} \Lambda_{pq} s_{1p} s_{2q} \quad , \qquad (p,q = x,y,z) \end{aligned} \tag{79}$$

with

$$\Lambda_{pp} = g^2\beta^2(r_{12}^2 - 3p_{12}^2) \, r_{12}^{-5} \quad , \tag{80}$$

$$\Lambda_{pq} = -3g^2\beta^2(p_{12}q_{12}) \, r_{12}^{-5} \quad . \tag{81}$$

The matrix elements of the spin Hamiltonian can be factorized into the product of a radial integral over the space wavefunctions and one-electron spin-operators acting on the spin wavefunctions.

$$\begin{aligned} H_{ij} &= \langle\phi_k\Theta_{1i}| \sum_{p,q} \Lambda_{pq} s_{1p} s_{2q} |\Phi_k\Theta_{1j}\rangle \\ &= \sum_{p,q} \Omega_{pq} \langle\Theta_{1i}|s_{1p}s_{2q}|\Theta_{1j}\rangle \quad . \end{aligned} \tag{82}$$

By a suitable choice of coordinates, the tensor quantity Ω_{pq} can be reduced to its principal axes. As a result, the expression for the spin Hamiltonian operator simplifies to:

$$\mathcal{H}_{ss} = \Omega_{xx} s_{1x} s_{2x} + \Omega_{yy} s_{1y} s_{2y} + \Omega_{zz} s_{1z} s_{2z} \quad . \tag{83}$$

We are interested at this point in going from the one-electron spin-operators to the operators for the state, namely $\underline{S} = \underline{s}_1 + \underline{s}_2$. In Appendix 2 we detail the transformations required to arrive at the final expression for the spin-Hamiltonian, namely:

$$\mathcal{H}_{ss} = D\left[S_z^2 - \frac{1}{3}\underline{S}^2\right] + E\left[S_x^2 - S_y^2\right] \quad . \tag{84}$$

In order to derive the matrix elements of the S operator over the spin wavefunctions, we make use of the relations of Table 7 and finally arrive at the secular determinant for the zero-field splitting

$$\begin{vmatrix} \frac{1}{3}D-W & 0 & E \\ 0 & -\frac{2}{3}D-W & 0 \\ E & 0 & \frac{1}{3}D-W \end{vmatrix} = 0 \quad , \tag{85}$$

with solutions

$$W_1 = -\frac{2}{3}D \quad ,$$

$$W_{2,3} = \frac{1}{3}D \pm E \quad . \tag{86}$$

Observed values of D and E generally range from 0 cm^{-1} to ~0.2 cm^{-1}.

TABLE 7. EFFECT OF SPIN-ORBIT OPERATORS ON TRIPLET WAVE-FUNCTIONS [RESULTS TO BE MULTIPLIED BY $(\frac{1}{2})^{\frac{1}{2}}\hbar$]

State / Operator	T_1^{+1}	T_1^0	T_1^{-1}
S_x	T_1^0	$T_1^{+1} + T_1^{-1}$	T_1^0
S_y	iT_1^0	$-iT_1^{+1} + iT_1^{-1}$	$-iT_1^0$
S_z	$(2)^{\frac{1}{2}}T_1^1$	0	$(2)^{\frac{1}{2}}T_1^{-1}$

In the presence of an external magnetic field, assuming an isotropic $\hat{g}$-factor, the spin-Hamiltonian becomes:

$$\mathcal{H}_s = \beta\underline{H}\cdot\hat{g}\cdot\underline{S} + D\{S_z^2 - \tfrac{1}{3}\underline{S}^2\} + E\{S_x^2 - S_y^2\}$$

$$= g\beta H(\ell S_x + mS_y + nS_z) + D\{S_z^2 - \tfrac{1}{3}\underline{S}^2\} + E\{S_x^2 - S_y^2\}, \qquad (87)$$

with ℓ, m and n being the cosine directors of the external magnetic field H. The modified secular equation is:

$$\begin{vmatrix} \frac{1}{3}D+g\beta Hn-W & (2)^{-\frac{1}{2}}g\beta H(\ell-im) & E \\ (2)^{-\frac{1}{2}}g\beta H(\ell+im) & -\frac{2}{3}D-W & (2)^{-\frac{1}{2}}g\beta H(\ell-im) \\ E & (2)^{-\frac{1}{2}}g\beta H(\ell+im) & \frac{1}{3}D-g\beta Hn-W \end{vmatrix} = 0 \quad . \qquad (88)$$

Its expression is simplified by choosing the magnetic field in the direction of one of the principal axes of the molecule.

VII.D. Magnetophotoselection

In cases where the radiative transitions $^1S_0 \rightarrow {}^1S_1$ which populate the triplet level have well-defined polarization (in-plane polarization for $\pi \rightarrow \pi^*$ transitions, and out-of-plane polarization for $n \rightarrow \pi^*$ transitions), it is possible to simplify the EPR pattern observed for molecules imbedded in a frozen glassy matrix (34).

When the radiation promoting the triplet-state population is unpolarized, three EPR doublets are observed, corresponding to $\Delta M_S = \pm 1$ transitions, and due to molecules with one molecular axis parallel to the direction of the applied magnetic field $\underline{H}$. If the exciting light in a $\pi \rightarrow \pi^*$ absorption is polarized in a direction parallel to the applied magnetic field, it will still be possible to observe the EPR signal from those molecules whose in-plane axes are parallel to $\underline{H}$. For $\underline{E} \perp \underline{H}$, only the molecules with out-of-plane axis parallel to $\underline{H}$ will be active in EPR. Therefore the assignment of EPR signals to triplet components becomes more specific. The scheme in question is outlined in Fig. 10.

VII.E. Optically-Detected Magnetic Resonance (ODMR)

The triplet-level splitting is normally less than 0.3 cm^{-1}; that is far smaller than the homogeneous line-widths of the phosphorescent emission originating from the individual triplet components. It is

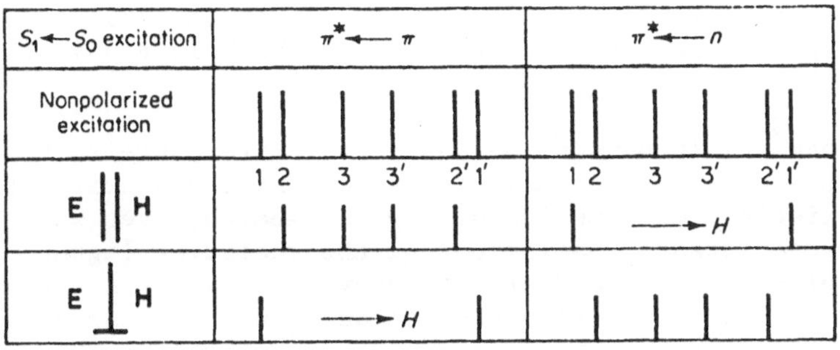

$S_1 \leftarrow S_0$ excitation	$\pi^* \leftarrow \pi$	$\pi^* \leftarrow n$
Nonpolarized excitation		
E ∥ H	1 2 3 3′ 2′ 1′	1 2 3 3′ 2′ 1′ → H
E ⊥ H	→ H	

Fig. 10. Schematic representation of the $\Delta M_S = \pm 1$ transitions in the ESR spectra of triplet states in glassy matrices. The triplet states are populated via the $S_0 \rightarrow S_1$ absorption, followed by intersystem crossing. E is the electric vector of the exciting light; H is the applied magnetic field [Ref. (31), p. 366].

therefore very difficult to detect optically the splitting of the triplet level. When the sample temperature is so low that the spin-lattice relaxation between the triplet components is negligible, the phosphorescent emission will be determined by the level population and the radiative decay-rates of the individual triplet components. If microwave power is applied under these conditions, so as to change the population of pairs of sublevels, a change should be detected in the phosphorescent emission. Results of this technique will be discussed in Section X, on protein triplets.

VIII. TRIPLET-STATE SENSITIZATION OF LANTHANIDE EMISSION IN METALLO-ORGANICS

In view of its long lifetime, the triplet state is potentially an effective reservoir of excitation energy in molecular systems. This excitation energy can be utilized "in situ" for biphotonic processes, to provide the molecule with excitation energy well in excess of that corresponding to the wavelength of the exciting radiation, or it can be released to nearby species, creating new triplet states in different chemical entities. Alternatively, the triplet energy can be transferred to an inorganic ion and cause it to emit. This particular property was actively investigated in recent years as likely to lead to so-called "rare-earth chelate" lasers.

In the mid-1960's, when the search for new laser materials was at its peak, a very determined effort was made in many laboratories to develop a liquid-phase rare-earth laser. One basic problem with rare-earth ions as laser materials is the intrinsic difficulty in efficient excitation of a material that has narrow, low absorption-cross-section transitions both in the visible and near UV.

It had been known in the analytical literature that metallo-organic or chelate complexes of rare-earths could emit into the characteristic line emission of the rare-earth, even when excited in the broad intense absorptions of the chelating ligand. If we also recollect the effect of both heavy and paramagnetic ions (like Gd^{3+}) in quenching the fluorescence and promoting the phosphorescence of complexes of dibenzoylmethane, we can anticipate the main pathways of the excitation transfer from the chelate to the rare-earth ions. On the basis of the observations on Gd^{3+}, it was expected that the excitation energy would be diverted mainly to the triplet level of the chelating agent and hence to the rare-earth ions. The details of the latter process were not quite clear. The ligand triplet had to be higher in energy than the emitting rare-earth level, but it was not established whether the lowest triplet was mainly responsible for the transfer and whether some of the rare-earth excited levels were preferential acceptors.

In the mid-1960's a sustained effort was made at GTE Labs, then in Bayside, New York, to develop a liquid-phase rare-earth chelate laser based on this approach (35). One basic drawback in the overall scheme greatly complicated the development work. The central rare-earth ion not only favored the $^1S \rightarrow {}^3T_1$ intersystem crossing in the ligand, but also enhanced the subsequent deactivation of the triplet down to the ground state. This effect is revealed in Table 6 by the very short phosphorescence decay time of the Gd complex, even at 77°K. The rare-earth emission in the chelates decreased rapidly with the increase in temperature over 77°K. As a result, the chelate solutions had to be kept at low temperatures; voids and bubbles formed in the very viscous solutions and had to be eliminated by mounting the cell windows over movable quartz pistons. Laser action on Eu^{3+} emission was finally observed by Lempicki and Samelson (35). The thresholds for lasing, though, were very high, i.e., flash energies of 1500J to 1600J were needed, and the lasing was very inefficient. Finally this effort was discontinued and shifted to what appeared to be a more promising approach, namely that of rare-earth ions dissolved in aprotic solvents (36).

The utilization of the emission from rare-earth chelates in display devices has been suggested very recently by Japanese workers (37, 38). Thin cells of chelate solutions, containing in one case Eu^{3+} and in another case Tb^{3+}, are excited by UV light. Voltage pulses are applied to the cells and lead to the reduction of Eu^{3+} to its divalent, nonemitting state or, in the case of Tb^{3+}, to the formation of a similarly nonemitting Tb^{4+} species. The net result is a modulation of the visible emission from these cells.

IX. INTERSYSTEM CROSSING IN LASER DYES

We consider now in greater detail the effects of singlet-triplet mixing in a class of complex molecules that has attracted considerable attention in laser technology, both as laser generators and as down-converters of laser radiation with wide tuning range. We are referring specifically to those highly absorbing materials (dyes) that have high enough quantum yield for fluorescence to warrant their use in cw or pulsed dye-lasers.

In dye-laser operation the presence of triplet-state population is deleterious on several grounds: it reduces the quantum yield for fluorescence; it tends to depopulate the ground state under excitation of long duration; and it introduces parasitic absorption of laser radiation, because of intense triplet-triplet absorption.

These effects can be visualized by a computer simulation of dye-laser operation (39). Differential equations describing the decay rates for the laser-cavity photon-density, for the population inversion in the singlet state and for the triplet-state population are solved by means of Runge-Kutta double-precision subroutines. The rate equations for the singlet and triplet population and for the cavity photon-density are:

$$\frac{dY_1(x)}{dx} = Y_1(x)[\alpha Y_2(x)-1] - \frac{\alpha}{\delta} Y_1(x)Y_3(x) + AY_2(x) , \tag{89}$$

$$\frac{dY_2(x)}{dx} = \gamma^{-1}\left(\frac{\ell n2}{\pi}\right)^{\frac{1}{2}} \exp\left\{-\left[\frac{x}{\gamma} (\ell n2)^{\frac{1}{2}}\right]^2\right\} - \frac{\alpha}{\gamma}\left(\frac{\ell n2}{\pi}\right)^{\frac{1}{2}} Y_1(x)Y_2(x) - Y_2(x) \left(\frac{1}{\beta} + \frac{1}{\varepsilon}\right) , \tag{90}$$

$$\frac{dY_3(x)}{dx} = \frac{Y_2(x)}{\varepsilon} - \rho Y_3(x) . \tag{91}$$

The kinetic equations are expressed as a function of time, defined in cavity-lifetime units, namely $x = t/t_c$, with

$$t_c = n_r\ell/(1-R)c . \tag{92}$$

The dye is assumed to be excited by a pump pulse with a Gaussian shape in time, and symmetric with respect to the time origin:

$$W(t) = W_{max} \exp[(-t/T_1)(\ell n2)^{\frac{1}{2}}]^2 \quad . \tag{93}$$

The quantity $Y_1(x) = q(t)/t_c W_{max}$ is a time-dependent photon-density;

$$Y_2(x) = (n_{S*} - n_S)/N \cong n_{S*}/N \quad , \tag{94}$$

is the singlet population-inversion normalized to the total number of pump photons, namely:

$$N = \int_{-\infty}^{\infty} W(t)dt \quad , \tag{95}$$

and

$$Y_3(x) = n_T/N \quad , \tag{96}$$

is the normalized triplet-population.

The parameters in the rate equations define the fluorescent lifetime ($\beta = \tau_f/t_c$); the pump halfwidth ($\gamma = T_1/t_c$); the ratio of singlet-singlet absorption to triplet-triplet absorption at the laser frequency ($\delta = \varepsilon_{SS*}/\varepsilon_{TT*}$); the intersystem crossing rate ($\varepsilon = 1/k_{S*T}t_c$); the reduced triplet-quenching rate ($\rho = t_c/\tau_p$) and the ratio of pump photons to the threshold population inversion ($\alpha = N/n_0$).

The solutions of the rate equations provide the time development of the quantities of interest, such as instantaneous triplet-state population, population inversion, the shape and intensity of the lasing pulse, and the overall efficiency of the lasing process. The latter is expressed as the ratio of the laser-cavity photon-density to the total number of pump photons.

Fig. 11 shows that for very long triplet-state lifetimes and fast intersystem crossing rates, the laser pulse is practically extinguished at the peak of the pump power because of the steady increase in triplet-state population and in concomitant population inversion required to sustain laser oscillation. If the triplet lifetime is reduced to 300 ns, the laser pulse begins to follow the shape of the excitation pulse (Fig. 12). The predicted efficiency begins to look acceptable and continues to rise if only the triplet lifetime can be further reduced (Fig. 13a). We plot in Fig. 13b the predicted laser efficiency as a function of triplet lifetime for a fixed value of total pump photons, supplied by Gaussian pulses of progressively increasing duration. For triplet lifetimes bracketed

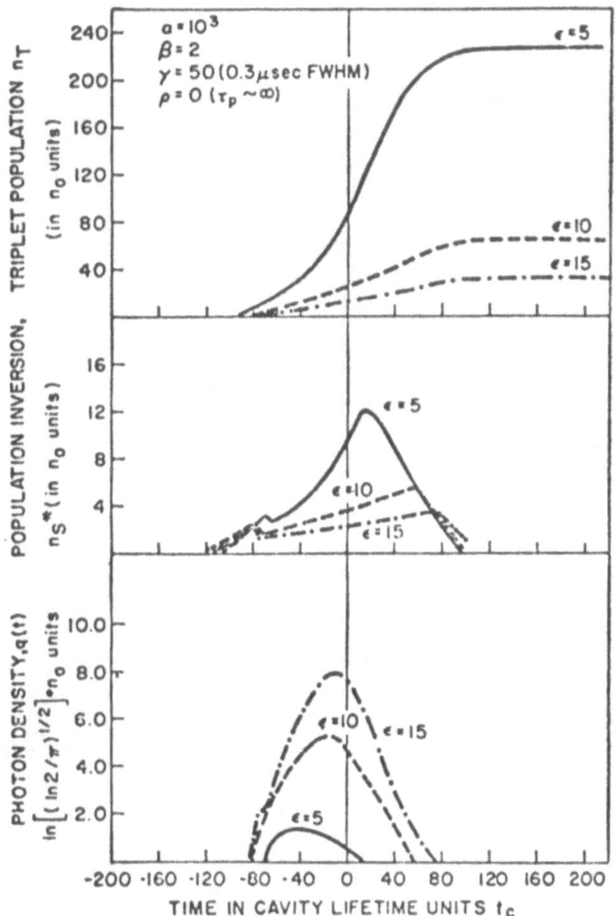

Fig. 11. Computer simulation of the effect of triplet-state population on dye-laser operation (39). The triplet lifetime is assumed to be infinite. The intersystem crossing rate is specified by the parameter $\varepsilon = 1/t_c K_{S^*T}$. Pump pulse of 0.3 μs FWHM. For fast intersystem crossing ($\varepsilon = 5$), the triplet-state population effectively extinguishes lasing (photon-density plot) near the peak of the pump pulse.

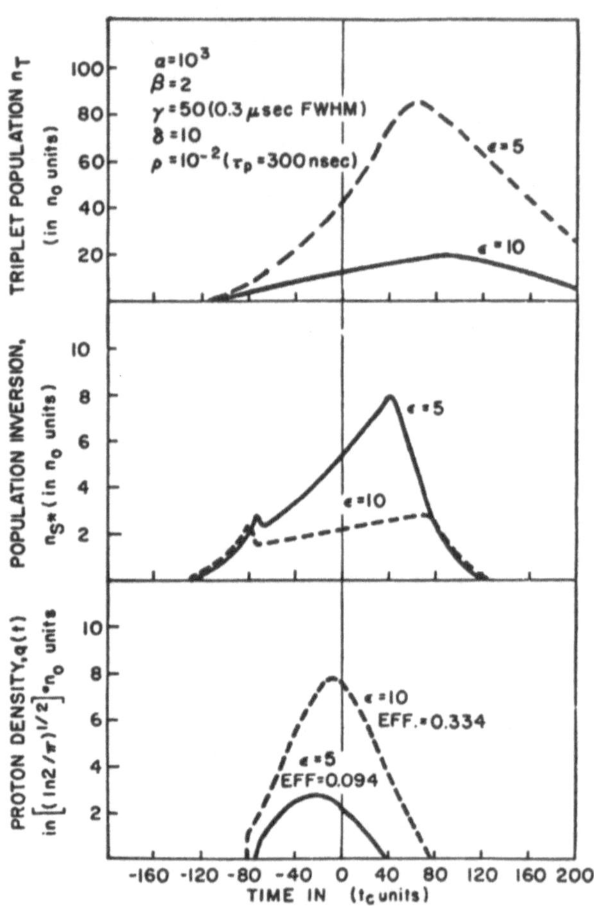

Fig. 12. Improved pulse-shape and efficiency of lasing action, following shortening of the triplet-state lifetime down to 300 ns (39). The values of the other parameters are unchanged from Fig. 11.

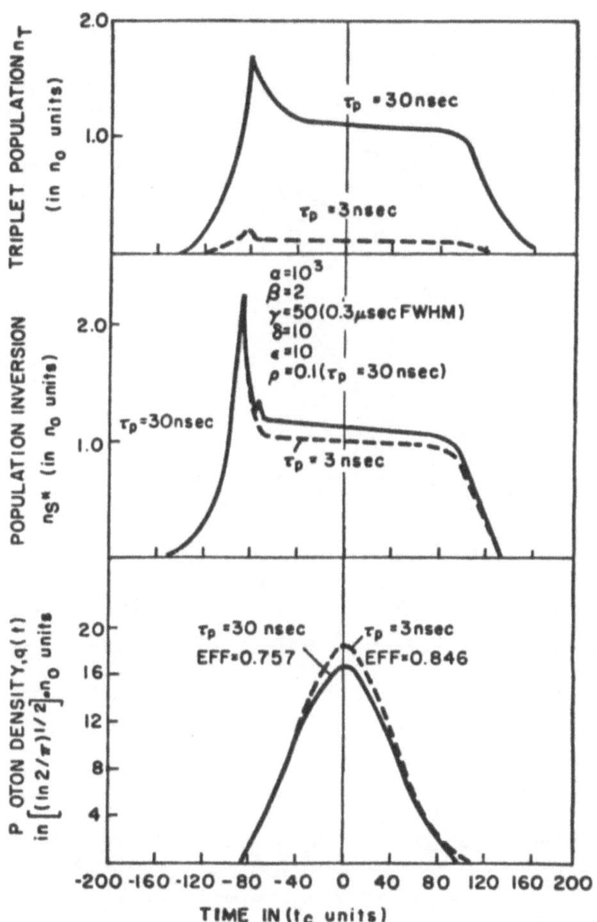

Fig. 13a. Further improvement in laser efficiency with reduction in triplet lifetime (39).

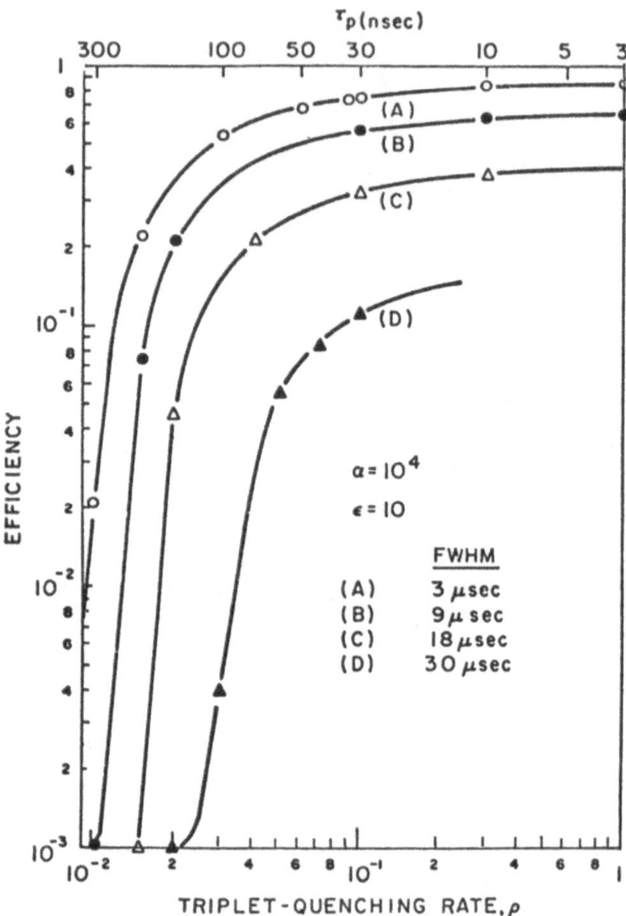

Fig. 13b. Predicted efficiency as a function of triplet lifetime and of pump duration, for a fixed value of the pump energy.

between the realistic values of 300 ns and 100 ns the laser efficiency is of the order of 10%, but very rapidly drops with the increased duration of the pump pulse.

The computer simulation demonstrates quantitatively the need for reducing the triplet-state lifetime and the intersystem crossing rate. The condition of reduced triplet lifetime was found to be particularly important under conditions of flashlamp excitation, if long-duration laser emission was the objective. The way to reduce the triplet lifetime is via introduction of suitable triplet quenchers. It has long been known from phosphorescence studies that oxygen can be very effective in quenching triplet-state population, so that the use of rigorously deoxygenated material is a prerequisite for any accurate triplet-lifetime determination. In effect, oxygen can be used as a triplet quencher in spite of its tendency to quench also the excited-singlet population and to promote photoreactions in dyes. Even more effective as triplet quenchers were aromatic hydrocarbons, such as cyclo-octatetraene and cycloheptatriene. Their addition improved the laser efficiency and made it possible to obtain laser emission 500 μs long, with the laser pulse following the shape of the excitation pulse (40). At these long emission times and high-energy flashlamp excitation, other limiting effects set in: especially important is the formation of optical inhomogeneity ("schlieren") in the dye solution at these long pump times.

An additional requirement for efficient laser operation is, as we have seen, the need to reduce the intersystem crossing rate. Heavy-atom substituents should be avoided, as exemplified by some fluorescein derivatives such as eosin, erythrosin and dithiofluorescein (Figure 14). In eosin the triplet yield is 76%, against the 3% yield for fluorescein. As an empirical rule, Drexhage (41) has proposed that high triplet yields are associated in the planar dye molecules with large circulation loops for electrons. As a way to block these circulation loops Drexhage proposes the replacement of typical oxygen bridges used to rigidize xanthene dyes, with a tetrahedrally coordinated carbon atom. In support of his model Drexhage reports reduced triplet formation in going from oxazine to carbazine (Fig. 14).

X. TRIPLET STATES IN BIOLOGICAL SYSTEMS

X.A. Proteins

The building blocks of proteins are amino-acids, of composition

$$H_3N^+ - \underset{\underset{\displaystyle R}{|}}{CH} - COOH \quad . \qquad (97)$$

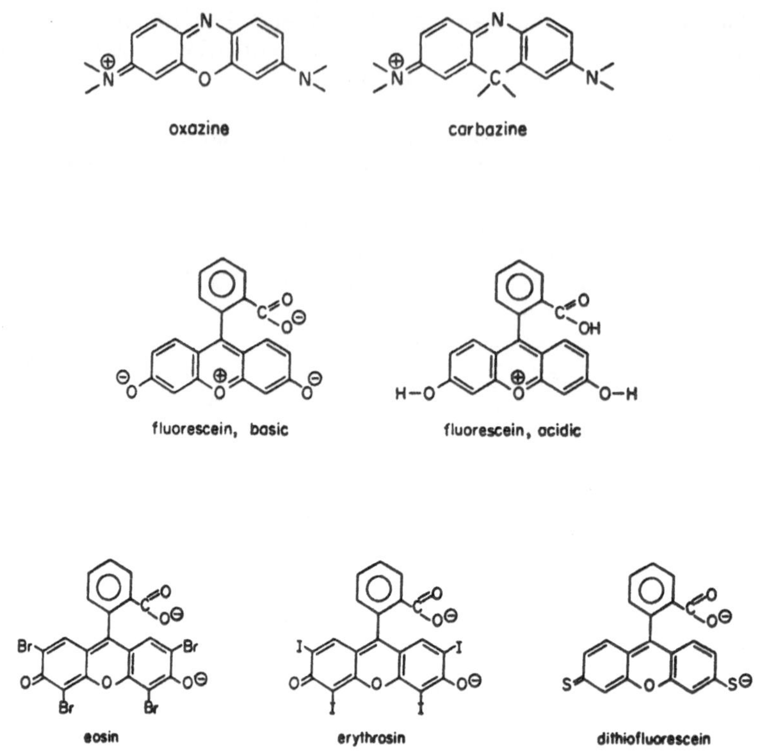

oxazine

carbazine

fluorescein, basic

fluorescein, acidic

eosin

erythrosin

dithiofluorescein

Fig. 14. Formulae for some of the dyes discussed in the text.

Polymerization of amino acids proceeds from the elimination of a water molecule between the amino group and the carboxylic group of adjacent amino-acid molecules. Proteins are naturally occurring amino-acid polymers and, in living systems, perform a variety of functions: catalysis and regulation of biochemical processes (enzymes); mechanical mobility (contractile proteins); defense mechanism (antibodies); and distribution of chemicals to points of utilization (transport proteins).

Proteins are built from combinations out of a set of about twenty amino acids, three of which are aromatic amino acids (Fig. 15). The latter play a dominant role in photoprocesses affecting proteins, and are mainly responsible for the protein fluorescence and phosphorescence properties. The three aromatic amino acids are phenylalanine (derived from benzene), tyrosine (derived from phenol) and tryptophan (derived from indole). The fluorescence from the three aromatic amino acids peaks at 282 nm (phenylalanine), 303 nm (tyrosine) and 348 nm (tryptophan). Their fairly structured phosphorescent bands are only detected at low temperatures (77°K) and occur in the blue region of the spectrum (Fig. 16). The pertinent emission parameters (42) are contained in Table 8. At 77°K the total quantum yield ($\phi_f + \phi_p$) is roughly unity for phenylalanine and tyrosine while it is only 0.89 for tryptophan, pointing to the existence in the latter of nonradiative losses even at 77°K. The phosphorescence decay times are quite long, of the order of seconds (Table 8). At temperatures between 170°K and 200°K the quantum yield for phosphorescence in the solvent medium commonly used (frozen 0.5% glucose solutions) drops very dramatically. The long triplet lifetimes at 77°K are quite suitable for the observation of a variety of photoprocesses, including EPR and optically-detected magnetic resonance (ODMR). In view of its extended lifetime, the triplet state is also responsible for the formation (on irradiation) of a variety of free radicals, and is an essential intermediate in photoionization and sensitization reactions.

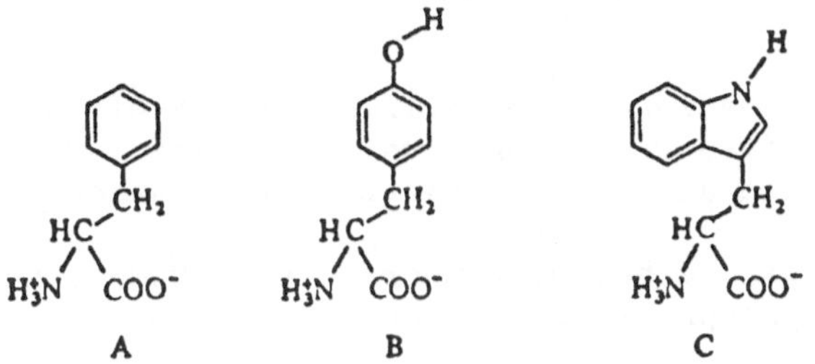

A B C

Fig. 15. Aromatic amino-acids commonly found in proteins.
A = phenylalanine (derived from benzene)
B = tyrosine (from phenol)
C = tryptophan (from indole)

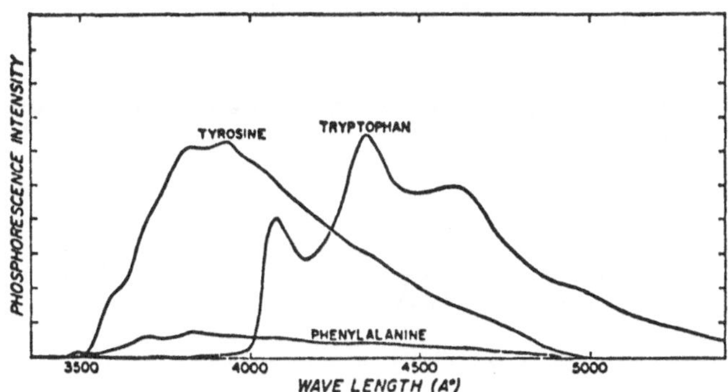

Fig. 16. Phosphorescence spectra of aromatic amino acids in 0.5% glucose at 77°K (42).

TABLE 8. SPECTRAL PARAMETERS FOR AROMATIC AMINO-ACIDS (42)

	Tryptophan		Tyrosine		Phenylalanine	
Absorption maxima (nm)	280	218	275	222	258	205
Molar extinction coefficient at maximum	6500	27,000	1290	8000	200	8500
Fluorescence maximum (nm)	330		300		290	
Fluorescence quantum yield (Φ_f)	0.72		0.47		0.41	
Phosphorescence maximum (nm)	435		395		385	
Phosphorescence quantum yield (Φ_f)	0.17		0.53		0.59	
Phosphorescence lifetime (sec)	5.8 6.65		2.7 2.90		5.5 7.70	

Detailed information on the triplet states of the aromatic amino acids has been derived from ODMR studies at 77°K and at liquid helium temperatures. Since at 77°K spin-lattice relaxation is still effective in equalizing the triplet-level populations, spin-alignment is obtained only at liquid helium temperatures. Under these conditions the steady state phosphorescent intensity is expressed by:

$$I_0 = S \Sigma_i k_i^r N_i^0 \quad , \tag{98}$$

with k_i^r the radiative rate of the triplet level i. If microwave radiation is used to saturate (equalize the population between) pairs of triplet components, the phosphorescent intensity will change by:

$$\Delta I(0) = \frac{1}{2} S (N_i^0 - N_j^0)(k_j^r - k_i^r) \quad . \tag{99}$$

This intensity change, under pulsed microwave excitation (fast passage) will decay as:

$$\Delta I(t) = \frac{1}{2} S \left(N_i^0 - N_j^0\right) \left(k_j^r e^{-k_j t} - k_i^r e^{-k_i t}\right) \quad , \tag{100}$$

hence the possibility of deriving the radiative decay rates of the individual triplet components. It is found in tryptophan that only the T_x level decays radiatively. Rate equations for the populations of the triplet components can also be written for the case when the spin-lattice relaxation is not negligible. Again, ODMR can be used to derive both the level total-decay rates and the spin-lattice relaxation rates between level pairs.

In moving now from the case of the individual amino-acids to their incorporation into proteins, it is found that the fluorescence and the phosphorescence properties of the proteins strongly resemble those of the constituent aromatic amino-acids. Phenylalanine emission is generally not detected, both because of the intrinsic low absorptivity of phenylalanine and because of energy transfer to tyrosine and tryptophan. Consequently, proteins are grouped into two classes (class A and class B) according to emission properties. Class A exhibits tyrosine emission, while class B has the tryptophan emission characteristics. In comparison with the corresponding amino-acid chromophores, phosphorescence decay times in proteins are generally shorter at 77°K. The triplet quenching in attributed to the presence in the protein of the disulfide linkages, with the possibility of electron transfer from the triplet levels to the disulfides.

Class A proteins have low quantum yields for phosphorescence. The quantum yield, and consequently the emission lifetime and other emission parameters, can be brought in line with those of the corresponding amino-acids by denaturing the proteins and reducing the disulfide linkages. The tryptophan phosphorescence in proteins, and especially the triplet splitting, is a sensitive function of the protein environment, as concluded from detailed ODMR studies. This makes tryptophan an "innate" probe of the protein structure.

As observed in the case of the amino-acid monomer, the triplet emission lifetime of tryptophan in proteins decreases from 4.2°K to 77°K. The intersystem process $T_1 \rightarrow S_0$ is a thermally activated process at 77°K, and the protein molecules are dynamically active systems even at these very low temperatures.

In proteins, it is possible to probe the population of triplet sublevels by a combination of flash excitation and fast passage ODMR. Application of this technique in lysozyme shows that the tryptophan triplet population arises from direct intersystem crossing from S_1 (no triplet-triplet energy transfer) and that the out-of-plane T_z level has a significantly lower pumping rate from the S_1 level.

X.B. Energy Transfer in Models of Biomembranes

Several simplified models of living cells have been proposed in the literature for the study of specific aspects of energy transfer. Cellular membranes in particular are important in their role of segregating reactants and thus favoring specific biochemical processes by bringing reactants into close proximity. We shall review two types of measurements based on these simplified models. In one case the singlet-singlet energy transfer, and specifically the effect of spatial distribution of donor-acceptors, is used as a structural probe, or molecular ruler, in the 1.5- to 7.5-nm range. In the second instance the membrane is used to bring an acceptor-donor pair closely together, thereby assuring a highly efficient energy transfer.

In the first study (43) a lecithin, phosphatidylcholine forms a vesicle, or globule, and contains on its surface membrane fluorescent molecules surrounded by suitable acceptors. The fluorescent intensity of the donor decays in time, according to Förster's theory, as:

$$I(t) = I(0)\, e^{-t/\tau_0}\, e^{-\sigma S(t)} \quad , \tag{101}$$

with τ_0 the donor fluorescent lifetime in the absence of acceptors, where σ is the surface density of acceptors, and $S(t)$ is the spatial energy-transfer term:

$$S(t) = \int_a^\infty \left[1 - e^{-(t/\tau_0)(R_0/r)^6}\right] 2\pi r dr \quad . \tag{102}$$

The energy transfer is assumed to take place in a two-dimensional surface, wherein a is the closest approach of the donor-acceptor pair. The quantity R_0 is the critical distance that appears in Förster's theory and is expressed, in Ångstrom units, as:

$$R_0 = (JK^2 Q_0 n^{-4})^{1/6} \times 9.79 \cdot 10^{-3} \quad , \tag{103}$$

with J the spectral overlap integral in $cm^3\ M^{-1}$; K^2 as a dipole-dipole orientation factor; Q_0 the quantum yield for fluorescence in the absence of acceptors; and n the refractive index of the medium. The overall efficiency of the energy transfer is given by:

$$E = 1 - (1/\tau_0) \int_0^\infty [F(t)/F(0)]\, dt \quad , \tag{104}$$

with F(0) the emission intensity following short-pulse excitation. The kinetics of the decay of donor fluorescence are seen to depend on the three parameters a, R_0 and σ. Therefore the analysis of the time decay of the emission can supply the value of the R_0 and σ parameters and can provide structural information on the membrane.

In the second type of studies (44) sodium dodecyl sulphate (SDS) is used to trap a triplet donor (naphthalene) inside a micelle. Molecules such as detergents and soaps, containing both polar and nonpolar groups and which are themselves water soluble, can combine and aggregate with similar compounds to form partially soluble complexes, with the hydrophobic end of the molecules filling a spherical cavity. SDS can form these micelles and can trap in their interior naphthalene, while a rare-earth ion, such as Tb^{3+}, can be attached to the charged surface of the micelle. While naphthalene in aqueous solutions interferes with and quenches the energy transfer from ketone triplets to Tb^{3+}, in the micelle configuration it proves to be very efficient in sensitizing Tb^{3+} emission. Since the energy transfer from the singlet level of naphthalene to Tb^{3+} is spin-forbidden, and since the fluorescence decay of naphthalene is unaffected by the presence of Tb^{3+}, one can safely conclude that the energy transfer in the micelle occurs from the naphthalene triplet. Escabi-Perez, et al., provide additional support (44) for this conclusion. The naphthalene decay time for phosphorescence in the micelle is 10^{-5}s, and this is also the build-up time for Tb emission. Using diffusion theory and a 2-nm estimate for the radius of the micelle, the time for the migration of the naphthalene triplet to the surface can be estimated from diffusion theory as:

$$t = \frac{3\pi\eta\chi^{-2}}{kT} , \qquad (105)$$

as being 262 ns. In the formula χ^{-2} is the mean path-length and η is the microscopic viscosity. The 262 ns thus derived are only a brief interval in the lifetime of the naphthalene triplet (10^{-5}s). In the micelle configuration, and thanks to the presence of naphthalene, the Tb emission intensity increases by two orders of magnitude. Escabi-Perez and coworkers (44) speculate that this type of segregation in simple membranes may lead to efficient devices for solar-energy conversion.

REFERENCES

1. G. W. Robinson and R. P. Frosch, J. Chem. Phys. 37, 1962 (1962).
2. A. E. Douglas, J. Chem. Phys. 45, 1007 (1967).
3. G. B. Kistiakowsky and C. S. Parmenter, J. Chem. Phys. 42, 2942 (1965).
4. A. T. Vartanian, Izv. Akad. Nauk SSSR 3, 341 (1938).
5. G. W. Robinson in The Triplet State, (A. B. Zhalan, ed.), Cambridge University Press, p. 213 (1967).
6. N. A. Borisevich, Excited States of Complex Molecules in the Gaseous Phase, Nauka i Tekhnika, Minsk (1967).
7. B. S. Neporent, Izv. Akad. Nauk SSSR 15, 533 (1951).
8. V. L. Levshin, Dokl. Akad. Nauk SSSR 1, 474 (1935).
9. B. I. Stepanov, Dokl. Akad. Nauk SSSR 112, 839 (1957).
10. N. A. Borisevich, loc. cit., p. 235
11. B. I. Greene, R. M. Hochstrasser and R. B. Weisman, J. Chem. Phys. 71, 544 (1979).

12. M. L. Goldberger and K. M. Watson, Collision Theory, John Wiley and Sons, New York (1964).
13. J. M. Ziman, Elements of Advanced Quantum Theory, Cambridge University Press (1969).
14. K. F. Freed in Radiationless Processes in Molecules and Condensed Phases, (F.K. Fong, ed), Springer Verlag, Berlin, p. 23 (1976).
15a. R. Kubo and Y. Toyozawa, Prog. Theor. Phys. 13, 160 (1955).
15b. Y. Toyozawa, Dynamical Processes in Solid State Optics, (R. Kubo and Y. Toyozawa, eds.), Benjamin, New York, p. 104 (1967).
16. W. Siebrand in The Triplet State, (A.B. Zhalan, ed.), Cambridge University Press, p. 31 (1967).
17a. R. Englman and J. Jortner in Proceedings of the International Conference on Luminescence, (F. Williams, ed.), North-Holland Publishing Company, Amsterdam, p. 134 (1970).
17b. S. Fischer, J. Chem. Phys. 53, 3195 (1970).
17c. F. K. Fong and M. M. Miller, Chem. Phys. Letters 10, 408 (1971).
18. B. S. Neporent in Luminescence of Crystals, Molecules and Solutions, (F. Williams, ed.), Plenum Press, p. 31 (1973).
19. J. C. Hsieh and E. C. Lim, J. Chem. Phys. 61, 736 (1974).
20. S. F. Fischer, A. L. Stanford and E. C. Lim, J. Chem. Phys. 61, 582 (1974).
21. N. A. Borisevich, Excited States of Complex Molecules in the Gaseous Phase, Nauka i Tekhnika, Minsk (1967).
22. H. F. Hameka in The Triplet State, (A. B. Zahlan, ed.), Cambridge University Press, p. 1 (1967).
23. A. Terenin, Acta Physicochim. U.R.S.S. 18, 210 (1943).
24. G. N. Lewis and M. Kasha, J. Amer. Chem. Soc. 66, 2100 (1944).
25. D. F. Evans, Nature 176, 777 (1955).
26. C. A. Hutchison and B. W. Mangum, J. Chem. Phys. 29, 952 (1958).
27. R. B. Cundall, R. J. Fletcher and D. G. Milne, J. Chem. Phys. 39, 3536 (1963).
28. R. B. Cundall, A. S. Davies and K. Dunnicliff in The Triplet State, (A. B. Zahlan, ed.). Cambridge University Press, p. 183 (1967).
29. D. S. McClure, J. Chem. Phys. 17, 905 (1949).
30. S. P. McGlynn, R. Sunseri and N. Christodouleas, J. Chem. Phys. 37, 1818 (1962).
31. S. P. McGlynn, T. Azumi and M. Kinoshita, in Molecular Spectroscopy of the Triplet State, Prentice-Hall, Englewood, New Jersey, P. 308 (1969).
32. S. P. McGlynn, T. Azumi and M. Kinoshita, in Molecular Spectroscopy of the Triplet State, Prentice-Hall, Englewood, New Jersey, p. 308 (1969).
33. S. P. McGlynn, T. Axumi and M. Kinoshita, in Molecular Spectroscopy of the Triplet State, Prentice-Hall, Englewood, New Jersey, p. 331 (1969).
34. P. Kottis and R. Lefebvre, J. Chem Phys. 41, 3660 (1964).
35. A. Lempicki and H. Samelson, in Lasers, Vol. 1, (A. K. Levine, ed.), M. Dekker, New York, p. 181 (1966).

36. A. Heller and A. Lempicki, Appl. Phys. Lett. 9, 198 (1966).
37. S. Sato and Y. Itasaka, Jap. J. Appl. Phys. 18, 1295 (1979).
38. S. Sato, Jap. J. Appl. Phys. 18, 1399 (1979).
39. R. Pappalardo, H. Samelson and A. Lempicki, J. Appl. Phys. 43, 3776 (1972).
40. R. Pappalardo, H. Samelson and A. Lempicki, IEEE J. Quantum Electronics 6, 716 (1970).
41. K. H. Drexhage, in Dye Lasers, (F. P. Schäfer, ed.), Springer Verlag, Berlin, p. 144 (1973).
42. A. H. Maki and J. A. Zuclich, in Topics in Current Chemistry, Vol. 54, Springer Verlag, Berlin, p. 115 (1975).
43. B. K. Fung and L. Stryer, Biochem. 17, 24, 5241 (1978).
44. J. R. Excabi-Perez, F. Nome, and J. H. Fendler, J. Amer. Chem. Soc. 99, 24, 7749 (1977).

APPENDIX 1

We have by definition that:

$$(E - K)|\Psi\rangle = V|\Psi\rangle \quad , \tag{A.1.1}$$

$$(K - E_o)|\Psi_o\rangle = 0 \quad , \tag{A.1.2}$$

$$|\Psi\rangle = |\Psi_o\rangle + |\phi\rangle \quad , \tag{A.1.3}$$

$$\langle\Psi_o|\phi\rangle = 0 \quad . \tag{A.1.4}$$

We want to express $|\phi\rangle$ by means of $|\Psi\rangle$, so as to obtain $|\Psi\rangle$ simply as a function of $|\Psi\rangle$. We operate on the left in (A.1) with the projection operator $P = 1 - |\Psi_o\rangle\langle\Psi_o|$, and take advantage of its commuting with K. By definition P^o is orthogonal to Ψ_o. The result is:

$$(E - K)|\phi\rangle = P\, V|\Psi\rangle \quad . \tag{A.1.5}$$

If we noe operate on the left in (A.1) with $M = |\Psi_o\rangle\langle\Psi_o|$ we obtain:

$$|\Psi_o\rangle(E - E_o)\langle\Psi_o|\Psi\rangle = |\Psi_o\rangle\langle\Psi_o|V|\Psi\rangle \quad . \tag{A.1.6}$$

Hence

$$E - E_o = \langle\Psi_o|V|\Psi\rangle \quad . \tag{A.1.7}$$

The expression for $|\Psi\rangle$ becomes:

$$|\Psi\rangle = |\Psi_o\rangle + (E - K)^{-1} PV|\Psi\rangle \quad , \tag{A.1.8}$$

where in general $(E - K)^{-1}|\Theta\rangle$ is expressed in a basis of eigenstates $|\psi_n^o\rangle$ of K as:

$$\sum_n a_n \frac{1}{E - E_n^o} |\Psi_n^o\rangle \quad . \tag{A.1.9}$$

By iteration we have the Brillouin-Wigner series:

$$|\Psi\rangle = |\Psi_o\rangle + (E-K)^{-1}PV(|\Psi_o\rangle + (E-K)^{-1}PV[|\Psi_o\rangle + (E-K)^{-1}PV\{\cdots\}]). \tag{A.1.10}$$

APPENDIX 2

Starting from the expression for the spin-Hamiltonian:

$$\mathcal{H}_{ss} = \Omega_{xx}s_{1x}s_{2x} + \Omega_{yy}s_{1y}s_{2y} + \Omega_{zz}s_{1z}s_{2z} \, , \tag{A.2.1}$$

we consider the matrix elements of $\Omega_{xx} + \Omega_{yy} + \Omega_{zz}$:

$$\langle\Phi_k|g^2\beta^2(r_{12}^2 - 3x_{12}^2 + r_{12}^2 - 3y_{12}^2 + r_{12}^2 - 3z_{12}^2)|\Phi_k\rangle \, , \tag{A.2.2}$$

which are identically zero, since

$$r_{12}^2 = x_{12}^2 + y_{12}^2 + z_{12}^2 \quad .$$

We redefine Ω_{pp} so as to multiply both the x and y components of the one-electron spin-operator by the same factor. This is made possible by the transformation:

$$\Omega_{xx} = \frac{1}{2}(\Omega_{xx} - \Omega_{yy}) - \frac{1}{2}\Omega_{zz} \quad , \tag{A.2.3}$$

$$\Omega_{yy} = -\frac{1}{2}(\Omega_{xx} - \Omega_{yy}) - \frac{1}{2}\Omega_{zz} \quad , \tag{A.2.4}$$

with the result that the Hamiltonian operator becomes:

$$\mathcal{H}_{ss} \equiv \mathcal{H}_S = \frac{1}{2}(\Omega_{xx} - \Omega_{yy})(s_{1x}s_{2x} - s_{1y}s_{2y}) + \frac{1}{2}\Omega_{zz}(3s_{1z}s_{2z} - \underline{s}_1 \cdot \underline{s}_2).$$

We now want to introduce the spin-operators for the state, namely $\underline{S} = \underline{s}_1 + \underline{s}_2$ and $S_p = s_{1p} + s_{2p}$. On the basis of the relations

$$s_{ip}^2 |\alpha\rangle = \frac{1}{4}|\alpha\rangle \qquad s_{ip}^2 |\beta\rangle = \frac{1}{4}|\beta\rangle \quad , \tag{A.2.5}$$

and

$$s_{1p}s_{2p} = \frac{1}{2}S_p^2 - \frac{1}{4} \qquad \underline{s}_1 \cdot \underline{s}_2 = \frac{1}{2}\underline{S}^2 - \frac{3}{4} \quad , \tag{A.2.6}$$

we derive

$$\mathcal{H}_{ss} = \frac{1}{2}(\Omega_{xx} - \Omega_{yy})(\frac{1}{2}S_x^2 - \frac{1}{4} - \frac{1}{2}S_y^2 + \frac{1}{4}) + \frac{1}{2}\Omega_{zz}(3\frac{1}{2}S_z^2 - 3\frac{1}{4} - \frac{1}{2}\underline{S}^2 + 3\frac{1}{4})$$

$$= \frac{1}{4}(\Omega_{xx} - \Omega_{yy})(S_x^2 - S_y^2) + 3\frac{1}{4}\Omega_{zz}(S_z^2 - \frac{1}{3}\underline{S}^2) \quad . \tag{A.2.7}$$

Taking the matrix elements of the operator $\mathcal{H}_S$ between the states Φ_k, we introduce the ZFS parameters D and E as:

$$D = \langle\Phi_k \left| \frac{3g^2\beta^2}{4} \frac{r_{12}^2 - 3z_{12}^2}{r_{12}^5} \right| \Phi_k\rangle$$

$$E = \langle\Phi_k \left| \frac{3g^2\beta^2}{4} \frac{y_{12}^2 - x_{12}^2}{r_{12}^5} \right| \Phi_k\rangle \quad . \tag{A.2.8}$$

MULTIPHONON PROCESSES, CROSS-RELAXATION AND UP-CONVERSION IN ION-ACTIVATED SOLIDS, EXEMPLIFIED BY MINILASER MATERIALS

F. Auzel

Centre National d'Etudes des Télécommunications
196, Rue de Paris 92220 BAGNEUX (France)

ABSTRACT

In the first part, the basic concentration-independent processes such as multiphonon decay and sidebands are discussed, then the concentration-dependent effects of energy transfer, cross-relaxation, and up-conversion are considered both from a microscopic and macroscopic point of view in an attempt to present a unified view of the energy transfer problem in insulating solids.

In the second part, these processes are exemplified by the important role they play as radiationless processes governing the behavior of the standard as well as the newly considered stoichiometric (self-activated) laser materials.

I. INTRODUCTION

Most of the impetus given to the understanding of the processes involved in ion-activated solids for the last fifteen years are due to the development of new fluorescent light sources, either coherent (laser) or incoherent (fluorescent lamp), and the need to improve them. Such sources must first of all have a good efficiency, regardless of all other specifications. Usual basic limitations to a high quantum efficiency are by definition the different nonradiative processes which are precisely the focus of this fourth school in Erice.

Until recently one had to consider ions as dopants added as activators into a host solid; now one also considers solids with a high concentration of activators ($>10^{21}$ cm^{-3}), even to the point of having them as constituents of the solid. This evolution can be traced back to the need for micrometric-size light sources such as, for example, minilasers for optical communications, displays with up-converting phosphors, or fluorescent inks for marking recognition. In this last case the useful dimension of the light source is only the thickness of the inked print! For such applications one is involved with the so-called stoichiometric or self-activated materials in which interactions between activators usually provide new paths for nonradiative processes to take place. High concentration may also be involved when sensitization of an activator has to be considered.

In this article, I shall first discuss the basic phenomena involved in nonradiative processes, beginning with the concentration-independent ones, such as multiphonon decay, then going on to the concentration-dependent ones such as energy transfer, cross-relaxation, and up-conversion. Secondly, in order to illustrate these processes I shall give examples taken from current research in stoichiometric minilaser materials.

II. CONCENTRATION-INDEPENDENT PROCESSES OR ONE-CENTER PROCESSES

I shall deal with the nonradiative and radiative processes in rare-earth ions. They can usually be considered in the first approximation as independent of the interactions between activator ions, that is, as one-center processes. However in some peculiar cases they could also be concentration-dependent, as I shall point out in the discussion.

II.A. Multiphonon Nonradiative Probabilities

Once an ion has been excited (the way in which it has been excited does not matter), it can lose energy nonradiatively in making a transition from the excited level to the one just below it. Experimentally it is found that the quantum efficiency of an excited state can be lower than expected from the one-phonon interaction for a given energy gap, even at low concentration (that is, without any possibility for energy transfer to sinks to take place). A good description of this situation is given by the relation between quantum yields of levels and their difference in energy from the next lower level (1) (2). Monochromatic excitation has shown (3) that energy decay was effectively proceeding as in Fig. 1., by cascade. This comes from the fact that the nonradiative transition probability (W_{NR}) is then well described with respect to the energy difference ΔE between two consecutive levels by an exponential law:

$$W_{NR}(\Delta E) = W_{NR}(0)\ \exp - \alpha_{NR}\Delta E.$$

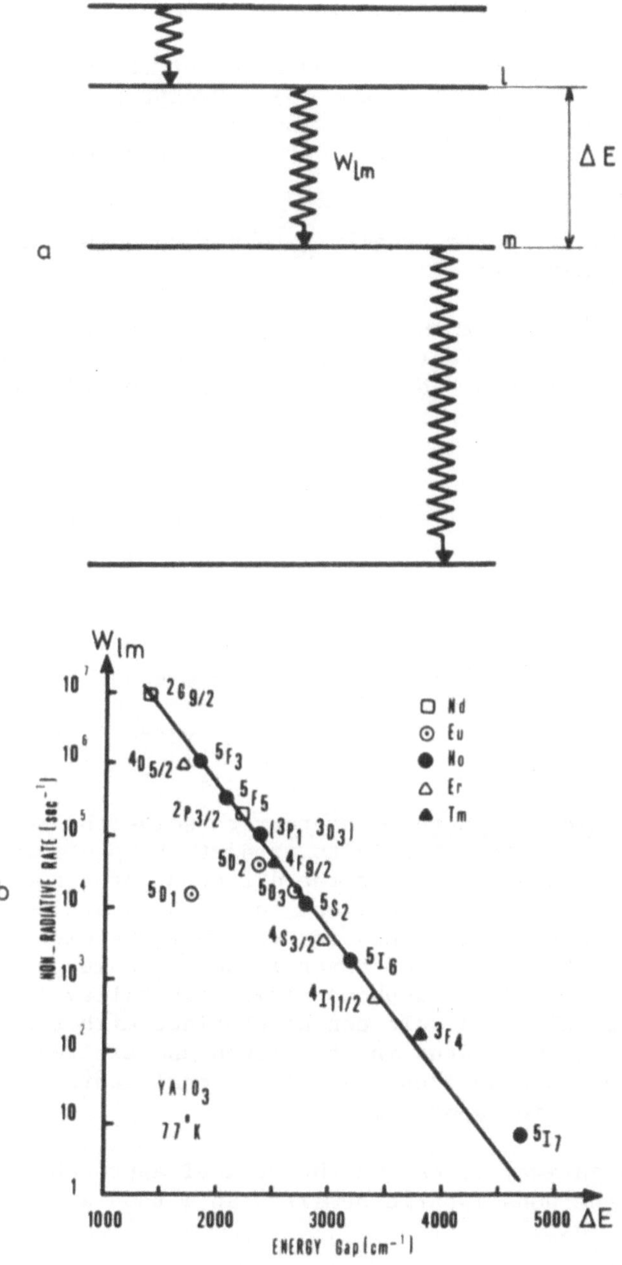

Fig. 1. Multiphonon nonradiative decay: a) a sequential "ladder"-type process ; b) an example of the experimental law (from M.J. Weber, Phys. Rev. 138, 54 (1973)).

Here, ΔE is larger than the highest phonon energy of the matrix, which is the condition for a multiphonon process to be considered.

Such exponential laws are found also for molecules (4) and for deep centers in semiconductors (5). Except in a few cases where ΔE is of the order of the highest vibrational energy, no selection rule is found with respect to the set of quantum numbers of the levels.

Experimentally W_{NR} is usually obtained through one of the three following methods:

i) If W_{NR} is larger than W_R, the radiative transition probability, then a direct measurement of W_{NR} is obtained by a lifetime measurement for the level under consideration.

$$W_{NR} \simeq 1/\tau \quad . \tag{1}$$

ii) When W_{NR} is smaller than W_R, then W_R is first estimated from an absorption measurement and W_{NR} is obtained by:

$$W_{NR} = \frac{1}{\tau} - W_R \quad . \tag{2}$$

iii) Assuming the validity of the rate-equation model, which shall be discussed in section III.A, one can, by measuring intensity ratios solve the resulting system of equations for W_{NR}.

1) Theoretical Aspects. In order to solve the N-phonon nonradiative decay problem, that is to predict the probability of such a process, one needs to consider the different processes which allow an N-phonon jump. If wave functions for phonons are considered as pure harmonic states, being the same whatever the electronic state of the impurity, then jumps of several quantum numbers are forbidden in the first order and the N-phonon jump probability is zero. Removal of this selection rule can be obtained either by considering different vibrational states in the ground and excited electronic states or by considering identical vibrational states but in an Nth-order perturbation approximation.

This straightaway gives two theoretical approaches called, respectively, the nonadiabatic Hamiltonian method and the Nth-order method (6).

2) The Nonadiabatic Hamiltonian Method ($S_o \neq 0$). It was first pointed out by Frenkel (7) that if vibrational states are allowed to differ with the considered electronic state either by normal coordinates or frequency, changes by more than one quantum number

can be obtained in a first-order perturbation leading to an N-phonon process. One should note that the same effect could be also obtained by anharmonic effects in the vibrations, which are usually neglected.

The following standard approximations are accepted in the literature in order to proceed far enough with the calculations so as to obtain formulas comparable with experimental results:

- Applicability of the "Fermi golden rule" for the calculation of the transition probability per unit time. This approximation is valid as long as first-order time-dependent perturbation can be used, that is for small electron-lattice interactions. For strong interactions and when the two considered adiabatic electronic states are nearly degenerate, a more general method using a correlation-function approach should be used (8) (9).

- Use of true adiabatic wave functions as given by the Born-Oppenheimer approximation as opposed to crude Born-Oppenheimer (CBO) functions.

- "Condon approximation." This approximation is based on the assumption that adiabatic electronic wave functions exhibit a weak dependence on the nuclear coordinates. Then one can factor the relevant matrix elements into an electronic part and a vibrational part. Since we shall be mainly concerned with relative rather than absolute decay rates, we are somewhat justified in using this approximation which has received the favour of many papers. It is more precise than the C.B.O. approximation since some averaging is done in computing the electronic matrix elements (10).

- Vibrational states between ground and excited electronic states differ only by their normal coordinates and not by their frequency. This corresponds to the linear approximation for the interaction energy which is correct if this interaction is weak. In the linear approximation only the equilibrium position for the adiabatic electronic state is changed, not the vibrational frequencies. This insures also harmonicity of the vibration, that is vibrations are normal modes, described by harmonic oscillator wave functions.

- Interaction with only one phonon mode. This approximation is usually necessary at the end of the calculation to obtain some usable results. If such an approximation is made at the beginning, a one-dimensional model is valid as in the case of vibronic absorption (11).

With such approximations, the transition probability for a nonradiative N-phonon process between two electronic states i and f is found by Fermi's golden rule to be:

$$W(i\to f) = \frac{2\pi}{\hbar} \sum_{v'v''} P_{v'} \sum_{s} R_s(f,i)^2 \left| < X_{fv''_s} \left| \frac{\partial}{\partial Q_s} \right| X_{iv'_s} > \right|^2$$

$$\times \prod_{j\neq s} \left| < X_{fv''_j} \left| X_{iv'_j} \right> \right|^2 \delta(E_{fv''} - E_{iv'}). \qquad (3)$$

(These symbols are similar to the ones used in B. Di Bartolo's article in this volume.) The so-called "promoting" term:

$\left| <fv''_s \left| \frac{\partial}{\partial Q} \right| iv'_s> \right|$ comes from the nonadiabatic Hamiltonian and corresponds to the usually neglected term in the Born-Oppenheimer approximation.

The term:

$<X_{fv''_j} \left| X_{iv'_j} \right>$ is the usual Franck-Condon factor corresponding to the so-called "accepting" modes.

From this point on, the problem lies in handling the summation over v' and v" for the density of states weighted by $P_{v'}$. The summation has been first obtained by Huang and Rhys (12) as applies to F-centers by series expansion of the harmonic-oscillator wave functions with displaced coordinates, use of a recursion formula between integrals, and careful combinative analysis of useful terms for an N-phonon process.

Their result is the following (12):

$$W_N(i\to f) = \frac{h^2}{\omega_m}\Big\{ \frac{1}{2} Z^2 \left[\bar{n} R_{N+1} + (\bar{n}+1) R_{N-1}\right] + |Y|^2 \left[(\bar{n}+\tfrac{1}{2})^2 + \tfrac{1}{2}\bar{n}(\bar{n}+1)\right] R_N - |Y|^2 (\bar{n}+\tfrac{1}{2}) \times \left[\bar{n} R_{N+1} + (\bar{n}+1) R_{N-1}\right] + \frac{1}{4}|Y|^2 \left[\bar{n}^2 R_{N+2} + (\bar{n}+1)^2 R_{N-2}\right]\Big\} \delta(E_f - E_i), \qquad (4)$$

where

$$R_N = \exp\left[-(2\bar{n}+1) S_o\right] \left(\frac{\bar{n}+1}{\bar{n}}\right)^{\frac{1}{2}N} I_N\left[2S_o\sqrt{\bar{n}(\bar{n}+1)}\right] \qquad (5)$$

$$Z^2 = \left(\frac{\omega_m}{\hbar}\right) \sum_s R_s^2(f,i) \tag{6}$$

$$Y = \left(\frac{1}{\hbar\omega_m}\right) \sum_s R_s(f,i)\ (A''_s - A'_s). \tag{7}$$

$$S_o = \frac{\hbar}{2\omega_m^3} \sum_s (A''_s - A'_s)^2 , \tag{8}$$

in which ω_m is the frequency of the mode s and has been assumed to be the same for each s and to be constant whatever the electronic state is. I_N is the Bessel function of order N with imaginary argument.

The occurrence of orders N+1, N-1, N, N+1, N+2, gives rise to useful recurrences which can be used for effective computations. (See W.H. Fonger's article in this volume.)

Another method in handling the summations over v' and v'' is the "generating function method" of Lax-Kubo-Toyosawa which gives essentially the same results, with the same approximations (13), (14). This last method is more general and does not rely upon expansion techniques for harmonic-oscillator wave functions. Lax (13) introduced the method by using eq. (3) and the integral definition (15) for the delta function which contains the energy explicitly as a variable:

$$\delta(E_{iv'} - E_{fv''} - E) = \frac{1}{2\pi\hbar} \int_{-\infty}^{+\infty} \exp\left[i(E_{iv'} - E_{fv''} - E)\frac{t}{\hbar}\right] dt. \tag{9}$$

Then eq. (3) can be written in the form

$$W_{i\to f}^{(E)} = \frac{1}{\hbar^2} \int_{-\infty}^{+\infty} G(t) \exp\left[-\frac{iEt}{\hbar}\right] dt , \tag{10}$$

which is the definition for the inverse Fourier transform of $W_{i\to f}^{(E)}$ (14). G(t) has been called the generating function of $W_{i\to f}$ by analogy with moment - generating functions from distribution theory in statistics.

Let F(x) be a distribution function for a random variable x . Its moment-generating function is defined as :

$$M(t) = \int_{-\infty}^{+\infty} e^{tx}\ dF(x).$$

The moments of order i at the origin being defined as:

$$\alpha_i = \int_{-\infty}^{+\infty} x^i \, dF(x),$$

are generated by M(t) by :

$$\alpha_i = \left(\frac{d^i M}{dt^i}\right)_{t=o} .$$

The different moments i = 0,... ∞ uniquely define the distribution F(x). So a general advantage of the generating function is to obtain directly the moments of the line-shape function for $W_{i\to f}$, for example, without needing to take the inverse Fourier transform of G(t).

But in the case of small coupling, as we shall see later, the line shape is so asymmetrical (it is the exponential gap law) that its moments are not usually considered and the inverse Fourier transform integral of eq. (10) can be estimated by expanding G(t) in a series in which only the N-phonon term is retained and found by simple Fourier inversion of single-frequency components, giving delta functions for each N (16). The end result of Miyakawa and Dexter by this method is for 0 K :

$$W_N(i\to f) = \left(\frac{2\pi}{\hbar}\right) R^2 \left(1 - \frac{N}{S_o}\right)^2 \left(\frac{S_o^N}{N!}\right) \delta(\Delta E - N\hbar\omega_m), \tag{11}$$

in which R is equivalent to the matrix element Y of Huang and Rhys (eq. (7)) and ΔE is the energy gap.

By using the Stirling formula ($N! \simeq \sqrt{2\pi N}\,(N/e)^N$) eq. (11) is transformed approximately into an exponential law

$$W_N(i\to f) = \left(\frac{2\pi}{\hbar}\right) R^2 \left(1 - \frac{N}{S_o}\right)^2 \exp(-\alpha\Delta E)\, \delta(\Delta E - N\hbar\omega_m), \tag{12}$$

with the exponential parameter given by :

$$\alpha = (\hbar\omega_m)^{-1} \left(\ln (N/S_o) - 1\right). \tag{13}$$

To find the inverse of G(t), another method can be used which involves a "steepest descent" approximation (17) (18) (19) for the integration in the inverse Fourier transform, which is given by

$$\int_{-\infty}^{+\infty} \exp\left[f(t)\right] dt \simeq \exp\left[f(t_s)\right] \times \left[\frac{-2\pi}{f''(t_s)}\right]^{1/2} , \tag{14}$$

with t_s defined by $f'(t_s) = 0$ (the "saddle point"). The exponential law arises directly from the term $\exp\left[f(t_s)\right]$, explicitly found to be (17) (18) (19):

$$\exp\left\{-\Delta E/\hbar\omega_m \left[\ln(N/S_o) - 1\right]\right\}, \tag{15}$$

which is the same result as that obtained by Stirling's approximation.

There is a good reason for this: eq. (14) corresponds to Laplace's method for obtaining an asymptotic expansion of functions defined by integrals (20). This method states that the major contribution to the value of the integral occurs in the vicinity of the point where $f(t)$ is maximum ($f'(t_s) = 0$).

Expanding $f(t)$ around $t = ts$, one finds (for $\alpha < t < \beta$) :

$$\int_\alpha^\beta \exp\left[f(t)\right] dt \simeq \int_{ts}^{t_s+\varepsilon} d\varepsilon \left\{\exp\left[f(t_s)\right] x \exp\left[\varepsilon f''(t_s)\right] x \exp\left[\frac{\varepsilon^2}{2} f''(t_s)\right]\right\};$$

as $\varepsilon \to o$:

$$\int_\alpha^\beta \exp\left[f(t)\right] dt \simeq \exp\left[f(t_s)\right] \int_{t_s}^{t_s+\varepsilon} \exp\left[\frac{\varepsilon^2}{2} f''(t_s)\right] d\varepsilon .$$

Extending the integration to $-\infty$ and $+\infty$ yields :

$$\int_{-\infty}^{+\infty} \exp\left[f(t)\right] dt \simeq 2 \exp\left[f(t_s)\right] \left[\frac{-\pi}{2f''(t_s)}\right]^{1/2} \tag{16}$$

in which the well-known value of the integral of a gaussian has been used. Eq. (16) is just found to be eq. 14. Now let us apply eq. 14 to the following definition of a factorial (21):

$$\int_o^\infty t^n e^{-t} dt = n! .$$

The saddle point is given by :

$$f'(t) \equiv \frac{n}{t} - 1 = o \quad \text{and} \quad t_s = n$$

$$f''(t) \equiv -\frac{n}{t^2}$$

$$\text{giving : } \int_0^\infty t^n e^{-t} dt \simeq \left[n^n e^{-n}\right] \left[\frac{-\pi n}{-2}\right]^{1/2} .$$

Therefore: $n! \simeq (2\pi n)^{\frac{1}{2}}(n^n e^{-n})$, which is just Stirling's formula. In others words one cannot claim as in Ref. (19) that the saddle-point method is more precise than the use of Stirling's formula. Mathematically the same approximation is used in both cases, and the equivalence of the exponential gap law is not fortuitous. More details on the "saddle point" method will be found in Dr. R. Englman's article in this volume.

The ($N\mp1$) and ($N\mp2$) phonon terms are found in Kubo and Toyosawa (14), Diestler (19), and Miyakawa and Dexter (16) where it is clearly shown that the generating function is in fact the product of two generating functions, one giving rise to a two-phonon process mostly, and the other an N-phonon process (14).

A more general advantage of the use of generating functions is that even in more general cases and without approximations, the different summations which arise, as in eq. (3) for instance, can be presented in a closed form inolving traces of products of density matrix elements (13) (14) (16).

The generating function should not be confused with the generator function (22) sometimes used in the same instance (23) to generate the Hermite polynomials involved in the harmonic-oscillator wavefunctions.

a. <u>Discussion of the Results</u>. In order to make comparison easier we shall first discuss only the low temperature case ($\bar{n} \ll 1$). From eq. (4) we get, neglecting constants,

$$W_N(i\rightarrow f) \propto R_N \left(\frac{3}{4} Y^2\right) + R_{N-1} \left(\frac{Z^2-Y^2}{Y}\right) + R_{N-2} \left(\frac{Y^2}{4}\right) . \qquad (17)$$

This equation shows an interesting feature: the N-phonon process involves also (N-1) and (N-2) phonon processes simultaneously. This feature though neglected by Huang and Rhys in their own result is general and is due to the action of the promoting modes. It can be physically viewed in the following manner:

At low temperature, in which case energy cannot be absorbed from the lattice, in order to breakdown the adiabaticity, the promoting modes take their energy from the only available energy, the energy gap. This reduces the number of phonons involved by at most two; the action of the promoting and accepting mode are simultaneous.

The Bessel function in eq. (5) for the N-phonon term can be approximated for two separate cases (14), $S_o > 5$ and $S_o < 5$ (16).

- Small coupling ($S_o < 5$), low temperature. ($x << N$ and $N > 0$). Writing for simplification :

$$x = 2 S_o \sqrt{\bar{n}(\bar{n}+1)} ,$$

one has

$$I_N(x) \simeq \frac{(x/2)^N}{N!} . \tag{18}$$

- Strong coupling ($S_o > 5$), high temperature. ($x >> |N|$ and $N^2 >> 1$) :

$$I_N(x) = \left(\frac{2\pi}{N}\right)^{1/2} \exp\,(x - N^2/2x) . \tag{19}$$

Since we are interested only in the small coupling case we shall consider the first limiting case only.

From it, one has a simple recursion between the N-phonon, (N-1)-phonon and (N-2)-phonon terms in eq. (4) at low temperature :

$$R_N \simeq \exp\left(- S_o\right)\bar{n}^{-N/2}(S_o\bar{n}^{1/2})^N/N! = \exp\left(- S_o\right)S_o^N/N! \tag{20}$$

This form of R_N is the so-called "Pekarian" function (24) and

$$R_{N-1} = R_N\, N/S_o$$
$$R_{N-2} = R_N\, N(N-1)S_o^2, \tag{21}$$

from which eq. (17) is written

$$W_{i\to f} \propto Y^2 \left(1 - \frac{4}{3}\frac{N}{S_o} + \frac{N}{S_o^2} + \frac{N^2}{3S_o^2}\right)\exp - S_o . S_o^N/N! \tag{22}$$

In the experimental case we shall further consider (Ln^{3+} ions), one has usually (see Ref. 6) $S_o \simeq 0.05$ and also N is a number between 2 and 10, so that one can write

$$W_{i\to f} \propto N^2/S_o^2 \;.\; S_o^N/N! \tag{23}$$

In this approximation (small coupling, multiphonon, low temperature) the results of Huang and Rhys are equivalent to the one of Miyakawa and Dexter as given by eq. (11) in which

$$\left(1 - \frac{N}{S_o}\right)^2 \simeq \left(\frac{N}{S_o}\right)^2 , \tag{24}$$

for $N > S_o$.

Workers who have either considered promoting matrix elements as constants (17) or have not considered them (5) (22) will find, of course, only the N-phonon term. That is, they do not find a preexponential term of order $(N/S_o)^2$ equivalent to the one in eq. (12).

To summarize, we can describe the situation by the following:

The nonradiative transition probability is found to be the product of two terms: one, due to the "promoting modes," is at most a two-phonon term, which in considering a N-phonon process gives rise to an (N-2) phonon term; the second, due to the "accepting modes" and which are in the form of Franck-Condon integrals, gives rise to an N-phonon term. In the literature two approaches are found:

i) Matrix elements for promoting modes are assumed constant or ignored. This is equivalent to considering only the Franck-Condon integrals and to considering only the N-phonon term leading to the α parameter given in eq. (5); (12) (17) (22).

ii) Matrix elements for promoting modes are included in a preexponential function, this gives an exponential parameter identical to i). (8), (16), (19).

(Note: By N-phonon processes we understand as in Huang and Rhys that N is the net number of created phonons.)

b. <u>The Temperature Dependence</u>. If we consider eq. (4), as given by Huang and Rhys, the temperature dependence is very complicated since it is contained in the phonon occupancy number,

$$\bar{n} = (\exp(\hbar\omega_m/kT) - 1)^{-1} ,$$

which arises several times explicitly in eq. (4) and also implicitly in eq. (5) in the form

$$I_N\left[2\, S_o\sqrt{\bar{n}(\bar{n}+1)}\right] .$$

First we look at the temperature dependence for the N-phonon term R_N using the series definition for I_N :

$$R_N = e^{-(2\bar{n}+1)S_o}\,\frac{(\bar{n}+1)^{N/2}}{\bar{n}^{-N/2}}\sum_{s=o}^{\infty}\frac{\left[S_o^2\,\bar{n}(\bar{n}+1)\right]^{s+N/2}}{s!\,(N+s)!} \qquad (25)$$

$$= e^{-(2\bar{n})S_o} \frac{S_o^N e^{-S_o}}{N!} (\bar{n}+1)^N \sum_{s=o}^{\infty} \frac{N!\, S_o^{2s} \bar{n}^s (\bar{n}+1)^s}{s!\,(N+s)!}$$

$$= e^{-S_o} \frac{S_o^N}{N!} (\bar{n}+1)^N e^{-2\bar{n}S_o} \left[1 + \frac{S_o^2\, \bar{n}(\bar{n}+1)}{(N+1)} + \ldots \right] . \quad (26)$$

For $S_o < 1$, and not too high a temperature, we can retain only the first term; that is, for a "Pekarian function" at 0°K we associate a temperature dependence of the form

$$(\bar{n} + 1)^N e^{-2\bar{n}S_o} \quad .$$

As a further simplification, we may neglect the variation with temperature of the term $(N/S)^2$.

This can be justified by the fact that in eq. (4) for $\bar{n} > 1$, the term which is retained is of the order of a term independent of temperature:

$$\frac{\bar{n}^2}{(\bar{n}+1)^2} \left(\frac{N}{S_o^2}\right)^2 \simeq \left(\frac{N}{S_o}\right)^2 ,$$

so that we have finally :

$$W_{i\to f} \propto \left(\frac{N}{S_o}\right)^2 \frac{e^{-S_o}}{N!} (\bar{n}+1)^N e^{-2\bar{n}S_o} \quad . \quad (27)$$

The last two factors give the variation with temperature proposed by Fong et al. (25). In fact for small values of S_o, $\exp{-2S_o\bar{n}}$ can be neglected ($\simeq$ 4% for LaF_3 at 400°K) and one is left with the usual dependence $(\bar{n}+1)^N$ as experimentally verified, for instance, by Riseberg and Moos (26).

3) <u>The Nth-order Method (S_o = o)</u>. Instead of considering different vibrational states in the ground and excited electronic state of the impurity, let us now assume that they are identical. Then S_o = o and a multiphonon transition is forbidden in first order because transitions between harmonic-oscillator states are allowed only one at a time. An N-phonon transition is only obtained now through an m-order perturbation method due to a ℓ-order variation in the static electric field at the impurity with N = m+ℓ (6) (27). This situation has been investigated by Kiel for the cases of Cr^{3+} and Pr^{3+} (28), following an approach originated by Orbach for spin-lattice relaxation problems.

Instead of considering only the static field in the Hamiltonian of the considered impurity, one considers its Taylor series expansion about the equilibrium ion position. That is :

$$H_{\text{crystal field}} = H_o + \sum_i \frac{\partial V}{\partial Q_i} Q_i + \frac{1}{2} \sum_{i,j} \frac{\partial}{\partial Q_i} \frac{\partial V}{\partial Q_j} Q_i Q_j$$

$$+ \dots \frac{1}{n!} \sum_{i..n} \frac{\partial}{\partial Q_i} \dots \frac{\partial V}{\partial Q_n} \times Q_i \dots Q_n , \qquad (28)$$

Q_i being the ith normal mode coordinate. H_o is the static field Hamiltonian which for rare earth ions can break down the 4f-4f parity selection rule. One can then understand how vibrations modulating the crystal field V can be coupled to the electronic transitions.

Applying time-dependent perturbation theory, an N-phonon process linking two electronic levels can be obtained by taking contributions from all terms of total order N.

Initial and final wave functions of the whole system are considered to be Crude Born-Oppenheimer (CBO) states :

$$|a\rangle \prod_i |n_i\rangle \text{ and } |b\rangle |n_i+1\rangle |n_k+1\rangle \dots \Pi' |n_i\rangle$$

this assumes adiabaticity since electronic wave functions $|a\rangle$ and $|b\rangle$ and vibrational wave functions $|n_i\rangle$ are considered to be independant. This is clearly different from the nonadiabatic case of section II A, where electronic states depend parametrically on lattice coordinates giving real Born-Oppenheimer states and $S_o \neq o$. The probability per unit time for transition between two electronic states $|a\rangle$ and $|b\rangle$ by a N-phonon process is (27) :

$$W(N) = \frac{2\pi}{\hbar} \sum_{\underbrace{i..j}_{N},m_1..m_{N-1}} \Bigg[|\langle n_i+1|Q_i|n_i\rangle|^2 \dots |\langle n_j+1|Q_j|n_j\rangle|^2$$

$$\times \frac{|\langle b| \frac{\partial V}{\partial Q_i} |m_{N-1}\rangle|^2 \dots |\langle m_1| \frac{\partial V}{\partial Q_j} |a\rangle|^2}{(E_{m_{N-1}} + \hbar\omega_i + \dots + \hbar\omega_{j-1} - E_a)^2 \dots (E_{m_1} + \hbar\omega_i - E_a)^2}$$

$$\times \quad g(\omega_i) \dots g(\omega_j) \, \delta\,(E_b - E_a + \hbar\omega_i \dots \hbar\omega_j) \Bigg]$$

$$+ \ldots + \frac{2\pi}{\hbar} \sum_{\substack{i\ldots j \\ N}} \left[\left(\frac{1}{(N-1)!}\right)^2 \left| \langle b| \frac{\partial}{\partial Q_i} \ldots \frac{\partial}{\partial Q_j} |a\rangle \right|^2 \right.$$

$$\times \; |\langle n_i+1|Q_i|n_i\rangle|^2 \ldots |\langle n_j+1|Q_j|n_j\rangle|^2$$

$$\left. \times \; g(\omega_i) \ldots g(\omega_j)\; \delta\,(E_b - E_a + \hbar\omega_i \ldots \hbar\omega_j) \right] , \qquad (29)$$

where $|m_1\rangle \ldots |m_{N-1}\rangle$ are electronic intermediate states, and $g(\omega_i) \ldots g(\omega_j)$ are the phonon densities of states for $\omega_i \ldots \omega_j$. Since intermediate terms are generated by all other possible cross-products, eq. (29) cannot be put in closed form due to the difficulties in summing over numerous intermediate states and all relevant lattice modes.

To go further in the calculations, one can either consider only the Nth-order perturbation with first-order field or the first-order perturbation with N^{th}-order field (28) (29). However, one can approximate a solution by considering the ratio of two consecutive terms. In both cases, one finds that if $\frac{\partial v}{\partial Q}$ is small then

$$\frac{W(N)}{W(N-1)} = Y \quad , \qquad (30)$$

with Y nearly a constant as small as 0.05 (28). This is the central result of the Nth-order method. One then considers the iterative properties of eq. (30), which shows that we are considering a series in which each succeeding term is smaller by the characteristic constant number Y. Calculating W(N) from W(0) yields:

$$W(N) = W(0)\; Y^N = W(0) \quad e^{N \log Y} \qquad (31)$$

If we replace N by the ratio of the energy gap ΔE to some single effective phonon energy $\hbar\omega$:

$$W(N) = W(0)\; \exp{-\alpha\, \Delta E} \quad , \qquad (32)$$

with

$$\alpha = \frac{1}{\hbar\omega} \; \log\,(1/Y) \quad , \qquad (33)$$

in which W(0) and Y are characteristic of the matrix and of its interaction with the considered impurity.

Basically this is the phenomenological development that has also been used by McGill (30) for multiphonon absorption of pure crystals and therefore for a different physical process.

In view of this, a specific form of the theory leading to an exponential behaviour does not mean that the associated physical mechanism is the one responsible for the multiphonon process considered. As long as an iterative method is involved, the only test for any theory would be to predict values of Y, as given by experiments, from "ab initio" calculations. This has yet to be performed.

In comparison, the nonadiabatic Hamiltonian method bears more physical significance than the Nth-order method.

As for the temperature dependence, it is derived in the same way for both N^{th} order methods.

From eq. (29), the temperature dependence occurs through the products of normal mode matrix elements for one-phonon processes:

$$\underbrace{|\langle n_i+1|Q_i|n_i\rangle|^2 \ \dots \ |\langle n_j+1|Q_j|n_j\rangle|^2}_{N \text{ factors}} \ . \tag{34}$$

Using the phonon creation a_i^+ and annihilation a_i^- operators defined by: (31)

$$\begin{aligned} a_i^+|n_i\rangle &= \sqrt{n_i+1}\ |n_i+1\rangle \quad \text{and} \\ a_i^-|n_i\rangle & \quad \sqrt{n_i}\ |n_i-1\rangle \ , \end{aligned} \tag{35}$$

we have

$$Q_i = \sqrt{\hbar/2M\omega_i}\ (a_i^+ + a_i^-) \ ,$$

with n_i being the average occupation number, and n_i=(exp $h\omega_i/kT$-1)$^{-1}$. As a result,

$$\langle n_i+1|Q_i|n_i\rangle = \sqrt{\hbar/2\pi\omega_i}\ \ \langle n_i+1|a_i^+ + a_i^-|n_i\rangle = \sqrt{\hbar/2\pi\omega_i}\ \sqrt{n_i+1} \ , \tag{36}$$

so that the product given by eq. (34) is proportional to

$$\underbrace{(n_i+1) \ \dots \ (n_j+1)}_{N \text{ factors}} \ , \tag{37}$$

which is maximum for $\omega_i = \omega_j$ leading to the temperature-dependent factor $(n+1)^N$ experimentally verified (26) (27).

II.B. Radiative Transition Probabilities

Generally, nonradiative transitions are experimentally studied through radiative transitions, as shown for instance by eqs. (1) and (2). So, in order to obtain information on radiationless processes, useful relations for radiative transition probabilities are recalled. The basic physics behind them has been given in B. Di Bartolo's article. Two types of transitions, either in emission or absorption, have to be considered: electronic and vibronic.

1) Electronic Transitions: Judd's Theory. In this case we address ourselves to transitions between two purely electronic states $|i\rangle$ and $|f\rangle$. From Fermi's Golden Rule, the transition probability is proportional to the matrix element for the electric-dipole operator (of odd parity), which is nonzero only if $|i\rangle$ and $|f\rangle$ have opposite parity (Laporte's Rule). In transition metals and lanthanide ions, for which most of the transitions take place within 3d or 4f electronic configurations, the transition is then forbidden in first order. This explains the small oscillator strengths found ($f \simeq 10^{-5}$, 10^{-6}) which are of same order of magnitude as the allowed magnetic-dipole transitions (even-parity operator).

For rare-earth ion absorption intensities, Judd assumes that the observed transitions are due to different parity admixtures in $|i\rangle$ and $|f\rangle$ from odd-parity static crystal-field terms (32). This results in better wave functions for the description of the electronic states:

$$\langle I| = \langle i| - \sum_k \frac{\langle i|v|k\rangle\langle k|}{E(i)-E(k)}$$

$$|F\rangle = |f\rangle - \sum_k \frac{|k\rangle\langle k|v|f\rangle}{E(f)-E(k)} ,$$

$|k\rangle$ being a state of another configuration of opposite parity $(n'\ell')$ to the one of $|i\rangle$ and $|f\rangle$.

Assuming the energy denominators to be approximately equal and constant (ΔE), the oscillator strength of an electronic transition within the configuration f^n is found to be:

$$f = \frac{1}{\chi} \sum_{\lambda=2,4,6} T_\lambda \nu \, |\langle f^n \psi J \| U^\lambda \| f^n \psi' J'\rangle|^2 \tag{38}$$

with $\chi \simeq (n^2+2)^2/9n$ representing a correction for the local field as introduced by Dexter (33). T_λ are parameters involving the odd static crystal field terms and the radial part of the wave functions as shown by the theoretical expression for it :

$$T_\lambda = \frac{\chi 8\pi^2 m}{3h}(2\lambda+1)\sum_{k=1,3,5}(2k+1)\,B_k\,\Xi^2(k,\lambda)/(2J+1) \tag{39}$$

$$B_k = \sum_q (B_q^k)^2/(2k+1)^2$$

where B_q^k are the crystal field parameters :

$$V = \sum_{k,q} B_q^k\ C_q^k$$

and

$$\Xi(k,\lambda) = 2\sum_{n'\ell'}(2\ell+1)(2\ell'+1)(-1)^{\ell+\ell'}\begin{Bmatrix}1 & \lambda & k\\ \ell & \ell' & \ell\end{Bmatrix}\begin{pmatrix}\ell & \ell'\\ o & o & o\end{pmatrix}\begin{pmatrix}\ell' & k & \ell\\ o & o & o\end{pmatrix}$$

$$\times \langle n\ell|r|n'\ell'\rangle\ \langle n\ell|r^k|n'\ell'\rangle/\Delta\bar{E}\ .$$

$\langle|r|\rangle$ and $\langle|r^k|\rangle$ are the radial integrals for one electron.

The (2J+1) denominator in eq. (39) gives an average value assuming all (2J+1) crystal field splitting to have a natural excitation, that is to be equally populated. In some cases this can lead to an error of 50 % for large ground-state splitting (34).

$\langle||U^J||\rangle$ is the reduced matrix element for U^λ the sum over all the electrons of the single-electron tensor $u^{(\lambda)}$ for which :

$$\langle n\ell||u^\lambda||n'\ell'\rangle = \delta(n,n')\ \delta(\ell,\ell')$$

For practical purposes, the T_λ are taken as adjustable parameters which, by a least-squares fitting, describe a set of absorption probabilities between different J levels.

Experimentally, oscillator strengths are pure numbers obtained through integration of absorption coefficients :

$$f = \frac{m\,C_o^2}{\pi e^2\,N\chi}\int \alpha(\sigma)\,d\sigma \tag{40}$$

where N is the number of absorbing centers
C_o the speed of light
$\alpha(\sigma)$ the spectral absorption coefficient.
σ the wave number
χ the local field correction.

It should be noticed that oscillator strengths are by definition characteristic of one independent impurity or center, hence the N denominator.

Through Einstein's relation connecting the spontaneous transition probability (A) and the induced transition probability (B)

$$A_{if} = \frac{8\pi h\nu^3}{c_o^3} B_{if} \qquad \text{and} \qquad g_i B_{if} = g_f B_{fi} \tag{41}$$

with g_i, g_f being level degeneracies, one can obtain the radiative lifetime of a level f, using the knowledge of A_{if}:

$$\tau_{of} = \frac{1}{A_f} = \frac{1}{\sum_i A_{if}} \tag{42}$$

where $\sum_i$ includes all the emission processes originating from level f.

Then the quantum efficiency of level f is simply:

$$\eta = \frac{\tau_f}{\tau_{of}} \quad \text{where } \tau_f \text{ is the measured lifetime of f.}$$

From this the total nonradiative probability is obtained by:

$$W_{NR_f} = \frac{1}{\tau_f} - \frac{1}{\tau_{of}} \quad . \tag{43}$$

The B_{fi} and A_{if} transition probabilities are easily obtained when only two levels are considered, $|i\rangle$ being the ground state; it is much more difficult when several states $|i\rangle$ are involved.

Experimentally it is hard to obtain absorption between excited states and in practice one makes use of rate equations, assuming their applicability. Another useful method (34) is to realize that in eq. (38), T_λ parameters are valid for all other f-f transitions for a given set of hosts and ions whereas only $\langle||U^\lambda||\rangle$ differs for a transition between excited states and can be computed from the orbital part of the wave function describing the levels. The procedure is then the following:

- Measure the absorption of all possible f-f oscillator strengths from the ground state.

- obtain T_λ from these results.

- calculate $\langle||U^\lambda||\rangle$ using intermediate-coupling wave functions.

-from eq. (38) obtain transition strengths between any set of levels and then all A_{if}.

- calculate τ_{of} and compare it with the τ_f experimentally measured.

- obtain the quantum efficiency and the nonradiative transition probability from level $|f>$.

2) Vibronic transitions. When states are not described by pure electronic wave functions but by Born-Oppenheimer states then mixed-nature transitions, partially vibrational and partially electronic (called vibronic), are found in absorption, excitation or emission.

As for nonradiative transitions, vibronic transitions can be separated into two classes, according to the zero or nonzero value of the Huang-Rhys coupling parameter. This gives rise to, respectively the so-called M and Δ processes (35).

(a) The M Process ($S_o = o$). In the preceding chapter, we saw how the odd term of the static electric crystal field induces electronic transitions. In the same manner vibrations can produce odd terms in the dynamic crystal field giving rise to a phonon-forced electric-dipole transition.

As for multiphonon nonradiative N^{th}-order methods, the crystal field is expended in a series:

$$V = \sum_{k,q} \left[B^k_q + \sum_i \frac{\partial B^k_q}{\partial Q_i} Q_i \right] C^{(k)}_q \quad ,$$

where Q_i are the normal-coordinates, and the wave functions under consideration are CBO ones.

Again the contribution to the oscillator strength has the form:

$$f \propto \sum_{\lambda=2,4,6} T'_\lambda \; \nu \; |<||U^\lambda||>|^2$$

with a T'_λ having a B_k given by the dynamical part of the crystal field :

$$B'_k = \sum_{q,i,n,n'} \left| \frac{\partial B^k_q}{\partial Q_i} \right|^2 |<n|Q_i|n'>|^2 g(\omega_i)/(2k+1)^2 \; .$$

This means that the experimental T_λ takes into account the vibronic contribution for an M process, and no information about the electron-phonon coupling can be gained. When there is inversion symmetry at the ion site, the only contribution comes from the M process. This is the most favorable situation for experimental observation. Besides this one-phonon vibronic transition, N-phonon processes could also be considered by taking the Nth-order term in the field expansion.

(b) The Δ Process ($S_o \neq o$). By analogy with the nonradiative transitions of the non-adiabatic Hamiltonian, the Δ process corresponds to a shift in the equilibrium position in the lattice coordinates of the electronic states. In other words, vibrations are different in the two electronic states.

Experimentally, four extreme cases may be distinguished for the general problem of electron-lattice coupling (36) related to sideband appearance via a Δ process.

α) Small change in mass and spring constants and weak electron-lattice coupling. This is, for example, the case in triply ionized rare earth halide matrices where there is usually a restriction to one-phonon sidebands.

β) Small change in mass and spring constants and strong electron-lattice coupling. This is the case, for instance, in some divalent rare earths such as Sm^{2+} in CaF_2 (also with one-phonon sidebands).

γ) Large change in mass and spring constants and strong coupling. This is the case in some divalent rare earths and also in F centers in KCl which show multiphonon sidebands up to 10th-order.

δ) Large change in mass and spring constants and weak coupling in which case only limited vibronic line structure is to be expected. This is the case when the doping impurity strongly alters the phonon modes and there is the possibility for localized modes.

Since triply ionized rare earth ions (Ln^{3+}) in usual matrices for lasers or summation-of-photon action by energy transfer (37) appear to fall in class α, it can be understood easily why very little or no experimental work on multiphonon sidebands of Ln^{3+} can be found in the literature (37).

Yet multiphonon sidebands seem to play a role in multiphonon-assisted energy transfer as shall be seen later in III.A.

With the same approximations and notation as in section II.A, one can write for the absorption probability

$$W_{A_{i\to f}}(E) = \frac{2\pi}{\hbar} \sum_{v'v''} P_{v'} |\langle fv''|eR|iv'\rangle|^2 \, \delta(E_{fv''}-E_{iv'} - E),$$

which is the "Golden Rule" expression for the electric-dipole operator eR.

Using adiabatic wave functions and the "Condon approximation" one gets :

$$W_{A_{i\to f}}(E) = \frac{2\pi}{\hbar} |\langle\Phi_f|eR|\Phi_i\rangle|^2 \sum_{v'v''} P_{v'} \prod_j |\langle X_{fv''_j}|X_{iv'_j}\rangle|^2 \tag{44}$$

$$\times \, \delta(E_{fv''} - E_{iv'} - E).$$

Comparing this equation with eq. (3), one sees clearly the analogy with nonradiative decay coming from the "accepting modes" terms. The essential difference is of course the absence of "promoting modes" terms since the radiative transition between electronic states is assumed to be permitted for the electronic dipolar transition contrary to the M-process case. Another difference is that nonradiative transitions occur in the limit of zero excitation energy (E = 0).

To calculate the double summation, the same methods as in II.A are used: either the initial method of Huang and Rhys (12) or the generating function method (14), yielding the same results with the same approximations.

With the previous notation one gets from (12), for example:

$$W_{i\to f}(E) = \frac{2\pi}{\hbar} |\langle M\rangle|^2 R_N \delta(E_f - E_i - E) \quad .$$

As done previously in eq. (26), one can write for temperature T:

$$R_N \simeq e^{-S_o} \frac{S_o^N}{N!} (\bar{n} + 1)^N \quad . \tag{45}$$

N! can also be approximated by Stirling's formula, giving the exponential gap law with the same α parameter as found for nonradiative decay.

The $(\bar{n}+1)^N$ factor has an obvious physical interpretation in terms of induced and spontaneous emission of phonons. If only absorption of phonons is considered it should be replaced by $(\bar{n})^N$.

Note on the Pekarian Function and the Energy-Gap Law. Since the function $e^{-S_o} S_o^N/N!$ is the key result for multiphonon vibronic sidebands as well as for the gross features of nonradiative decay, it should be interesting to try to interpret its physical meaning.

As underlined by Huang and Rhys (12) in their paper, S_o is by far the most important constant in the theory. It is now usually called the "Huang-Rhys-Pekar coupling constant." As shown from its definition (eq. (8)), it is directly linked to the amount by which the normal coordinates are displaced by the electron-phonon coupling. Following Toyozawa, S_o can be viewed also as the number of phonons emitted or absorbed in the transition (38).

The "Pekarian function" is in fact a normalized Poisson distribution and it could be viewed in the following way.

Let p be the probability for one of a set of S equivalent modes to jump to the next state. p^N is the probability of having N simultaneous jumps in the same direction for N modes. But the number of different ways to choose the N simultaneous jumping modes among the total set of S is:

$$C_S^N = \frac{S(S-1)\ \dots\ (S-N+1)}{N!} \quad .$$

That is, the total probability P_t for N jumps among a set of S modes is:

$$P_t^{(N)} = \frac{p^N S(S-1)\dots(S-N+1)}{N!} \simeq \frac{(p\ S)^N}{N!} \quad ,$$

for S >> N. The term pS represents the average number of modes which have been involved in a jump, that is $pS = S_o$, and, in the case of 0^oK

$$P_t^{(N)} = \frac{S_o^N}{N!} \quad .$$

Now the total probability for all N-processes is equal to unity and we have:

$$P_t = \sum_N k \ \frac{S_o^N}{N!} = k\ e^{S_o} = 1 ,$$

that is

$$P(N) = \frac{e^{-S_o} S_o^N}{N!} ,$$

the required Poisson distribution.

Now at temperature T, absorption and emission of phonons have to be simultaneously considered in such a way that

$$N = k - \ell ,$$

where k is the number of phonons emitted and ℓ the number of phonons absorbed. Then (11)

$$P_\ell(N,T) = e^{-S_o(\bar{n}+1)} \frac{S_o^k (\bar{n}+1)^k}{k!} \frac{e^{-S_o \bar{n}} S_o^\ell (\bar{n})^\ell}{\ell !} , \quad (46)$$

and in vibronic absorption for instance :

$$P_t(N,T) = \sum_\ell \frac{e^{-S_o(2\bar{n}+1)} S_o^\ell (\bar{n})^\ell S_o^{N+\ell} (\bar{n}+1)^{N+\ell}}{\ell ! (N+\ell)!}$$

$$= e^{-S_o(2\bar{n}+1)}$$

$$\times \sum_\ell \frac{S_o^{2\ell+N} (\bar{n})^\ell (\bar{n})^{N/2} (\bar{n})^{-N/2} (\bar{n}+1)^{N+\ell} (\bar{n}+1)^{N/2} (\bar{n}+1)^{-N/2}}{\ell ! (N+\ell)!}$$

$$= e^{-S_o(2\bar{n}+1)} \frac{(\bar{n}+1)^{N/2}}{(\bar{n})^{N/2}} \sum_\ell \frac{\left[S_o^2 \bar{n}(\bar{n}+1)\right]^{\ell+N/2}}{\ell ! (N+\ell)!} . (47)$$

Then from eq. (25), $P_t(N,T) \equiv R_N$ of Huang and Rhys.

From this, the small-coupling approximation of eq. (26) is equivalent even at temperature T to considering only phonon emission by $(\bar{n}+1)^N$ or only absorption by $(\bar{n})^N$. This possibility arises from the rapid decrease of P(N) with the number of phonons. This is shown in Fig. 2.

According to the value of S_o and assuming that energy can be varied only by discrete equal values for N (one phonon frequency), P(N) is represented as is Fig. 2. For a small coupling (S_o = 0.1), the envelope of the distribution is nearly exponential and represents the gap law we are interested in. For medium (S = 1) and strong (S = 10) coupling the distribution becomes more symmetrical and approaches the usual Gaussian found for such cases (16).

Since this simple reasoning underlying "Pekarian" functions can be applied to any phenomenon in which multiphonon jumps are involved, one can now understand why when small coupling is considered, exponential gap laws are found whatever the process is. The precise knowledge of the phenomena comes as a modification to the statistics involved giving the gross features.

II.C Discussion

All the processes above, either radiative or nonradiative, electronic or vibronic, have been essentially considered to be taking place in one center independently of other nearby centers. Is this assumption still valid when a high concentration of ions is involved?

The answer for rare earth ions is yes, most of the time.

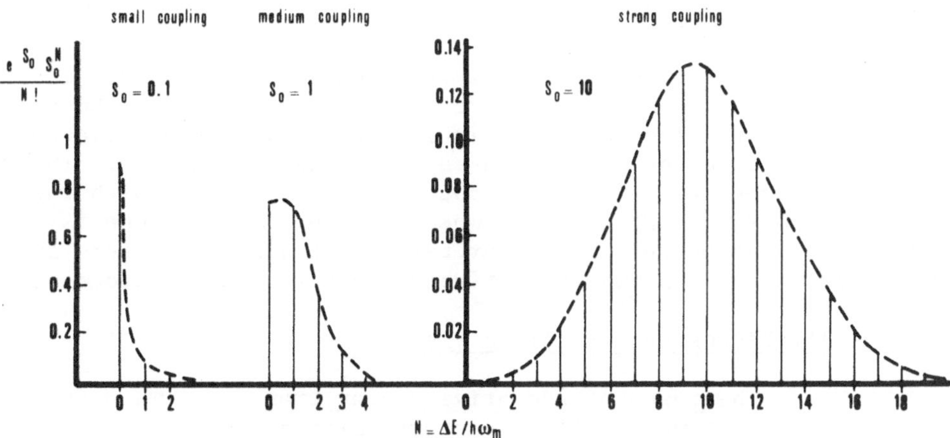

Fig. 2. The "Pekarian function" for different coupling parameter (S_o) values.

However, two cases of concentration-sensitive electron-phonon coupling have been reported recently for Eu^{3+} (39) (40) where a doubling of S_o is found as one proceeds from 1% to 100% concentration. This can increase the nonradiative transition probability by an order of magnitude and is a new possibility to be considered for a self-quenching we proposed (40).

As for radiative transitions, when the concentration is increased for transition metal ions, new lines of increasing intensity appear (41). For Cr^{3+}, they are attributed to exchange-coupled pairs, with the absorption probability increasing as the square of the concentration. Clearly in these cases we are not dealing any more with one-center processes.

In the rare-earth case, pair spectra have also been observed in trichlorides (42), but oscillator strength estimates indicate values as small as 10^{-10} (43), negligible with respect to one-center values (10^{-6}).

For the Nd^{3+} case in stoichiometric materials, we shall dwell somewhat more on that point in section IV.

III. CONCENTRATION-DEPENDENT PROCESSES OR MULTICENTER EFFECTS

When the concentration of active ions is increased, long before the appearance of new lines due to pairs or modifications in radiative transition probabilities, a migration of energy between the centers is found. We are going to study this now, assuming that the multiphonon decay and the radiative transition remain one-center processes.

III.A. Energy Transfer

Energy transfer occurs in a system when absorption and emission do not take place within the same center. It may occur without charge transport. Then one may distinguish between radiative and nonradiative, resonant and phonon-assisted energy transfer. Theoretical approaches start from a microscopic point of view with the interaction between two centers, whereas experimentally one is faced with a macroscopic result averaged over all the centers in the sample. Fig. 3 shows the different energy transfer processes.

1) The Microscopic Point of View

a. The resonant radiative energy transfer. When the transfer is radiative, real photons are emitted by the sensitizer ions and are then absorbed by any activator ions within a photon travel distance. Because of this fact, such transfer depends on the shape of the sample.

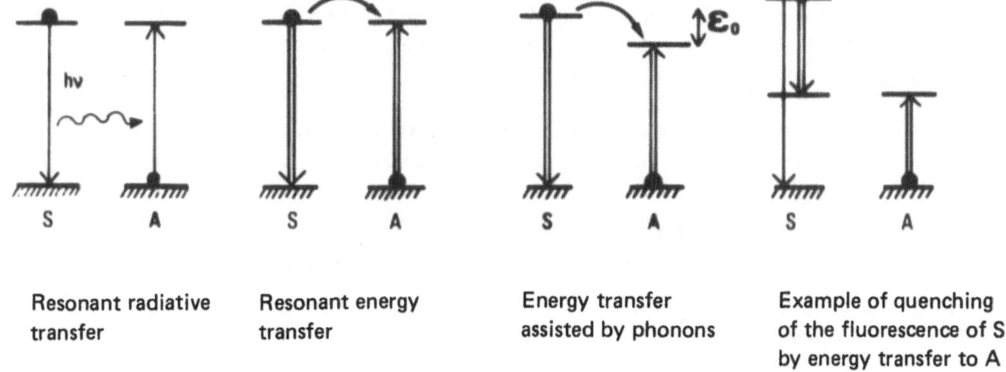

Fig. 3. Different types of energy transfer to an activator in its ground state.

Moreover, according to the degree of overlap between the emission spectrum of the sensitizer (S) and the absorption spectrum of the activator (A), the structure of the emission spectra of the sensitizer will change with activator concentration. Since photons are emitted anyway, the sensitizer lifetime is independent of the activator concentration. These three facts are the criteria used to distinguish between radiative and nonradiative resonant energy transfer. The probability for such transfer is given by the product of the emission probability at the sensitizer and the absorption probability at the activator.

For a frequency ν one has :

$$P_{SA}(\nu) = A_{21_S} g_S(\nu) \times B_{12_A} g_A(\nu) \rho_S(\nu) , \tag{48}$$

where $g_S(\nu)$ and $g_A(\nu)$ are the lineshape functions for emission of S and absorption of A. A_{21_S} and B_{12_A} are Einstein coefficient for S and A.

For each emitted photon, the photon density at the activator is given by the photon flux at distance R of the activator from the sensitizer :

$$\rho_S(\nu) = \frac{h\nu}{C\, 4\pi R^2} . \tag{49}$$

From the well-known relation for the absorption cross section with Einstein coefficients :

$$\sigma(\nu) = \frac{h\nu}{C} B_{12}\, g_A(\nu) \tag{50}$$

we obtain:

$$P_{SA}(\nu) = A_{21_S}\, g_S(\nu)\ \sigma(\nu)/4\pi R^2 . \tag{51}$$

Considering σ_A, the integrated cross-section :

$$\sigma_A = \int \sigma(\nu)\, d\nu ,$$

one can write $\sigma(\nu) = \sigma_A\, g_A(\nu)$

or
$$P_{SA}(\nu) = A_{21_S}\ \sigma_A\ g_S(\nu) g_A(\nu)/4\pi R^2 \tag{52}$$

which by integration gives :

$$P_{SA}(R) = \frac{\sigma_A}{4\pi R^2}\ \frac{1}{\tau_S} \int g_S(\nu) g_A(\nu)\, d\nu \tag{53}$$

where τ_S is the sensitizer lifetime. The integral represents the spectral overlap between A and S. It should be noted that the distance dependence goes as R^{-2}. Such resonant radiative transfer may permit long-range energy diffusion between identical ions and gives rise to photon trapping effects of the same type as the one observed a long time ago in gases (44). Trapping effects increase the apparent experimental lifetime and τ_s has to be measured on thin and lightly doped samples. These effects are particularly strong in Cr^{3+} and Yb^{3+} (45) (46).

b. The resonant nonradiative energy transfer. Let us consider the simple case of two ions, each with one excitable electronic state separated from its electronic ground state by nearly equal energy. With suitable interaction between the two electronic systems, the excitation will jump from one ion to the other before one is able to emit a quantum of fluorescence. The mutual interactions are the Coulomb interactions of the Van der Waals type between the two ions. Förster (47), who first treated such a case theoretically by quantum-mechanical theory, considered the dipole-dipole interactions. He assumed that the interaction is strongest if for both transitions electric-dipole transitions are permitted (48). The interaction energy is then proportional to the inverse of the third power of the interionic distance and the transfer probability is given by:

$$P_{SA} = \frac{2\pi}{\hbar} \left| < S\ A^o \mid H_{SA} \mid S^o\ A > \right|^2 \rho_E \quad ; \tag{54}$$

H_{SA} = Interaction Hamiltonian
ρ_E = density of states provided by the vibrational motion contributing to the line-broadening of the transition.

which is proportional to the inverse sixth power of the separation. The wavefunctions to be considered for the matrix element describe an initial state of the system with the sensitizer in its excited state and the activator in its ground state, the final state having the sensitizer in its ground state and the activator in its excited state.

Therefore, the transfer probability can be written as

$$P_{SA} = \frac{1}{\tau_s}\left(\frac{R_o}{R}\right)^6 \quad , \tag{55}$$

where τ_s is the actual lifetime of the sensitizer excited state, including multiphonon nonradiative decay, and R_o is the critical transfer distance for which excitation transfer and spontaneous deactivation of the sensitizer have equal probability. R_o can be written as (49):

$$R_o^6 = \frac{3}{64\,\pi^5}\;\frac{1}{\varepsilon^2}\;\frac{\sigma_A\,\eta_s^{\circ}}{\bar{\nu}^4}\int_o^{\infty} g_s\,(\nu)\quad g_A\,(\nu)\quad d\nu \qquad (56)$$

where ν is the wavenumber, $g_A(\nu)$, $g_S(\nu)$ normalized shape functions η_S^o the quantum efficiency of the sensitizer in the absence of the activator, ε the refractive index, $\bar{\nu}$ the average wavenumber of the transition, and the integral represents the energy overlap between the emission in the sensitizer and the activator absorption. However, Dexter pointed out (50) that this theory should be extended to include higher multipole and exchange interactions. In fact, for an isolated atom, one can consider the transition probability as decreasing as $(a_o/\lambda)^{2n}$ where a_o is the Bohr radius, λ the wavelength, and n an integer.

Whereas, in an energy transfer process with a dependence on near-zone interactions, the transition probabilities drop off as $(a_o/\rho)^{2n}$ where ρ is the separation of the interacting ions. ρ can be as much as three orders of magnitude smaller than λ, so that the energy transfer effect tends to be more pronounced in systems with forbidden transitions (50). This holds true for ions with transitions forbidden to first order, such as transition metal and lanthanide ions.

The energy transfer probability for electric multipolar interactions can be more generally written as (50):

$$P_{SA}(R) = \frac{(R_o/R)^s}{\tau_s}\ , \qquad (57)$$

where s is a positive integer taking the following values:

s = 6 for dipole-dipole interactions
s = 8 for dipole-quadrupole interactions
s = 10 for quadrupole-quadrupole interactions.

It should be noted that, for dipole-dipole interactions, differences between radiative and nonradiative resonant transfer lay essentially in that for radiative transfer there is no critical R_o depending only upon concentration. The variation goes as R^{-2} instead of R^{-6}, and the sensitizer lifetime does not depend upon the distance R.

Now, to be able to calculate effectively $P_{SA}(R)$, eq. (57) is not very useful, because R_o cannot be easily obtained theoretically. Applying Racah's tensorial methods straightaway at the beginning of the calculation of Dexter, eq. (57), permits one to develop calculations analogous to Judd's theory of radiative transitions. The case of the multipolar interactions was treated in this way by Kushida (51) and calculations for magnetostatic, and exchange interactions were performed in our group (52).

We start with eq. (54). The probability of a resonant energy transfer between two ions linked by some H(R) interaction, can be written as

$$P_{SA}(R) = \frac{2\pi}{\hbar}\frac{1}{g_S^* \; g_{A^o}} \sum_{IF} |\langle S^*A^\circ|H(R)|S^\circ A^*\rangle|^2 \int \gamma_S(E)\Gamma_A(E)\,dE, \quad (58)$$

where g_S^* ($g_A^{*\circ}$) is the degeneracy of the S^* (A°) level, $\gamma_S(E)$ ($\Gamma_A(E)$) is the normalized shape function of emission (absorption) spectrum and $\sum_{IF}$ represents a sum over all possible transitions that contribute to the transfer.

We now introduce the result (52) that for equivalent electrons (4f electrons for instance) electrostatic, magnetostatic and exchange (without change of spin) Hamiltonians H(R) can be written as a combination of tensorial operators acting on the S ion, the A ion and their relative coordinates (R,Ω) ;

$$H(R) = \sum_{\ell_1 \ell_2 \Lambda} \alpha_{\ell_1 \ell_2 \Lambda}(R) \left[\left(U^{(\ell_1)}(S) \otimes U^{(\ell_2)}(A) \right)^{(\Lambda)} \otimes C^{(\Lambda)}(\Omega) \right]^{(0)} \quad (59)$$

where $U^{(\ell)}$ are tensorial operators as already seen in II.B. and $C^{(\Lambda)}$ are Racah spherical harmonics. When introduced into eq. (58), this expression of H(R) leads to :

$$P_{SA}(R) = \frac{2\pi}{\hbar}\frac{I}{g_S^* \; g_A^\circ} \sum_{\ell_1 \ell_2} |C_{\ell_1 \ell_2} \langle S^*||U^{(\ell_1)}||S^\circ\rangle\langle A^\circ||U^{(\ell_2)}||A^*\rangle|^2 \quad (60)$$

with

$$I = \int \gamma_S(E)\,\Gamma_A(E)\,dE$$

$$|C_{\ell_1 \ell_2}|^2 = (2\ell_1+1)^{-1}(2\ell_2+1)^{-1}\sum_{\Lambda}(2\Lambda+1)^{-1}\,|\alpha_{\ell_1 \ell_2 \Lambda}(R)|^2 \quad (61)$$

R = distance from S to its first neighbour A ion.

This expression of the transfer probability has following advantages :

- Radial and orbital parts have been separated.

- Only a few reduced matrix elements need be calculated. They are the same ones for the three interactions we consider (for any interaction leaving spins unchanged).

- Comparison between two different interactions can be made through comparison of $\alpha_{\ell 1 \ell 2}\Lambda(R)$, that is to say, $C_{\ell 1 \ell 2}$ coefficients. They are independent of the states involved in the transfer and we call them $E_{\ell 1 \ell 2}$, $M_{\ell 1 \ell 2}$ and $X_{\ell 1 \ell 2}$ for electrostatic, magnetostatic and exchange interactions, respectively.

- Forced electric-dipole transitions, as calculated by Judd's method, can be included in eq. (60).

- This expression gives also a single mathematical form regardless of the interaction, which is a convenient result.

The somewhat complicated expressions for the different $C_{\ell 1 \ell 2}$ of 4f electrons are given in (52). But we can note that:

- For electrostatic interaction ($E_{\ell 1 \ell 2}$) the $\ell_1 = 1$ $\ell_2 = 1$ term, corresponding to dipolar-dipolar interaction, being zero in first order, necessitates the introduction of Judd's T_λ parameters. The $E_{\ell 1 \ell 2}$ values are between $E_{22} \simeq 30\ cm^{-1}$ for quadrupole-quadrupole intensities to $E_{66} \simeq 3 \times 10^{-1}\ cm^{-1}$, but all contain some dipole-dipole part due to the T_λ.

- For magnetostatic interactions ($M_{\ell 1 \ell 2}$) only terms with $\ell_i = 1,3,5$ are nonzero and have the following order of magnitude: $M_{11} \simeq 1\ cm^{-1}$; $M_{55} \simeq 2 \times 10^{-7}\ cm^{-1}$.

- For exchange interactions ($X_{\ell 1 \ell 2}$) we have $1 \leq \ell_i \leq 6$, giving estimates of 1 to $10^{-1}\ cm^{-1}$ for the coefficients.

These results show that exchange or magnetostatic interactions can be found in cases of small dipole-dipole and quadrupole electrostatic interactions if the matrix elements allow them.

c. The phonon-assisted nonradiative energy transfer. If now we consider two ions with excited states of different energies, the probability for energy transfer should drop to zero where the overlap integral $\int g_S(\nu) g_A(\nu)\, d\nu$ vanishes. However, it is found experimentally that energy transfer can take place without phonon-broadened electronic overlap provided that the overall energy conservation is maintained by production or annihilation of phonons with energies approaching $k\theta_d$, where θ_d is the Debye temperature of the host matrix (53). Then for small energy mismatch ($\simeq 100\ cm^{-1}$) energy transfer assisted by one or two phonons can take place (54) (55). However, in energy transfer between rare earths, energy mismatches as high as several thousand reciprocal centimeters are encountered. This is much higher than the Debye cutoff frequency found in normally encountered hosts, so that multiphonon phenomena have to be considered here.

This was done by Miyakawa and Dexter in a general theory of multiphonon processes which arose from the necessity of obtaining a theoretical estimation of the probability of energy transfer in up-conversion processes which are the object of section III.C (16). They showed that it is also legitimate to write the probability of energy transfer in the form of eq. (54), where $\rho(E)$ is taken as S_{SA} the overlap of the lineshape functions for emission in ion S and absorption of ion A, including the phonon sidebands in the lineshape. It is necessary to consider each partial overlap between the m-phonon emission line shape of ion S and the n-phonon absorption lineshape of ion A. This mathematical assumption has been given recently a physical meaning by our demonstration of the existence of multiphonon sidebands for trivalent rare-earth ions, even in such a case of very small electron-phonon coupling (37). Fig. 4 gives an example of excitation through a multiphonon sideband and Fig. 5 shows the exponential behavior.

Using a generating-function method along the same lines as for vibronic-sideband studies (II.B), S_{SA} can be expressed as follows (16):

$$S_{SA} = \sum_N e^{-(S_{oS}+S_{oA})} \left[\frac{(S_{oS}+S_{oA})^N}{N!}\right] \alpha_{SA}(0,0;E)\ \delta(N,\Delta E/\hbar\omega), \quad (62)$$

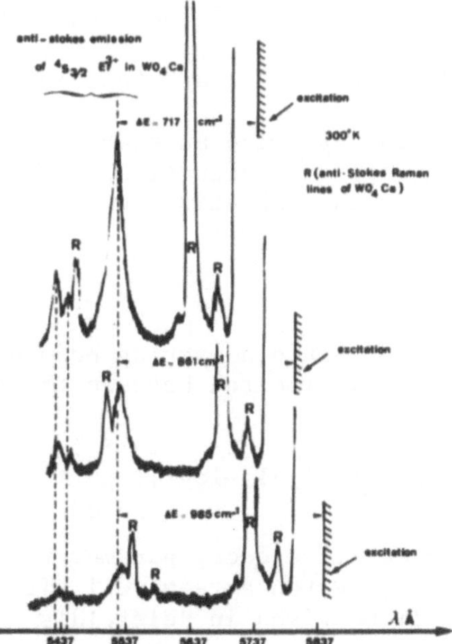

Fig. 4. Multiphonon anti-Stokes sideband excitation of $^4S_{3/2}(Er^{3+})$ for different excitation mismatches (ΔE) (from Ref. 37).

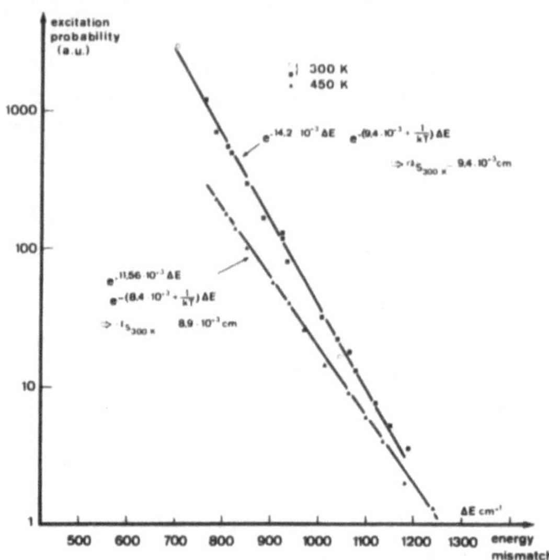

Fig. 5. Example of multiphonon anti-Stokes excitation exponential gap law for $LaF_3:Er^{3+}$ $(^4S_{3/2})$ (from Ref. 37).

where S_{oS} and S_{oA} are the respective lattice coupling constants for the ions S and A, N the order of the multiphonon process with $N = \Delta E/\hbar\omega_m$, ΔE the energy mismatch between both ions, and $\hbar\omega_m$ the phonon cutoff frequency. σ_{SA} (0,0;E) is the zero-phonon overlap integral between S and A. A "Pekar" function is still found.

The expression for S_{SA} with an energy mismatch of ΔE for small S_o constant and for an occupation number $\bar{n} = (\exp(\hbar\omega/kt)-1)^{-1}$ not exceeding unity at the operating temperature, can be approximated with Stirling's formula by:

$$S_{SA}(\Delta E) = S_{SA}(0)\ e^{-\beta\Delta E} \tag{63}$$

where $S_{SA}(0)$ is the zero-phonon overlap between S and A in the case where there is no energy mismatch between the two ions. β is given by:

$$\beta = (\hbar\omega)^{-1}(\log N/S_o(\bar{n}+1)-1)-\log(1+\frac{S_{oA}}{S_{oS}}) \equiv \alpha - \gamma\ , \tag{64}$$

involving α the nonradiative decay parameters. This exponential dependence on energy mismatch agrees well with experiment (56). Examples of this will be given in Reisfeld's article in this volume.

This expression is particularly useful since we have seen that the multiphonon relaxation rate and the multiphonon sidebands are described by an exponential function with respect to energy mismatch with an exponential parameter involving also S_o. Measurement of one of the three processes can give insights on the other processes.

2) The Macroscopic Case of Energy Transfer in Real Samples. Up to this point we have been dealing with the microscopic case of two ions interacting with one another. To discuss the case of real macroscopic samples with many ions and to obtain a link with experimental facts, a statistical analysis of the energy transfer is necessary.

We have then to think about the overlap integrals that we have seen arise in all transfer between two ions. In the microscopic case we are sure that the lineshapes involved can be only due to some homogeneous broadening even for transfer between two identical ions in different lattice sites.

In the macroscopic case, we can measure absorption and emission spectra taking into account all broadening processes averaged over the whole sample, for instance the inhomogeneous broadening process due to emission and absorption at centers in different lattice sites. Then the overlap integral measured experimentally from the usual spectra is a measure in excess of the real overlap since we take into account emission and absorption of centers at any distances, even those which cannot interact. The error is the largest for the processes at shortest interacting distance (exchange), and so is certainly negligible for radiative transfer, since photons can travel a much larger distance than the spread of the spatial disorder. The error is also smaller for systems with small inhomogeneous broadening having centers in only one type of lattice site, that is, without disorder.

Fluorescence line-narrowing techniques could give some idea about the homogeneous part of an emission line, but the statistical analysis for the whole sample should still be performed. The inhomogeneous case will be discussed again later. Supposing only a sensitizer-activator interaction, an average transfer efficiency can be calculated (48) (50). This was studied in some detail by Inokuti and Hirayama (57). They considered the number of activators situated at random in a sphere around a sensitizer such that the activator concentration is constant when the volume of the sphere and the number of activator ions considered goes to infinity. Then the average probability for transfer from one sensitizer to any acceptor is:

$$W_{SA} = N_A \int_{R_{min}}^{\infty} P_{SA}(R)\ 4\pi R^2\, dR \quad . \tag{65}$$

Introducing eq. (65) into the expression for the intensity emitted by all sensitizers, each with different activator neighborhood, they obtained the following relation for the intensity decay of the emission of the sensitizer surrounded by many activators:

$$I\,(t) = \exp\left[-\frac{t}{\tau_{SO}} - \Gamma\left(1-\frac{3}{s}\right)\frac{C}{C_o}\left(\frac{t}{\tau_{SO}}\right)^{3/s}\right], \quad (66)$$

where τ_{so} is the decay constant of the sensitizer in the absence of activators; c is the activator concentration; c_0 is the critical activator concentration; and s the parameter of the multipolar interaction.

The comparison between experimental decay and this theoretical expression has been widely used to determine the index of the multipolar interaction involved (45) (58) (59). This theory is valid only when there is no sensitizer-to-sensitizer transfer or activator-to-activator transfer. This formulation, therefore, has to be modified for high concentrations of sensitizers and activators. Then rapid energy migration between sensitizers or between activators is possible, because of the perfect resonance conditions. The general result is complicated (60) but, for large t, I(t) decays exponentially (61):

$$I(t) = \exp\left(-\frac{t}{\tau_s} - \frac{t}{\tau_D}\right). \quad (67)$$

Then, two cases can be distinguished:

a) Diffusion-limited case. In this case spontaneous decay of excited sensitizers, diffusion among sensitizers and energy transfer between sensitizers and activators are of about the same order of magnitude.

For sufficiently long times and dipole-dipole interactions one has (61) (62) (63):

$$\frac{1}{\tau_D} = 4\pi D N_A \rho \quad ; \quad (68)$$

ρ is a length defined by: $\rho = 0.68\ (C/D)^{1/4}$; D is the diffusion constant; N_A the activator concentration; C is the sensitizer-activator energy transfer constant, such as $C = (R_o)^s/\tau$, and R_o is the critical transfer distance of eq. (57).

For this diffusion-limited case, the product $D\rho$ is found to depend linearly upon sensitizer concentration. One has from (62):

$$D = (8/4\pi)\ C_{SS}/R^4 \quad ,$$

C_{SS} being the sensitizer-sensitizer transfer constant. For limited diffusion we take R as the average distance between sensitizers:

$$R = (3/4\pi)^{1/3}\ N_S^{-1/3} \tag{69}$$

therefore:

$$D\rho = 2C^{\frac{1}{4}}\ C_{SS}^{\ 3/4}\ N_S \ , \tag{70}$$

which has been experimentally verified (64). One has:

$$\tau_D^{-1} = V\ N_S\ N_A \tag{71}$$

with: $V = 8\pi C^{\frac{1}{4}}\ C_{SS}^{\ 3/4}$.

b) Fast-diffusion case. For high sensitizer concentration, the diffusion rate can be faster than spontaneous sensitizer decay or sensitizer-activator energy transfer. The limiting step is no longer diffusion and D appears to saturate with increasing donor concentration; each activator experiences the same excited sensitizer neighbourhood. We take R as the minimum distance between sensitizers as permitted by the lattice ($R = R_{min}$). Therefore:

$$D = (8/4\pi)\ C_{SS}/R^4_{min} \quad . \tag{72}$$

$D\rho$ is now a constant with respect to N_S.

One has

$$\tau_D^{-1} = UN_A \quad , \tag{73}$$

with U a constant depending on the type of interaction as discussed earlier in this section through eq. (60).

Another approach to the macroscopic case is the use of the well-known rate equations which deal with the population of ions in a given state. This was used as a phenomenological approach in studies of lasers. In the same manner, rate equations were derived for the energy transfer between Yb^{3+} and Er^{3+} (65), (66), (67). The applicability of those equations in relation to the Inokuti and Hirayama statistics has been discussed by Grant (68). Based on the first principles of quantum statistics, Grant shows that the average transition probability which enters into the rate equation, provided diffusion is sufficiently fast, depends on powers of the concentration reflecting the number of types of coupled particles.

The basic result of Grant is that the energy transfer probability is proportional to the activator concentration:

$$W_t = UN_A \tag{74}$$

this result is the same as obtained by fast-diffusion studies (eq. 73).

The practical interest in considering diffusion is that the decays are again exponential, as when ions are not interacting. This validates the use of rate equations. Further, the constant U can be estimated from Dexter's theory (50) or from our unified result whatever the type of multipolar or exchange interaction: eq. (60). Comparison between calculated values for U permits one to predict the type of multipolar interaction involved.

3) Anderson's Theorem. We have seen that the passage from the preceding microscopic view to macroscopic theories was strictly valid only in the case of homogeneous broadening for $g_A(\nu)$ and $g_S(\nu)$, the activator and sensitizer lineshapes. For the inhomogeneous case, Anderson has proven directly from quantum statistics the following general theorem (69) applying to any transport by quantum jumps: If the interaction Hamiltonian H_{SA} falls off at large distances faster than R^{-3} and if the average value of $\langle H_{SA} \rangle$ is less than a critical value H_C of the order of magnitude of ΔE_{inhomo} (the inhomogeneous energy spread), then transport of energy is forbidden. This is the so-called Anderson localization. Since the R^{-3} interaction is the dipole-dipole case, all other interactions should follow this theorem when $\langle H_{SA} \rangle < \Delta E$. However, this theorem does not take into account any contact with an external thermal reservoir, which does exist at room temperature for a real sample. Then thermal processes may actually control the transport processes.

Physically this theorem simply states that energy transport is not possible whenever the interaction (H_{SA}) on one site is too small to compensate the energy mismatch ΔE_{inhomo} of a nearby site (see Fig. 6).

Outside the region of validity of this theorem, Anderson predicts the following behavior: for the dipole-dipole case and $\langle H_{SA} \rangle \simeq \Delta E_{inhomo}$, transport occurs by very long paths involving virtual jumps. Experimental evidence of this theorem for transition metals, rare-earth ions, and organic systems is still being debated. (70) (71) (72).

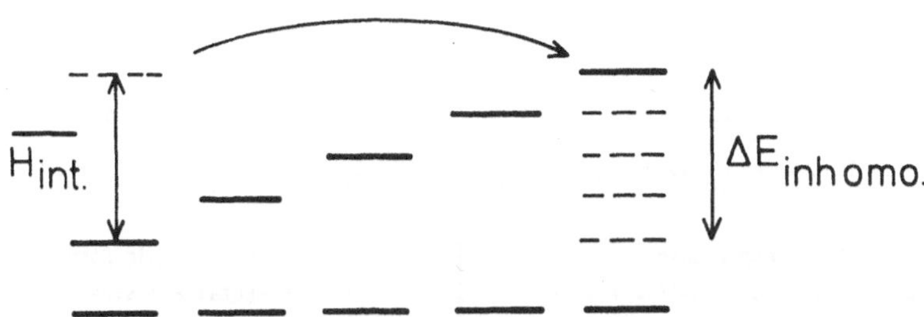

Fig. 6. Physical meaning of Anderson's theory. Energy transfer forbidden for: $\bar{H}_{int} < \Delta E_{inhomo}$ and $\bar{H}_{int}$ α R^{-s} and s > 3 (R = distance between sites).

At any rate, it is experimentally found for rare-earth ions that between about 77°K and 300°K, overlap integrals measured from emission and absorption spectra even with an inhomogeneous part give some reasonable account of the different energy transfers occurring in rare-earth ions (73) (74) (75). Several reasons appear to explain this:

i) In practical systems, high rare-earth concentration is considered (> 10^{19} cm^{-3}). Then one may have $H_{SA} \gtrapprox \Delta E_{inhomo}$ and Anderson's localizaion, applying essentially at low concentration, becomes negligible.

ii) Whatever the process, energy at high rare-earth concentration migrates in such a way that on a time scale of tens of μs, site-selective pulsed excitation gives the same inhomogeneous spectra that is obtained under CW broadband excitation, even in glasses where the inhomogeneous broadening is the largest (76).

Then one can state that overlap-integral measurements are roughly valid on a time scale > 1 μs. This could explain why in NdP_5O_{14} no apparent spectral migration is found for delays > 1 μs after a pulsed excitation (77), migration being on a shorter scale than 1 μs.

TABLE I

SUMMARY OF THE DIFFERENT CASES OF ENERGY TRANSFER PROCESSES

A Microscopic Studies followed by macroscopic statistics :

Microscopic Case (theoretical one sensitizer-one activator case).		Macroscopic Case (Experimental all sensitizers-all activator case).
- Resonant - Radiative : $P_{SA}^{(R)} = \frac{\sigma_A}{4\pi R^2 \tau_S} \int g_A(\nu) g_B(\nu) d\nu$ - Non Radiative : $P_{SA}^{(R)} = \frac{(R_0/R)^s}{\tau_s}$; (s = 6,8,10). with : $R_0^s \propto \int g_A(\nu) g_B(\nu) d\nu$	Intermediate Statistics : Average probability for one sensitizer → all acceptors	- Without diffusion among sensitizers : non exponential decay : $I(t) = \exp\left[\frac{-t}{\tau_S} - \Gamma(1-\frac{3}{s}) \frac{C}{C_0} (\frac{t}{\tau_S})\right]^{3/s}$
- Non Resonant : Non Radiative : $P_{SA}^{(R)} = \frac{(R_0/R)^s}{\tau_s}$; (s = 6,8,10) with : $R_0^s \propto e^{-\beta\Delta E}$	→ $W_{SA} = N_A \int P_{SA}(R) 4\pi R^2 dR$ strictly valid for $g_A(\nu)$ and $g_B(\nu)$ homogeneous	- With diffusion among sensitizers : (long time approximation) exponential decay : $I(t) = \exp\left(\frac{-t}{\tau_S} - \frac{t}{\tau_D}\right)$ "Diffusion limited" : $\tau_D^{-1} = V N_S N_A$ "Fast diffusion" : $\tau_D^{-1} = U N_A$ (rate equation case).

B Direct microscopic quantum statistics :

- rate equation case (Grant) → $\tau_D^{-1} = U N_A$
- inhomogeneous brodenning case (Anderson's localization)
  - → no diffusion for $\langle H_{SA} \rangle < \Delta E_{inhomogeneous}$ (without thermal effects) and $\langle H_{SA} \rangle \propto R^{-s/2}$ for S >6
  - → diffusion if thermal effects (one and two-phonons).

A summary of the different cases of energy transfer is given in Table I.

III.B. Cross-Relaxation

This term usually refers to all types of energy transfer occurring between identical ions. In such a case the same kind of ion is sensitizer and activator.

As shown in Fig. 7, cross relaxation may give rise to the diffusion process already considered between sensitizers when the involved levels are identical or to self-quenching when they are different. In the first case there is no loss of energy wheareas in the second case there is a loss or a change in the energy of the emitted photons.

Theoretically the same treatment is valid as for the more general case of energy transfer, however, experimentally it may be more difficult to distinguish between sensitizers and activators. So any of the microscopic processes discussed in III.A may happen with a maximum overlap when an identical couple of levels are involved. From the macroscopic point of view, the diffusion-limited case predicts from eq. (71):

$$\tau_D^{-1} = VN^2 \tag{75}$$

for $N_S \equiv N_A = N$

and in the fast-diffusion case:

$$\tau_D^{-1} = UN \tag{76}$$

we shall illustrate these cases in Nd^{3+} laser materials in section IV.

III.C. Up-Conversion by A.P.T.E. Processes

Until the mid-1960's, all energy transfers considered between rare-earth ions were of the type shown in Fig. 7, where the sensitizer ion is in one of its excited states while the activator is in its ground state. This type of energy transfer explains sensitized fluorescence as well as concentration quenching. In 1966 I reported evidence (78) for energy transfer via the somewhat different process shown in Fig. 8. In these cases, both the activator and sensitizer ions are in their excited states prior to energy transfer.

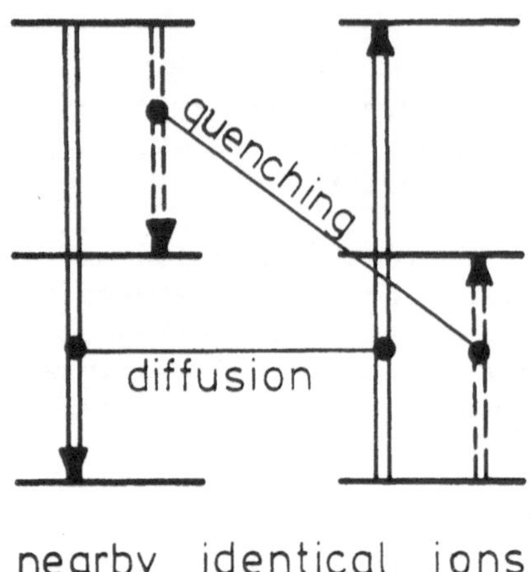

Fig. 7. Two cross-relaxation processes: diffusion and quenching.

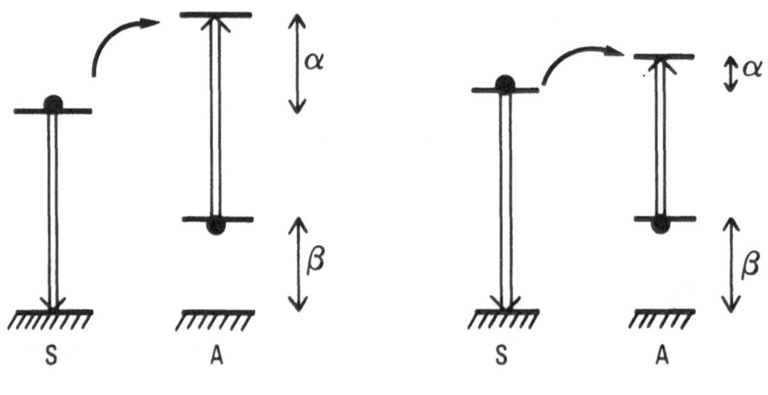

Resonant case

$\beta - \alpha = 0$

Phonon assisted case

$\beta - \alpha = \varepsilon_0$

Fig. 8. Energy transfer to an activator in one of its excited states in a resonant and non-resonant case: the elementary A.P.T.E. process.

The probability calculations previously described apply to energy transfer between excited states (65) and rate equations are particularly useful to describe them (66) (67). From section III. A, this is an a posteriori proof of the important role played by fast diffusion among sensitizers in such processes.

Now if one considers a level scheme in ladder form for the activator of the type shown in Fig. 8, a multiple sequential up-conversion process may take place which we called an APTE process (Addition de Photons par Transfert d'Energie); this is possible provided a direct emission from an intermediate level is less probable than a transfer to an upper level; this stresses the need for long-lived intermediate states. Then a direct relaxation to the ground state may give rise to spontaneous emission of anti-Stokes photons.

Other kinds of up-conversion exist also giving anti-Stokes spontaneous emission, namely: the cooperative sensitization (79) and the cooperative luminescence (80). These processes have already been reviewed elsewhere (65) and at Erice (81) so I shall deal only with the essential features of them.

All these processes being basically nonlinear: their efficiency increases with the excitation intensity. A comparison of them, normalized for input power, is given in Table II.

TABLE II

COMPARISON BETWEEN DIFFERENT MULTIPHOTON UPCONVERTING PROCESSES (65)

Mechanisms	efficiency μ (cm^2/W)	Material
Two sucessive transfers	$\sim 10^{-3}$	YF_3 : Yb:Er
Two-step absorption	$\sim 10^{-5}$	SrF_2: Er
Cooperative sensitization	$\sim 10^{-6}$	YF_3 : Yb:Tb
Second harmonic generation	$\sim 10^{-11}$	KDP
Two-photon absorption	$\sim 10^{-13}$	CaF_2: Eu^{2+}

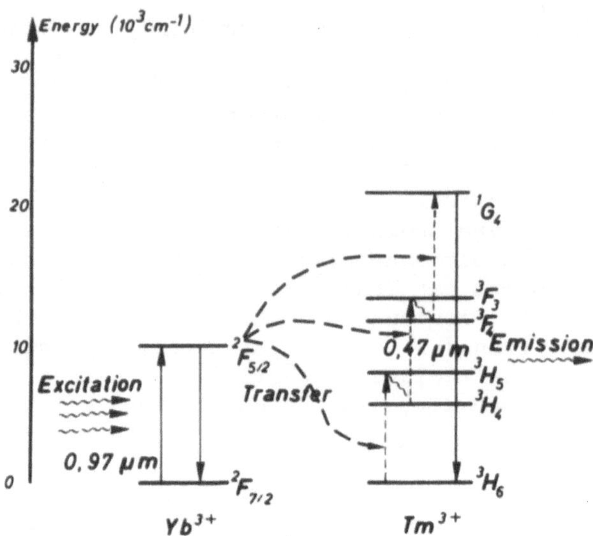

Fig. 9. IR-to-blue 3-photon A.P.T.E. process by Yb^{3+} - Tm^{3+} coupling.

1) <u>Sequential Energy Transfer</u>. A practical example of such a process for the summation of three infrared photons into one photon is presented in Fig. 9. This is a typical case where phonon-assisted energy transer is involved. Up to five sequential summations have been so obtained (Fig. 10).

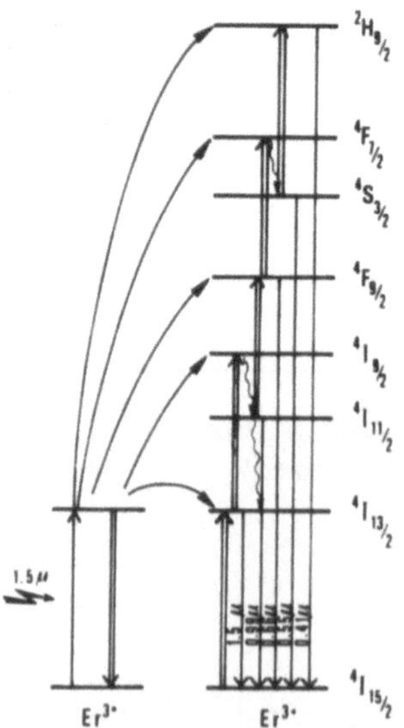

Fig. 10. IR-to-UV 5-photon A.P.T.E. process by Er^{3+} - Er^{3+} coupling.

As already pointed out, rate equations apply because fast diffusion is generally involved between sensitizers (Yb^{3+}). In fact radiative resonant transfer is involved partially (46) (82).

The rate equations give the rate of change of the number of particles in a given energy state and they can be written in the following manner :
For the population of the ith level of the activator we have :

$$\frac{dA_i}{dt} = S_2 \sum_{j=1}^{i-1} A_j \alpha_{ji} - S_2 \sum_{j=i+1}^{n} A_i \alpha_{ij}$$
$$+ \sum_{j=i+1}^{n} A_j \omega_{ji} - \sum_{j=1}^{i-1} A_i \omega_{ij}$$
$$+ S_1 \sum_{j=i+1}^{n} A_j \beta_{ji} - S_1 \sum_{j=1}^{i-1} A_i \beta_{ij}$$
$$+ \sum_{j=i+1}^{n} A_i A_j \gamma_{ji} - \sum_{j=1}^{i-1} A_i A_j \gamma_{ij} \quad . \qquad (77)$$

And for the population of the excited level of the sensitizer (Yb^{3+}) :

$$\frac{dS_2}{dt} = S_1 \sigma \phi - S_2 \left(\omega_s + \sum_{j=i+1}^{i} \sum_{i=1}^{n-1} \alpha_{ij} A_i \right)$$
$$+ S_1 \sum_{j=2}^{n} \sum_{i=1}^{n-1} \beta_{ji} A_j \quad . \, (78)$$

Since the total concentration of activators (A) and sensitizers (S) is constant we also have :

$$\sum_{i=1}^{n} A_i = A \quad \text{and} \quad S_1 + S_2 = S \quad .$$

σ is the cross section for absorption of the sensitizer and ϕ is the incident IR photon flux. The following assumptions have been made.

i) The sensitizer has only two levels 1 and 2, as in the case of Yb^{3+}, with populations S_1 and S_2, level 2 being the excited state.

ii) The probabilities α_{ij}, β_{ij}, ω_{ij} are independent of sensitizer and activator concentrations.

α_{ij} is the transfer rate from sensitizer to activator;
β_{ij} is the back-transfer rate from activator to sensitizer;
γ_{ij} is the activator-to-activator transfer rate;
ω_{ij} represents the spontaneous deactivation rate for radiative and/or nonradiative transitions; subscripts i and j represent the levels of the activator involved in the transfer.

iii) Stimulated emission, direct absorption in the activator, as well as sensitizer-sensitizer transfer have been neglected. Even with these simplifications, the equations are still somewhat intractable and some further simplifications are made in order to compare their results with experiment.

First, one can deal with the steady-state case (65) and set all derivatives to zero. One can also reduce the number of levels involved, dealing only with the most obvious and neglecting their Stark splitting. For instance in Tm^{3+}, only 6 levels are considered ($3H_6$, $3H_4$, $3H_5$, $3F_4$, $3F_3$, $1G_4$) (82).

In the calculations of the transfer rates, estimates of the spectral overlap integral have to be used since transitions between excited states do not generally appear in absorption spectra. When transfer is phonon-assisted, the emission of phonons for a given temperature is more probable than an absorption. This gives unidirectional energy transfer because of the reduced back transfer. One can write in populating a level i from j:

$$\alpha_{ij} = V_{ji} + U_{ji} \qquad \text{for the direct-transfer rate}$$

$$\beta_{ji} = V_{ij} + U_{ij} \qquad \text{for the back-transfer rate}$$

with $U_{ij} = V_o \exp - \beta \, \Delta E_{ij,Yb}$

and $V_{ij} = U_o \exp - \delta \, \Delta E_{ij,Yb}$,

where β is the exponential parameter defined in III.A and:

$\delta \simeq \beta + 1/kT$; with $\Delta E_{ij,Yb}$ the energy mismatch between the energy difference of levels i,j and the two levels of $Yb^{3+}(^2F_{5/2}, ^2F_{7/2})$.

The back-transfer probability is reduced with respect to the direct transfer by a factor:

$$\exp - \Delta E/kT \quad .$$

Solving for A_i gives the following general result: the number of activators in an excited state is proportional to the nth power of Φ for an n-transfer process.

2) Cooperative Pair Effects and Excited-State Excitation. In addition to the stepwise energy transfer described previously, another hypothesis was advanced (79), (83) for the anti-Stokes fluorescence of the Er^{3+} and the Yb^{3+}-Tm^{3+}-doped compounds, namely, cooperative pair effects. Among them, we shall consider those which may lead to up-conversion: the cooperative sensitization of luminescence, cooperative luminescence, cooperative energy transfer and photon absorption. By "cooperative" it is understood that more than one impurity center takes part in the elementary process of either sensitization (or energy transfer) or luminescence.

(a) The Cooperative Sensitization of Luminescence. This phenomenon is in fact the reverse of the one which was predicted by Dexter (84). Here the excitation of one ion can produce the excitation of two other ions. The energy accumulated by two excited ions is given in one transfer to one ion which then reaches an excited state from which it can relax to the ground state. Fig. 11 gives the energy scheme of this basic interaction where ions A and B, which may be identical, are in their excited states a and b and ion C is in its ground state 0. The final state is the one where A and B are in their ground states 0 and C is in its excited state c. Conservation of energy requires that $\varepsilon_a + \varepsilon_b = \varepsilon_c$, so that the photon emitted by ion C will be the energy sum of the two photons absorbed by ions A and B. This is, in some way, analogous to the two triplet-exciton fusion into one singlet exciton observed in anthracene (85).

In order to have a nonvanishing matrix element for the interaction between these ions, one has to consider the second order terms in the multipole-multipole interaction or the effect of electronic overlap exchange, which is negligible at least in the fluoride hosts. This has been discussed in detail by Miyakawa and Dexter (67). In that case, the cooperative transfer probability per unit time is (67):

$$P_{coop} = \frac{2\pi}{\hbar} |D_{abo}{}^{ooc}|^2 \rho E^{coop} \qquad (79)$$

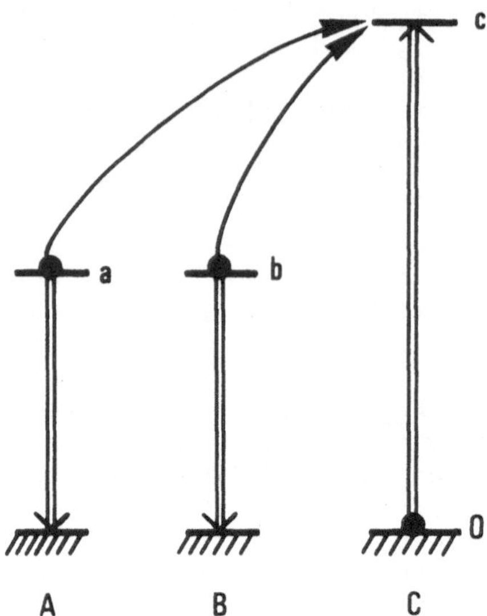

Fig. 11. Energy scheme for cooperative sensitization of ion c by ions A and B.

with the assumption that ions A and B are identical and where ρE^{coop} is the phonon density of states which can be written in the form of an overlap integral of the absorption line shape $S_c(E)$ of ions C with a convolution $S_{AB}(E)$ of the emission line shape of ions A and B, i.e., $\rho E^{coop} = \int dz S_c(z + E) S_{AB}(z)$. This line shape takes into account the intensity distribution of the phonon sidebands. $D_{abo}{}^{ooc}$ is the matrix element linking the system initial and final states: A and B in states a,b; C in state c. It can be expressed as a sum of products of matrix elements for interaction between two of the three ions, over virtual states of ion C, those two-ion matrix elements being, in turn, developed in a multipole expansion, as seen in Section III.A. In this calculation, it is assumed that the interaction between ions is sufficiently weak so that within a short time interval with respect to the emission lifetime of the final state, the wave function of the system loses its phase memory, and the transfer of energy and the emission of light proceed independently. Furthermore, it is also required that the interaction be such that molecular states of the two or three ions are not involved, since experimentally the transition energies in cooperative phenomena remain unchanged from the sum of the individual excitation energies (79), (80), (86).

(b) Cooperative Luminescence and Cooperative Energy Transfer with Photon Absorption. The cooperative luminescence is the emission in a single process of one photon from two excited interacting ions. Fig. 12 gives the energy scheme for such a process. This was observed by Nakazawa and Shionoya (80) between two Yb^{3+} ions. The effect is the reverse of the cooperative absorption experimentally shown by Varsanyi and Dieke (86) and discussed by Dexter (87). A nonvanishing transition probability is obtained even in the absence of overlapping wave functions with the use of first-order perturbation theory to obtain the wave functions of the two-ion system taking into account their electronic interactions. It is then found that the matrix elements for a transition from a state of the system with two ions in their excited states $|a,b\rangle$ to a state with two ions in their ground state $|0,0\rangle$ or vice-versa, with emission or absorption of one photon conserving the energy of the system, may be divided into four parts, two of which are symmetrical with respect to the ions considered. They can be written

$$-\sum_{a'\neq 0} \frac{(a'|e\,R_A|a\rangle\langle 00|H'|a'b\rangle}{\varepsilon_{a'} + \varepsilon_b} - \sum_{a'\neq 0} \frac{\langle 0|e\,R_A|a'\rangle\langle a'0|H'|ab\rangle}{\varepsilon_{a'} - \varepsilon_a - \varepsilon_b} \tag{80}$$

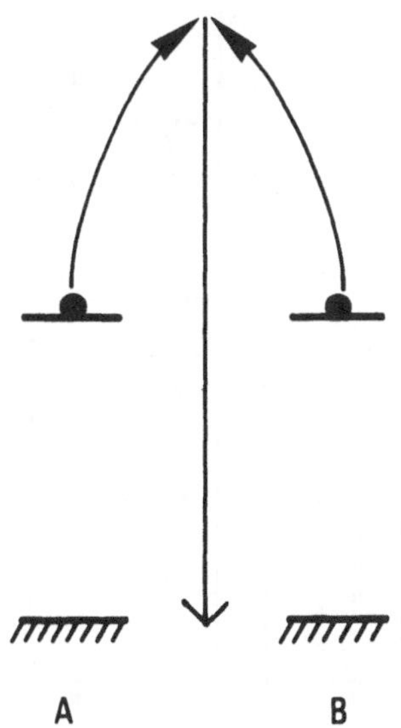

Fig. 12. Energy scheme for cooperative luminescence of ions A and B.

and their correspondents with the permutation $a \to b$ and $b \to a$. ε'_a, ε_b, and ε_a are the energies of states $|a'\rangle$, $|b\rangle$, $|a\rangle$, with respect to the ground levels $|0\rangle$. $e\,R_A$ is the perturbation associated with the electromagnetic field of the emitted or absorbed photon; H' is the Coulombic interaction between the two ions which, as seen from energy transfer, may be developed in a multipole expansion. Here again it is found that for this interaction, the degree of forbiddenness is in powers of (a_o/ρ) instead of (a_o/λ) for the radiation interaction part.

The physical significance of (80) is described in Fig. 13; the upper part of the figure corresponds to the first term: the process starts from the state $|a,b\rangle$ by emission of a photon $h\nu_\varepsilon$ to the virtual state $|a'b\rangle$ (or $|ab'\rangle$); it then proceeds nonradiatively through the interaction H' between the two ions to the final state $|00\rangle$. However, to conserve energy and if a photon is to be emitted, we have $h\nu < \varepsilon_a + \varepsilon_b$. Therefore, this term should be less important than the second, which is depicted in the lower part of the figure; interaction between the two ions takes place first with a transition from the state $|a,b\rangle$ to $|a', b\rangle$ (or $|a,b'\rangle$) and from that virtual intermediate state the emission of an anti-Stokes photon $h\nu = \varepsilon_a + \varepsilon_b$ is obtained, the system relaxing to state $|0,0\rangle$.

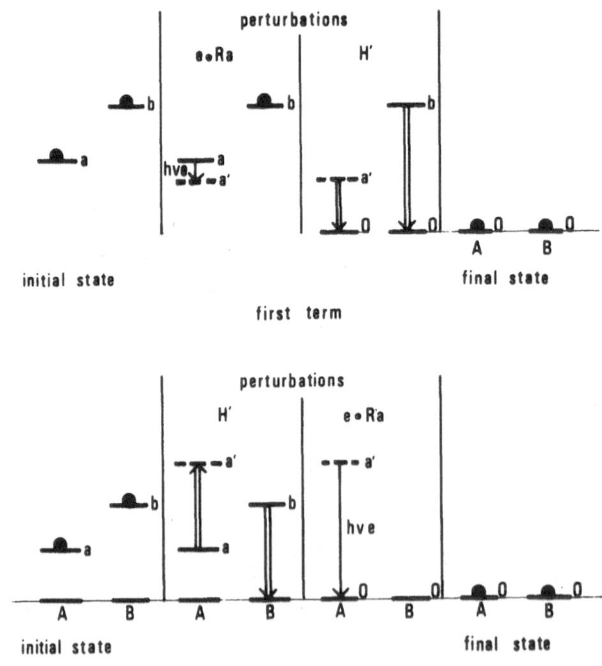

Fig. 13. Physical significance of the two terms of eq. (80)

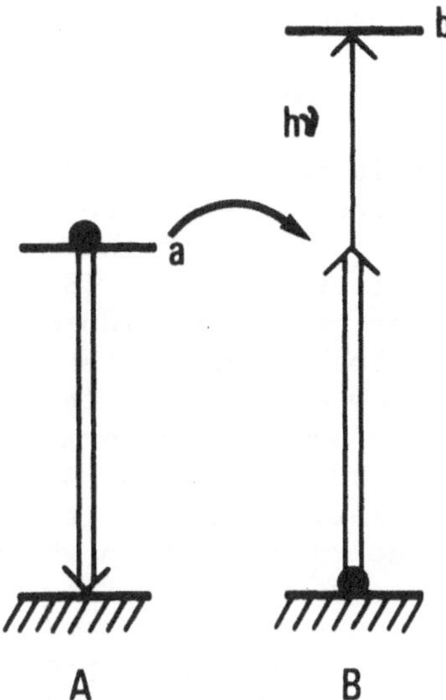

Fig. 14. Cooperative energy transfer and simultaneous photon absorption.

Closely linked to that phenomenon is the cooperative energy transfer with simultaneous photon absorption predicted by Altarelli and Dexter (88) which is shown in Fig. 14. Ion A is in state |a>; then ion B absorbs cooperatively a photon hν and reaches state |b>, so that $h\nu = \varepsilon_b - \varepsilon_a$. Once excited, state |b> can relax emitting one anti-Stokes photon.

As for cooperative luminescence,one can use the first-order perturbation theory but the initial state of the system is now |a,0>, the final state being |0, b>, which gives, when calculating the transition matrix element under hν, four nonvanishing terms analogous to (80) showing the role played by initial states.

When considering ion B alone (88), this process should be 10^4 to 10^5 times more efficient than two-photon absorption and should not be overlooked when using a laser source for hν.

(c) Discussion of the Results. The four following criteria have been used to distinguish between multiple successive transfers and cooperative phenomena;

1) the respective energy level positions and their differences;
2) the power law dependence of emission versus excitation;

3) the power law dependence of emission versus sensitizer concentration;
4) the rise and decay times of the emission.

Although these criteria permit us to distinguish between a single-ion and a two-ion process, the experimental results are the same whether the two-ion process is of the multiple successive transfer type or of the cooperative type (89). For instance, for the 2-transfers case of the green emission of Yb^{3+}-Er^{3+}, the emission law is quadratic both with excitation and Yb^{3+} concentration; the rise time can show a delay and the decay time is about $\tau_{Yb}/2$. This is because the populating term of the final state in a double-photon multiple-transfer is $\alpha_{12}\alpha_{23}\ A_1S_2^2$, whereas in a double-photon cooperative sensitization process it is αS_2^2. This is the principal difference in the rate equation where α, α_{12}, and α_{23} are found to be independent of the concentration of A or S.

In order to ascertain which type of process is involved, one can practically rely on energy difference considerations, which gives foresight as to which transfer is the most likely to occur.

This is true for Yb^{3+}-Tb^{3+}, Yb^{3+}-Ho^{3+}, Yb^{3+}-Yb^{3+}, where cooperative sensitization or fluorescence has been recognized beyond any doubt (37).

The basic difference between the two processes is in the transition selection rules or cross sections, which are highest for the process which uses the most real intermediate states. The physical interpretation of this basically is that photons travel fast so that their energies are not available in the medium for a very long time, as compared to the excitation energy in real excited states (90). This situation is summarized in Table II.

IV. APPLICATION TO STOICHIOMETRIC Nd^{3+} MINILASER MATERIALS

Now that low-attenuation optical fibers in the 1.05 μm region are available, some new interest arises for Nd^{3+} laser materials (91). But the conditions are somewhat different from the well-developed Nd^{3+}-doped YAG lasers. For a Nd^{3+} laser source to compete with semiconductor lasers, the following characteristics are necessary (91):

- overall length should be less than one centimeter
- laser threshold should be obtained for a transverse pumping from semiconductor diodes.

The compactness requirement, as well as transverse pumping, dictate a high concentration for Nd^{3+}. This comes from the fact that compactness means a high gain per unit length, that is, a high number of active ions per unit volume. As for transverse pumping, easily manufactured planar LEDs can provide high power density only when close coupling with index matching is used.

The active region in the single crystal has to be of about the same dimension as the high brilliance region of the pumping diode junction. That means that absorption of pumped light has to take place within less than 100 μm. This condition implies a high concentration of absorbing ions.

A difference with APTE up-conversion materials (92) is that the same type of ions (Nd^{3+}) contributes both to absorption and emission, so that a separate optimization of absorption and emission is no longer possible. What is tried is to reduce the constraint upon the emission optimization by using materials with small self-quenching.

IV.A. The Self-Quenching Limitation of Quantum Yield. Different Hypotheses

To obtain a high Nd^{3+} concentration ($>10^{21}$ cm^{-3}), Nd^{3+} has to be one of the matrix constituents. Such materials are usually called "stoichiometric" (93) or more appropriately "self-activated" (94) materials. But unfortunately, for any luminescence systems a high activator concentration generally leads to poor quantum efficiency for emission; this is the pervading effect of self-quenching due to the ion-ion interactions discussed in section III. Figs. 15 and 16 present a typical behaviour for fluorescence intensities and lifetimes of different materials versus Nd^{3+} concentration.

The levels of Nd^{3+} involved in pumping, emission and self-quenching are recalled in Fig. 17; pumping occurs by emission of $Al_xGa_{1-x}As$ diodes into absorption bands of ${}^4I_{9/2} \rightarrow {}^2H_{9/2}$, ${}^4S_{5/2}$ at 0.8 μm. A laser transition takes place at ≈1.06 μm (${}^4F_{3/2} \rightarrow {}^4I_{11/2}$) or at 1.35 μm for ${}^4F_{3/2} \rightarrow {}^4I_{13/2}$. As shown in Fig. 15, self-quenching is usually studied through ${}^4F_{3/2}$ lifetime variations as the Nd^{3+} concentration is increased, because the lifetime is an easily obtained parameter related to quantum efficiency for all emissions from a given level j by:

$$\eta_j = \frac{\tau}{\tau_{oj}} , \tag{81}$$

where τ is the lifetime effectively measured for ${}^4F_{3/2}$; τ_{oj} is the purely radiative lifetime of ${}^4F_{3/2}$ which can be linked to transition probabilities by eq. 42. Then a concentration quenching rate is defined as:

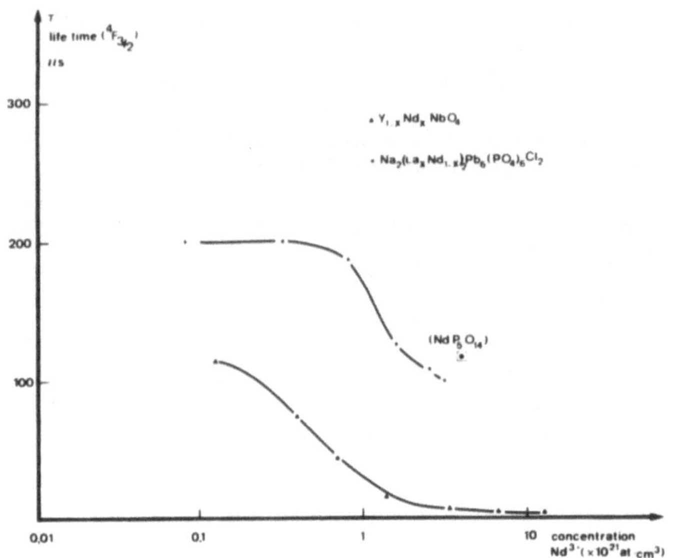

Fig. 15. Typical behaviour of emission intensity of $^4F_{3/2}(Nd^{3+})$ for strong ($NdNbO_4$) and small-quenching ($Na_2Nd_2Pb_6(PO_4)Cl_2$ materials in powder form. Reference points are given for for NdPP and YAG.

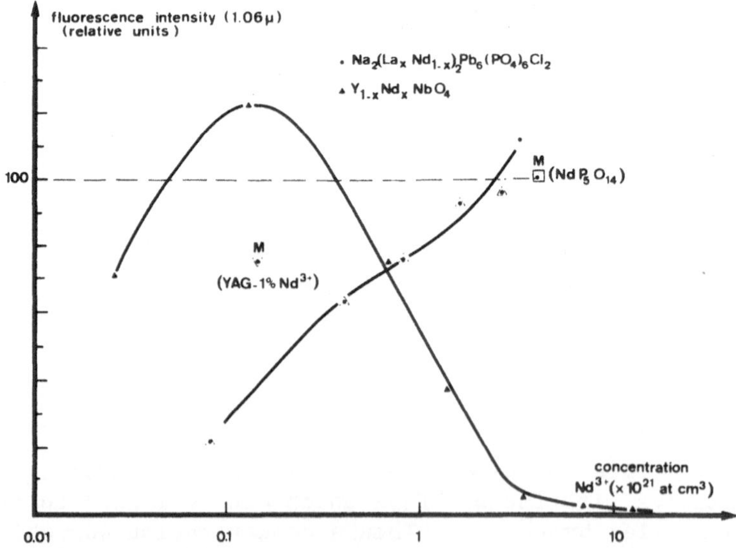

Fig. 16. Lifetime behaviour of $^4F_{3/2}(Nd^{3+})$ versus concentration for same sample as in Fig. 15.

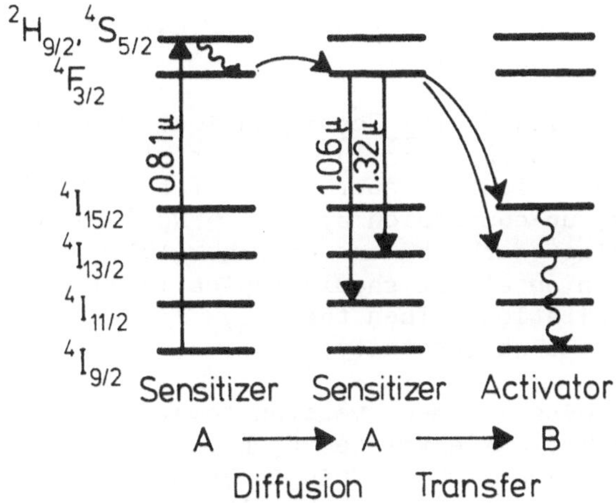

Fig. 17. Energy levels of Nd^{3+} ions involved in diode-excited laser emission and quenching.

$$R_Q = \tau^{-1} - \tau_{oj}^{-1} \quad . \tag{82}$$

Such quenching is also studied by fluorescence intensity measurements on powders (92) and by photoacoustic spectroscopy (95) (96), which is the subject of Dr. A. Rosencwaig's article in this volume.

1) C.F.O.M. Hypothesis. If one assumes τ_o to be independent of the Nd^{3+} concentration (C_{nd}) the variations of $\tau(C_{nd})$ are a good representation of those of $\eta(C_{nd})$. This point may be questioned (93) because for high Nd^{3+} concentration there can be some overlapping between the 4f configuration of one ion and the 5d configuration of its next neighbour. From this consideration, Danielmeyer proposed (97), that the lifetime decrease with concentration could be the indication of some transition probability enhancement.

As we have seen in section II.B, 4f-4f transitions of trivalent rare-earth ions are forbidden in first order by Laporte's Rule. The transitions are the result of parity mixing between 4f and 5d configurations coming from the odd crystal-field terms. In analogy with this effect the above said enhancement would come from crystal field overlap mixing of pairs (C.F.O.M.). By oscillator strength measurements as a function of concentration in $La_{1-x}Nd_xP_5O_{14}$(NdLaUp), we failed to show any probability increase within experimental error (98) (see Fig. 18). From this, one can safely consider that C.F.O.M. is not operative, at least in NdP_5O_{14}(NdPP) and that the lifetime decrease follows the quantum yield decrease. However in transition-metal ions some increase in transition probability with concentration is found (99).

2) Up-Conversion (APTE) Recombination. The APTE process was discovered while studying Er^{3+} laser materials and was looked for as a detrimental side effect of energy transfer in laser materials (34). Similarly, up-conversion by $Nd^{3+} \rightarrow Nd^{3+}$ could lead to self-quenching according to the scheme of Fig. (19). As seen in section III.C such an effect should typically have a nonlinear behavior with excitation. Then the $^4F_{3/2}$ lifetime should be sensitive to high excitation intensity.

In YAG, assuming an up-conversion towards $^2G_{9/2}$, 10^5 W/cm^3 of pumping intensity was estimated to be necessary for such an effect to be operative (100). This is beyond the usual operating limits. For NdPP, exciting more than 10% of the ions has led to the observation of up-conversion quenching in the initial part of the $^4F_{3/2}$ decay (101). By analogy with the Auger effect for X-ray emission, such APTE up-conversion recombinations are sometimes called "Auger-like" recombinations. Of course the still less probable cooperative effects can be completely neglected.

3) Exciton Annihilation. When energy migrates quickly between Nd^{3+} ions through resonant energy transfer involving the $^4F_{3/2}$ and $^4I_{9/2}$ levels, the excitation can be assumed to be delocalized all over the sample. In YAG doped with only 1% of Nd^{3+}, it was experimentally found that the excitation is delocalized over about 750 different ions during the $^4F_{3/2}$ lifetime (100). One can regard this delocalized excitation as an exciton. If one of these ions is really a quenching center (any other ion or defect to which the energy may migrate), the excitation (exciton) is annihilated. The excitonic model could be still more appropriate for stoichiometric materials because the whole host could be involved in the delocalization. However in NdPP, migration seems to occur only between identical sites (102) as shown by the existence of spatial diffusion without spectral diffusion. Diffusion in itself not being a recombination loss process, the exciton annihilation is more probable at the point where there are more defects. Experimentally it was found that the $^4F_{3/2}$ lifetime is shorter near the sample surface indicating some surface exciton annihilation. A recent systematic study of the surface recombination due to different adsorbed gases indicates, however, a limited quenching of ≈3% (102).

4) Concentration-Enhanced Electron-Phonon Coupling. As pointed out in section II, multiphonon decay is generally independent of concentration. However, we recently have shown that S_o, the Huang-Rhys coupling parameter, could double between 1% and 100% concentration for Eu^{3+} (40). This could increase the multiphonon nonradiative

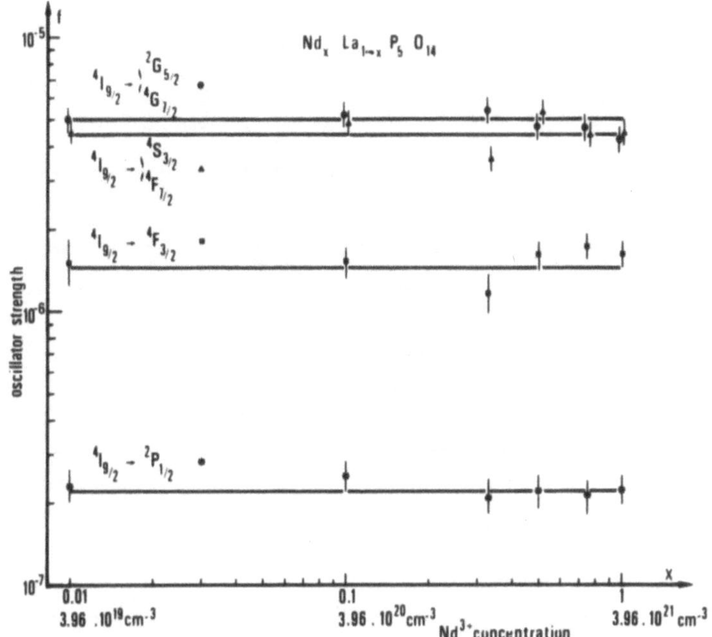

Fig. 18. Constancy of Nd^{3+} oscillator strengths with concentration in a stoichiometric material (NdPP).

decay by one order of magnitude without the need for the energy transfer described in section III. Up to now such effects have been found only for Eu^{3+} and cannot be generalized without further experiments (39) (40).

5) Cross-Relaxation Processes. The four processes above produce only moderate effects for Nd^{3+} and cannot explain why in YAG, for instance, concentration quenching is important ($R_Q \simeq 10^5 s^{-1}$) even at $\simeq 10^{20}$ cm^{-3} whereas in some other materials such as NdPP it is not ($R_Q \simeq 5x10^3 s^{-1}$). The strong self-quenching of Nd^{3+} in $Na_{0.5}Gd_{0.5}WO_4$:Nd (104), LaF_3:Nd (105), YAG:Nd (106) or glass (107) has been well-known for a long time to be due to the following cross-relaxation energy transfer:

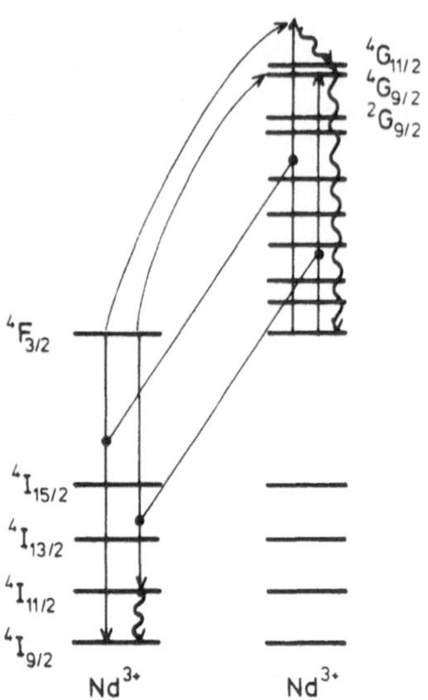

Fig. 19. Up-conversion self-quenching of Nd^{3+} by an A.P.T.E. process.

$$Nd_A^{3+}(^4F_{3/2}) + Nd_B^{3+}(^4I_{9/2}) \longrightarrow 2Nd^{3+}(^4I_{15/2}) \tag{83}$$

as depicted in Fig. 17. Such transfer is rather of the nonradiative resonant type (see III.A). When energy-matching for resonance is too bad, phonon-assisted energy transfer may take place in addition to eq. (83) according to:

$$Nd_A^{3+}\ (^4F_{3/2}) + Nd_B^{3+}\ (^4I_{9/2}) \longrightarrow 2\ Nd^{3+}\ (^4I_{13/2}) + N_1\ h\omega_m \tag{84}$$

or

$$Nd_A^{3+}\ (^4F_{3/2}) + Nd_B^{3+}\ (^4I_{9/2}) \rightarrow Nd_A^{3+}(^4I_{13/2}) + Nd_B^{3+}\ (^4I_{15/2}) + N_2 h\omega_m \tag{85}$$

Of course, at high concentration, energy migration is also involved and one can view cross-relaxation as an exciton annihilation process. Because of the rapid multiphonon nonradiative relaxation between levels above $^4F_{3/2}$, the most likely energy path for diffusion is:

$$Nd_A{}^{3+}(^4F_{3/2}) + Nd_B{}^{3+}(^4I_{9/2}) \longrightarrow Nd_A{}^{3+}(^4I_{9/2}) + Nd_B^{3+}(^4F_{3/2}). \quad (86)$$

Since transition probabilities involved in the above relations for resonant transfer are not expected to vary a lot from matrix to matrix (108), the role played by the overlap integral should be most important in comparing strong- and small-quenching materials. This result comes directly from eq. (56) of section III.A.

Table III presents some previously unpublished results of current work in our laboratory, showing effectively that measured overlap integrals ($^4F_{3/2} \rightarrow {}^4I_{15/2}$ and $^4I_{9/2} \rightarrow {}^4I_{15/2}$) are systematically smaller for small-quenching materials.

On the other hand, the overlap for the $^4F_{3/2} \rightarrow {}^4I_{9/2}$ diffusion transition is much larger and about constant. From this it can be inferred in our view that small-quenching or strong-quenching materials behave in a way predicted by known energy-transfer processes without resorting to new ones.

In the case of NdLaPP and other small-quenching materials (see Fig. 16), the quenching rate R_Q is found to be:

$$R_Q = KN \quad (87)$$

where N is the Nd^{3+} concentration. This is the same form as eq. (76) for the "fast-diffusion case."

In the case of YAG, glasses and strong-quenching materials, quenching is found to be (see Fig. 20):

$$R_Q = kN^2 \quad , \quad (88)$$

which is the form of eq. (75) for the diffusion-limited case. Such a form has also been found for NdPP (109). This discrepancy can be linked with the following result: For $K_3Nd(PO_4)_2$(KNP), at concentration $<3.2 \times 10^{20}$ cm^{-3}, a law in kN^2 is found whereas for $N > 3.2 \times 10^{20}$ cm^{-3} the law in KN predominates (110).

This may be ascribed to a passage from limited diffusion to fast diffusion with increasing concentration (94). Also as seen in III.A, the distinction between the two regimes, for fixed concentration, is in relation to the strength of the cross-relaxation

TABLE III

Nd^{3+} spectral overlap integral for diffusion ($^4F_{3/2}$, $^4I_{9/2}$) and quenching ($^4F_{3/2}$, $^4I_{15/2}$, $^4I_{9/2}$) of strong and small self-quenching materials.

Materials	Maximum concentration (cm^{-3})	Quenching rate at maximum concentration R_Q (s^{-1})	Cross-relaxation overlap	
			Quenching (Kayser^{-1})	Diffusion (Kayser^{-1})
$NdNa(WO_4)_2$	5.45 10^{21}	2.4x10^5	6.5 10^{-4}	15 10^{-4}
YAG:Nd	.8 10^{21}	1 x 10^5	4 10^{-4}	15.7 10^{-4}
NdP_5O_{14} (Nd P.P)	3.96 10^{21}	5.2 10^3	0.65 10^{-4}	11.6 10^{-4}
$Na_2Nd_2Pb_b(PO_4)Cl_2$ (CLAP)	3.4 10^{21}	6 x 10^3	1.4 10^{-4}	18.3 10^{-4}

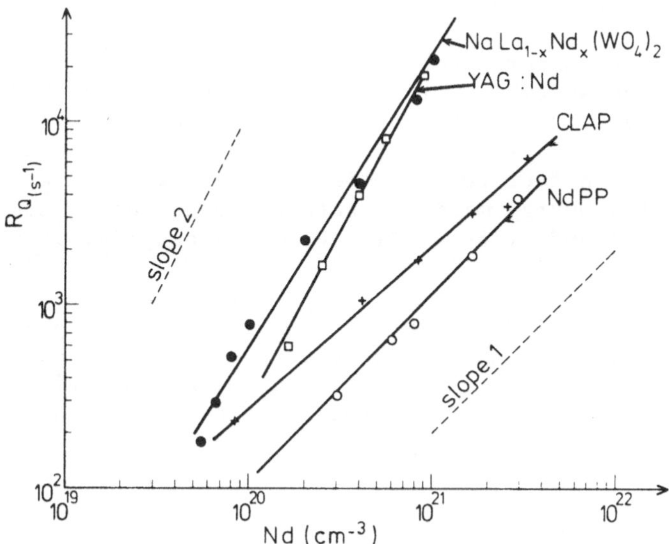

Fig. 20. Self-quenching rate for high- and low-quenching materials. Points for YAG and NdPP taken from Ref. (94).

energy transfer with respect to the diffusion one: a strong quenching overlap would give a "limited diffusion," whereas a small overlap would give a "fast diffusion." But for a particular crystal growth, if the relative probabilities for diffusion and quenching are modified whatever the reason, we think one could expect a change in the expected diffusion behavior.

The results of Table III for overlaps are in good agreement with the concentration power law behavior of Fig. 20.

At any rate, the linear or the square law does not reflect the type of interactions, though this argument is sometimes found in the literature to reject or assume cross-relaxation, dipole-dipole interactions, or exchange. Observed laws are only of a statistical origin, the type of interaction giving the magnitude of the constant. For the same reason, the shortest distance between ions does not seem to be the most important parameter, though so generally admitted in the literature (111). Proof is given by the results of the following section.

IV.B The Role of the Crystal Field and Necessary Criteria for Nd^{3+} Minilaser Materials

In order to understand which is the principle parameter involved in the spatial overlap, let us consider the Nd^{3+} "free-ion" states (Fig. 21). The positions of such states are obtained from the "center of gravity" of the experimental crystal-field states (112). A fact to be noticed is that the energy difference (e_2-e_1) for the energy mismatch between energy differences of the $(^4I_{15/2}, {}^4I_{9/2})$ and $(^4F_{3/2}, {}^4I_{15/2})$ level couples, is about constant (75).

For NdF_3 one finds $(e_2-e_1) = -\ 440\ cm^{-1}$; for Nd_2S_3 one has $(e_2-e_1) \simeq -\ 426\ cm^{-1}$, though these compounds are at each end of the nephelauxetic scale (113), between which the variation in level position is largest. On the average (e_2-e_1) can be considered

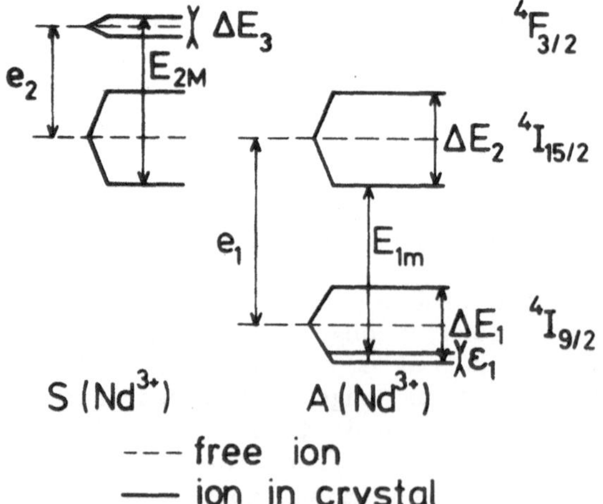

Fig. 21. Free-ion and crystal-field levels of Nd^{3+} involved in self-quenching of $^4F_{3/2}$.

as an invariant with value $\simeq -430\ cm^{-1}$. The negative sign indicates that energy transfer could not take place in the "free-ion" state. Overlap is due essentially to the crystal field splitting compensating the energy mismatch. Inside a crystal, the new energy difference to be considered is $E_{2M} - E_{1m}$ giving the non self-quenching condition:

$$E_{2M} - E_{1m} < 0 \quad . \tag{89}$$

Assuming the center of gravity of the crystal states is about at the center of the crystal field maximum splitting, one has from Fig. 21 the following nonoverlapping condition (75):

$$\Delta E_2 - \frac{\Delta E_1}{2} + \frac{\Delta E_3}{2} < 380\ cm^{-1} \quad , \tag{90}$$

with ΔE_1; ΔE_2; ΔE_3 the Stark splitting for, respectively, $^4I_{9/2}$; $^4I_{15/2}$; $^4F_{3/2}$.

Now we look for a simple link between the splitting of cross-relaxation involved levels and crystal field parameters, whatever the site symmetry.

The crystal states are given by the eigenvalues of the matrix:

$$< f^3\ SLJJ_z \left| V \right| f^3\ SL'J'J'_z > \tag{91}$$

where V is the crystal-field site potential:

$$V = \sum_{k \neq 0,q,i} B_q^k\ (C_q^k)_i \tag{92}$$

already seen in II.B.

By comparing the experimental results to the eigenvalues of the crystal potential matrix, experimental values of B_q^k are obtained. For rare-earth ions k = 2,4,6 and q has 2k+1 values such that $|q| < k$; the number of different B_q^k may be as high as 15 for low-symmetry sites and as small as 2 for cubic sites. This precludes at first sight a crystal field strength comparison for different site symmetries.

However eq. (92) can be considered as the definition of a function of an Hilbert space described by spherical harmonics. In such a space a norm can always be defined. Then we shall compare norms of functions v, even when some of them are in a two-dimensional space.

So we introduce a crystal field strength scale parameter as:

$$N_v = \left[\sum_{k\neq 0,q} \frac{4\pi}{2k+1} (B^k_q)^2 \right]^{\frac{1}{2}} \simeq \left[\sum_{k\neq 0,q} (B^k_q)^2 \right]^{\frac{1}{2}} \tag{93}$$

where $(4\pi/2k+1)^{\frac{1}{2}}$ is the normalization factor which can be neglected in a first approximation except when $k \neq 4,6$ only. N_v is found to describe well the maximum splitting for different site symmetries as long as the considered B^k_q in N_v are the same as those necessary for the degeneracy removal of a given J level. This is when: $2J \geq k$ for $k = 2,4,6$.

Experimentally the splitting of $^4I_{9/2}$ is found (Fig. 22) to be well described by:

$\Delta E_1(^4I_{9/2}) \simeq 0.23\ N_v$ (correl. coeff = 0.988)

also $\Delta E_2(^4I_{15/2}) \simeq 0.27\ N_v$ (correl. coeff = 0.988)

$\Delta E_3(^4F_{3/2}) \simeq 0.026\ N_v$ (correl. coeff = 0.64) .

The lower correlation coefficient for $^4F_{3/2}$ is explained by the violation of the inequality $2J \geq k$. Taking:

$$N_v = \left[\sum_{k=2,q=0,2} (B^k_q)^2 \right]^{\frac{1}{2}} \tag{94}$$

would increase correlation to 0.92 but would reduce generality. ΔE_3 being smaller than ΔE_1 and ΔE_2, this point is neglected.

The nonoverlap condition, eq. (90) is reduced then to:

$\Delta E_1 < 470\ \text{cm}^{-1}$

or $N_v < 1800\ \text{cm}^{-1}$,

indicating that small-quenching materials are to be searched for among small crystal-field materials. As shown in Fig. 22, this well describes as a whole the materials known from the literature.

Though a general theory of melting points of solids (Tm) does not exist up to now, we have been looking for a relation between Tm and N_v. We found (75):

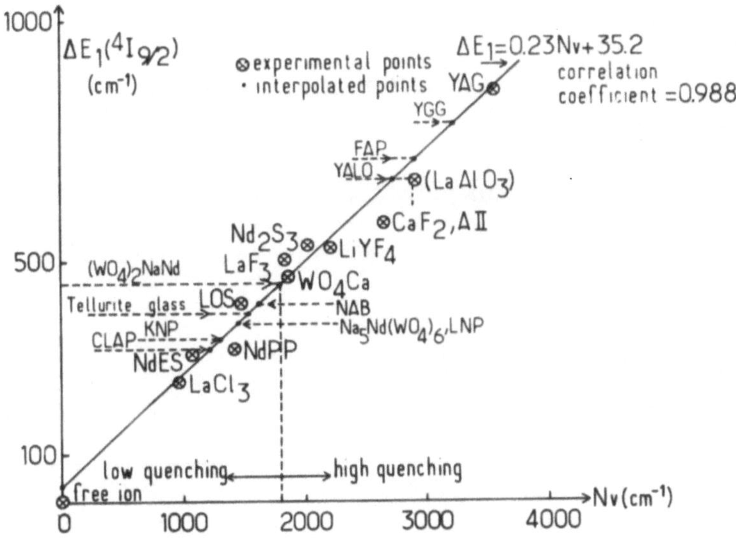

Fig. 22. Maximum splitting of $^4I_{9/2}(Nd^{3+})$ versus scalar crystal-field strength (N_v), showing the low-quenching region.

$$Tm \simeq 0.48\ N_v + 300 \qquad \text{(corr. coeff. 0.88).}$$

Translation of the criterion upon N_v shows that small quenching would be obtained for (see Fig. 23) :

$$Tm < 1200°C \quad ;$$

the large intercept (300 cm^{-1}) would correspond to the B^o_o term not contained in N_v.

In another approach, the lattice energy obtained from scratch hardness measurements indicates equivalently that small quenching should be obtained for materials with small bond strength U/z (114):

$$U/z < 500 \text{ kcal mole}^{-1} \text{ per bond.}$$

For a given cohesive energy (U), the number of bonds (z) should be large. This can explain why all minilaser materials known to date have a large number of component ions (4 on the average) except $NdCl_3$ which is of very low cohesive energy (≃ 680 kcal $mole^{-1}$).

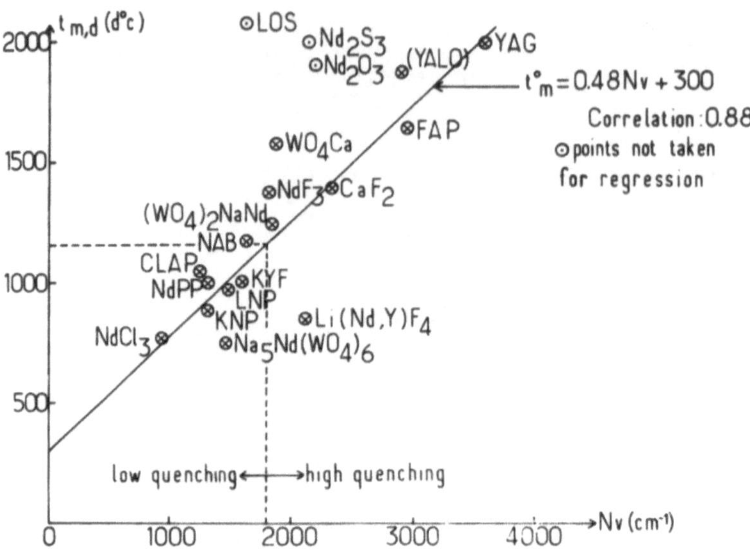

Fig. 23. Melting or decomposition temperature versus crystal-field strength (N_v), showing low-quenching region.

IV.C. Prediction of Self-Quenching Behavior of Other Ions: Eu^{3+}, Tb^{3+}, and Pr^{3+}.

Concentration quenching of $Eu^{3+}(^5D_o)$ and $Tb^{3+}(^5D_4)$ levels are known to be small in any matrix, due to the large energy gap to the next lower level (115), (116) and for this reason cannot be considered as typical for concentration-quenching studies.

On the other hand, $Eu^{3+}(^5D_1)$ and $Tb^{3+}(^5D_4)$ are, respectively, quenched through the following energy transfers (see Fig. 24):

$$Eu^{3+}(^5D_1) + Eu^{3+}(^7F_o) \longrightarrow Eu^{3+}(^5D_o) + Eu^{3+}(^7F_3 \text{ or } ^7F_2);$$

$$Tb^{3+}(^5D_3) + Tb^{3+}(^7F_o) \longrightarrow Tb^{3+}(^5D_4) + Tb^{3+}(^7F_o \text{ or } ^7F_1) .$$

"Free-ion" energy differences (e_2-e_1) involved in such transfers are readily found to be positive except for $^5D_1 \rightarrow ^7F_3$ for which $(e_2-e_1) \simeq -120$ cm^{-1}.

The splitting of $^7F_3$ is found to be described by $\Delta E(^7F_3) \simeq 0.112\, N_v$ and $\Delta E(^5D_1)$ is taken as $\simeq 30$ cm^{-1} (2J = 2, case); this gives a condition for the field: $N_v < 890$ cm^{-1}. From Fig. 22 this condition is never realized for the usual host, even $LaCl_3$. It means that small-quenching inorganic materials are not likely to

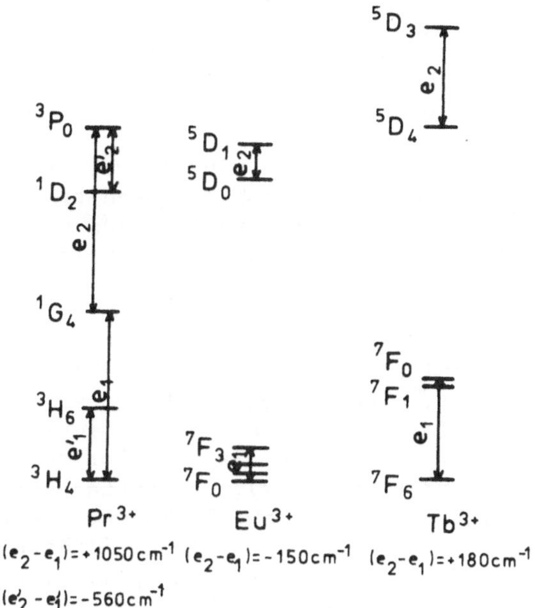

Fig. 24. Free-ion levels of Pr^{3+}, Eu^{3+}, Tb^{3+} involved in the self-quenching of $^3P_o(Pr^{3+})$, $^5D_1(Eu^{3+})$, $^5D_3(Tb^{3+})$.

be discovered for $Eu^{3+}(^5D_1)$ and $Tb^{3+}(^5D_3)$ within the limits of our description. This may explain why EuP_5O_{14} and TbP_5O_{14} show very little or no emission from $^5D_1$ and $^5D_3$, though they were claimed to be small-quenching materials (117), (118) (in fact for (^5D_o) and $(^5D_4)$ which, as said above, is not pertinent).

A similar prediction is also obtained for $Pr^{3+}(^3P_o)$ (Fig. 24). From the energy-level scheme, likely cross-relaxations are:

$$Pr^{3+}(^3P_o) + Pr^{3+}(^3H_4) \rightarrow Pr^{3+}(^3H_6) + Pr^{3+}(^1D_2) \quad \text{with} \tag{95}$$

$(e_1' - e_2') \simeq -560\ cm^{-1}$ and $Pr^{3+}(^3P_o) + Pr^{3+}(^3H_4) \rightarrow 2Pr^{3+}(^1G_4)$ + phonons, (96)

for which an average "free-ion" energy difference is found to be positive: $(e_2 - e_1) = +\ 1050\ cm^{-1}$. This last self-quenching is always likely to take place with the assistance of the emission of phonons, the energy of which depends on the crystal-field splitting. This prediction is confirmed. Recent experiments with stoichiometric materials known to give small quenching for Nd^{3+}, such as $K_5PrLi_2F_{10}$, have shown a strong room-temperature quenching for Pr^{3+} (^3P_o) (119).

However, this last case is not so straightforward: the phonon-dependent process of eq. (95) would have to be compared with the probability of the process of eq. (96) which could behave as for $Nd^{3+}(^4F_{3/2})$. We propose to relate this to an "abnormal" behavior of PrP_5O_{14} recently found (120).

In this material with a large vibrational frequency of the P-O bond, selection rules for one-phonon assisted transfer (121) could inhibit the transfer. Whereas in the fluoride case, the lower vibrational frequency would relax selection rules through a higher-order multiphonon process.

In summary, for a resonant condition, several numerical criteria have been given for splitting, crystal field, cohesive energy per bond and melting point. The method can be generalized to any ion, but quenching properties have to be analyzed level by level and cannot be generalized to other levels, ions or matrices. The strong quenching of $Tb^{3+}(^5D_3)$ (117), for instance in TbP_5O_{14}, or of $Pr^{3+}(^3P_o)$ in $K_5PrLi_2F_{10}$ in which the Tb-Tb and Pr-Pr distance could not be much shorter than the Nd-Nd distance in NdP_5O_{14} and $K_5NdLi_2F_{10}$, is an additional proof of the primary role of resonance over the shortest-distance one (though it is the hypothesis generally advanced up to now (111)).

Of course, a resonance criterion cannot give by itself any information about absolute quantum yield; low site symmetry and small optimum phonon coupling conditions have also to be considered (75) in order to obtain fully "necessary and sufficient" criteria for small-quenching materials.

V. CONCLUSION

Multiphonon processes and energy transfer, which are the basic phenomena involved in radiationless processes such as nonradiative decay, cross-relaxation and up-conversion, have been reviewed. Though the former are mainly concentration-independent, as opposed to the latter, a common philosophy can be drawn from the review of their theoretical treatment: The experimental behavior, through the observed relative law, often bears more statistical than physical meaning. The physics behind them is globally given by a scale factor providing absolute values. The application of the knowledge of such processes to the research of small self-quenching materials permits one to give a simple rule of thumb: If (e_2-e_1) is $\gtrsim 0$, no small-quenching materials are likely to be found. If $(e_2 - e_1) < 0$, small-quenching materials are likely to be found for materials with crystal-field splittings of order smaller than (e_2-e_1). From this, one can conclude that there is no small self-quenching material per se, as was thought when NdP_5O_{14} was proposed. This has been very recently confirmed by the strong self-quenching of $PrP_5O_{14}(^1D_2)$ (120) and $K_5PrLi_2F_{10}$ (119).

However, to go beyond the gross features, I think more experimental and theoretical research is needed in the understanding of the respective role played by the inhomogeneous and homogeneous part of the spectral overlaps involved in energy transfer.

REFERENCES

1 K.H. Hellwege, Ann. de Phys. 40, 529 (1942).

2 G.E. Barasch and G.H. Dieke, J. Chem. Phys. 43, 988 (1965).

3 F. Varsanyi, in Quantum Electronics, (P. Grivet and N. Bloembergen, eds) Dunod, p. 787 (1964).

4 S. Murata, C. Iwanaga, T. Toda, and H. Kokubun, Chem. Phys. Lett. 13, 101 and 15, 152 (1972).

5 N.F. Mott, E.A. Davis and R.A. Street, Phil. Mag. 32, 961 (1975).

6 F. Auzel, in Luminescence of Inorganic Solids (B. Di Bartolo, ed.), Plenum, p. 67 (1978).

7 J. Frankel, Phys. Rev. 37, 17 (1931).

8 F.K. Fong, Theory of Molecular Relaxation, John Wiley and Sons (1975).

9 S. Fisher, J. of Chem. Phys., 53, 195 (1970).

10 K.K. Rebane, Impurity Spectra of Solids, Plenum Press (1970).

11 D. Curie, Optical Properties of Ions in Solids, NATO N°8, Plenum, p. 71 (1975).

12 K. Huang and A. Rhys, Proc. Roy. Soc. A.204, 406 (1950).

13 M. Lax, J. Chem. Phys. 20, 1752 (1952).

14 R. Kubo and Y. Toyozawa, Progr. Theor. and Phys. 13,160 (1955)

15 W. Heitler, The Quantum Theory of Radiation, Oxford Press, p. 66 (1954).

16 T. Miyakawa and D .L. Dexter, Phys. Rev. B1, 2961 (1970).

17 R. Englman and J. Jortner, Mol. Phys. 18, 145 (1970).

18 K.F. Freed, Radiationless Processes in Molecules and Condensed Phases (F.K. Fong. ed.) Springer-Verlag (1976).

19 D.J. Diestler, Radiationless Processes in Molecules and Condensed Phases, (F.K. Fong ed.) Springer-Verlag (1976).

20 A. Erdelyi, Asymptotic Expansions, Dover Pub. Inc. (1956).

21 Handbook of Chemistry and Physics (C.D. Hodgman, ed.) The Chemical Rubber Pub. Co. p. 267 (1953).

22 N. Robertson and L. Friedman, Phil. Mag. 33,753 (1976).

23 L. Shiff, Quantum Mechanics, Mc Graw Hill, p. 70 (1949)

24 J.J. Markham, Rev. of Modern Phys., 31, 956 (1959).

25 F.K. Fong, S.L. Naberhuis, and M.M. Miller, J. Chem. Phys. 56, 4020 (1972).

26 L.A. Riseberg and H.W. Moos, Phys. Rev. 174, 429 (1968).

27 H.W. Moos, J. of Lum. 1,2, 106 (1970).

28 A. Kiel, Quantum Electronics, (P. Grivet and N. Bloembergen eds). Dunod, p. 765 (1964).

29 W.E. Hagston and J.E. Lowther, Physica 70, 40 (1973).

30 T.C. Mc Gill, Optical Properties of Highly Transparent Solids, (S.S. Mitra and B. Bendow, eds), Optical Physics and Engineаring series, Plenum, p.3. (1975).

31 B. Di Bartolo, Optical Interactions in Solids, Wiley, New-York (1968).

32 B.R. Judd, Phys. Rev. 127, 750 (1962)

33 D.L. Dexter, Solid State Physics, 6, 353 (1958).

34 F. Auzel, Ann. Telecom. 24, 199 (1969).

35 T. Miyakawa, Luminescence of Crystals, Molecules and Solutions (R. Williams, ed.) Plenum, p. 394 (1973).

36 W.E. Bron and W. Wagner, Phys. Rev. 139 A, 233 (1965).

37 F. Auzel, Phys. Rev. B13, 2809 (1976).

38 Y. Toyozawa, Dynamical Processes in Solid State Optics, (R. Kubo, ed.). Benjamin p. 90 (1967).

39 T. Hoshina, S. Imanaga and S. Yokono, J. of Lum. 18/19,88 (1979.

40 F. Auzel, G.F. De Sá and W.M. de Azevedo, "Dynamical Processes in the Excited States of Ions and Molecules in Solids Conference", Madison, Wi (1979). To be published in Bull. Ann. Phys. Soc.

41 G.F. Imbusch, Luminescence of Inorganic Solids (B. Di Bartolo ed.). Plenum, p. 155 (1978).

42 G.H. Dieke and E. Dorman, Phys. Rev. Lett. 11, 17 (1963).

43 E. Dorman, J. Chem. Phys. 44, 2910 (1966).

44 H.G. Kuhn, Atomic Spectra, Longmans, p. 390 (1962)

45 F. Auzel, Ann. Telecom. 24, 363 (1969).

46 F. Auzel and D. Pecile, C.R. Acad. Sc. Paris B277, 155 (1973)

47 T. Förster, Ann. Phys. (Germany) 2, 55 (1948).

48 T. Förster, Radiat. Res. (Supplement) 2, 326 (1960)

49 D. Curie, Luminescence Cristalline, Dunod, Paris (1960)

50 D.L. Dexter, J. Chem. Phys. 21, 836 (1953)

51 T. Kushida, J. Phys. Soc. Jpn. 34, 1318 (1973)

52 J.F. Pouradier and F. Auzel, J. de Phys. 39, 825 (1978)

53 J.D. Axe and P.F. Weller, J. Chem. Phys. 40, 3066 (1964)

54 R. Orbach, Optical Properties of Ions in Crystals (H.M. Crosswhite and H.W. Moos, eds). Intersciences, p. 445 (1967)

55 R.Orbach, Optical Properties of Ions in Solids, (B. Di Bartolo, ed.). Plenum, p. 355 (1975)

56 N. Yamada, S. SHionoya, and T. Kushida, J. Phys. Soc. Jpn. 32, 1577 (1972)

57 M. Inokuti and F.Hirayama, J. Chem. Phys. 43, 1978 (1965)

58 E. Nakazawa and S. Shionoya, J. Chem. Phys. 47, 3211 (1967)

59 L.G. Van Vitert, J. of Lum. 4, 1 (1971)

60 M. Yokota and O. Tanimoto, J. Phys. Soc. Jpn.,22, 779 (1967)

61 M.J..Weber, Phys. Rev. B4, 2932 (1971)

62 N. Krasutsky and H.W. Moos, Phys. Rev. B8,1010 (1979)

63 N.V. Artomonova, C.M. Briskim, A.L. Burshtein, L.D. Zusman and A.G. Skleznen, Sov. Phys. JETP 35, 457 (1972)

64 E. Okamoto, M. Sekita, and H. Masui, Phys. Rev. B11, 5103 (1975)

65 F. Auzel, Proc. IEEE 61, 758 (1973)

66 R.A. Hewes and F.F. Sarver, Phys. Rev. 182, 427 (1969)
67 T. Miyakawa and D.L Dexter, Phys. Rev. B1, 70 (1971)
68 W.J.C. Grant, Phys. Rev. 109, 648 (1958)
69 P.W. Anderson, Phys. Rev. 109,1492 (1958).
70 D.C. Ahlgren and R. Kopelman, J. Chem. Phys. 70, 3139 (1979)
71 S. Chu, H.M. Gibbs, A. Passner, and S. Geshwind, "Dynamical Processes in the Excited States of Ions and Molecules in Solids Conference" Madison, Wi. (1979), to be published in Bull. Am. Phys. Soc.
72 J. Koo, L.R. Walker, and S. Geschwind, Phys. Rev. Lett.35 1669 (1975)
73 R. Reisfeld, Structure and Bonding, 30, 65 (1976)
74 J.P. Budin, J.C. Michel, and F. Auzel, J. Appl. Phys. 50, 641 (1979)
75 F. Auzel, Mat. Res. Bull. 14, 223 (1979)
76 S.A. Brawer and M.J. Weber, Appl. Phys. B19, 32 (1979)
77 J.M. Flaherty and R.C. Powell, Phys. Rev. B 19, 32 (1979)
78 F. Auzel, C.R. Acad. Sci. (Paris), 262,1016 (1966) and C.R. Acad. Sci. (Paris), 263, 819 (1966)
79 V.V. Ovsyankin and P.P. Feofilov, Sov. Phys. JETP Lett. 4, 317 (1966)
80 E. Nakazawa and S. Shionoya, Phys. Rev. Lett. 25, 1710 (1970)
81 R.K. Watts, Optical Properties of Ions in Solids (B. Di Bartolo, ed). Plenum, p. 337 (1975)
82 F. Auzel and D. Pecile, J. of Lum. 11, 321 (1973)
83 P.P. Feofilov and V.V. Ovsyankin, Appl. Opt. 6, 1828 (1967)
84 D.L. Dexter, Phys. Rev. 108, 630 (1957)
85 R.G. Kepler, J.C. Caris, P. Avakian, and E. Abrahamson, Phys. Rev. 10, 400 (1963)
86 F. Varsanyi and G.H. Dieke, Phys. Rev. Lett. 7, 442 (1961)
87 D.L. Dexter, Phys. Rev. 126, 1962 (1962)
88 M. Altarelli and D.L. Dexter, Opt. Commun. 2, 36 (1970)
89 F. Auzel, CNET Internal Report PEC 55 (1970) unpublished.

90 D.L. Dexter, "Luminescence of Crystals, Molecules and Solutions" (F. Williams ed.). Plenum, p. 57 (1973).

91 J.P. Noblanc, Appl. Phys. (Germ). 13, 211 (1977) and J.P. Budin, M. Neubauer and M. Rondot I.E.E.E. J. of Quantum electr. QE 14 831 (1978)

92 F. Auzel, Proceeding of the 2nd International School on Semiconductor Optoelectronics, CETNIEWO 6-14 mai 1978, Wiley, (to be published 1980).

93 H.G. Danielmeyer, Solid State Physics XVI, Pergamon, p. 253 (1953)

94 A.A. Kaminskii, Luminescence of Inorganic Solids, (B. Di Bartolo, ed.) . Plenum, p. 511 (1978)

95 F. Auzel, D. Meichenin, and J.C. Michel, J. of Lum. 18/19, 97 (1979)

96 R.S. Quimby and W.M. Yen, Optics Letters, 3, 181 (1978)

97 G.H. Danielmeyer, J. of Lum. 12/13, 715 (1976)

98 F.Auzel, IEEE J. of Quantum Electron. Q.E. 12, 258 (1976)

99 G.F. Imbusch, Luminescence of Inorganic Solids (B. Di Bartolo, ed). Plenum, p. 155 (1978)

100 H.G. Danielmeyer and M. Blätte, Appl. Phys. (Germ) 1,269 (1973)

101 M. Blätte, H.G. Danielmeyer, and R. Verich, Appl. Phys. (Germ) 1, 275 (1973)

102 J.M. Flaherty and R.C. Powell, Phys. Rev. B19, 32 (1979)

103 R.C. Powell, "Dynamical Processes in the Excited States of Ions and Molecules in Solids Conference". Madison, Wi (1979) to be published in Bull. Am. Phys. Soc.

104 G.E. Peterson and P.M. Bridenbaugh, J.O.S.A. 54, 644 (1964).

105 C.K. Asawa, and M. Robinson, Phys. Rev. 141, 251 (1966)

106 H.G. Danielmeyer, Advances in Lasers, Vol. IV ; Decker (1975)

107 J. Chrysochoos, J. Chem. Phys. 61, 4596 (1974)

108 F. Auzel and J.C. Michel, C.R. Acad. Sci. (Paris) B279,187 (1975)

109 W. Strek, C. Szafranski, E. Lukowiak, Z. Mazurak and

B. Jerowska, Phys. Stat. Solid. 41, 547 (1977)

110 H.Y. Hong and S.R. Chinn, Mat. Res. Bull. 11, 421 (1976)

111 S.R. Chinn, H.Y Hong, and J.W. Pierce, Laser Focus, p. 64 (May 1976)

112 G.H. Dieke, Spectra and Energy Levels of Rare-Earth Ions in Crystals, Interscience Publishers,(1968).

113 P. Caro and J. Derouet, Bull. Soc. Chim. France N°1,46 (1972)

114 F. Auzel, The Rare-Earths in Modern Science and Technology. Vol. II, (G.J. Mc Carthy and J.B. Gruber, eds). Plenum, to be published .

115 L.G. Van Uitert and R.R. Soden, J. Chem. Phys. 32, 1687 (1960)

116 G.E. Peterson, and P.M. Bridenbaugh, J.O.S.A. 53, 1129 (1963)

117 B. Blanzat, J.P. Denis, and J. Loriers, Proc. 10th R.E. Conf. 2, 1170 (1973)

118 C. Brecher, J. Chem. Phys. 61, 2297 (1974)

119 A. Lempicki and B.C. Mc Collum, J. of Lum. 20, 291 (1979)

120 H. Dornauf and J. Heber, J. of Lum. 20, 271 (1979).

RADIATIONLESS PROCESSES IN LUMINESCENT MATERIALS

G. Blasse

Physical Laboratory, State University
P.O. Box 80.000, 3508 TA Utrecht, The Netherlands

ABSTRACT

Many luminescent materials are nowadays well known for their application. A very important property of such a material is the quantum efficiency of its luminescence which should be as high as possible. Nonradiative processes of importance are energy transfer (migration) between centres and nonradiative transitions between energy levels of a luminescent centre. In this series of lectures we will give (a) a short survey of energy migration and its importance for the efficiency of luminescent materials. (b) a qualitative approach to nonradiative transitions based upon the simple single-configurational coordinate diagram. Broad band as well as narrow line emission will be considered. Rules of thumb will be proposed which make the prediction of efficient luminescent materials possible. (c) a summary of the approach proposed by Struck and Fonger to nonradiative transitions. This approach covers broad band as well as narrow line emission, it justifies the use of the single configurational-coordinate model for many purposes (by considering multiple-coordinate and multiple-frequency models) and makes model calculations possible. The results of some calculations will be shown.

I. INTRODUCTION

Luminescent materials have nowadays a wide field of application. The more well-known examples are the fluorescent lamps and the cathode-ray tubes. One of the most important requirements for a luminescent material is a high value of the light output of the phosphor. All energy absorbed by the phosphor which is not emitted as radiation (luminescence) is dissipated to the crystal lattice by radiationless processes. This explains why these processes have always been a subject of many investigations in the phosphor field. In the early years, however, not much was known about their nature. Later qualitative considerations became available which were useful in the search for new luminescent materials. Only during recent years it has become possible to apply methods with a quantitative character to understand the radiationless processes better. It is the purpose of this chapter to discuss radiationless processes in luminescent materials along these lines.

Let us first have a look at the efficiency of some luminescent materials. In Table 1 we have given the radiant efficiency of some cathode-ray phosphors. This radiant efficiency is defined as the ratio of the emitted luminescent power to the power of the electron beam falling on the phosphor. It is clear that the radiant efficiency is far from 100%. In fact materials with efficiencies exceeding 20% are rare. The losses are here due to three factors, viz.: (a) the back-scattering of exciting electrons, (b) the ratio of the mean energy for production of electrons and holes to the band-gap energy (which amounts roughly a factor of 3), and (c) the efficiency of the processes in the luminescent centre itself. Using quantitative estimates for these values the theoretical maximum efficiency is estimated to be some 20% (2). Here we neglect the influence of possible killer centres (i.e. centres where nonradiative processes dominate). In this chapter we will not deal further with the efficiency of cathode-ray phosphors (see also the contribution by F. Williams to this course).

In Table 2 we have given the quantum efficiency of some photoluminescent phosphors. The quantum efficiency is the ratio between the number of emitted quanta to the number of absorbed quanta. Note that even a conversion factor like the quantum efficiency is far from 100% for commercial phosphors. The success of the Ca-halophosphate in fluorescent lamps is certainly not due to its quantum efficiency which is only some 70%. These figures illustrate

TABLE 1. RADIANT EFFICIENCY OF SOME STANDARD PHOSPHORS ISSUED BY THE NBS UNDER CATHODE-RAY EXCITATION (1).

number	sample	radiant efficiency (%)
1020	ZnS-Ag	21
1021	Zn_2SiO_4-Mn	8
1022	ZnS-Cu	11
1023	(Zn,Cd)S-Ag	19
1026	$CaWO_4$-Pb	3

TABLE 2. QUANTUM EFFICIENCY OF SOME PHOSPHORS UNDER ULTRAVIOLET EXCITATION (1, 3, 4, 5).

sample	quantum efficiency (%)
$CaWO_4$-Pb	75
Zn_2SiO_4-Mn	68
$Ca_3(PO_4)_2$-Tl	56
$Ca_5(PO_4)_3(F,Cl)$-Sb,Mn	71
Gd_2O_3-Eu^{3+}	52
YVO_4-Eu^{3+}	70
$Sr_2P_2O_7$-Eu^{2+}	65

the interest in the factors which determine the value of the quantum efficiency, especially if one considers the fact that the greater part of all luminescent materials emit with even lower efficiency. As an example we mention some well-known examples: the efficiency of La_2O_3-Eu^{3+} is considerably lower than that of Gd_2O_3-Eu^{3+}; $CaWO_4$ luminesces efficiently at room temperature, $SrWO_4$ does not luminesce at all.

All data mentioned above have been measured at room temperature. The efficiencies of the commercial materials increase hardly upon lowering the temperature. Nevertheless it is clear that the thermal quenching temperature of luminescence is also an important property in connection to the quantum efficiency. In this chapter we will deal exclusively with quantum efficiencies and thermal quenching temperatures under ultraviolet excitation. The advantage of such an approach is that we know which energy level of which ion (or group of ions) we are exciting.

It will be clear that, if we are attempting to explain and predict quantum efficiencies of luminescent materials in general, our approach will be approximative. Nevertheless it is possible to come to some general conclusions which are very helpful in predicting efficient new phosphors.

II. HISTORICAL DEVELOPMENTS

Since the end of the thirties, various models have been proposed to explain the presence or absence of luminescence. These models are based on what is termed the configurational-coordinate diagram. For a full account see ref.6 or 7. We shall start by considering this type of diagram (Fig.1). The potential energy of the

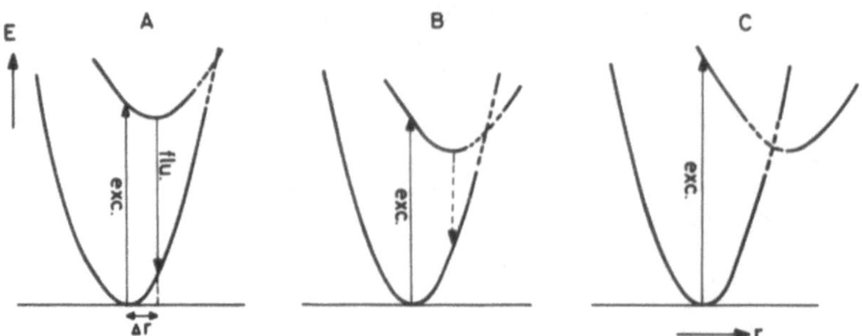

Fig. 1. Configurational coordinate diagrams. A: Mott-Seitz model, B: Dexter model, C: Seitz model (see text).

luminescent centre MO_n in the crystal lattice is plotted as a function of the configurational coordinate r. In the symmetric vibrational mode the quantity r represents the distance M-O. At higher temperature, higher vibrational levels may be occupied.

Due to the absorption of radiation of the appropriate wavelength the centre can be raised to an excited state. Since the equilibrium distance r_e of the excited state will not in general be equal to that of the ground state, and since the centre may be at different vibrational levels, this transition can correspond to a fairly broad absorption band.

Once in an excited state, the system will relax towards the equilibrium configuration of the excited level by dissipating heat. From this level the system returns to the ground state, thereby emitting radiation. The emission too, therefore, may consist of a broad band. Line emission is found in the case of small Δr as, for example, in the case of rare-earth ions. The emission generally lies at a lower energy than the absorption. This displacement of emission with respect to absorption is known as the Stokes shift.

From the configurational-coordinate diagram in Fig. 1A we can now understand why the emission will be quenched at higher temperature. If the luminescent centre is in the equilibrium configuration of the excited state, it may also, as a result of thermal activation, occupy a vibrational level situated near the point of intersection of the curves representing the excited and ground states (activation energy ΔE). Having arrived here, the centre will return non-radiatively to the equilibrium configuration of the ground state, dissipating heat in the process.

With the aid of the simple model in Fig. 1A (the Mott-Seitz model) we can therefore explain;

a) the broad-band character of the emission and absorption of many centres;

b) the Stokes shift of the emission;
c) the temperature dependence of the emission.

If now the equilibrium configuration of the excited state lies outside the curve of the ground state, then after excitation the intersection point of both curves is reached before the above-mentioned equilibrium configuration, and the system relaxes non-radiatively to the ground state (Fig.1C). No emission is then possible. This is the model which Seitz (8) proposed to explain the absence of luminescence in certain cases.

Dexter et al. (9) proposed a different model. This shows that under less rigorous circumstances non-radiative transitions to the ground state may occur (Fig.1B). The characteristic feature of the situation is that the intersection point of the two curves lies below the level reached after excitation. When, after excitation, the system relaxes to the equilibrium configuration of the excited state, the intersection point of the two curves is passed. Here too, a temperature-independent, radiationless return to the ground state can take place.

We learn from these models that the difference Δr between the equilibrium configuration of the excited state and that of the ground state must be small if luminescence is to occur.

III. ENERGY TRANSFER

In the model sketched in the section above the nonradiative losses occur in the luminescent centre itself. Before continuing this discussion we draw attention to the fact that nonradiative losses may also occur elsewhere in the lattice. This can only be of importance if the excitation energy of the luminescent centre is transferred to other centres. The theory of energy transfer is nowadays well known and we shall illustrate the relevant phenomena only by some examples. For the theory the reader is referred to refs.6, 7 and 10.

Let us start with the important phosphor YVO_4-Eu^{3+}. Undoped YVO_4 does not luminesce at room temperature, but emits an efficient blue luminescence at lower temperatures (11, 12). The absence of the blue emission at higher temperatures cannot be due to radiationless processes in the vanadate group, since the diluted system $YP_{1-x}V_xO_4 (x < 0.2)$ luminesces efficiently up to some 500 K. In the concentrated YVO_4 system, however, rapid energy migration among vanadate groups occurs at not too low temperatures. The critical distance for this energy transfer process amounts to some 8 Å, which is considerably longer than the shortest V-V distance. The migrating energy is finally captured by a killer centre so that no luminescence is observed (concentration quenching).

If we introduce Eu^{3+} ions in this lattice on the Y^{3+} sites, the Eu^{3+} ions can compete succesfully with the killer centres in trapping the migrating excitation energy. As a consequence the YVO_4-Eu^{3+} system emits an intense red Eu^{3+} luminescence upon excitation into the vanadate groups at not too low temperatures. This shows how the excitation energy of a vanadate group can be lost for the luminescence from that vanadate group itself.

At lower temperatures the situation becomes different. Energy transfer among the vanadate groups becomes inactive, because it is a thermally activated process. This is due to the Stokes shift of the emission relative to the excitation (Fig.2). As a consequence, below some 200 K the unactivated YVO_4 shows its vanadate luminescence. But the YVO_4-Eu^{3+} behaves not very differently. Since the VO_4^{3-}-Eu^{3+} transfer proceeds only over small distances (~ 4 Å), the emission of YVO_4-Eu^{3+} below 200 K consists mainly of vanadate emission.

The situation in YVO_4-Eu^{3+} can therefore be characterized by stating that at low temperatures the emission originates from the same centres which were directly excited, but at higher temperatures efficient transfer carries the excitation energy to traps (either killer sites or Eu^{3+} ions), and only at very high temperatures non-radiative processes in the intrinsic centres themselves are responsible for the loss of luminescence.

This is not generally the case as can be illustrated nicely on Sm^{3+}-activated scheelites (13, 14). In $CaWO_4$-Sm^{3+}, $CaMoO_4$-Sm^{3+} and $PbWO_4$-Sm^{3+} at low temperatures the situation is similar to that in

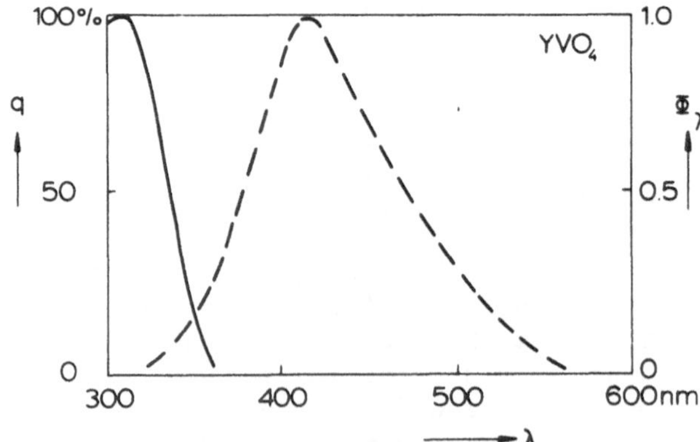

Fig.2. Excitation (left) and emission (right) spectrum of YVO_4 at liquid nitrogen temperature.

YVO_4-Eu^{3+}: no energy transfer, no radiationless processes. Upon increasing the temperature energy transfer and radiationless processes in intrinsic centres start at roughly the same temperature in $CaWO_4$. In $CaMoO_4$, however, we observe first radiationless processes and only at still higher temperatures there is some indication for energy migration. In $PbWO_4$-Sm^{3+} (and also $PbMoO_4$-Sm^{3+}) the different processes occur in the same sequence as in YVO_4-Eu^{3+}. This is schematically represented in Fig.3.

A completely different situation occurs if the energy migration does not need thermal activation. This is the case, for example, in compounds of Mn^{2+}, U^{6+} or RE^{3+}.

In MnF_2, for example, there is rapid energy migration through the Mn^{2+} sublattice, even at the lowest temperatures (7, 15). The emission is only for a very small part from the intrinsic Mn^{2+} ions. The migrating energy is trapped by those Mn^{2+} ions which have a slightly different energy level scheme due to the presence of impurities or defects. In MnF_2 the emission at low temperatures originates from Mn^{2+} ions which have Mg^{2+} ions as neighbours. The Mg^{2+} concentration is only in the ppm range.

In Ba_2CaUO_6 the uranate excitation energy migrates through the lattice, even at low temperatures. The highly efficient emission is from uranate groups near regions of disorder in this ordered perovskite (16) (see Fig.4). In $MgUO_4$ the luminescent efficiency remains low down to very low temperatures (17). Here also efficient

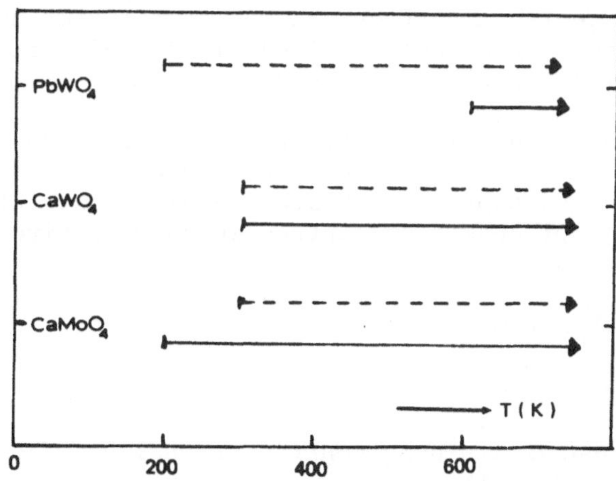

Fig.3. Schematic representations of the temperature regions in which energy migration (dotted line) and thermal quenching (full line) occur for $PbWO_4$, $CaWO_4$ and $CaMoO_4$. In the low temperature region no temperature-dependent radiationless processes occur.

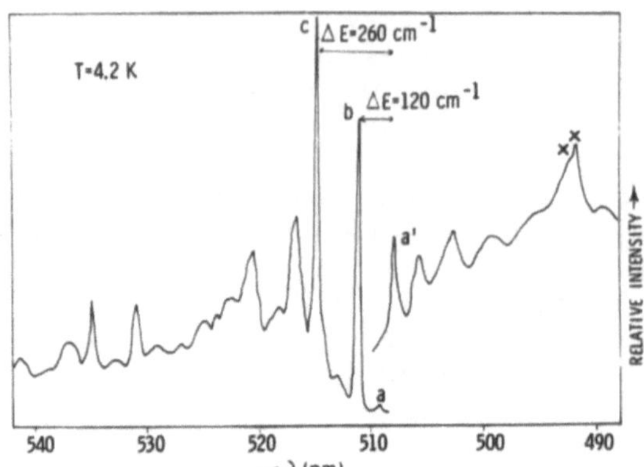

Fig.4. Part of the emission and excitation spectrum of Ba_2CaUO_6 at 4.2 K. Note that the emission zero-phonon lines a, b and c do not coincide with the excitation zero-phonon line a'.

energy migration occurs. The emission originates from different uranium traps dependent on temperature (see Fig.5), but the killer sites are competing with these uranium traps for the migrating excitation energy. The killer centres in this case are U^{5+} centres.

Concentrated rare earth compounds also belong to this class. Consider, for example, NdP_5O_{14}. It has been proposed that in this compound rapid energy migration occurs (spatial but nonspectral) which ends at surface quenching sites (18). This matter is discussed by F. Auzel elsewhere in this volume.

We have shown in this section that radiationless losses in luminescent materials may be due to the occurrence of energy transfer. In the remaining part of this chapter we will restrict ourselves to radiationless processes in centres which were directly excited by the incident radiation.

IV. QUALITATIVE APPROACH

IV.A Two Parabolae Models

The two parabolae single configurational coordinate model has been given in section II. Two very important parameters which determine the influence of radiationless processes are the offset between the two parabolae (Δr in Fig.1) and the energy difference between their respective minima. Let us consider these two parameters separately and start with the offset Δr.

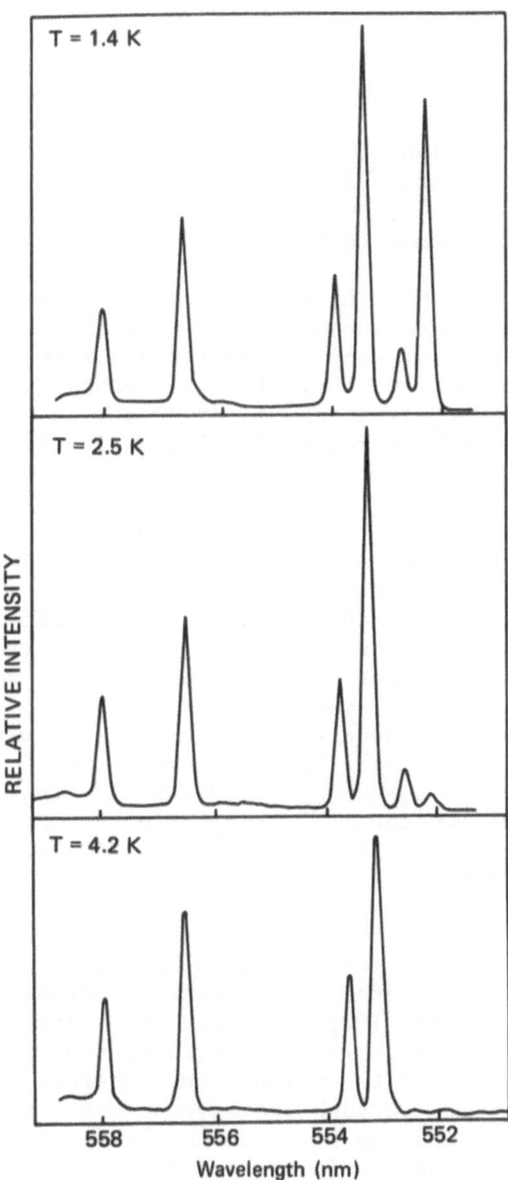

Fig.5. Emission spectra of $MgUO_4$ at different temperatures indicating emission from different uranate traps as a function of temperature.

Representative examples of centres with very small offset are the rare earth ions as far as we consider their $4f^n$ energy levels. Radiationless transitions between these levels have been studied intensively. It is very important by how many lattice phonons the energy difference between the two levels can be matched. If this number is high (say above 5), the probability of the radiationless transition becomes low. If the energy difference is large, one can be sure of efficient luminescence with high quenching temperature. Well-known examples are the $^5D_0$ emission of the Eu^{3+} ion, the $^5D_4$ emission of the Tb^{3+} ion and the $^6P_{7/2}$ emission of the Gd^{3+} ion. For further treatment we refer to R. Reisfeld's contribution to this volume.

Representative examples of centres with large offset are F-centres, tungstates and vanadates. Due to the large offset their emission and absorption spectra consist of broad bands without vibrational structure and their thermal quenching temperature is relatively low. Roughly speaking the quenching temperature is the lower, the larger the Stokes shift (19), which is determined by the value of the offset.

In order to predict the chemical composition of efficient luminescent materials we are looking for some general rules which indicate how the offset is determined by the crystal lattice. An important remark in this connection is that stiff lattices will imply relatively low values of Δr (20), so that they will constitute suitable host lattices for our purpose. It is a well-known fact that compounds like silicates, borates and phosphates are often used as host lattices for efficient luminescent materials (for example: Zn_2SiO_4-Mn^{2+}, $BaSi_2O_5$-Pb^{2+}, $BaSi_2O_5$-Eu^{2+}, $Ca_3(PO_4)_2$-Tl^+, $Ca_3(PO_4)_2$-Ce^{3+}, $Sr_2P_2O_7$-Eu^{2+}, $Cd_2B_2O_5$-Mn^{2+}). The sharp decrease of the quenching temperature in the series $CaWO_4$, $SrWO_4$, $BaWO_4$ and, similarly, in the series $LiYSiO_4$-Tb^{3+}, $LiYGeO_4$-Tb^{3+} can also be explained in this way.

A more refined Δr effect has been reported for the uranium luminescence in ordered perovskites A_2BWO_6 (21). The luminescent centre is a UO_6^{6-} octahedron. This regular octahedron is coordinated by six B^{2+} ions in the [100] direction and eight A^{2+} ions in the [111] direction (Fig.6). This means that, seen from the uranium, the B^{2+} ions are behind the oxygen ions, whereas the A^{2+} ions are located in directions through the oxygen ions. The thermal quenching temperature of the uranium emission does not depend on the radius of the A^{2+} ion, whereas it increases if the radius of the B^{2+} ion decreases. In view of the location of the A and B ions it is in fact expected that expansion of the luminescent centre upon excitation cannot be counteracted effectively by the A^{2+} ions, and that the B^{2+} ions have to receive this expansion. The larger the B^{2+} ion, the larger can be the expansion, i.e. Δr.

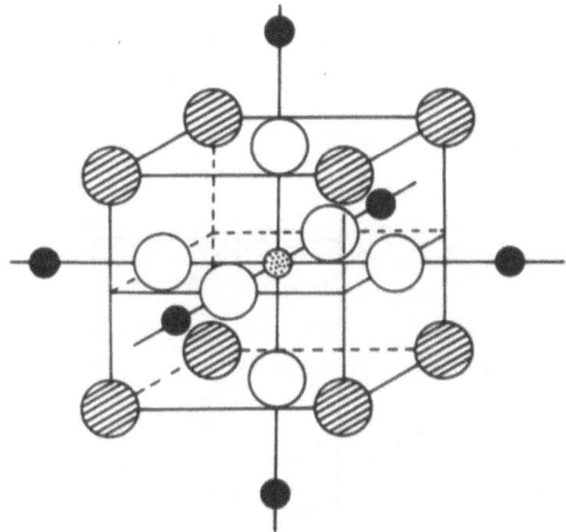

Fig.6. The ordered perovskite structure of compounds A_2BWO_6. Open circles O, hatched circles A, black circles B, dotted circle W.

These simple and rough rules connecting Δr with the nature of the host lattice can be extended further and seem to have a general validity, so that they can be used to predict host lattices for efficiently-emitting phosphors (20).

Let us now turn to the energy difference between the two parabolae. A qualitative inspection (Fig.7) shows already that the thermal quenching temperature of the luminescence will decrease, if the energy difference between the parabolae becomes smaller (for equal offset). This is corroborated by the general experience that broad band blue emission is a more general phenomenon than broad band red emission. A more detailed example is the difference between the tungstate and the molybdate luminescence in the scheelite structure. In view of the strong analogy between these two centres (i.e. equal Stokes shift and equal half width of the emission bands), we may assume that Δr will be the same for both centres. The lower quenching temperature in the case of $CaMoO_4$ is then ascribed to its lower energy emission (see Table 3). A more detailed example is the case of the octahedral niobate luminescence (19, 20). Here we find a relation between the value of the thermal quenching temperature of the luminescence and the spectral position of the absorption band. The samples with absorption at high energies show a considerably higher quenching temperature of their luminescence (Fig. 8).

Let us close this section by noting the work of Bartram and Stoneham (22) who were able to explain why the F centre luminesces

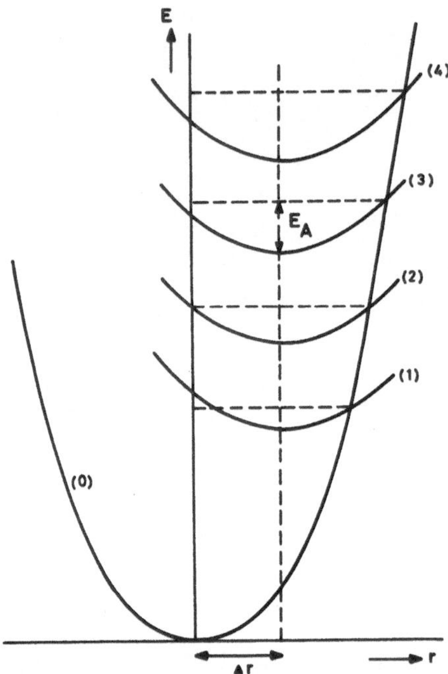

Fig.7. Illustration of the relation between the position of the absorption band and the quenching temperature of the emission. Δr is fixed. The shape of the excited parabolae does not change. Note the increase of the activation energy E_A from (1) to (4).

TABLE 3. SOME LUMINESCENT PROPERTIES OF $CaMoO_4$ and $CaWO_4$.

sample	Stokes shift	Half width emission band	Maximum emission band	Thermal quenching temperature
$CaMoO_4$	16.000 cm^{-1}	3500 cm^{-1}	19.000 cm^{-1}	350 K
$CaWO_4$	16.000 cm^{-1}	3500 cm^{-1}	24.000 cm^{-1}	425 K

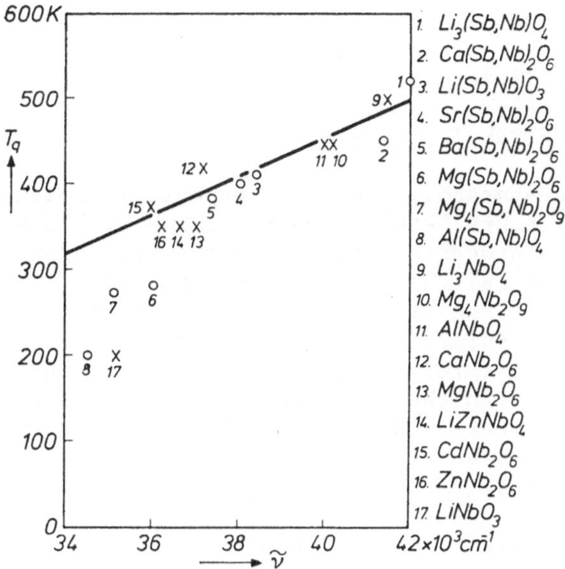

Fig. 8. Quenching temperature T_q of the niobate luminescence in various host lattices vs. the position of the maximum of the excitation band.

in some alkali halides, but not in others by using the Dexter criterium (9). They use a parameter Λ which is related to the offset of the parabolae (Λ = excited-state lattice-relaxation energy/optical-absorption energy). Some examples are given in Table 4).

TABLE 4. PREDICTION OF THE OCCURRENCE OF LUMINESCENCE OF F CENTRES IN ALKALI HALIDES (22).

Host	Λ*	luminescence		Host	Λ*	luminescence	
		predicted	observed			predicted	observed
NaI	0.384	no	no	NaCl	0.260	yes	yes
NaBr	0.375	no	no	KI	0.231	yes	yes
LiCl	0.371	no	no	KF	0.189	yes	yes
LiF	0.323	no	no	NaF	0.175	yes	yes
				RbI	0.211	yes	yes
				RbF	0.173	yes	yes

* condition for luminescence $\Lambda \lesssim 0.25$

We conclude that the very simple two-parabolae model is able to explain and predict in a very rough way the importance of radiationless processes. An extension to a three-parabolae model offers even more possibilities as will be shown now.

IV.B Three-Parabolae Models

A three-parabolae model is especially suitable if the emission transition occurs between two parabolae with a small offset, but the nonradiative processes via a higher third parabola with considerably larger offset (Fig.9): emission occurs via CD, the radiationless processes via CFGA (i.e. thermally activated). This approach has become very popular for trivalent rare earth ions. The parabolae with the small offset relate to energy levels of the $4f^n$ configuration. the parabola with the large offset relates to a charge-transfer state or a $4f^{n-1}5d$ state. But the model is more general. It has, for example, also been used to explain the radiationless processes in uranate centres.

The processes responsible for nonradiative losses in the Eu^{3+} ion upon excitation into the charge-transfer (c.t.) state have been elucidated mainly by Struck and Fonger (23). The first indication that the c.t. state of Eu^{3+} plays a role in the luminescence

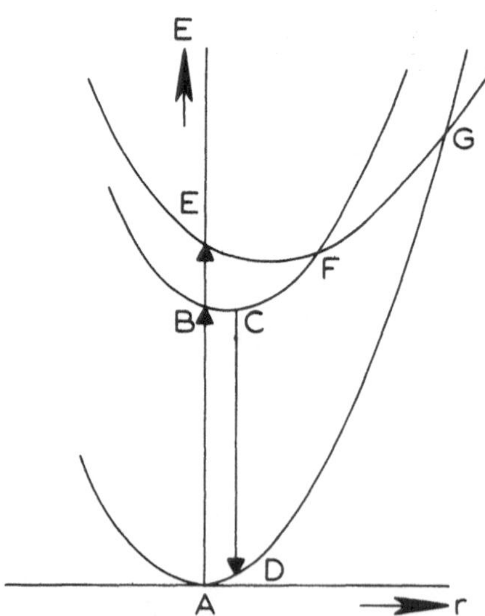

Fig.9. Three-parabolae configurational coordinate model to illustrate quenching of the CD emission via the large offset parabola.

quenching process was the fact that there is a relation between the spectral position of the first c.t. band of Eu^{3+} and the quenching temperature and room-temperature quantum-efficiency of the luminescence under excitation into the c.t. band (20). Bril and co-workers (24) showed that at room temperature the luminescence quantum efficiency for Eu^{3+} in $YAl_3B_4O_{12}$ amounts to 35% for excitation into the c.t. band and to 100% for excitation into the narrow 4f levels. It is a simple task to show that in a simple configurational coordinate model the quenching temperature of the luminescence and the cross-temperature quantum efficiency decrease, if the position of the c.t. band is at increasingly lower energy. This was mentioned above.

The picture became more clear by the work of Struck and Fonger on temperature quenching of trivalent lanthanides in the oxysulfides (23, 25). In host lattices like Y_2O_2S and La_2O_2S the c.t. band of the Eu^{3+} ion is situated at about 30.000 cm^{-1}. This is lower than in the greater part of the oxides due to the lower electronegativity of sulfur. Struck and Fonger observed direct feeding of the excited $^5D(4f^6)$ levels of Eu^{3+} by the c.t. state, but also ^5D quenching via the c.t. state. They used a configurational coordinate diagram as given in Fig. 10. The important fact is that, although the c.t. state lies well above the emitting ^5D states in the absorption and excitation spectra, its Franck-Condon shifted minimum lies relatively low (somewhere near $^5D_3$). As a consequence crossovers from ^5D levels to the c.t. state are possible.

The direct contact between the c.t. state and the ^5D levels is shown by direct feeding of the ^5D levels by the c.t. state. If the Eu^{3+} ion in Y_2O_2S is excited into the $^5L_7$ level, emission is observed from $^5D_3$, $^5D_2$, $^5D_1$ and $^5D_0$. The same emission spectrum is observed for excitation into the $^5D_3$ level. If excitation is into the c.t. state, i.e. at higher energies, emission occurs only from $^5D_2$, $^5D_1$, and $^5D_0$ in the same ratio as if the excitation had occurred into the $^5D_2$ level. This means that c.t. excitation skips the $^5L_7$ and $^5D_3$ levels and feeds directly the $^5D_2$ level. In La_2O_2S-Eu, where the c.t. band is at still lower energy, the excited c.t. state feeds directly the $^5D_1$ level for about two-thirds and the $^5D_2$ level for about one-third.

The temperature dependence of the ^5D emissions in Y_2O_2S-Eu^{3+} for excitation into the c.t. state has also been studied. Although the total emission intensity is practically temperature-independent the separate ^5D emissions quench sequentially in the order $^5D_3$, $^5D_2$, $^5D_1$, $^5D_0$ with increasing temperature. For La_2O_2S-Eu^{3+} the same sequence has been found, but the corresponding quenchings occur at lower temperatures. These quenchings are due to thermally promoted transitions from the ^5D levels to the c.t. state followed by return crossovers to lower ^5D states. In the oxysulfides the c.t. state is low enough to allow such transitions. The crossover rates for

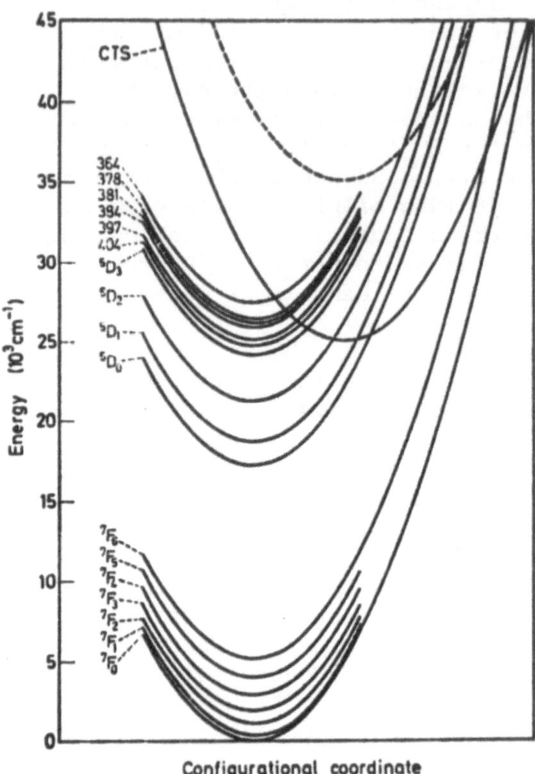

Fig.10. Configurational coordinate model for the Eu^{3+} ion. The small offset parabolae relate to the $4f^6$ states, the large offset parabolae to the charge transfer state (full line in oxysulfides, broken line in oxides).

c.t. state → ^5D levels are estimated to be 10^{11}-10^{12} sec^{-1}, so that the absence of luminescence from the c.t. state is understandable.

It will be clear that, if the c.t. state is at higher energy, these phenomena will no longer be observable. The present example illustrates the importance of low-lying charge-transfer states in nonradiative transitions.

It has been shown that in the case of the isoelectronic Sm^{2+} ion ($4f^6$) similar processes occur. The compound $BaCl_2$-Sm^{2+} for example, shows at 4.2 K mainly $^5D_1$ emission. This emission is quenched at about 130 K and the $^5D_0$ emission takes over. At still higher temperature the $^5D_0$ emission diminishes and is replaced by broad-band luminescence, due to emission from the 5d configuration (26). The 5d level plays here the same role as the charge-transfer state in the case of Eu^{3+}, the difference being that in the case of Sm^{2+} the interfering configuration level shows also luminescence. It is possible to show that in the case of $BaCl_2$-Sm^{2+} the $^5D_0$ level is populated from the 5d configuration which acts as an intermediary in the thermal quenching of $^5D_1$.

Also the uranate emission (UO_6^{6-}) is thermally quenched by a higher excited state as shown by the data in Table 5 (27). It is once again striking how well the simple model works. Finally we would like to mention some experiments using high pressure, because they are of interest in this connection.

IV.C High-Pressure Studies

From our simple model we expect that under high pressure the luminescence efficiency and the quenching temperature will increase if the offset is positive (expanded excited state), but that they will decrease if the offset is negative (contracted excited state). It is interesting in this connection to cite some results obtained by Drickamer's group (28).

The tungstate and molybdate centre have definitely a positive offset (excitation into antibonding orbitals). For these compounds the quenching temperature increases with increasing pressure. The Eu^{2+} ion, however, has a negative offset. This follows from the experiments, but also from theoretical considerations. In fact it was found that the quenching temperature of the Eu^{2+} luminescence decreases with increasing pressure (Fig. 11).

TABLE 5. SOME LUMINESCENCE DATA ON U^{6+}-ACTIVATED PHOSPHORS (OPTICAL DATA IN 10^3 cm^{-1}).

Host lattice	Maximum excitation band*	Maximum emission band**	Difference between previous columns	$T_{1/2}$
$Y_3Li_2Te_2O_{12}$	31	17.5	13.5	540 K
Mg_3TeO_6	31.5	18.7	12.8	500 K
Li_4WO_5	29.5	19.2	10.5	440 K
Ba_2MgWO_6	27	19.5	7.5	350 K
$BaNa_{0.4}W_{0.6}O_3$	24	19.2	5	< 100 K

* AE in Fig. 9. **CD in Fig. 9.

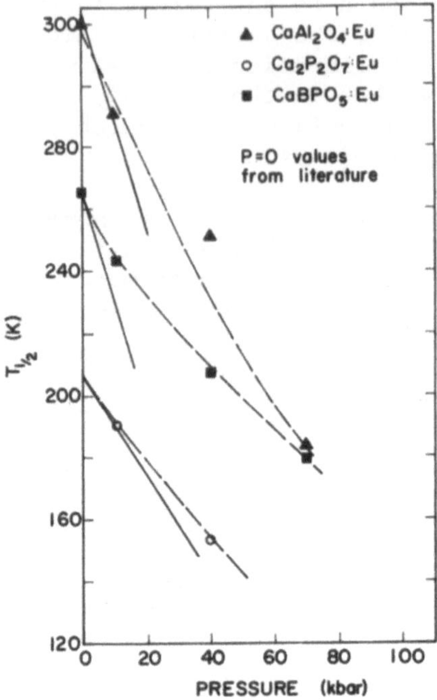

Fig.11. Quenching temperature of Eu^{2+} luminescence as a function of pressure.

It is also interesting to note that for organic compounds there exists an $\eta^{2/3}$ dependence of luminescence intensity on viscosity. This can also be explained by a simple two parabolae model assuming a less curved parabola in the excited state if pressure is applied (29).

Finally we note that, if La_2O_2S-Eu^{3+} is studied under pressure, the charge-transfer state shifts to higher energy (30). As a consequence the amount of $^5D_3$ emission from the Eu^{3+} ion increases considerably with pressure as would be expected from the Struck and Fonger investigations reported above.

We conclude that the simple configurational coordinate model can explain very reasonably a lot of luminescent phenomena in which radiationless processes play a role. This is surprising in view of the simplicity of the model (harmonic approximation, one vibrational mode). Up till a few years quantitative calculations were not very feasible. In the next section we will sketch the developments in this field during recent years which have been greatly influenced by the work of Struck and Fonger (see also the contribution of the latter to this course).

V. QUANTITATIVE APPROACH

V.A General

Struck and Fonger (31) have given a unified model of the temperature quenching of narrow-line and broad-band luminescence using a quantum-mechanical single-configurational coordinate (QMSCC) model. Their model is based upon Condon's quantum-mechanical handling of the Franck-Condon principle. Calculations were made possible by evaluating the vibrational overlap integrals exactly using the Manneback recursion formulae. Parameters for this model are depicted in Fig.12. In this figure the horizontal coordinate is the single configurational coordinate Q, the vertical coordinate is the total energy E of the system. The parabolae u and v are the potential energy wells of the electronic ground state and the relevant excited state, respectively. The u and v parabolae quantum numbers, wavefunctions, and phonon energies are called n, u_n, $\hbar\omega_u$ and m, v_m, $\hbar\omega_v$, respectively. a_{uv} measures the Franck-Condon (FC) offset

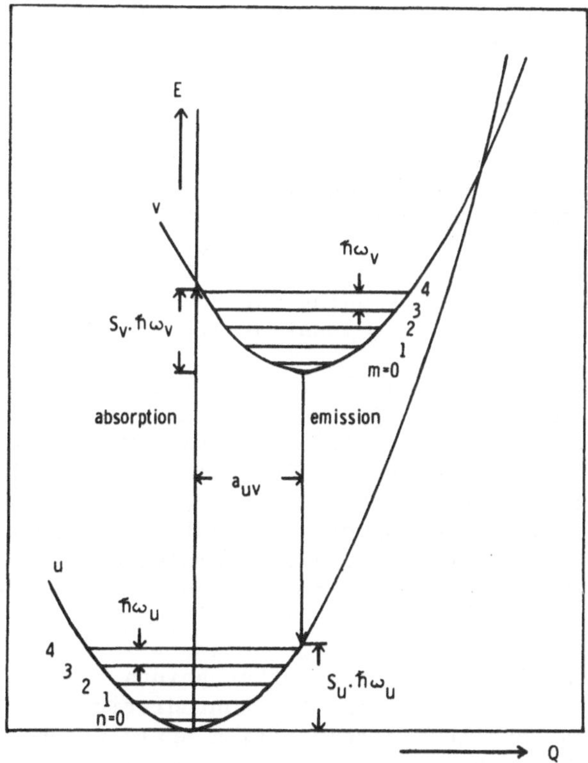

Fig.12. Single configurational coordinate diagram indicating the nomenclature used.

and is expressed in terms of S_u and S_v: $a_{uv}^2 = 2(S_u + S_v)$. $S_u \cdot \hbar\omega_u$ and $S_v \cdot \hbar\omega_v$ represent the relaxation energies after emssion and absorption, respectively. The QMSCC model is taken in the thermal Condon approximation; i.e. the initial vibrational states are in thermal equilibrium and the $v_m \rightarrow u_n$ rate is proportional to the squared overlap integral $< u_n | v_m >^2$.

Thus the radiative rate R_{nm} and the nonradiative rate N_{nm} are

$$R_{nm} = R_{uv}\ (1-r_v) r_v^m < u_n | v_m >^2 \quad \text{, and}$$
$$N_{nm} = N_{uv}\ (1-r_v) r_v^m < u_m | v_m >^2 \qquad (1)$$

where R_{uv} and N_{uv} are the electronic parts of the transition integral, and $r_v = \exp(-\hbar\omega_v/kT)$ is the Boltzmann factor. The $(1-r_v)r_v^m$ are the thermal weights, the $< u_n | v_m >^2$ the Franck-Condon factors.

The electronic factors are not investigated and taken constants. The thermal Franck-Condon weights vary enormously with temperature and parabola offset.

For radiative transitions, the total rate is obtained by summing R_{nm} over all initial and final states, so that

$$\sum_{n,m=0}^{\infty} R_{nm} = R_{uv}. \qquad (2)$$

For nonradiative transitions the total rate is obtained by summing N_{nm} over all nearly resonant v_m, u_n states. For the moment the nuclear operator driving the nonradiative transitions is taken 1 (see also below). In contradiction with many models on nonradiative transitions it is possible to take account of unequal force constants in the ground and excited state by using the Manneback angle θ ($\tan^4 \theta = k_v/k_u$, where k are the force constants).

In this way it is possible to calculate the temperature dependence of nonradiative processes for two-parabolae systems with small as well as with large offset and for three-parabolae systems. Also the quantum efficiency as a function of temperature can be evaluated. In section V.B we present the results of a number of model calculations which support the qualitative approach given in section IV. In section V.C we give examples of calculations on real systems where the experimental results could be fit using the QMSCC parameters.

First we will consider some other points.
A. The nuclear operator driving the nonradiative transitions is $\partial/\partial z_v$ or even a complicated combination of 1, z_v and $\partial/\partial z_v$. In ref.32 it has been shown that the Condon-approximation overlap integrals $< u_n | v_m >^2$ can be easily replaced by the linear and derivative operators $< u_n | z_v | v_m >^2$ and $< u_n | \partial/\partial z_v | v_m >^2$, respectively.

It turns out that quenching differences for the different operators are too small to be distinguished experimentally. The dominant factors in the calculation of thermal quenching remain the parabola offset and the thermal Franck-Condon weights. The nature of the nuclear operator driving the nonradiative transitions is not of very large importance.

B. For a localized centre the Huang-Rhys-Pekar(HRP) model and the single configurational coordinate model with equal force constants give the same formula for the absorption and emission band shapes (33). This is surprising because in the HRP model there is coupling with $\sim 10^{22}$ cm^{-3} lattice nuclear coordinates. In ref.34 it has been shown that this result can be understood as follows.

In the SCC model the normalized probability for a v → u optical transition accompanied by the generation of p_u phonons is

$$W_{p_u} = \sum_{m=m_0}^{\infty} (1-r_0) r_0^m < u_n | v_m >^2 \quad , \qquad (3)$$

where m_0 is the larger of 0 and $-p_u$. It can be shown that W_{p_u} satisfies a recursion formula, viz.

$$S_0 < m >_v W_{p_u+1} + p_u W_{p_u} - S_0 < 1+m >_v W_{p_u-1} = 0 \ (35). \qquad (4)$$

From this formula it can be derived that the W_{p_u} distribution has a reproductive property (34), viz. if p_{u1} and p_{u2} are two independent W_{p_u}-distributed variables, then $p_{uT} = p_{u1} + p_{u2}$ is also a W_{p_u}-distributed variable satisfying the recursion formula.

This means that the W_{p_u} distribution describing the optical-band shape in the SCC model (for equal force constants) possesses the reproductive property that two such distributions in independent coordinates with a common frequency combine to give a resultant which is also a W_{p_u} distribution. If we now turn to a centre characterized by offset parabolae in N independent coordinates, it can be shown by induction that the weight W_{p_u} attached to the set of all transitions in which a combined total of $p_u = p_{u1} + p_{u2} + \ldots + p_{uN}$ phonons is generated, satisfies the W_{p_u} recursion formula with $S_0 = S_{01} + S_{02} + \ldots + S_{0N}$. In the HRP model N is the number of host lattice modes. The distribution governing the number of phonons accompanying an optical transition in the HRP model must satisfy the recursion formula and has the W_{p_u} distribution as has the SCC model. In the HRP model the parameter S_0 is the sum of $10^{22} \cdot S_{0i}$ with each S_{0i} of the order of 10^{-22}.

C. We have now extended the simple SCC model to a multiple-coordinate model. The only approximation used is the single $\hbar\omega_0$ value. Recently Struck and Fonger (36) have derived the nuclear factors in the rates of transition for a multiple-coordinate model where all coordinates have a common phonon energy $\hbar\omega_0$. For the Condon approximation operator 1, the nuclear factor is the HRP W_p function (see above). For the operators $z(=0_+)$ and $\partial/\partial z(=0_-)$ the nuclear factors $W_{p,0_\pm}$ for $N \geqslant 2$ coordinates are $M_{p,0\pm} = (1-\gamma)W_{p,0\pm} + \gamma L_p$, where $W_{p,0\pm}$ are the nuclear factors for one coordinate, L_p is a combination of W_p functions and γ is a parameter between 0 and 1.

The largest differences between single- and multiple-coordinate nuclear factors $W_{p,0\pm}$ and $M_{p,0\pm}$ occur for S large. $W_{p,z}$ vs p has a mode at p = S. M_z has this mode filled in by γL_p. Absorption bands showing the double maximum $W_{p,z}$ behaviour are not known, so that only a description with $M_{p,z}$ with γ near to 1 can describe these. So would models with different phonon energies (see below).

$W_{p,\partial/\partial z}$ vs p has also a double maximum for S large and at low temperatures. The centre cusp (near p = S) is filled in with increasing temperature. For $M_{p,\partial/\partial z}$ the cusp is filled in by γL_p at every temperature. The $\partial/\partial z$ operator is expected for nonradiative transitions. These transitions are hard to study as a function of p, since multiple transitions with different energy gaps $p\hbar\omega_0$ are needed.

The nonradiative rate and its temperature dependence can be inferred from the thermal quenching of luminescence. Calculations in ref.36 show that this is not a promising approach to distinguish between the SCC and multiple-coordinate models. So in conclusion the SCC model can be used for representative model calculations and for the explanation of experimental results.
D. Struck has recently developed a multi-frequency model (37). He obtained the nuclear factors for K equally spaced phonon energies. These factors are combinations of K single-$\hbar\omega$ nuclear factors given above. The distributions calculated in this model are very similar to the SCC model distributions, except that some fine structure appears. Applied to typical thermal quenchings of luminescence, the multi-frequency model is only under very uncommon circumstances distinguishable from the SCC model $W_{p,0}$ and the single-$\hbar\omega_0$-model $M_{p,0}$.
E. Fong and coworkers have also derived general formulae for nonradiative processes driven by the derivative nuclear operator (38). It can be shown that his formula is the $\partial/\partial z$ operator rate in the equal-force-constants SCC model. To do this W_p should be written as a combination of five W_p functions by applying the recursion formula and the W_p functions should be Stirling approximated (39).

From this section we conclude that the simple SCC model can be reliably used to describe thermal quenching of luminescent centres, even if the constant operator is used. All further refinements do not yield essentially different results. For this reason model calculations may give us a very good feeling of the factors which influence thermal quenching of luminescent centres.

The Struck and Fonger method to calculate nonradiative transition probabilities has been critized recently by Englman. The reader is referred to his book on "Nonradiative decay of ions and molecules in solids" (North Holland, 1979). He points out that to improve the significance of these results the inter-state vibronic coupling would have to be included in the Hamiltonian, especially near the crossing point of the parabolae. The reason for this is that here the coupling cannot be regarded as perturbational. Englman also suggests that the density of phonon states must be used in order to establish significant transition rates.

The interesting point of Englman's criticism is that he (together with Barnett) has made calculations where these effects were included (46). These calculations were performed on the Cr^{3+} ion with the aim to find the ratio of the amount of ^2E and $^4T_2$ emission as a function of temperature. The parameters concerned were the offset between the parabolae of these two excited states and their energy difference. The coupling between the two potential curves (W) was also varied. The cause of this coupling may be, among others, spin-orbit coupling (intersystem crossing) or breakdown of the Born-Oppenheimer approximation (vibronic effects). In their calculation W is discussed in terms of the spin-orbit coupling mechanism.

These calculations yield the admixture of the ^2E state in the vibronic levels of the $^4T_2$ state. The nonradiative decay rate increases, if the amount of admixture increases. Using these results the temperature dependence of the emission can be found. This gives the following information:

(a) In the low-temperature region ($T < 0.5\ \hbar\omega$) the ^2E content of the emission rises for large and is constant for small offset. The rise is due to occupation of the second vibrational $^4T_2$ level.
(b) In the high-temperature region ($T > \hbar\omega$) the ^2E emission quenches due to thermal repopulation of the $^4T_2$ states with a higher radiative probability.

These calculations have been largely overlooked, but illustrate nicely the importance of interstate coupling. For our purpose, however, the Struck and Fonger method is accurate enough.

Let us now turn back to the simple QMSCC model as applied by Struck and Fonger. In section V.B we will review a number of model calculations on hypothetical luminescent materials. In section V.C, finally, we will consider several examples of fitting experimental results with the QMSCC model.

V.B. Model Calculations

In one of their first papers on the SCC model (31) Struck and Fonger presented results of a calculation on the thermal quenching of some representative luminescent centres. These were performed for the equal force constant case and $\hbar\omega_0$ was given the value 400 cm^{-1}. For large offset cases a_{uv} = 5 (S_0 = 6.25), for small offset cases a_{uv} = 0.6 (S_0 = 0.09).

Consider first the case of fast bottom crossover, i.e. the v → u transition in Fig.13. The crossover is about 1000 cm^{-1} above

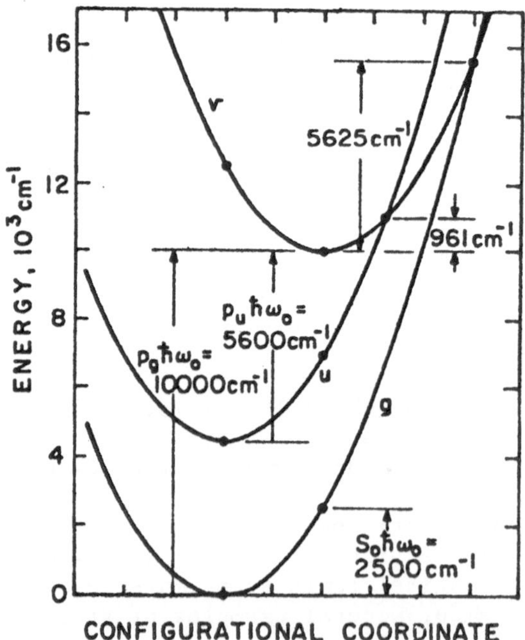

Fig.13. g, u and v parabolae used in model calculations (see text).

the minimum of the v parabola. Excitation in v feeds u. The $v \rightarrow u$ feeding has the rate $N_{uv}W_{p_u}$. Further we take $h\nu_{zp,vu} = p_u\hbar\omega_o = 5600\ cm^{-1}$ (see Fig. 13) and $R_{ug,v} = 10^{-6}s^{-1}$ and $N_{uv} = 10^{13}s^{-1}$.

Results are given in Fig. 14. Tabulated results can be found in ref.31. In the Mott-Seitz model (Fig.1) we expect radiative emission at low temperatures which is thermally quenched at higher temperatures. The Struck and Fonger calculation shows that there is not any radiative emission at all. In Fig. 14 we find the squared vibrational overlap as a function of m, the vibrational quantum number in the excited parabola. Its dependence on m is relatively weak. Also has been drawn the nonradiative rate obtained by the product of the squared overlap, the thermal weight and the electronic factor. It is immediately clear that the nonradiative rate, W_{nr}, is orders of magnitude larger than the radiative rate, W_r.

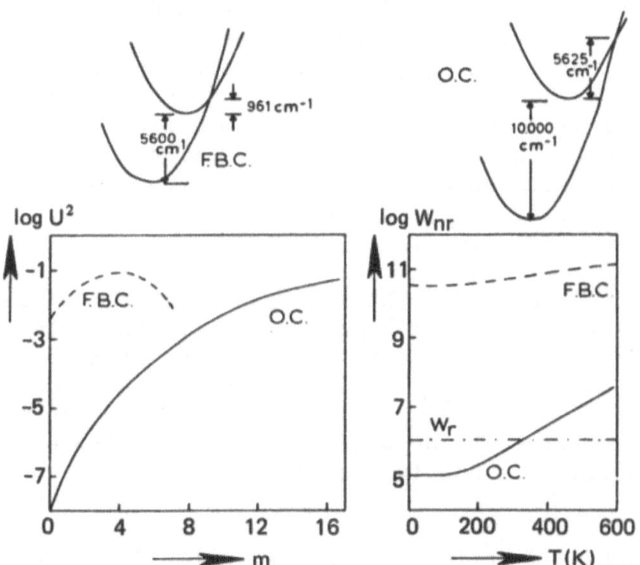

Fig.14. Top left: typical example of fast bottom crossover (F.B.C.); top right: typical example of outside crossover (O.C.) Bottom left: the dependence of the squared vibrational overlap (u^2) as a function of m, the vibrational quantum number in the excited state (see Fig. 12), for F.B.C. and O.C. Bottom right: the temperature dependence of the nonradiative transition rate (W_{nr}) for F.B.C. and O.C. in comparison with the radiative rate (W_r). Numerical values from ref. 31.

Note the weak temperature dependence of W_{nr} and the fact that W_{nr} cannot be described with the Mott-Seitz expression $\exp(-\Delta E/kT)$.

So this system does not yield any luminescence at all. It is nevertheless of great importance in the luminescence process of many well-known luminescent systems. It describes, for example, the feeding of the emitting ^5D levels of the Eu^{3+} ion by the charge-transfer state in a large number of compounds (see below). Other examples are the feeding of the emitting ^2E level of Cr^{3+} in Al_2O_3 by the $^4T_2$ level and the feeding of the emitting level in the octahedral UO_6^{6-} centre by the level to which an allowed absorption transition corresponds.

The second result taken from the Struck and Fonger calculations relates to the case of "outside crossover",i.e. the crossover is some 10,000 cm^{-1} above the minimum of the higher parabola (for example the $v \rightarrow g$ case in Fig.13). This case yields quite different results. Note the strong dependence of the squared vibrational overlap on the vibrational quantum number of the excited state, m. Note also that, even for m = 0, this overlap does not vanish completely. Fig.14 gives also the temperature dependence of the radiationless transition W_{nr}. An important observation is the fact that even at 0 K the ratio of W_{nr} and W_r is about 0.1 in the present calculation. This means that the quantum efficiency at 0 K is less than 1 in spite of the high value of ΔE. Since the data used in this calculation are typical of practical situations, this shows that the quantum efficiency of efficient phosphors at low temperatures is not necessarily 100%.

It is also interesting that the mean vibrational level at which the radiationless transition occurs is considerably lower than the vibrational level corresponding with ΔE, even at high temperatures. This mean level is 1.7 (at 259 K), 3.5 (at 374 K) and 6.1 (at 547 K), whereas the crossover is at m = 14. In contradiction with the Mott-Seitz model transitions near m = 14 are negligible. The downward transition is poorly described with a single activation energy.

The situation described by this calculation is typical for many broad-band emitting phosphors, e.g. vanadates, tungstates, and compounds activated with Pb^{2+}, Bi^{3+}, Ce^{3+} and Eu^{2+}.

The SCC model was also able to describe nonradiative processes between two small-offset parabolae and some other cases (31).

Another model calculation was presented recently by Bleijenberg and Blasse (40). They considered a hypothetical, red-emitting phosphor with a two-parabolae SCC diagram with a large offset. R_{uv} and N_{uv} were taken to be 10^4 and 10^{14} s^{-1}, respectively. Further θ= 45°, $\hbar\omega_o$=500 cm^{-1}, a_{uv}=7.746 and p_u=50. The distance between the parabolae minima was 25000 cm^{-1}, This phosphor, then, has a temperature quenching temperature $T_{\frac{1}{2}}$ = 450 K. By changing one of the parameters the dependence of $T_{\frac{1}{2}}$ upon that parameter could be evaluated (Fig.15). This calculation reveals some of the trends mentioned before in section IV.

Fig.15a shows that $T_{\frac{1}{2}}$ increases if the energy difference between the parabola minima increases. This agrees with the experimental results for the octahedral niobate group (Fig.8). Fig.15c shows that small variations in the Franck-Condon offset influence the luminescence efficiency drastically. The variations applied here can easily be expected for one and the same luminescent ion in different host lattices (compare Fig.1).

Fig.15b indicates that lattices with low vibrational frequencies provide suitable hosts for efficient luminescent materials as is well

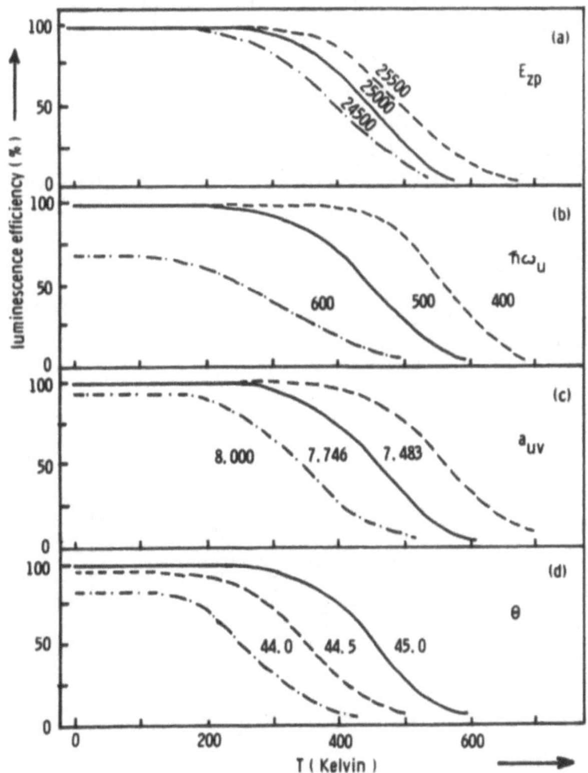

Fig.15. Temperature dependence of the luminescence intensity of a model phosphor system. (a) energy difference between parabola minima (E_{zp}) varied; (b) $\hbar\omega_u$ varied; (c) offset a_{uv} varied; (d) θ varied.

known from common practice.

A variation of the Manneback angle θ also influences the temperature quenching of luminescence. It should be realized that especially a_{uv} and θ cannot be varied independently. Large a_{uv} will imply low θ and therefore low quenching temperature.

This calculation gives an approximate idea of the nature of the factors influencing the luminescence efficiency. They can be used as rules of thumb in predicting and explaining efficient luminescent materials.

In closing this section we note that the θ variation is a way to mimic the effects ofan harmonicity which may seriously influence the rate of nonradiative processes (41).

V.C Calculations on Actual Cases

First we consider the case of ruby (Al_2O_3-Cr^{3+}) for which Fonger and Struck (42) have calculated the temperature dependence of the radiative and nonradiative transitions. The SCC diagram is shown in Fig.16. There are smaller-offset states (^2E, $^2T_1$ and $^2T_2$) and larger-offset states ($^4T_2$ and $^4T_1$). The parameters required for the calculations were obtained by fitting U_{p_u} distributions to absorption spectra. The phonon frequency required for this purpose amounts to

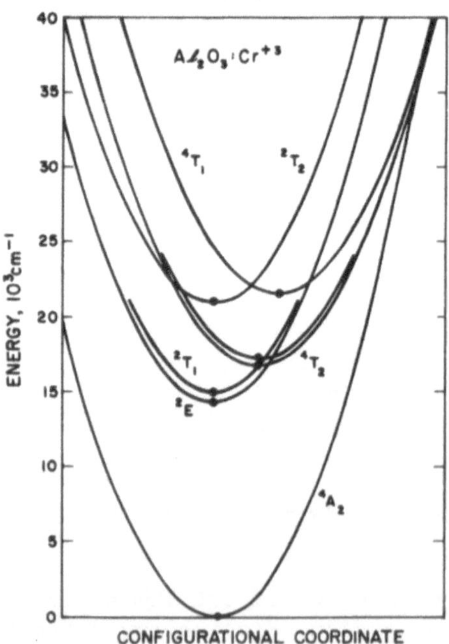

Fig.16. SCC diagram of the Cr^{3+} states in ruby.

500 cm^{-1} for the broad absorption bands and to 400 cm^{-1} for the narrow lines.

Further all nonradiative rates were calculated (Table 4) using the same parameters. N_{uv} was always taken 10^{14} s^{-1}. If the transition was slow (< 10^{10} s^{-1}), however, the phonon energies were increased to give smaller p numbers and faster rates. The computed rates range from 10^{13} to 10^{-27} s^{-1}. The 40 orders of magnitude variation indicates that the model makes a first order analysis of relaxation.

For example, excitation into $^4T_1$ goes to $^2T_2$, $^2T_1$, and ^2E by fast-bottom crossovers, excitation into $^2T_1$ to ^2E by fast small-offset multiphonon emission. From ^2E radiationless processes to $^4A_2$ can be neglected, and excitation goes radiatively to $^4A_2$ or, after thermal excitation, radiatively or nonradiatively via $^4T_2$ to $^4A_2$.

Fig. 17 shows the experimental and calculated total quantum efficiency of ruby as a function of temperature. It is clear that

TABLE 6. CALCULATED NONRADIATIVE RATES FOR RUBY INCLUDING PARAMETERS. RATES ARE FOR 0 K AND WOULD BE HIGHER AT HIGHER TEMPERATURES. FOR TRANSITIONS FROM $^2T_2$ SEE ORIGINAL LITERATURE (42).

Transition	θ	a_{uv}	$\hbar\omega_f$ (cm^{-1})	rate (s^{-1})*
$^4T_1 \rightarrow {}^2T_2$	44°	5.230	536	8.4×10^{11}
$\rightarrow {}^4T_2$ π	45°	1.297	833	4.7×10^{9}
$\rightarrow {}^4T_2$ σ	45°	1.297	833	8.6×10^{8}
$\rightarrow {}^2T_1$	44°	5.230	536	2.4×10^{12}
$\rightarrow {}^2E$	44°	5.230	536	1.3×10^{12}
$\rightarrow {}^4A_2$	44°	4.051	893	2.7×10^{4}
$^4T_2\sigma \rightarrow {}^2T_1$	44°	3.526	536	2.0×10^{13}
$\rightarrow {}^2E$	44°	3.526	536	1.5×10^{13}
$\rightarrow {}^4A_2$	44°	2.731	893	2.3×10^{3}
$^2T_1 \rightarrow {}^2E$	45°	1.095	536	2.2×10^{13}
$\rightarrow {}^4A_2$	45°	0.849	893	1.6×10^{-13}
$^2E \rightarrow {}^4A_2$	45°	0.849	893	2.5×10^{-12}

* smallest rate: $^2T_2 \rightarrow {}^4A_2$ 2.1×10^{-27} s^{-1}.

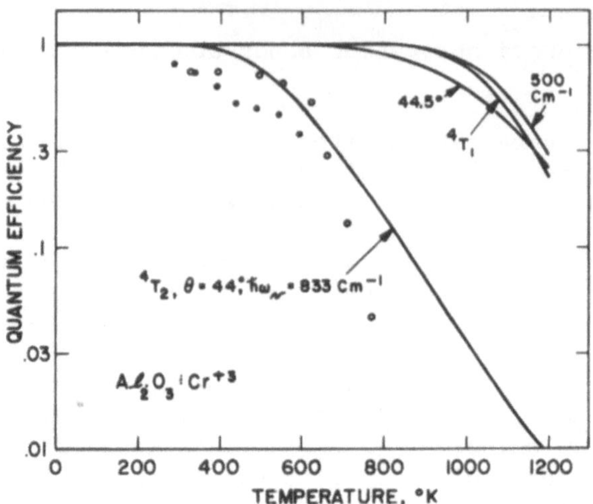

Fig. 17. Thermal quenching of the ruby total emission. Circles are experimental, full lines have been calculated.

a 833 cm^{-1} phonon is needed, because a 500 cm^{-1} phonon yields too high a quenching temperature. The quantum efficiency decreases above 450 K because of $^4T_2 \rightarrow {}^4A_2$ crossovers.

The other example treated by Struck and Fonger (43) is the case of the Eu^{3+} luminescence in Y_2O_2S and La_2O_2S mentioned above (compare Fig. 10). Here it was possible to fit the charge-transfer absorption band shapes and the quenching of the ^5D emissions using the set of parameters given in Table 7 (see Fig. 18). The reader is reminded of the fact that this quenching occurs via the charge-transfer state, i.e. $^5D \rightarrow CTS \rightarrow {}^5D$ quenching.

In addition the CTS $\rightarrow {}^5D$ feeding fractions could be calculated with these parameters (see Table 8 for an example).

TABLE 7. PARAMETERS FOR QMSCC MODEL FITS FOR $Y_2O_2S-Eu^{3+}$ AND $La_2O_2S-Eu^{3+}$ (43)

	Y_2O_2S		La_2O_2S
u	$^5D_0$	$^7F_0$	$^7F_0$
v	$^5D_1$	CTS	CTS
θ	45°	44°	44°
a_{uv}	0.346	7.58	8.32
$\hbar\omega_u (cm^{-1})$	435	295	295
$\hbar\omega_v (cm^{-1})$	435	275	275

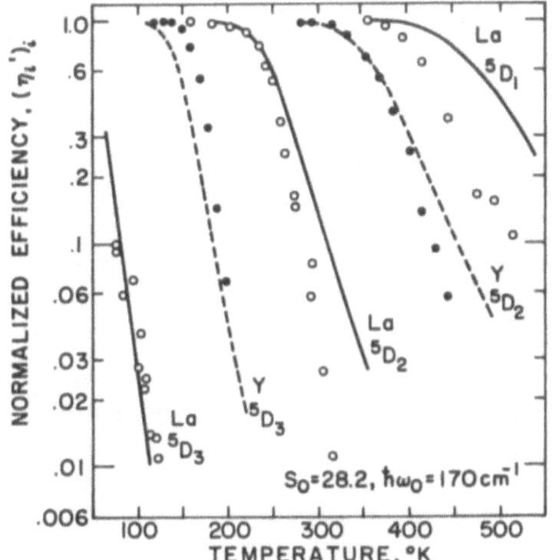

Fig.18. Quenching of the ^5D emissions of Y_2O_2S-Eu^{3+} and La_2O_2S-Eu^{3+}. Circles are experimental, lines have been calculated with the parameters of table VII.

In his contribution to this volume, Dr. Fonger gives another example dealing with Yb^{3+} in which energy transfer is also included.

In the meantime other groups have also worked with these calculations. Two examples are given here. The first relates to the luminescence of Ni^{2+} in $KZnF_3$ (44). Upon excitation into the $^3T_{1g}$ level the Ni^{2+} ion shows green, red and infrared emission. In view of their equal decay time the green and red emission originates from the same level ($^1T_{2g} \rightarrow {}^3A_{2g}$ and $^1T_{2g} \rightarrow {}^3T_{2g}$, respectively). The infrared emission corresponds to the $^3T_{2g} \rightarrow {}^3A_{2g}$ transition. The parameters for an SCC diagram were obtained from the temperature

TABLE 8. OBSERVED AND CALCULATED CTS → ^5D_i FEEDING FRACTIONS AT 77 K FOR La_2O_2S-Eu^{3+}. α_i IS DEFINED AS (CTS → ^5D_i)/(CTS → $^5D_{3,2,1,0}$).

	α_3	α_2	α_1	α_0
observed	0.02	0.30	0.65	0.04
calculated (θ=44°)	0.00	0.31	0.63	0.06
calculated (θ=45°)	0.00	0.02	0.91	0.07

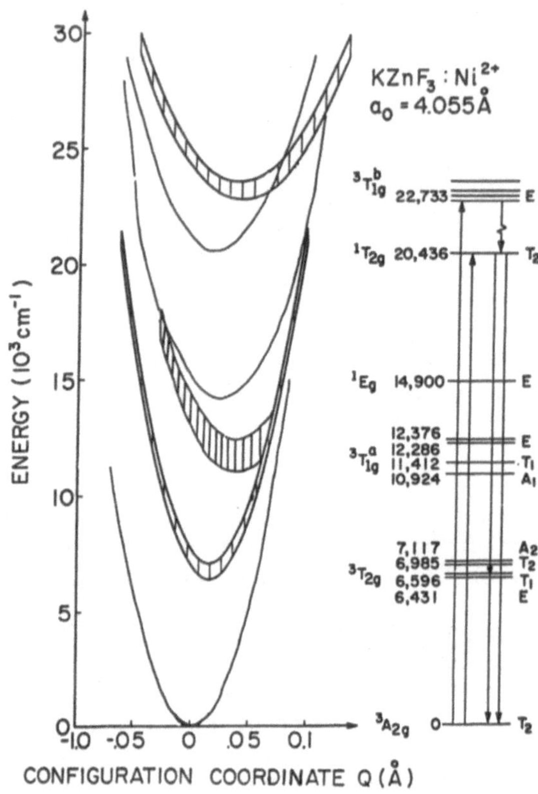

Fig.19. Proposed SCC diagram for Ni^{2+} in $KZnF_3$ (after a Struck and Fonger analysis). The parabola for the spin-orbit split levels are assumed to be the same and have been cross-hatched.

dependence of the decay times. They are as follows:for the $^1T_{2g}$ quenching $p_u = 20$, $a_{uv} = 3.25$, $\hbar\omega_v = 279\ cm^{-1}$, $\theta = 45^o$ and $N_{uv} = 10^{13}\ s^{-1}$ and for the $^3T_{2g}$ quenching $p_u = 18$, $a_{uv} = 2.61$, $\hbar\omega_v = 357\ cm^{-1}$, $\theta = 44.9^o$ and $N_{uv} = 10^{13}\ s^{-1}$. The resulting SCC diagram is presented in Fig.19. This strongly suggests that the ^1E_g level is the interactive level for nonradiative transitions from the $^1T_{2g}$ level.

Finally we mention an example from our own group, viz. a calculation of the thermal quenching of the uranate luminescence by Bleijenberg and Breddels (45). The special point of this calculation is that it relates to a series of compounds so that it can be seen how the parameters vary as a function of the host lattice. In section IV.A the relevant system has already been mentioned, viz. the uranium-activated ordered perovskites A_2BWO_6-U. Thermal quenching occurs via a large off-set third parabola (see Fig.9) and depends strongly on the choice of A (= Ca, Sr, Ba) and B (= Mg, Ca, Sr, Ba) (see Fig.20). All these curves and these optical spectra could be fit (45) by the parameters given in Table 9. Only for Ba_2MgWO_6-U we need parameter values which deviate from the trend in the series. To fit the data another parabola is needed. In view of the complicated energy level scheme of the uranate complex this is in itself not too surprising.

The results of this procedure can now be compared with our earlier qualitative approach. It is clear that indeed the offset is determined by the B ions (in the same way as predicted) and not by the A ion. However, other parameters are of influence too. Most striking in this aspect is the role of ΔE, but also the variations of θ and $\hbar\omega_g$ is seen to be necessary to fit the experimental results. For further details the reader is referred to the original literature.

VI. CONCLUSION

The field of nonradiative processes in luminescent materials has been studied intensively during the last one or two decades. There is a qualitative understanding of the factors which influence the luminescence efficiency and thermal quenching. From this knowledge rules of thumb can be deduced which are very helpful in predicting new and efficient luminescent materials. The QMSCC model introduced by Struck and Fonger offers the possibility to check the qualitative approach by performing calculations. The results of calculations on model systems confirm the rough model which was available at that moment and give a good feeling of the physical processes which are of importance in the radiationless transitions concerned. Fitting experimental results on existing phosphor systems to the QMSCC model extends the possibilities of the use of the simple configurational coordinate model considerably. Earlier it was possible to explain many luminescent phenomena qualitatively by

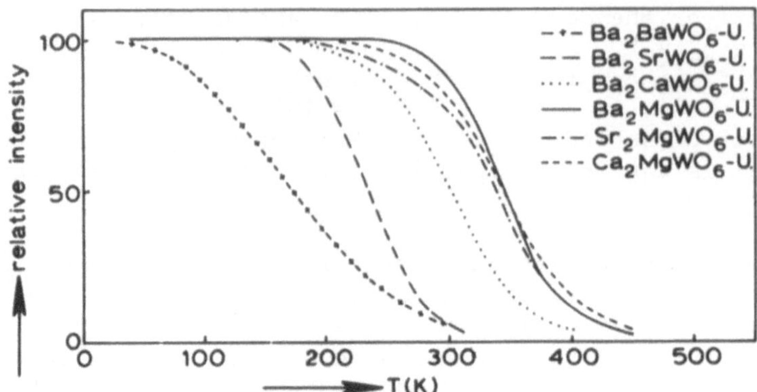

Fig. 20. Quenching of the uranate emission in the ordered peroyskite systems A_2BWO_6-U.

TABLE 9. PARAMETERS FOR QMSCC MODEL FIT FOR PHOSPHORS A_2BWO_6-U (45). DATA RELATE TO THE TRANSITION FROM THE LARGE-OFFSET PARABOLA TO THE GROUND STATE PARABOLA.

	$h\nu_0$ [1)]	$h\nu_{zp}$ [2)]	θ	a_{uv}	ΔE [3)]	$h\omega_g$ [4)]
Ca_2MgWO_6-U	29380	21775	44.23	10.000	1475	330
Sr_2MgWO_6-U	29310	21700	44.23	10.000	1500	330
Ba_2MgWO_6-U	27200	24200	45.00	6.030	4000	330
	29260	21650	44.23	10.000	1650	330
Ba_2CaWO_6-U	29040	21365	44.23	10.000	1165	330
Ba_2SrWO_6-U	28800	20850	44.17	10.587	925	320
Ba_2BaWO_6-U	28820	20520	44.08	10.860	870	310

[1)] maximum and [2)] zero-phonon energy of the excitation band relating to the large-offset parabola (in cm^{-1}).
[3)] energy difference between the minima of the small- and large-offset excited parabolae (in cm^{-1}); i.e. the u and v parabolae in Fig. 13.
[4)] vibrational frequency in the ground state (in cm^{-1}).

this model, but a real confirmation was difficult: quantitative calculations were practically impossible and the "elasticity" of the model made many explanations feasible. Nowadays all kinds of hypothesis in this model can be checked. Surprisingly enough, experience has taught that the fitting is rather critical which usually makes a unique physical solution possible. Although the exact physical nature of the nonradiative processes may remain obscure, it is possible to have a general picture of the processes of importance and the factors influencing them.

REFERENCES

1. A. Bril, in Kallman and Spruch, Luminescence of Organic and Inorganic Materials, Wiley, New York, 479 (1962).
2. See e.g. G.F.J. Garlick, Chapter 12 in Luminescence of Inorganic Solids (P. Goldberg, ed.), Academic Press, New York (1966).
3. A. Bril and W.L. Wanmaker, J. Electrochem. Soc. 111, 1363 (1964).
4. A. Bril, W.L. Wanmaker and J. Broos, J. Chem. Phys. 43, 311 (1975).
5. G. Blasse, W.L. Wanmaker and J.W. ter Vrugt, J. Electrochem. Soc. 115, 673 (1968).
6. B. Di Bartolo, Optical Interactions in Solids, Wiley, New York (1968).
7. G.F Imbusch, Chapter 1 in Luminescence Spectroscopy (M.D. Lumb, ed.), Academic Press, New York (1978).
8. F. Seitz, Trans. Faraday Soc. 35, 74 (1939).
9. D.L. Dexter, C.C. Klick and G.A. Russell, Phys. Rev. 100, 603 (1955).
10. R.K. Watts, in Optical Properties of Ions in Solids (B. Di Bartolo, ed.), Plenum Press, New York, 307 (1975). (The 1974 Erice course).
11. G. Blasse, J. Luminescence 1,2, 766 (1970).
12. C. Hsu and R.C. Powell, J. Luminescence 10, 273 (1975); G.E. Venikouas and R.C. Powell, J. Luminescence 16, 29 (1978).
13. J.A. Groenink, C. Hakfoort and G. Blasse, Phys. Stat. Sol.(a), in press.
14. J.A. Groenink and G. Blasse, J. Solid State Chem., in press.
15. J. Hegarty and G.F. Imbusch, Colloq. Int. C.N.R.S. 255, 199 (1977); B.A. Wilson, W.M. Yen, J. Hegarty and G.F. Imbusch, Phys. Rev. B19, 4238 (1979).
16. D.M. Krol and G. Blasse, J. Chem. Phys. 69, 3124 (1978).
17. D.M. Krol, K.P. de Jong and G. Blasse, J. Luminescence 20, 241 (1979).
18. J.M. Flaherty and R.C. Powell, Phys. Rev. B19, 32 (1979).
19. G. Blasse, in Luminescence of Inorganic Solids (B. Di Bartolo, ed.), Plenum, New York, 457 (1978). (The 1977 Erice course.)
20. G. Blasse and A. Bril, Philips Techn. Rev. 31, 303 (1970).
21. J.Th.W. de Hair and G. Blasse, J. Solid State Chem. 19, 263 (1976).
22. R.H. Bartram and A.M. Stoneham, Solid State Comm. 17, 1593 (1975).

23. C.W. Struck and W.H. Fonger, J. Luminescence 1,2, 456 (1970).
24. A. Bril, G. Blasse and J.A.A. Bertens, J. Electrochem. Soc. 115, 395 (1968).
25. C.W. Struck and W.H. Fonger, J. Chem. Phys. 52, 6364 (1970).
26. H.V. Lauer and F.K. Fong, J. Chem. Phys. 65, 3108 (1976).
27. G. Blasse, J. Electrochem. Soc. 124, 1280 (1977).
28. C.E. Tyner and H.G. Drickamer, J. Chem. Phys. 67, 4103 (1977); J. chem. Phys. 67, 4116 (1977).
29. L.A. Brey, G.B. Schuster and H.G. Drickamer, J. Chem. Phys. 67, 2648 (1977).
30. H.G. Drickamer, private communication.
31. C.W. Struck and W.H. Fonger, J. Luminescence 10, 1 (1975).
32. C.W. Struck and W.H. Fonger, J. Luminescence 14, 253 (1976).
33. See e.g. D. Curie, p.71 in ref.10.
34. W.H. Fonger and C.W. Struck, J. Luminescence 8, 452 (1974).
35. C.W. Struck and W.H. Fonger, J. Chem. Phys. 60, 1988 (1974).
36. C.W. Struck and W.H. Fonger, to be published.
37. C.W. Struck, to be published.
38. F.K. Fong, Theory of Molecular Relaxation, Wiley, New York (1975).
39. W.H. Fonger and C.W. Struck, J. Chem. Phys. 69, 4171 (1978).
40. K.C. Bleijenberg and G. Blasse, J. Solid State Chem. 28, 303 (1979).
41. M.D. Sturge, Phys. Rev. B8, 6 (1973).
42. W.H. Fonger and C.W. Struck, Phys. Rev. B11, 3251 (1975).
43. C.W. Struck and W.H. Fonger, J. Chem. Phys. 64, 1784 (1976).
44. M.V. Iverson and W.A. Sibley, J. Luminescence 20, 311 (1979).
45. K.C. Bleijenberg and P.A. Breddels, to be published.
46. R. Englman and B. Barnett, J. Luminescence 3, 37 (1970); B. Barnett and R. Englman, J. Luminescence 3, 55 (1970).

DEFECT STRUCTURES IN PHOSPHORS BY RADIATIONLESS TRANSITIONS

R. Grasser and A. Scharmann

I. Physikalisches Institut der Justus-Liebig-Universität
Heinrich-Buff-Ring 16, 6300 Giessen
Federal Republic of Germany

ABSTRACT

The purpose of this article is to emphasize the importance of defect production and defect motion in the radiationless deactivation of excited luminescent materials. The presentation is essentially confined to alkali halides and to oxygen-dominated phosphors. Both systems strongly interact with ionizing radiation fields. In the beginning, we discuss the formation of intrinsic defects by charge redistribution in halides and in oxygen-dominated phosphors. Athermal migration of pre-existing defects in solids, as a further radiationless channel is presented. A more detailed discussion is devoted to the radiolysis in halide compounds. In the last part of our paper, laser induced optical breakdown in dielectrics is briefly described.

I. INTRODUCTION

In this series of lectures we shall present mechanisms of defect production in inorganic phosphors by radiationless transitions and discuss, to some extent, the structure of defects formed during irradiation. According to the above statement the interacting radiation fields are restricted to those fields which only excite and ionize the crystal particles. Defect production by collision events in the initial stage is excluded from our considerations. In this connexion wide-band-gap substances as alkali halides and oxygen dominated phosphors are of main interest. Both classes of luminescent materials are highly sensitive to ionizing radiation.

The next chapter starts with a discussion of the creation of defects by redistribution of electrons in halides, tungstates, and related phosphors during irradiation with energetic electromagnetic radiation. Radiationless transitions occur in these systems as intermediate stages in the trapping of carriers at pre-existing capture centers and in the undistorted host lattice. The following chapter deals with athermal migration of defects in solids. In low temperature experiments recombination-enhanced motion of defects in crystals seems to be an important phenomenon in the interaction of optical radiation fields with solid state phosphors. The subject of the following chapter is radiolysis in halide compounds. This is the most widely and most thoroughly investigated problem within the scope of our topic. Nonradiative recombination of self-trapped excitons is the source of energy for the formation of separated F and H Centers. The final chapter contains a brief consideration of optical breakdown in dielectrics. Exposure of halide crystals or other optical materials to intense laser pulses leads to intrinsic optical damage in the form of fracture and/or melting.

For a more detailed study of radiolysis processes in ionic materials the reader is referred to the books (1,2) cited in the bibliography to this article. We have made no attempt to make this bibliography complete.

II. CREATION OF DEFECTS BY CHARGE REDISTRIBUTION IN LUMINESCENT MATERIALS DURING IRRADIATION WITH IONIZING RADIATION

A leading feature of the effect of ionizing radiation on solid phosphors is the formation of free carriers within the electronic system of the irradiated crystals. All crystals investigated by experimental physicists are necessarily real crystals. They are imperfect and contain impurities. The interaction of free carriers with these defects gives rise to three main groups of process:

1) Scattering
2) Capture with the emission of light
3) Capture without the emission of light.

Carrier capture without the emission of light can involve loss of energy to phonons or to other carriers. The phonon emission mechanism is usually dominant except at high carrier concentrations. It can be distinguished by its strong temperature dependence.

In general, charge redistribution results from the capture of free carriers. This trapping process is confined in many phosphor systems to defect sites in the crystal lattice. A class of luminescent materials, however, additionally shows the self-trapping process, where carriers become localized in the initially undisturbed crystal lattice. In the trapping of carriers radiationless transitions occur as intermediate stages. The radiationless transition to be regarded is from one electronic state to another.

Since all phosphors are impure, charge redistribution among pre-existing defects might be illustrated by any system of non-metallic materials. The effects are most easily identified in crystals doped with transition metal ions. A well-known example is ruby (Al_2O_3/Cr), where a fraction of the Cr^{3+} impurities is converted to Cr^{2+} and Cr^{4+} by electron capture and hole capture during irradiation. Analogous processes occur in semiconductors due to carrier redistribution among donors and acceptors. The changed charge states of defects are stable, at least through some temperature range.

We shall not discuss here charge redistribution only among pre-existing defects. The considerations in this chapter are limited to those charge redistribution processes in the course of which one kind of carriers is captured by pre-existing defects, while the companion carriers are trapped in initially undisturbed environments of the host lattice. The latter radiation-induced intrinsic defects produced by a self-trapping process control to a great extent the optical, electrical, and magnetical behaviour of a series of ionic phosphors. Self-trapped centers are not defects in the usual sense. They are electronic carriers which produce sufficient local distortion of the host lattice to get immobile. In this state they behave as if they were conventional defects.

By charge redistribution energy of the radiation field is stored within the irradiated crystal. Subsequent heating of the sample leads to a release of the trapped carriers and can produce thermally stimulated conductivity and thermoluminescence. Thereby, the solid fully recovers.

II.A. Mechanism of Defect Creation

The mechanism of defect creation by charge redistribution in luminescent materials during irradiation with ionizing radiation is self-trapping of electronic carriers. An obvious, however, nevertheless important initial condition for the self-trapping of a carrier (e.g. hole) is the capture of the companion carrier (electron) by a pre-existing conventional defect elsewhere in the crystal.

The self-trapping is an intrinsic effect. It is the consequence of the most general principle that a physical system rendered to itself tends towards a state of minimal energy. In halide crystals energy is gained by the self-trapping of holes through the formation of a X_2^- molecular ion from two adjacent halogen ions. As cited in (3), Gilbert has summarized the conditions for a self-trapping of carriers (4). He argues that the transition of a delocalized hole in the valence band in an undisturbed lattice to the self-trapped configuration may occur in three steps. Self-trapping should be possible if the sum of the change in total energy in these stages is negative. The three steps are: localization of the hole, polarization of the host lattice by the localized hole whilst the separation of the two halogen ions is kept fixed, relaxation of the X_2^- molecular ion to its optimum nuclear separation.

Localization of the hole is achieved by constructing a wave-packet from the valence band states of the undistorted lattice. To do this, a localization energy is needed which amounts to the difference between the mean energy in the band and the band extremum. It is about half the bandwidth. The sign of this energy is positive, since the localization procedure requires excited states of the hole

$$\Delta E_1 = E_{loc} > 0 \ . \tag{1}$$

Polarization of the lattice in the second stage gives a reduction in energy,

$$\Delta E_2 = E_{pol} < 0 \ . \tag{2}$$

Relaxation of the X_2^- molecular ion yields an amount of energy that consists of small terms owing to changes in the positions of neighbouring ions and a main contribution due to the bonding energy of the X_2^- molecular ion

$$\Delta E_3 \simeq E_{bond} < 0 \ . \tag{3}$$

On these assumptions, Gilbert has estimated that the hole in aklali halides should be self-trapped, the electron should not be. Data for the electron and the hole in KCl are summarized in Table 1. The high localization energy together with the small bonding energy prevent self-trapping of electrons in KCl crystals.

Stoneham points out that it is misleading to regard the self-trapping distortion as a Jahn-Teller effect. In spite of the fact

TABLE 1. DATA FOR THE ELECTRON AND THE HOLE IN KCl

	Hole	Electron
E_{loc}	0.3eV(valence band)	1.9eV(conduction band)
E_{pol}	-0.5eV	-0.3eV
E_{bond}	-1.5eV(Cl_2^- bond)	-0.3eV(K_2^+ bond)
Total	-1.7eV	+1.3eV

that the valence bands are usually degenerate, there is no direct influence.

The difference between electrons and holes in halide crystals is most obviously demonstrated for the exciton state. In this excited state configuration, which might be considered as another charge state of the X_2^- molecular ion, the localized self-trapped hole binds an electron in a relatively diffuse orbital.

In a recent paper Mott and Stoneham (5) have discussed a very interesting and, as it seems, a more comprehensive approach to the self-trapping of electronic carriers in solids. The authors use the concept of small self-trapped molecular polarons, to describe the features of the self-trapped hole in ionic crystals. To begin with, they compare the dielectric polaron, which occurs for electrons or holes in a polar lattice due to the Coulomb interaction of the carrier and the polar environment, with the molecular polaron in which self-trapping is caused by short-range interactions.

In the case of the dielectric polaron, it is generally assumed that the electron or hole produces for itself a potential hole of the form

$$-e^2/\varepsilon_{eff}\, r, \qquad \varepsilon_{eff}^{-1} = \varepsilon_\infty^{-1} - \varepsilon_o^{-1}. \tag{4}$$

ε_o and ε_∞ are the static and optic dielectric constants (6,7). On account of this Coulomb field, self-trapping is always possible. The linear dimension of the orbital of the self-trapped electron is of order

$$\hbar^2\varepsilon_{eff} \,/\, m_{eff}e^2. \tag{5}$$

m_{eff} = effective mass of the electron. Expression (4) only holds for large values of r compared to the radius given by eq. (5). If the orbital radius of the dielectric polaron is comparable to the lattice constant, the self-trapped carrier is called a small polaron. When the small polaron 'delocalizes' to a more extended orbital, it should experience only weak trapping. So, as m_{eff} decreases, it is to be expected to observe a continuous transition from small to large polaron behaviour. Due to the Coulomb interaction, self-trapping in the dielectric case should occur without any critical feature.

Molecular polarons should show quite another behaviour. Here, self-trapping is accompanied by the formation of a chemical bond between neighbouring atomic units of the crystal. The molecular polaron is only formed for sufficiently large effective mass of the carrier. If m_{eff} is too small, nothing happens. Contrary to the dielectric polaron, the molecular polaron only exists in the state of a small polaron. This critical change of behaviour is predicted by polaron theory (8,9,10).

A straightforward treatment of our problem, a discussion of the effect of the electron-lattice interaction on the electronic eigenstates of a deformable solid, characterized by a short-range electron-lattice interaction, is given in the theoretical approach performed by Emin (9,10). On the basis of a variational calculation of the eigenstates of the three-dimensional analogue of Holstein's Molecular Crystal Model, the author determines the conditions under which carrier self-trapping does or does not occur in this system. In Holstein's Molecular Crystal Model one considers a single excess electron placed in a periodic array of identical molecular units (11). With each molecule in the lattice a configurational coordinate q_i is associated which represents a distortion of the molecule from its equilibrium configuration (= displacement of the atoms of the molecule of the molecular crystal). In the total Hamiltonian of the system, the electron-lattice interaction energy (the change in the electron's potential energy as a result of a deviation of the molecules of the system from their equilibrium configuration) is assumed, consistent with the prior studies of the Molecular Crystal Model, to be a linear function of the configurational coordinates q_i. Apart from an arbitrary constant, the energy of an electron at the ith molecule upon which it resides is taken to have the short-range form

$$E(q_i) = - Aq_i, \tag{6}$$

where A is an electron-lattice coupling constant. Furthermore, for simplicity the electron is assumed to be coupled only to the optical-mode vibrations of the crystal.

The presence of a substantial short-range component of the electron-lattice interaction cannot be disregarded even in an ionic crystal.

As already shown by Toyozawa (8), the solution consists of two qualitatively distinct types of states which are possible eigenstates of the Molecular Crystal Model. In detail, one type of state is that of a weakly coupled electron-lattice system while the other is that characterizing a small molecular polaron. In the first case the system is adequately described by considering the electron-lattice interaction as a small perturbation on an otherwise uncoupled electron-lattice system. In the second situation the electron-lattice interaction must be treated ab initio.

The results of the calculation are illustrated in Fig. 1. The figure shows schematically the energy spectrum associated with

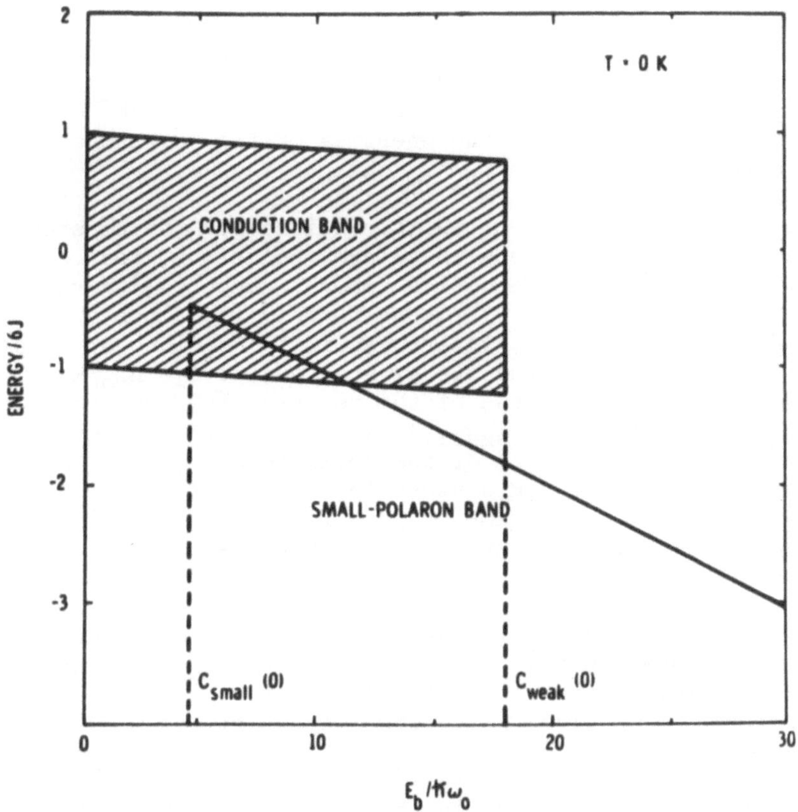

Fig. 1. One-electron energy spectrum at T = OK (from ref. 10).

the excess electron at T = OK. The energies of the spectrum corresponding to nearly uncoupled lattice states, the conduction band states, and the molecular (small) polaron states are plotted as a function of the electron-lattice coupling parameter, $E_b/\hbar\omega_o$. $E_b/\hbar\omega_o$ is a measure of the electron-lattice coupling strength. E_b denotes the molecular polaron binding energy, ω_o is the optical-mode frequency. The plot is derived for $6J/\hbar\omega_o = 10$, where J is the rigid-lattice electronic transfer integral. In a cubic lattice the conduction band width is 12J. The energy of the coupled system minus zero-point vibrational energy, $n\hbar\omega_o/2$, is measured in units of the rigid lattice band half-width, 6J. n is the number of molecules in the lattice. Since the degeneracy due to the geometric equivalence of a regular lattice is lifted when finite electronic overlap exists, the band of the self-trapped molecular polaron states is extremely narrow. The width is about $12J \exp(-E_b/\hbar\omega_o) \ll \hbar\omega_o$. As shown in Fig. 1, for $E_b/\hbar\omega_o > C_{weak}(0)$ only the molecular polaron states exist, while for $E_b/\hbar\omega_o < C_{small}(0)$ only the nearly conduction band states exist. For $C_{small}(0) \leq E_b/\hbar\omega_o \leq C_{weak}(0)$ both types of states coexist. In the coexistence region there is a shift of the energies of the two bands of states relative to one another as $E_b/\hbar\omega_o$ changes.

An essential feature of the conditions for the existence of the weak-coupling and molecular polaron solutions, respectively, is their dependence on the optical phonon number N, where N is the Bose factor, $(\exp(\hbar\omega_o/kT)-1)^{-1}$. Due to this fact, the two limiting values $C_{small}(T)$ and $C_{weak}(T)$ are explicit functions of temperature. For a comparison, the energy spectrum at $T = \hbar\omega_o/k = \Theta$ is shown in Fig. 2. An increase of the temperature from absolute zero results in a decrease of $C_{small}(T)$ while $C_{weak}(T)$ increases, thereby increasing the size of the coexistence region. The N-dependence and hence temperature-dependence of the critical values of the coupling parameter give rise to the possibility of an abrupt N-dependent appearance or disappearance of states from the energy spectrum of the carrier.

As in Fig. 1, $6J/\hbar\omega_o$ is taken to be 10. Here, however, the purely vibrational contribution to the energy of the coupled electron-lattice system is $n\hbar\omega_o(N+1/2)$. n is the number of molecules in the lattice. As before, the energy is measured in units of 6J.

Furthermore, a decrease in the adiabaticity parameter, $6J/\hbar\omega_o$, is associated with a decrease of the coexistence region until for extremely narrow electronic crystal bands, where $6J \lesssim \hbar\omega_o$, only one solution exists. Adiabaticity means that the excess carrier may be assumed to adjust to the instantaneous positions of the atoms due to a sufficiently rapid motion of the carrier compared to the motion of the relatively heavy atoms of the system. When $6J \lesssim \hbar\omega_o$, the solution changes its character continuously from nearly rigid-lattice to molecular polaron as the electron-lattice coupling strength is increased. In the opposite case, for $\hbar\omega_o \to 0$ with $M\omega_o^2$ finite, the coexistence region increases to include all nonzero values of the electron-lattice coupling strength. $M\omega_o^2$ is the lattice stiffness, M is an appropriate reduced atomic mass.

From the knowledge of the one-electron energy spectrum of the coupled system, characterized by a short-range electron-lattice interaction, a physical interpretation of the conditions for the occurrence of untrapped and of self-trapped eigenstates is possible. According to Heisenberg's uncertainty principle, a carrier associated with an energy band of halfwidth 6J can only be confined

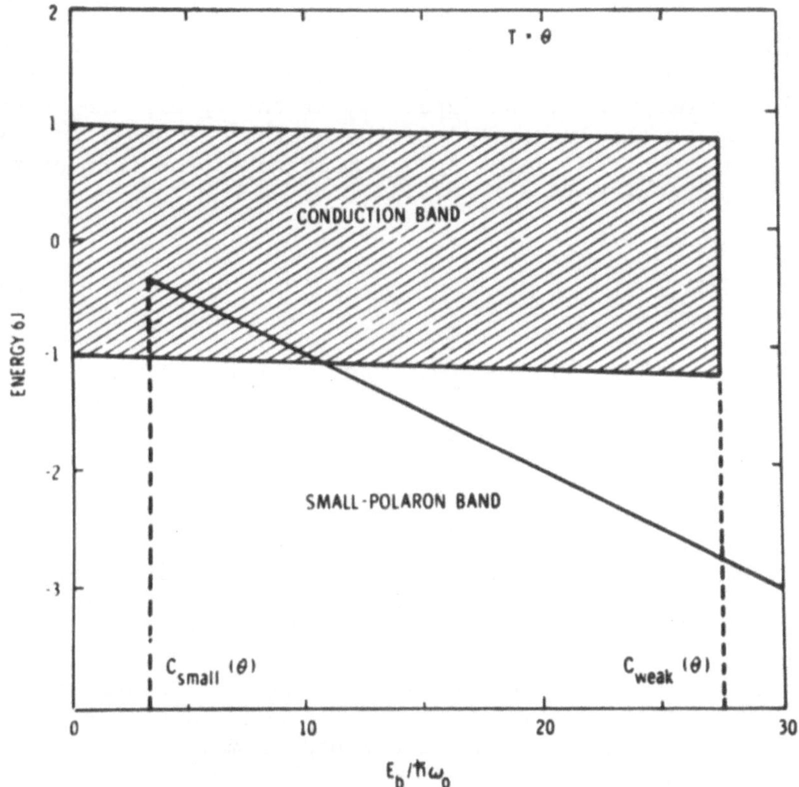

Fig. 2. One-electron energy spectrum at $T = \hbar\omega_o/k = \Theta$ (from ref. 10).

to a single site for a time

$$\tau \approx \hbar/6J. \tag{7}$$

τ is the time scale within which the carrier is able to force a substantial alteration of the positions of the atomic constituents associated with the occupied site. The force which the carrier exerts on the neighboring atoms of the occupied site is, due to eq. (6) within the linear approximation, the electron-lattice interaction constant A. Thus, during the time the carrier occupies a site the surrounding atoms are subject to a carrier-induced acceleration of the size

$$A\ /\ M. \tag{8}$$

A deviation of the motion of the neighbouring atoms from their motion in the carrier-free state produced by electron-lattice interaction is only not to be expected, if the specific carrier-induced change in the 'velocity' of these atoms is small compared with their carrier-free 'velocity'. A measure of the carrier-free 'velocity' of the atom is

$$\left(\frac{\hbar\omega_o(N + 1/2)}{M}\right)^{1/2} . \tag{9}$$

Hence, with eqs. (7), (8), and (9), the condition for a weak-coupling eigenstate is

$$\left(\frac{A}{M}\right)\left(\frac{\hbar}{6J}\right) < \left(\frac{\hbar\omega_o(N + 1/2)}{M}\right)^{1/2} . \tag{10}$$

This means that a carrier can escape self-trapping in a material, characterized by a short-range electron-lattice interaction, only, if it moves sufficiently rapidly between different sites to prevent the atomic constituents from being able to respond to its presence. All mechanisms that increase the time of stay of a charge carrier on a given site may destabilize the weakly coupled states and enhance the probability for the occurrence of molecular polaron states.

The fact that a system, characterized by short-range electron-lattice interaction only possesses rigid-lattice band states and molecular polaron states being essentially confined to a single unit cell, strongly depends on the dimensionality of the system (12,13). The results obtained by Emin and Holstein (13) from an adiabatic treatment of the ground state of a carrier placed in a deformable continuous medium are shown in Fig. 3. In this figure the curves of energy E(R) versus a scaling factor R are illustrated for systems of distinct dimensionality and for various models of the electron-lattice interaction. An investigation of a three-dimensional system in which the carrier interacts with the lattice solely via the long-range component of the electron-lattice interaction yields a solitary minimum of the E(R) curve as illustrated in Fig. 3(a). Here, the electron exists in a finite-radius bound state. The polaron (bound electron plus distortion) is always energetically stable. The same result holds for a carrier in a one-dimensional system interacting with the lattice solely via the short-range component of the electron-lattice interaction. In the

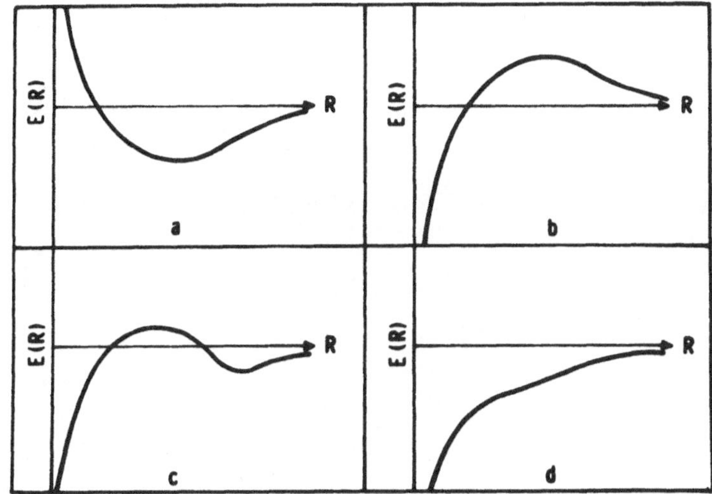

Fig. 3. E(R) curves (from ref. 13).

case of a carrier in a three-dimensional deformable system for which the electron-lattice interaction is short-range, the adiabatic theory gives the same solutions as those obtained by the variational calculations. As illustrated in Fig. 3(b), the function E(R) shows no finite-radius minimum. Minima occur at $R = \infty$, corresponding to an unbound electron in a rigid lattice (rigid lattice conduction band states), and $R = 0$, conforming to an electron self-trapped in an infinitely deep and infinitesimally localized deformation-induced potential well (molecular polaron states). There is no intermediate-range polaron (dielectric small polaron) in this case. An inclusion of the long-range component of the electron-lattice interaction along with the short-range component in a three-dimensional model can convert the nonpolaronic state to a finite-radius polaronic state as shown in Fig. 3(c). At sufficiently high values of the additional long-range component of the electron-lattice interaction, the finite-radius state shrinks in size finally collapsing into a molecular polaron state, Fig. 3(d).

The essential contribution of this adiabatic theory (13) is the result that it is possible for both types of polarons (dielectric polaron and molecular polaron) to coexist in the same three-dimensional system if the electron-lattice interaction is appropriate.

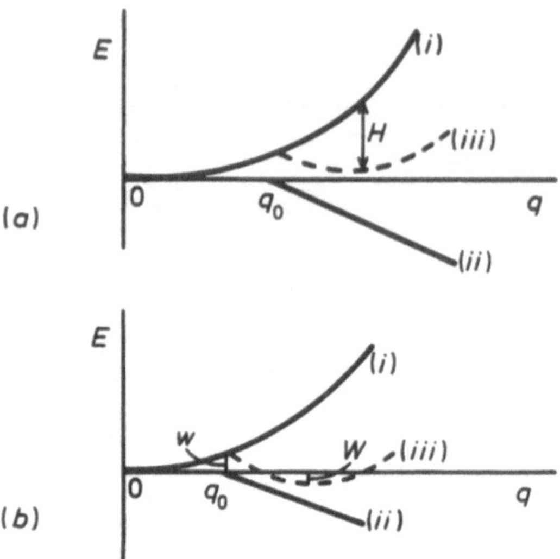

Fig. 4. Mechanism of self-trapping for a hole. (i) is the elastic energy, (ii) is the electronic energy, and (iii) is the total energy (from ref. 5).

Mott and Stoneham (5) have anewly analysed the results of all these calculations. Their version of argumentation is illustrated in Fig. 4. As before, q is the configurational coordinate. For a V_k center q denotes the change in the distance between the two halogens concerned. Curve (i) represents the elastic energy. To illustrate graphically the electronic part of the energy, one has to keep in mind that the short-range interaction is not of a Coulomb type. The potential well formed for the carrier is schematically shown in Fig. 5. In the three-dimensional case a potential well of depth V and radius a can only trap a carrier if the 'well size' is

$$2\ mVa^2/\hbar^2 > \frac{1}{4}\pi^2, \qquad (11)$$

where m is the mass of the carrier. This means that self-trapping occurs only for a finite value q_o of q. If it happens, the energy of the charge carrier will be lowered within the linear approximation as depicted in curve (ii) of Fig. 4. Thus, the total energy (iii) will pass through a minimum. Self-trapping of the carrier can only occur, if the energy value of the minimum drops below zero. This situation is shown in Fig. 4(b), where W denotes the binding

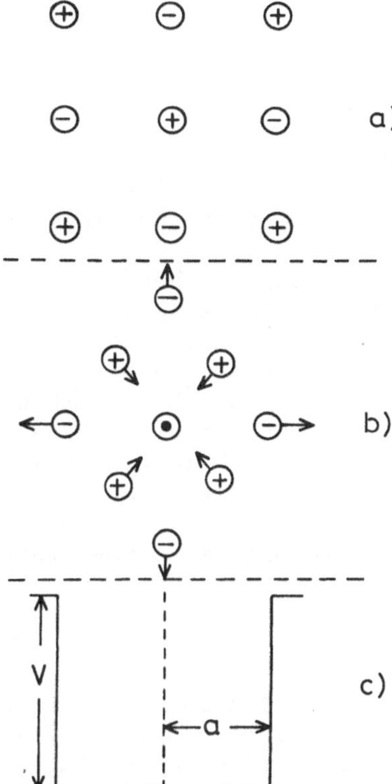

Fig. 5. Schematic representation of the potential well formed for an excess electron in a cubic ionic lattice. a) carrier-free lattice, b) readjustment of the positions of the ions in response to the excess electron on the central positive ion, c) formed potential well of depth V and radius a.

energy of the polaron. In the case plotted in Fig. 4(a) no polaron is created.

The essential point of the discussion of Mott and Stoneham (5) is the existence of the potential barrier w to self-trapping (Fig. 4(b)) and its consequences. For a carrier in thermal equilibrium at q = 0 self-trapping is possible, if the system surmounts the potential barrier w. At high temperatures the probability of this process will be

$$C \exp(-w/kT), \tag{12}$$

per unit time. At low temperatures the barrier can be overcome by tunnelling with the probability

$$C \exp(-w/\frac{1}{4} \hbar\omega). \tag{13}$$

ω is the vibrational frequency associated with the configurational coordinate q. C should be of order 10^{12} s^{-1}. The potential barrier w must lead to a delay in the self-trapping process.

Mott and Stoneham (5) assume that the self-trapping of a carrier in the state of a molecular polaron may be stimulated by large or small dielectric polarons, since, as mentioned above, dielectric and molecular polarons can exist together in the same system. For instance, this should be possible for holes in the ionic alkali halides. The authors expect that a dielectric polaron will be formed around a hole at 0 in Fig. 4 in a few multiples of 10^{-12} s, and that this will result in an increase of the effective mass of the hole (m in eq. (11)). The mechanism will decrease q_o in Fig. 4 and make self-trapping into the state of a molecular polaron easier.

II.B. Examples

Molecular polaron formation is not restricted to polar materials in which the classical coupling of a carrier to the optical modes of an ionic lattice may be large. Self-trapping occurs in both polar and nonpolar substances. It may be associated with an interaction of the carrier with both acoustic and optical-mode atomic displacements.

1. V_k center in alkali halides. The V_k center was discovered by Känzig (14). Early information on its electronic structure was derived from detailed EPR measurements on selected crystals (15) and optical investigations performed by Delbecq, Smaller, and Yuster (16).

The self-trapped hole, or V_k center, is probably the most important radiation produced defect in halide crystals. The lifetime of free holes in halides is extremely short, perhaps no more than 10^{-12} s. The hole becomes self-trapped forming a X_2^- molecular ion from two adjacent halogens. Bonding is achieved taking the p valence orbital into account. Fig. 6 shows the arrangement of the V_k center in the NaCl structure. The optical and paramagnetic spectra of the center are generally anisotropic. To good approximation they can be described in terms of <110> axial symmetry. The actual point group of the center is D_{2h}.

The molecular model regarding the V_k center as a X_2^- molecule replacing two neighbouring anions $(X^-)_2$ in the close-packed direction <110> for alkali halides greatly assists in under-

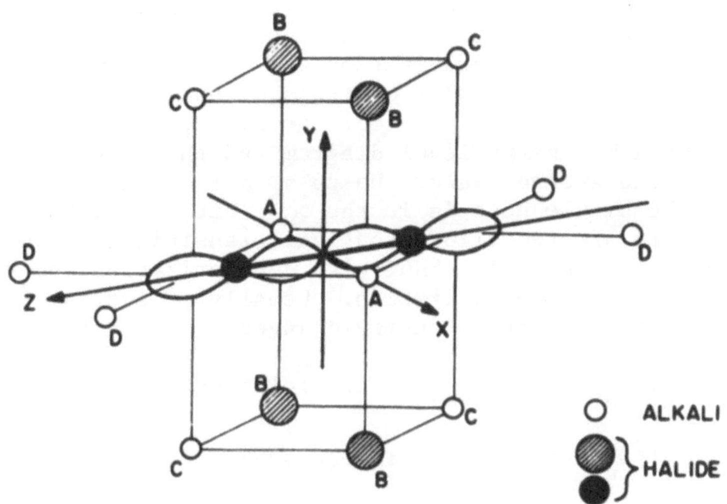

Fig. 6. V_k center in the NaCl structure (from ref. 2).

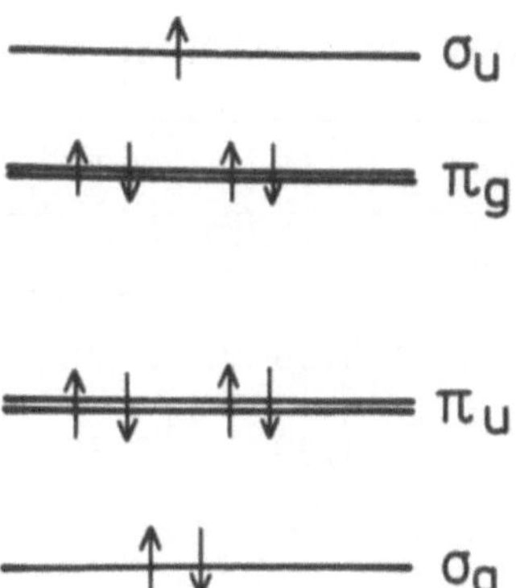

Fig. 7. MO Energy level diagram of the V_k center.

standing its properties. The model is strongly supported by experimental evidence. Many features of the V_k center are close to those of a free X_2^- molecule at the appropriate internuclear distance. In this description the crystal is considered only by defining the nuclear separation. Delocalization and distortion of the orbitals are neglected. The molecular orbitals of the V_k center are primarily constructed from the outer p electrons of the halogen X^- ions. Fig. 7 shows a MO energy level diagram and the occupancies of the orbitals in the ground state. Up to π_g all orbitals are fully occupied, the single hole is in the σ_u orbital. Without spin-orbit coupling there are two allowed optical transitions. These produce the ultraviolet absorption band with σ-polarization and the infrared band with π-polarization. Usually, the transitions are classified with respect to energy changes of the hole. They are:

$\sigma_u \rightarrow \sigma_g ({}^2\Sigma_u^+ \rightarrow {}^2\Sigma_g^+)$ ultraviolet band

$\sigma_u \rightarrow \pi_g ({}^2\Sigma_u^+ \rightarrow {}^2\Pi_g)$ infrared band.

Calculated (17) and experimentally measured peak values of the optical absorption bands of V_k centers in alkali halides are summarized in Table 2.

The calculated transition energies are found to be in reasonable agreement with, but somewhat smaller than, the experimental results. The agreement between experiment and theory is better for the alkali fluorides compared to the chlorides. Furthermore, the calculations yield a substantial reduction in the internuclear distance of the X_2^- molecule from that in an undistorted crystal. The reduction varies from about 20% for LiCl to 45% for RbF. The nuclear distance always lies between the free X_2^- separation and

TABLE 2. TRANSITION ENERGIES (eV) OF V_k CENTERS IN ALKALI HALIDES.

	LiF	NaF	LiCl	NaCl	KCl
Ultraviolet band					
ΔE_{exp}	3.48	3.38	3.15	3.28	3.39
ΔE_{theor}	3.30	3.30	2 69	2.65	2.37
Infrared band					
ΔE_{exp}	1.65				1.65
ΔE_{theor}	1.60	1.60	1.31	1.35	1.23

the X^--X^- separation for the perfect crystal. From ENDOR data, using the R^{-3} dependence of the anisotropic hyperfine constant, Daly and Mieher (18) estimate the displacements of the V_k ions for LiF and NaF. They obtain a reduction in separation of 27% for both LiF and NaF, confirming the calculated large distortions. For comparison, the predicted values are 25% for LiF and 35% for NaF.

The model of the V_k center as a X_2^- ion, where the crystalline environment only influences the separation of the component atomic units is qualitatively and quantitatively a very powerful tool to analyse its properties. Thus, the hyperfine interaction of the self-trapped hole with neighbouring ions indicates that the wave-function is close to that of the free molecular ion.

At a characteristic temperature (e.g. about 90 K for KI) the hole is no longer localized on a single X_2^- molecular ion by the self-trapping distortion and diffuses through the lattice (19). In the alkali halides the jumps to neighbouring sites involve reorientation of the V_k center axis by an angle of 60°.

2. Intrinsic hole center in oxygen-dominated phosphors. The class of oxygen-dominated phosphors consists of a rather large number of phosphors. We shall restrict our considerations to tungstates and molybdates crystallizing in the scheelite structure.

In the scheelite structure, space group C_{4h}^6, there are four molecules ABO_4 in the unit cell. The dimensions are $a = b \neq c$. The unit cell is plotted in Fig. 8. The crystals can be regarded as ordered arrays of BO_4^{2-} radicals ionically bonded to A^{2+} cations. The solids contain the polyatomic anions in the form of discrete and separated complexes. The covalently bonded tetrahedral oxyanions are slightly compressed along the c-axis leading to a reduction in symmetry from T_d to D_{2d}. MoO_4^{2-} groups are more covalent than WO_4^{2-} groups (20).

The scheelite type oxygen-dominated phosphors are wide-band-gap phosphors. This is illustrated in Fig. 9. The plot shows the reflection spectra of a nominally pure $CaWO_4$ crystal at room temperature (a) and at the temperature of liquid nitrogen (b). We have discussed these experimental results in terms of a semiempirical MO (molecular orbital) calculation of the isolated tetraoxo transition metal complex WO_4^{2-} (21).

The MO energy level diagram of the regular tetrahedral (T_d symmetry) WO_4^{2-} complex, as calculated with the self-consistent charge and configuration molecular orbital (SCCC-MO) method in (22), is shown in Fig. 10(a). The level diagram of Fig. 10(a) can serve as a basis for a mainly qualitative interpretation of the

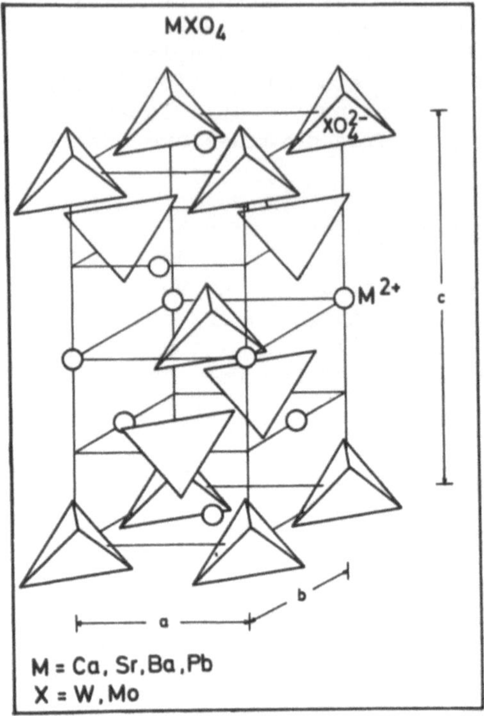

Fig. 8. Unit cell of scheelite type crystals ABO_4.

reflection spectra, only. With the exception of the well investigated MnO_4^- (23,24), where there is now some agreement on the ordering of the molecular orbitals, doubt still remains as to the ordering and energies of the orbitals in most other systems. The excited states scheme of the SCCC-MO method is very poor. A more comprehensive analysis of the optical spectra of tetraoxo transition metal complexes can be obtained by means of the Hartree-Fock-Slater discrete variational molecular orbital (HFS-DVM) method (25,26). With the above mentioned reservations in mind, we have assigned the first six main peaks (Fig. 9) to the first six charge transfer bands of the WO_4^{2-} oxyanion as summarized in Table 3.

TABLE 3. CHARGE TRANSFER BANDS OF THE WO_4^{2-} OXYANION, CORRESPONDING TO THE FIRST SIX MAIN PEAKS OF FIG. 9

Peak number	Transition	Peak energy (eV)
1	$t_1 \rightarrow 2e(^1A_1 \rightarrow {}^1T_2)$	5.9
2	$t_1 \rightarrow 4t_2(^1A_1 \rightarrow {}^1T_1)$	7.6
3	$t_1 \rightarrow 4t_2(^1A_1 \rightarrow {}^1T_2)$	8.3
4	$3t_2 \rightarrow 4t_2(^1A_1 \rightarrow {}^1T_2)$	9.4
5	$2t_2 \rightarrow 2e(^1A_1 \rightarrow {}^1T_1)$	11.7
6	$2t_2 \rightarrow 2e(^1A_1 \rightarrow {}^1T_2)$	12.3

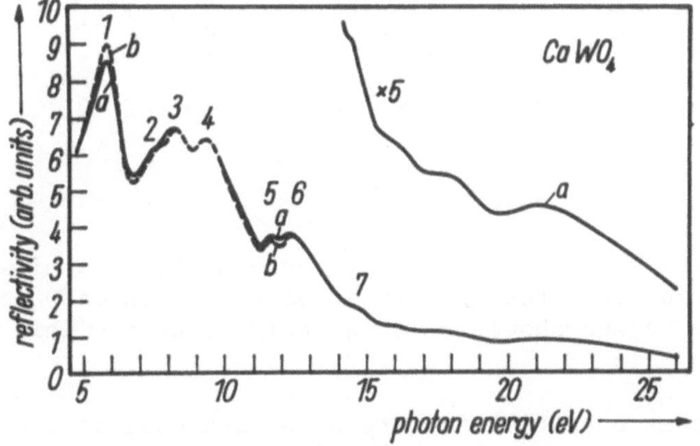

Fig. 9. Reflectivity of a nominally pure $CaWO_4$ crystal in arbitrary units versus photon energy (a) at room temperature, (b) at the temperature of liquid nitrogen (from ref. 21).

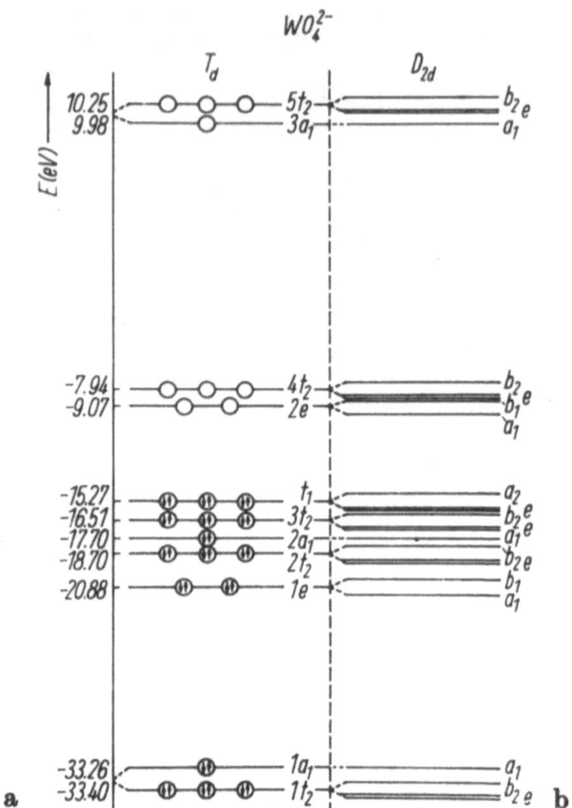

Fig. 10. Molecular orbital level diagram of the tungstate anion (a) for T_d symmetry, (b) for D_{2d} symmetry. The values of the orbital energies in T_d symmetry are taken from (22). The diagram (b) shows the level splitting in D_{2d} symmetry only qualitatively (from ref. 21).

In spite of all limitations, these optical results lead to the assumption that the electronic processes in $CaWO_4$ crystals in the valence region should be adequately treated within a localized description.

A further essential feature of oxygen dominated phosphors is the large Stokes shift (27). Thus, the luminescence band of $CaWO_4$ has its peak energy at about 2.84 eV. Luminescence measurements also support a description of the optical properties of these phosphors in a MO scheme of the oxyanions (28,29,30). Moreover, the experimental investigations point towards a strong electron-lattice interaction in this sort of materials.

Preliminary new studies of the photoconductivity of a nominally pure $CaWO_4$ crystal have failed. The measurements were performed using the electron synchrotron DESY as a light source. If $CaWO_4$ crystals should show photoconductivity at all, it would be very small. The experimental facts indicate narrow bands.

From all these results 1) isolated oxyanions
2) wide-band-gap materials
3) nearly localized states
4) strong electron-lattice interaction
5) narrow bands

one may expect that in these phosphors self-trapping of charge carriers should be a natural process. Actually, there exists conclusive experimental evidence for molecular polaron formation in tungstates and molybdates. Self-trapping occurs during X-rays or γ irradiation (ionizing radiation) of nominally pure and doped tungstate and molybdate crystals at sufficiently low temperatures. Irradiated crystals are coloured showing a radiation-induced, structured absorption from the infrared over the whole visible spectrum up to the fundamental absorption edge. Defects are produced by charge redistribution. A prominent defect, an intrinsic paramagnetic hole center, resulting from γ irradiation of undoped $CaWO_4$ at 77 K has been described for the first time by Zeldes and Livingston (31). This hole center has the g-values $g_x = 2.0065$, $g_y = 2.001$, $g_z = 2.0354$. The center reveals a small hyperfine interaction with two equivalent W^{183} nuclei (natural abundance 14.3 %) which is only partially resolved at temperatures $\geqq 78$ K. The hyperfine interaction is explained by the hole being equally shared between two adjacent tungstate tetrahedra (31,32).

Fig. 11 presents the EPR spectrum of the intrinsic hole centers in $CaWO_4$ at 4.2 K for an arbitrary direction of the magnetic field H relative to the crystal axes (33). Due to four magnetically distinct lattice sites for the hole center there are four major resonance lines, each accompanied by two pairs of satellite lines with an intensity of 1/12 of the central line. The satellite lines at 4.2 K result from hyperfine interaction of the hole with two nonequivalent W^{183} nuclei with the nuclear spin I = 1/2. The intensity ratio of 1:12 between the hyperfine and the central lines is caused by the natural abundance of 14.3 % W^{183} to 85.7 % tungsten nuclei with I = 0. Each line group can be described by a spin Hamiltonian

$$\mathcal{H} = \mu_B SgH + SA^{(1)}I_1 + SA^{(2)}I_2 \ , \qquad (14)$$

where the effective spin $S = \frac{1}{2}$, $I_1 = I_2 = \frac{1}{2}$ and μ_B is the Bohr magneton. $A^{(1)}$ and $A^{(2)}$ are the hyperfine tensors. The relatively

TABLE 4. PRINCIPAL VALUES OF THE COMPONENTS OF THE g-TENSOR AND THE HYPERFINE TENSORS $A^{(1)}$ AND $A^{(2)}$; $A^{(1)}$ AND $A^{(2)}$ WITHOUT SIGNS.

g	$A^{(1)}(10^{-4}T)$	$A^{(2)}(10^{-4}T)$
$g_x = 2.0089 \pm 0.005$	$A_1 = 6.25 \pm 0.20$	$A_1 = 14.41 \pm 0.20$
$g_y = 2.0119 \pm 0.005$	$A_2 = 5.37 \pm 0.20$	$A_2 = 15.56 \pm 0.20$
$g_z = 2.0234 \pm 0.005$	$A_3 = 6.46 \pm 0.20$	$A_3 = 13.27 \pm 0.20$

small nuclear Zeeman term has been omitted from eq. (14). The EPR data from an analysis of the experimental results obtained at 4.2 K (33) are summarized in Table 4.

The mean value $\bar{g} = 2.0147$ and the isotropic parts are $a^{(1)}_{iso} = (6.02\pm0.20) \times 10^{-4}T$ and $a^{(2)}_{iso} = (14.41\pm0.20) \times 10^{-4}T$. The smaller hyperfine interaction with a second tungsten nucleus, as shown by $A^{(1)}$, may be caused by a partial transfer of the hole to a neighbouring WO_4^{2-} tetrahedron. Edwards et al. (34) have discussed this mechanism for the system $CaMoO_4$/Nb. From the ratio of $A^{(1)}$ to $A^{(2)}$ for $CaWO_4$ at 4.2 K a probability of about 30% of the hole being at the adjacent oxyanion can be estimated.

Until now, by EPR measurements intrinsic hole centers are found in the scheelites $CaWO_4$, $SrWO_4$, $BaWO_4$ (35), $CaMoO_4$, and $SrMoO_4$. Although no intrinsic hole centers are detected in $CdMoO_4$, measurements on scheelite type $Cd_xCa_{1-x}MoO_4$ mixed crystals indicate their formation in this crystal, too (36). An optical proof of the self-trapped hole center in scheelites is extremely difficult due to strongly overlapping absorption bands. Above liquid nitrogen temperature the hole centers become unstable, their thermal decay being connected with the appearance of thermoluminescence and thermally stimulated current (32,35).

A reasonable mechanism for self-trapping of holes in scheelites seems to be a charge-transfer process, as proposed in (34), whereby charge-density is transferred from BO_4^{2-} to BO_4^- with no actual chemical bond formation. The self-trapping during irradiation with ionizing radiation may be described as follows:

$$BO_4^{2-} \xrightarrow[T \leqq 77\ K]{X\text{-rays},\ \gamma} BO_4^- + e^-_{trapped} \tag{15}$$

$$BO_4^- + BO_4^{2-} \xrightarrow[T \leqq 77\ K]{} (BO_4 - BO_4)^{3-} \tag{16}$$

The self-trapped hole center $(BO_4\text{-}BO_4)^{3-}$ represents a molecular polaron.

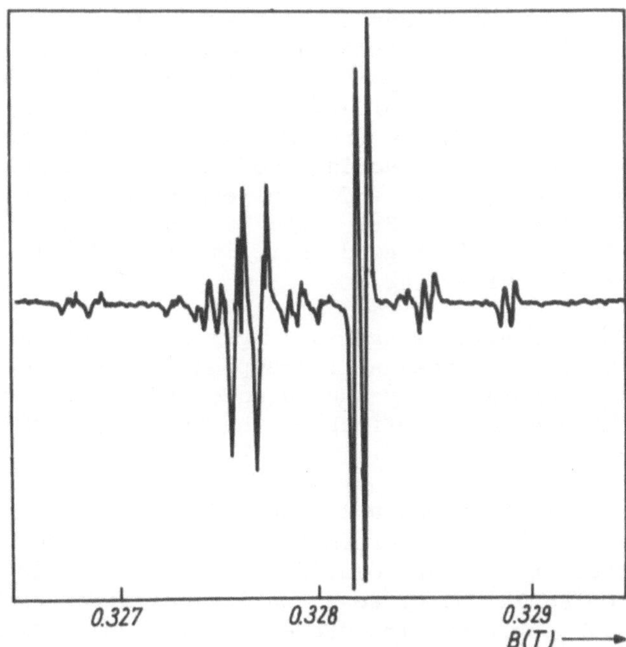

Fig. 11. EPR spectrum of the intrinsic hole centers in $CaWO_4$ at 4.2 K. Arbitrary direction of the magnetic field H to the crystal axes (from ref. 33).

According to the above mentioned features of the scheelites one has to expect that self-trapped electrons should also be possible in this class of crystals. Although stable WO_4^{3-} electron centers are found by EPR (31,35,37), until now there are no indications for an intrinsic nature of these centers.

Scheelites with valence states which are not solely determined by the oxyanions ($PbWO_4$, $PbMoO_4$, $CdMoO_4$) show quite another behaviour (36,38).

3. <u>Further self-trapped centers.</u> Self-trapped hole centers also have been observed in the alkaline earth fluorides (39,40,41). Spin resonance and optical data show that the centers have <100> orientation. The EPR results indicate the hole to be concentrated on two of the fluorine ions. It is generally assumed that the electronic properties of the self-trapped hole are those of an F_2^- molecular ion, and that the crystal environment merely changes the interatomic distance. This 'molecule in a crystal' model, already successfully used for the description of the V_k center in alkali halides, is strongly supported by measurements of the hole charge-density on ions neighbouring the F_2^- molecular configuration. The hyperfine interaction from the adjacent axial ions is small for CaF_2, less in SrF_2, and undetectable in BaF_2. Therefore, the hole is almost completely localized on the F_2^- molecular ion. After that, there are no detectable deviations from axial symmetry in the paramagnetic and optical spectra. This means that the effect of crystal fields on the molecular ion is negligible, since, unlike the F_2^- ion, the self-trapped hole center in alkaline earth fluorides does not have axial symmetry. Calculations of the optical and spin resonance properties of the self-trapped hole in CaF_2, SrF_2, and BaF_2 show that the 'molecule in a crystal' concept is a good approximation of this defect in alkaline earth fluorides (42).

In the silver halides AgCl and AgBr self-trapped holes are stabilized, too. A proposed self-trapping of the hole in the 4d cation shell is explained by the valence band structure of silver halides and a Jahn-Teller effect of the $4d^9$ configuration of the Ag^{2+} ion (43).

Self-trapped electrons are assumed to exist in orthorhombic sulphur crystals (44). Orthorhombic sulphur, a molecular crystal, is characterized by strong covalent bonds between the atoms of the S_8 ring molecule and by weak van der Waals forces binding these molecules together. Therefore, the intermolecular orbital overlaps are small and the molecular energy levels are perturbed only slightly by neighbour interactions. Narrow bands and molecular polaron formation should be natural consequences. There is experimental evidence (45) that self-trapped electrons are also formed in realgar (As_4S_4).

II.C. Concluding Remarks

The discourse given in this chapter shows that self-trapping rendered possible by charge redistribution during irradiation with ionizing radiation is a widespread mechanism of defect creation by radiationless transitions. Self-trapping of charge carriers is not confined to polar materials. The mechanism even works in Van der Waals crystals. The crucial condition for the formation of this kind of defect is a sufficiently large short-range component of the electron-lattice interaction. The extent to which the self-trapping effect occurs in any particular system cannot be estimated without further ado.

III. ATHERMAL MIGRATION OF DEFECTS IN SOLIDS

Quite recently significant experimental evidence has been gathered concerning changes in the macroscopic properties of, above all, elemental and compound semiconductors produced by recombination processes. These effects play a key role in the performance of many semiconductor devices. The changes in the properties are strongly bound to the presence of deep traps in the energy level spectrum of the materials. They are created by carrier capture or recombination processes at these deep states, whereby the processes must predominantly occur via radiationless transitions. A series of detailed investigations published in the last few years convincingly demonstrate that carrier capture or recombination processes at deep traps can enhance or induce dissociation and/or migration of the defect or impurity centers which generate the trapping states (46,47). The migration of pre-existing defect or impurity centers as well as the dissociation of pre-existing defect associates followed by a migration of a component defect may create new defect centers within the solid. The inverse process, an elimination of pre-existing defects, for example an annealing of implant damage, is also possible.

Deep defect states within the band gap of a given material decide whether or not these processes may occur. Suitable deep states possess those centers which tightly bind electronic particles. In semiconductors operating at room temperature and above, centers with binding energies $\gtrsim 0.5$ eV should not represent metastable trapping centers but are most likely to act as recombination centers. Perhaps transition metal point defects where trapping occurs into tightly bound d-states or lattice defects and their associates with one another and with chemical impurities have sufficiently large binding energies. Similar to transition metals, point defect vacancies and interstitials just as simple associates containing these elemental lattice defects may exist within the crystals in several charge states. All these centers are suitable candidates for the type of phenomena we are interested in here.

From combined electrical, optical, and magnetic measurements there is much information on the nature of these defects in elemental semiconductors. However, our knowledge about the identity of centers responsible for particular deep traps in the binary compounds as III-V and II-VI materials is very limited. From our own results obtained in the course of photoluminescence studies on annealed ultra-high pure ZnS crystals (48,49,50,51) we suppose that there are many unidentified centers in II-VI compounds consisting of simple lattice defects and their associates which produce essentially completely nonradiative recombination.

The phenomena related to carrier capture or recombination of carriers at defect sites, which are of interest here, concern the enhancement of atomic motion. In a soft mechanism recombination may merely change the electronic (exciton recombination) or charge (charge carrier recombination) state of the defect, producing a species of higher mobility. The probability that a defect will jump from one equilibrium position to another across the barrier E_m given by the host atoms is

$$\nu = \nu_o \exp(-E_m/kT). \qquad (17)$$

ν_o is a characteristic vibrational frequency of the order $10^{13} s^{-1}$. $\nu_o \simeq kT/h$, where h is Planck's constant. The barrier height E_m accounts for the interaction of the defect with the surrounding atoms. The interaction may include repulsive or attractive electronic forces. Therefore, E_m may change with the charge state of the migrating defect. A decrease of the activation energy E_m leads to an enhanced thermal motion. A faster diffusion rate may also result from an altered effective frequency ν_o. Examples for this mechanism are vacancies in Si which diffuse at very different rates in the four charge states V^+, V^o, V^-, and V^{--} (52) as well as anion vacancies in alkali halides (F',F,α). Moreover, in the alkali halides the motion of F centers in the ground and excited states are substantially different (53).

Besides the above mentioned mechanism, recombination may enhance the amplitude of local atomic vibrations, thereby enhancing the local probability of a jump. The resulting enhanced diffusion rate depends on the efficiency in excitation of the local modes in question and on the rate at which anharmonic interactions dissipate enery to other lattice modes. Under suitable conditions 'thermal spikes' may be produced by localized nonradiative energy dissipation which were postulated to create Frenkel defects for atoms with appropriately low displacement energy in the range of 1-2 eV (54). This mechanism is called the 'phonon kick'.

Violent adiabatic effects should be expected to occur at defect complexes. Defect associates with deep trapping states can be very efficient in nonradiative recombination, since the extended structure of these centers must be particularly sensitive to the charge state. When the charge state is changed through capture of a charge carrier strong electron-phonon interaction may produce a substantial change in a local lattice coordinate or even alter the symmetry of the center. During the following relaxation process energy is released locally which in extreme cases may be of the order of the band gap. In defect complexes the dissipation of the localized energy adiabatically produced on capture of a carrier at a deep trap may occur through structural changes, in which the form of the complex is altered by an ionic displacement promoted by the absorption of the localized energy. The structural change may be accompanied by a diffusion of a point defect. Structural changes of this type may only be observed at centers with a sufficiently strong electron-phonon interaction or at moderate coupling to a mode of exceptionally large vibrational energy where a few phonons take up a large amount of energy. These reactions caused by recombination are no longer enhanced thermal but rather induced athermal processes. It must be mentioned that besides the structural changes a purely electronic change can occur by ejection of an electronic particle from the center in a trap-related Auger process, especially in indirect gap materials.

A theoretical description of the recombination-enhanced motion of point defects in a quasi-equilibrium approach for the energy locally liberated at the center is given by Weeks et al. (55). The thermal bath and the localized recombination-produced contributions to the vibrational motions are independent. Therefore, the contributions of these two energy sources to the activation energy of motion are additive. The recombination enhanced activation energy E'_m may be written (55)

$$E'_m = E_T - E_r + E_R. \tag{18}$$

E_T = thermal activation energy, E_R = activation energy for the electron-hole recombination ($\sim$ 0.1 eV), E_r = recombination energy. Athermal diffusion processes occur within the quasi-equilibrium approximation when $E_r > E_T$. Furthermore, due to eq. (18) the recombination-enhanced motion must be faster in wider-band-gap materials, since there the recombination energies are larger. Recombination-enhanced and induced motion of defects should be a general phenomenon which is particularly essential in the wide-band-gap materials.

In the quasi-equilibrium approach it is assumed that the defect reaction occurs when the energy in some critical coordinate is raised above threshold during a fluctuation above a local equilibrium. The adiabatic reaction model discussed by Dean and Choyke (46) becomes formally identical to the quasi-equilibrium description when $E_r \gg E_T$. In this case the fluctuation effects are insignificant. The reaction occurs athermally before the critical coordinate loses the energy supplied directly from the radiationless recombination process. These adiabatic reactions are not within the scope of the quasi-equilibrium description.

The discussed effects can lead to the formation of more extended defects by a loss of the mobile component to other centers. It is assumed that such an irreversible loss-process promotes the climb of pre-existing dislocations creating dislocation networks. These extended defects are of central importance in the degradation of the efficiency of light emitting semiconductor devices.

III.A. Bourgoin-Corbett Mechanism

Bourgoin and Corbett (56) have discussed a mechanism for athermal motion of defects in solids. The authors start with the assumption that the equilibrium configuration of a defect is substantially altered when its charge state changes. The defect may prefer to occupy different sites in the lattice for different charge states. Then, by capture of electrons and holes alternately the defect athermally migrates through the crystal. The diffusion rate is given by the slower of the rates of capture of the two carriers.

The mechanism of athermal migration is illustrated in Fig. 12. The plot shows the potential energy of the defect as a function of its position in the lattice. In the charge state q = 0 the defect is in equilibrium at the bottom of the well at site H. For a motion of the defect in this charge state to equivalent sites the barrier at T must be overcome by thermal excitation. However, upon switching the charge state to q = 1 the defect, still located at H,

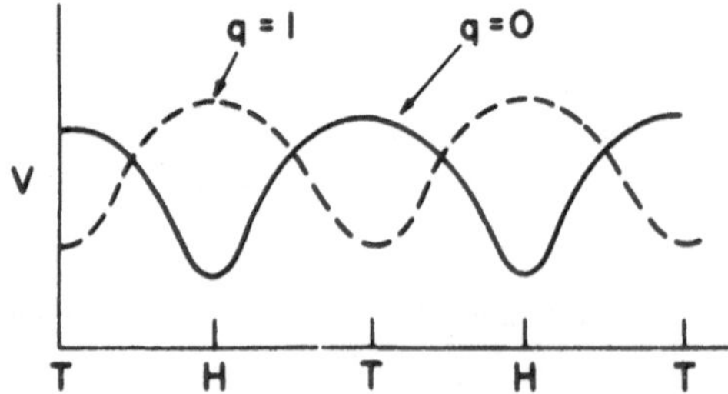

Fig. 12. Schematic diagram of the potential energy V of the defect for different charge states q versus its position in the lattice. T is a tetrahedral, H a hexagonal site (from ref. 56).

finds itself at the top of the potential curve for q = 1 and relaxes to the tetrahedral site T. Thus, an alternate change of the charge state results in an athermal motion of the defect.

To support this model, Bourgoin and Corbett (56) have considered the dependence of the equilibrium position of a defect on its charge state for a tetrahedral (T) and a hexagonal (H) site. The authors follow a calculation published by Weiser (57) on the ion energy for two sites (tetrahedral (5) and hexagonal (H) interstitial sites) in the diamond lattice. The interaction of a charged interstitial with its environment consists of an attractive and of a repulsive potential. The attractive part is produced by the interaction of the charged interstitial with the dipoles it has induced in the lattice, while the repulsive contribution is determined by the interaction of the interstitial with closed shells. Weiser shows that it is possible to calculate the stability by treating the difference in the attractive energies for the two sites and the difference in the repulsive energies for the two sites separately and then comparing these differences. As shown in Fig. 13, the difference in attractive energies has a q^2 dependence (ΔU_{POL} curve). q is the charge of the interstitial. The potential produced by a dipole is linear in the charge of the elements of the dipole. The difference in repulsive energies shows an approximately exponential behaviour with the ionic radius. This results in a decrease with increasing positive charge q (ΔU_{REP} curve). So the equilibrium site of the interstitial changes with the charge state q as shown in Fig. 13. At the cross-over the migration energy vanishes. This mechanism is used to explain displacement effects under irradiation at very low temperatures.

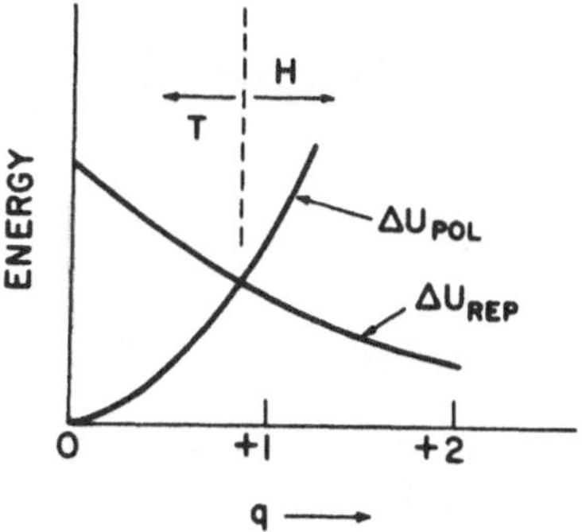

Fig. 13. Dependence of the equilibrium position of an interstitial on its charge state q (from ref. 56).

III.B. Photo-Induced Defect Reactions

As already mentioned above, recombination-enhanced and induced-defect processes should be particularly important in the wide-band-gap materials. Depending on the defect state, alkali halides exposed to ionizing radiation at low temperatures show several types of hole and electron centers which are produced simultaneously. In the case of hydrogen impurities, however, it is possible to create only one type of hole centers by UV irradiation of alkali halide crystals containing neutral hydrogen atoms in interstitial lattice positions (U_2 or H_i centers). U_2 centers can be produced by photo-dissociation of HH^- centers in alkali halides. By irradiation into the UV absorption band of U_2 centers in KCl at 4.5 K it is possible to destroy the U_2 band and to create two absorption bands due to U centers (H^- ions on Cl^- sites, H_s^-) and H centers (hole centers, Cl_2^- on Cl^- sites, $Cl_{2,s}^-$). The photodestruction of U_2 centers in KCl is described as (58,59)

$$H_i^o + Cl_s^- + Cl_s^- \xrightarrow{h\nu} H_s^- + Cl_{2,s}^- \quad . \tag{19}$$

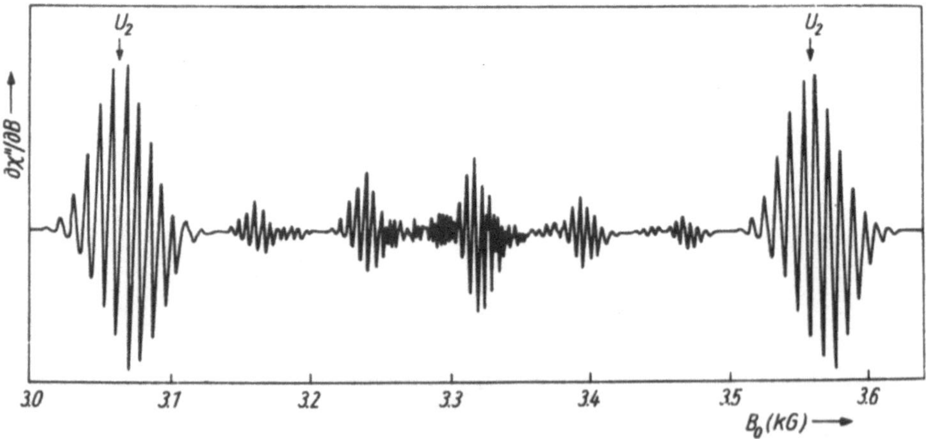

Fig. 14. EPR spectrum of a KCl crystal after partial photo-destruction of the U_2 band at 20 K. Outer 13-line-groups: U_2 centers; inner groups: H centers. $\vec{B}_o || \langle 100 \rangle$, ν_{EPR} = 9380 MHz (from ref. 60).

The U_2 decay and the H center formation have been investigated by EPR measurements (60). Fig. 14 shows the EPR spectrum after partial photodestruction of the U_2 centers at 20 K. The two outer line groups belong to U_2 centers. The inner lines are due to H centers. They possess for $\vec{B}_o || <100>$ seven nearly equidistant line groups. Each group consists of seven lines caused by hyperfine interaction with four Cl neighbour nuclei. Two of the H center line groups are hidden under the U_2 groups.

The experimental results convincingly demonstrate that recombination induced migration and dissociation effects on defects are important phenomena in the field of radiationless processes in solids.

IV. RADIOLYSIS IN HALIDE COMPOUNDS

In chapter II of our article we have discussed mainly intrinsic hole centers which are possible if the free companion electrons are trapped elsewhere in the lattice at defect sites. The mechanisms described in chapter III are confined to processes at point defects and defect structures. The topic of this chapter is related to the production of ionic defects in pure undoped materials. Here, the mechanism is the reaction of stable Frenkel defects in a perfect dynamic lattice by ionizing radiation. The effect is a more dramatic process than the formation of V_k centers. It is connected with a motion of lattice particles about finite distances.

IV.A. Background

Very early it was found that energetic electromagnetic (X-rays, γ-rays) or particle irradiation would colour simple halide crystals at temperatures at about room temperature and below. The colouration is due to the creation of defects which absorb light at wavelengths at which the undisturbed crystal is transparent. The intensity of the induced absorption bands is proportional to the concentration of definite defects. If the defects are produced by displacement of lattice particles from their normal sites through elastic collision events with incident projectiles, the number N of created defects per cm^3 of crystal is

$$N = N_o \sigma_d(E) \phi \ . \tag{20}$$

N_o is the atomic density, $\sigma_d(E)$ is the displacement cross section, and ϕ is the flux of incident particles. $\sigma_d(E)$ is a simple expression when the sample is thin enough so that the irradiating particles are of essentially constant energy throughout their traverse of the crystal. For an average displacement cross section of about 10 barns, eq. (20) shows that a 2-MeV electron can displace at best one lattice ion from its site to form a Frenkel pair. This calculational result is in strong contrast with experiment, a clear indication that an approximation which only takes account of collision processes is not sufficient to explain radiation damage in metal halide compounds. Thus, in alkali halides irradiated with electrons, many defects are formed per incident particle (61). Furthermore, Frenkel pair production is even possible by exciting a sample with UV light having an energy of the band gap. Defect creation in ionic crystals is very efficient and is most likely due to the conversion of electronic excitation energy into mechanical energy of lattice particles. These radiolysis processes also occur in nonpolar insulating materials and seem to govern wide regions of the photographic process and of photosynthesis.

IV.B. Simultaneous Creation of F and H Centers

The first general question with respect to radiolysis in halide compounds concerns the identity of the produced defects. In irradiated pure crystals at low temperatures where defect aggregation is negligible one generally may expect Schottky defects and/or Frenkel defects in the cation and/or anion sublattices. A mechanism for the formation of Schottky pairs consisting of cation and anion vacancies has been discussed in the literature (62). There, it is suggested that electronic excitation should deexcite at dislocation jogs, whereby positive and negative ion vacancies are alternately evaporated from the dislocation into the host lattice. Although one cannot state that any of the above mentioned defects are never created by ionizing radiation, the experiment convincingly shows that the produced defects are Frenkel defects in the halide sublattice. They are F-center interstitial-atom (F-H) pairs, for which the charge of each constituent is zero, or anion-vacancy interstitial-ion (F^+ or α-I) pairs. Overwhelming evidences that Frenkel defects, rather than Schottky pairs, are formed result from a lot of mechanical, thermal, electrical, magnetic, and optical measurements (1,2). Moreover, experimental studies indicate that the major fraction of Frenkel defects produced are formed in the environment of the perfect crystal.

The first serious attempt to understand how electronic excitation can produce vacancy-interstitial pairs in a halide crystal lattice was performed by Varley (63). The author assumed that irradiation creates multiply ionized halogen ions which, if the lifetime of their multiply charged state is long enough, will be ejected due to electrostatic forces into interstitial positions distant from the initial sites. A subsequent capture of electrons would then produce F center - H center pairs usually observed. The Varley mechanism only had a limited chance to survive. The production efficiences originally calculated by Varley (63) were much too low compared with the production rate for all Frenkel defects at liquid helium temperature. Efficiency measurements on KBr at 5 K performed with a beam of 50 kV X-rays showed that one vacancy is created for 2400 eV absorbed energy (64). Furthermore, in view of the very rapid formation of electron-hole pairs in halide crystals a Varley-like displacement process seems rather improbable. A mechanism of Klick (65) involving ionization of two neighbouring halogen ions also disappeared from more recent discussions.

Up to now, the most successful approach to an understanding of defect production in halide crystals by ionizing radiation is the concept of a single ionization mechanism. Electron-hole pairs rapidly formed have lifetimes of the order of 10^{-6} s and store energies of several electron volts. On the other hand in these crystals any electronic change is followed by a strong ionic relaxation. So the self-trapping of a hole leads to an appreciable shortening of the internuclear distance of the two participating adjacent halide lattice ions. Due to these facts, the recombination of an electron must have two main consequences: 1. a release of an amount of energy of several eV's and 2. the generation of an impulse pushing the two halogen ions towards their normal lattice sites. Whether or not in a definite halide crystal Frenkel defects are formed then depends on the released recombination energy relative to the energy required for displacement.

In the case of ionic crystals the displacement energy of an anion from its lattice site must be of the order of the lattice binding energy approximated by

$$E_{bind} = \alpha q^2/r_o \quad , \tag{21}$$

where α is the Madelung constant of the lattice, $q = Ze$ is the charge on the ions, and r_o is the distance between adjacent ions. The energy of an electron-hole pair is roughly equal to the energy of a molecule (for example NaCl) in the crystal:

$$E_{pair} = \alpha q^2/r_o - I + A \ . \tag{22}$$

I is the ionization potential of the cation and A is the electron affinity of the anion. These crude estimates yield that in alkali halide crystals recombination and displacement energies are of comparible size.

A consolidation of the single ionization model was achieved above all by UV absorption and luminescence measurements. Since Smakula's (66) investigations, as early as 1930, there was experimental evidence that UV photons could produce colouration in alkali halides. However, a connection of this colouration with a photon-induced creation of intrinsic defects was done only much later. It was shown that UV radiation could create much higher concentrations of F centers than the impurity concentrations in the samples (67). Moreover, the shapes of the F center formation curves were very similar to those obtained with X-rays (68). Studies on the correlation of the temperature dependence of the fundamental luminescence in KI with the temperature dependence of F center production clearly showed that the creation of F centers is connected with electron-hole recombination at self-trapped holes (69,70,71). It was found that just in the range where the fundamental emission disappeared, the F center formation efficiency increased by approximately an order of magnitude. UV light has since been shown to create F centers in several alkali halides. The anticorrelation in the temperature dependence of the fundamental luminescence and F center production has also been observed for different alkali halides. These experiments convincingly demonstrate that a single ionization mechanism of relatively high efficiency is in operation.

The next essential step in the verification of the single ionization mechanism was the experimental result that electron-hole recombination during irradiation at low temperatures resulting in fundamental luminescence was identical to recombination of electrons with trapped holes, for the appearance of intrinsic recombination-luminescent metastable states involving a self-trapped hole and a bound electron were responsible (72,73). This excited state configuration ($V_k e$) in ionic crystals is generally termed self-trapped or relaxed exciton. In the relaxed exciton states two neighbouring halogen ions are bound covalently in the same way that excited states of a diatomic rare gas molecule are bound. The relaxed exciton is the key to a better understanding of Frenkel defect production in alkali halides by ionizing radiation. The single

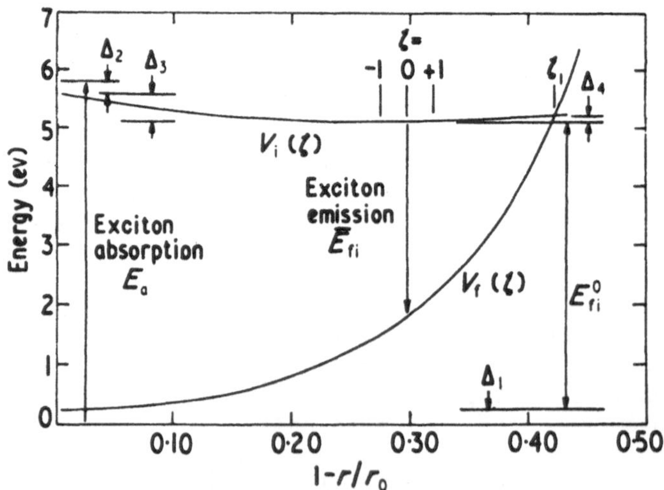

Fig. 15. Energies of electron-hole pair states in KI versus the relative separation of the two iodide ions (from ref. 71).

ionization mechanism becomes an excitonic mechanism. A Frenkel defect can be created by a radiationless transition of a relaxed exciton to the ground state. Through the nonradiative transition a large fraction of the band gap energy becomes available as kinetic energy of the two ions which may be sufficient to separate the ion from its lattice site. With the concept of the excitonic mechanism the problem is reduced enough to allow theoretical considerations.

A first more quantitative approach to F center production in alkali halides by electron-hole recombination is given by Pooley (71). Fig. 15 shows the potential curves of the ground state (calculated) and of the excited state (schematic) as a function of the relative separation r/r_o of the two halogen ions for KI. E_{fi}^{o} is the energy available for F center production. For defect creation the system has to surmount the barrier Δ_4. The model accounts for the fact that in KI thermally stimulated transitions from the metastable relaxed exciton state to the ground state occur with consequent F center formation at the expense of the luminescence. From results of a computational model Pooley (74) also concludes that the separation of the halogen ion from its lattice position is achieved by a replacement sequence along a <110> direction.

Further essential progress was obtained by Keller and Patten (75) using EPR technique. In this experiment electrons trapped at impurity sites in KCl were released at low temperature. The recombination of the electrons with holes trapped in the form of Cl_2^- or $ClBr^-$ centers caused H centers (halogen atom at an interstitial site) to be in proportion to the trapped hole centers destroyed. $ClBr^-$ self-trapped holes were found to be more effective in producing interstitials than was the intrinsic Cl_2^- hole center. It was the first experiment which persuasively demonstrated that electron-hole recombination produces Frenkel pairs.

The behaviour of alkali halides against ionizing radiation is mainly the consequence of a competition between luminescence and defect production. In this connection the electronic states of the relaxed exciton are of fundamental interest, since they initiate recombination luminescence as well as Frenkel defect formation. The relaxed exciton in alkali halides is a bound electron-hole pair localized in the perfect lattice according to intrinsic ionic relaxation. After excitation of an electron to a conduction band state, the lattice close to the hole distorts to a configuration similar to the V_k center (72,76,77). However, in the case of the relaxed exciton, the electron will screen the hole somewhat from its surrounding. The time scale of relaxation is of order of 10^{-13} s, while the lifetime of the excitation is $10^{-9} - 10^{-6}$ s. The excited electron recombines with the self-trapped hole, either directly from a band state or, more probably, from an exciton state, where the lattice subsequently returns to its normal configuration. The most essential feature of the relaxed exciton in alkal halides is the property that after electron trapping the hole remains nearly unperturbed and the self-trapping configuration is maintained. Thus, it is reasonably described as an electron bound to a V_k center - (V_ke).

The states of the relaxed exciton are discussed in terms of those of a diatomic rare gas molecule X_2^{2-} (78), the natural generalization of the X_2^- model of the V_k center. The ground state, which is unstable, has the electronic configuration

$$(\sigma_g\ np)^2\ (\pi_u\ np)^4\ (\pi_g\ np)^4\ (\sigma_u\ np)^2.$$

It is a ${}^1\Sigma_g^+$ state. The lowest relaxed exciton states result from the excited configuration.

$$\ldots\ldots\ldots\ (\sigma_u\ np)\ (\sigma_g(n+1)s).$$

These two states ${}^{1,3}\Sigma_u^+$ are bound states. The state ${}^1\Pi_u$, due to the electronic configuration

$$(\sigma_g\ np)^2\ (\pi_u\ np)^3\ (\pi_g\ np)^4\ (\sigma_u\ np)^2\ (\sigma_g(n+1)s)$$

is also of importance. These states allow us to describe the luminescence properties of the lowest triplet state of the relaxed exciton consistently. The recombination occurs with the emission of π-polarized and σ-polarized light. The π-band, observed in all alkali halides, has a long lifetime of about 10^{-6} s which decreases with increasing atomic number of the anions. It is attributed to the spin-forbidden transition ${}^3\Sigma_u^+ \rightarrow {}^1\Sigma_g^+$ made allowed by halogen spin-orbit coupling of the ${}^3\Sigma_u^+$ and ${}^1\Pi_u$ excited states. ${}^1\Pi_u$ mixes most strongly with ${}^3\Sigma_u^+$. The model accounts for all properties of the π-band. An assignment of the σ-band to a transition from the singlet ${}^1\Sigma_u^+$, corresponding to the triplet ${}^3\Sigma_u^+$, to the ground state ${}^1\Sigma_g^+$ is not entirely free from doubt. Blair et al. (79) suggest that in the case of the σ-band the X_2^{2-} model is not adequate and that charge transfer to the nearest cations must be considered. The σ-band occurs at higher energies and has a lifetime of about 10^{-9} s. Absorption and emission are well described by the Franck-Condon principle. The observed extreme Stokes shifts ($\sim$ 5 eV) are due to substantial lattice relaxation in the excited state associated with the self-trapping. Wood (77) argues that the lattice distortion near the exciton in alkali halides consists of two main parts, a symmetric relaxation about the anion (one) on which the hole is initially created and a asymmetric relaxation analogous to that of the V_k center. The author concludes that the asymmetric distortion produces the major part of the Stokes shift.

In recent years, detailed investigations were performed concerning the long-lived triplet state of the relaxed exciton in alkali halides (80,81,82). The results indicate that the hole is well-localized and the electron has a wavefunction very similar to that of the F center in its ground state. The quadrupole resonance of the two central halogen nuclei shows that the relaxed exciton is slightly displaced from the centrosymmetrical position between two lattice sites. With respect to the triplet states of the relaxed exciton, the electron and hole are substantially independent of each other. The spectral features are similar to those of analogous perturbed color centers. The optical transitions from the lowest triplet can be divided into two main types, consisting of hole-excitations analogous to those of the V_k center and electronic excitations analogous to those of the M center. The mentioned spectral characteristics are of special importance in modern discussions of the Frenkel defect production in alkali halides by ionizing radiation.

Quite recently, the combination of excited-state absorption spectroscopy (81) and extensive theoretical calculations has led to a very good understanding of the energy levels of the relaxed exciton in alkali halides. Excited-state absorption spectroscopy uses the techniques of time-resolved spectroscopy and electron-pulse irradiation. The theoretical studies consider the fact that there are two main types of excitation for a relaxed exciton. The 'hole transitions' involving mainly X_2^- ionic excitations are best treated using a Hartree-Fock method (83). The more extended states of the relaxed exciton, those states which are involved in the 'electronic transitions,' are well described by the pseudopotential method (84). Neither method gives a good recombination energy. The most important result of the pseudopotential calculation is that the σ (singlet) and π (triplet) luminescence bands derive from different orbital states, contrary to previous assumptions. Besides the known absorption band at 2 eV, an additional absorption band of the triplet relaxed exciton in the infrared range, predicted by the pseudopotential approach, has not been found experimentally (86). The best values for electron-hole recombination as well as for electron and hole excitation have been achieved using a semi-empirical self-consistent molecular orbital method (CNDO) for relaxed excitons in KCl (85). Advanced experimental techniques and detailed calculations have led to a significant improvement in the overall picture of the relaxed exciton in alkali halides.

The time-resolved investigations of colour center production in halide crystals (81,87) have yielded a series of very important facts. The absorption spectrum of a KBr crystal, induced by an electron pulse, is illustrated in Fig. 16 by the solid curve. The spectrum was measured at 50 ns after the end of pulsed irradition at 8K. The absorption peaks at 1.6, 2.06, 3.3, and 6.1 eV are attributed to bands due to triplet excitons, F centers, H centers, and α centers, respectively. The F band region has been reduced to 1/4. The absolute optical densities for the transient and stable spectra

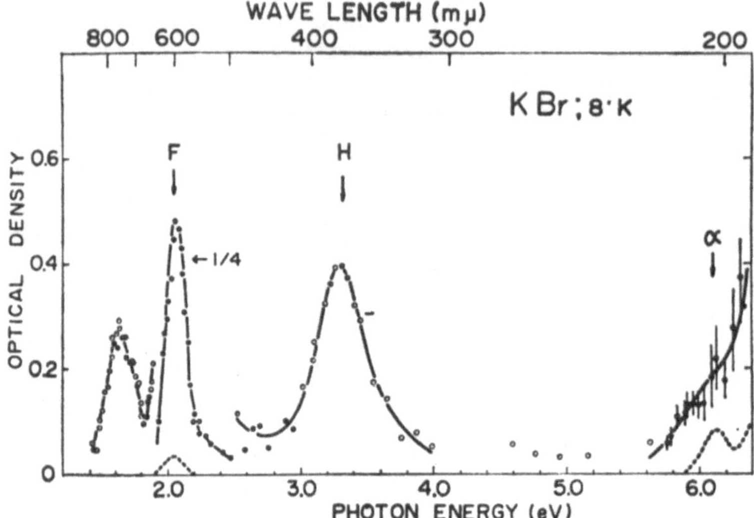

Fig. 16. Transient (solid curve) and stable (broken curve) absorption produced in KBr by an electron pulse (from ref. 87).

are not directly comparable. The band denoted H also includes relaxed exciton absorption. Fig. 16 shows that the transient F center production efficiency is clearly higher than that obtained by steady-state measurements. From the transient results one derives a production rate for F center formation of about 100 eV per center, a very reduced value compared to steady-state investigations. After that, the primary defects produced are pairs of H and F centers (87,88). Furthermore, due to the experimental results it is generally accepted that some excited states of the relaxed excitons serve as the precursor for the defect formation, however, with the exception of the lowest bound state of the relaxed exciton from which the long-lived π-luminescence is emitted. Itoh and Saidoh (89) suggest that Frenkel defect creation proceeds through an excited hole state. It is also found that the formation time of F centers in their ground state is about 11 ps in KCl at 25 K (88). Damage is produced very rapidly.

After it has become evident that Frenkel defect production in alkali halides results from a radiationless decay of relaxed excitons, the most serious problem arises concerning the separation of the vancancy and the interstitial. If the ($V_k e$) decays via a nonradiative crossover, as illustrated in Fig. 15, the available kinetic energy must be equally shared between the two halogen ions between which the bond has been broken. This, however, is a hard drawback to the original Pooley mechanism (74) which states that focused collisions along a close-packed line (<110>) of like ions is most favourable to produce a separation between the vacancy and the interstitial. Calculational estimates on F^+-I pairs indicate that it is highly improbable to produce a displacement with less than about 10 eV (90). For a displacement sequence a substantial asymmetry in the energy partition between the two halogen ions is necessary. In the study of Smoluchowski et al. (91) asymmetry is considered to some degree through introduction of zero-point vibrations. In a recent experiment Hall et al. (92) have shown that the recombination of electrons with highly oriented V_k centers in fact creates H centers, but that there is no polarization in the induced H band. The authors conclude that if H centers do separate from F centers via correlated <110> collision sequences the direction of the sequence is not maintained to its end.

For modern theories on Frenkel defect production and on separation of the vacancy and the interstitial in alkali halides by ionizing radiation two experimental results, the very rapid creation of F centers in their ground state and the fact that a state or narrow band of states above the lowest triplet state of the ($V_k e$) act as precursors to F-H pairs, are of basic importance. In contrast to the model of Pooley (74) which is of a dissociative type, in re-recent theoretical investigations translational motion of the relaxed exciton is discussed. Itoh et al. (85) have used the CNDO method to calculate the energies for different stages in the evolution of F and H centers from relaxed excitons. A particularly striking result of this calculation is a repulsive F-H interaction at the nearest-neighbour spacing, compared with an attractive interaction at larger spacing. It ensures the separation of the F and H center and creates a barrier against F-H recombination. Fig. 17 illustrates the evolution of the ionic positions with time. In the first stage optical excitation creates an exciton whose hole component undergoes self-trapping. A prompt decay of the F-H pair follows, because of the repulsive interaction at nearest-neighbour spacing. If the F-H pair evolves to the next-nearest-neighbour spacing, the pair has about 2 eV less energy than at infinite separation. The authors assume that this barrier to further separation of the F-H pair can be overcome if the H center is formed in an excited state. The special feature of this theory (85), but also its critical point, is that the hole component is proposed to be also excited.

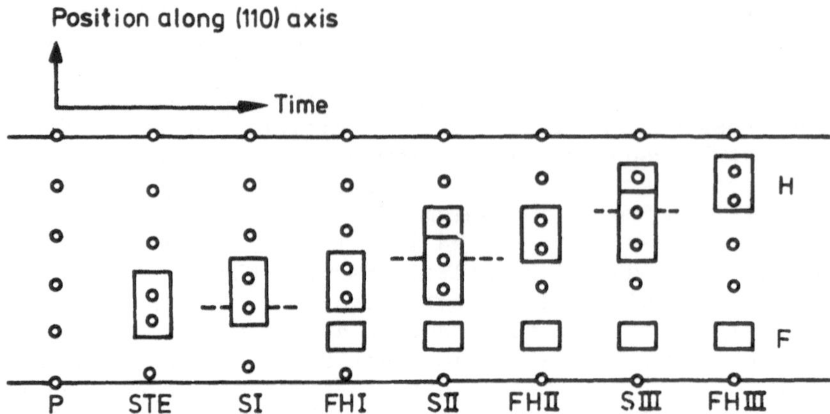

Fig. 17. Formation mechanism of a neutral vacancy (F center) plus a neutral interstitial (H center) from a relaxed exciton. P: Perfect <110> close-packed row. STE: Relaxed exciton with the hole in its lowest state. SI, SII, SIII: Saddle points, midway between lattice sites. FHI, FHII, FHIII: F-H center pairs at increasing distances (from ref. 85).

In the model of defect production given by Toyozawa (93), the author starts from excited states in which the electronic component of the (V_ke) alone is excited. Toyozawa suggests that the translational motion is a consequence of the adiabatic instability of the lowest (V_ke) state A_{1g}. The instability originates from a mixing with the second lowest state B_{3u} in the presence of a vibrational mode Q_2 of the center of mass of the X_2^- molecular ion. The terms A_{1g} and B_{3u} result from an electronic orbital notation appropriate to D_{2h} point symmetry. The mixing of the A_{1g} and B_{3u} states means that the two adiabatic potential surfaces do not cross and further, due to increased mixing the Q_2 mode may become unstable. The Q_2 mode instability, together with the stretching mode Q_1 of the X_2^- molecular ion, squeeze out the X_2^- in the <110> direction. The distorted electron wave function settles on the vacant lattice site resulting in a F center. Leung and Song (94) propose a revised model which includes an additional mode The mode takes into account the relaxation of other ions around the X_2^- molecular ion. The inclusion of the mode yields improved results. Kabler and Williams (95) also present a model in which a relaxed exciton becomes an F-H pair through a relaxation process that preserves the σ bond between the two halogen ions on which the hole is localized. The authors state that the initial ionic motion between relaxed exciton and F+H is not necessarily confined to a close-packed direction in alkali halides.

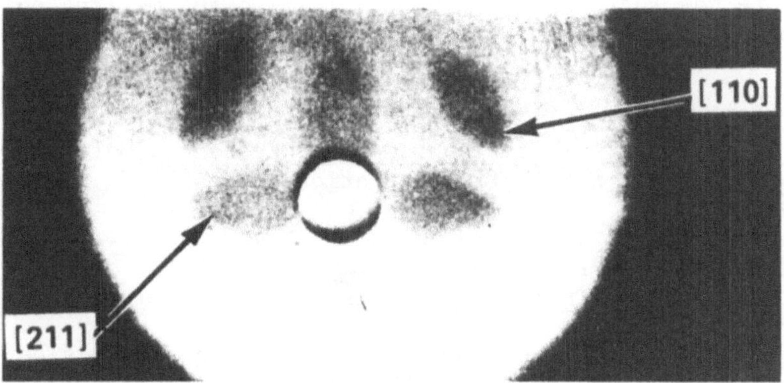

Fig. 18. Halogen emission pattern obtained from KCl crystals during exposure to intense ruby laser pulses (from ref. 96).

IV.C. Experimental Facts

In a recent experiment (96) multiphoton-induced directional emission of halogen atoms and ions from fcc KCl, KBr, and NaCl single crystals has been observed. The crystals were irradiated with strong ruby laser pulses. The striking feature of this experimental result is the emission of particles even in the <211> direction. The authors suppose that nonradiative decay of (V_ke) and subsequent replacement collision chains account for this effect. If this explanation holds, the result would be a clear example of the radiationless decay of the relaxed exciton in which the motion is not confined to the <110> direction.

V. OPTICAL BREAKDOWN IN DIELECTRICS

Up to now we have discussed mechanisms of defect creation by radiationless transitions which led to more or less simple effects. The mechanisms give rise to point defects and in the worst case to dislocation networks. In this chapter we briefly mention an extremely violent effect of defect formation in insulators by nonradiative processes. The effect is called optical breakdown.

Optical breakdown in insulating materials can be produced by very intense laser pulse excitation. During exposure of a NaCl crystal for instance to a pulse of high-intensity laser light F center bleaching, avalanche ionization (97), and multiphoton absorption (98) produce a very strong increase of the number of free electrons in the conduction band. By the inverse bremsstrahlung effect the free electrons receive energy from the optical field. The lattice is heated by electron-phonon collisions. Damage occurs when the temperature has increased to a value close to the melting point of the material.

REFERENCES

1. M. N. Kabler, in Radiation Damage Processes in Materials (C. H. S. Dubuy, ed.) Noordhoff International Publishing, Leyden (1975).
2. J. H. Crawford, Jr. and J. M. Slifkin, Point Defects in Solids, Vol. 1, Plenum Press, New York-London (1972).
3. A. M. Stoneham, Theory of Defects in Solids, Clarendon Press, Oxford (1975).
4. T. L. Gilbert, Lecture notes for the NATO Advanced Study Institute, Ghent, 1966, unpublished.
5. N. F. Mott and A. M. Stoneham, J. Phys. C: Solid State Phys. 10, 3391 (1977).
6. L. Landau, Phys. Zeitschr. D. Sowjetunion 3, 664 (19733).
7. N. F. Mott and R. W. Gurney, Electronic Processes in Ionic Crystals, Oxford University Press, Oxford (1940).
8. Y. Toyozawa, Prog. Theor. Phys. 26, 29 (1963).
9. D. Emin, Phys. Rev. Lett. 28, 604 (1972).
10. D. Emin, Adv. Phys. 22, 57 (1973).
11. T. Holstein, Ann. Phys. 8, 325 and 343 (1959).
12. A Sumi and Y. Toyozawa, J. Phys. Soc. Japan 35, 137 (1973).
13. D. Emin and T. Holstein, Phys. Rev. Lett. 36, 323 (1976).
14. W. Känzig, Phys. Rev. 99, 1890 (1955).
15. T. Castner and W. Känzig, J. Phys. Chem. Solids 3, 178 (1957).
16. C. J. Delbecq, B. Smaller, and P. H. Yuster, Phys. Rev. 111, 1235 (1958).
17. A. N. Jette, T. L. Gilbert, and T. P. Das, Phys. Rev. 184, 884 (1969).
18. D. F. Daly and R. L. Mieher, Phys. Rev. 183, 368 (1969).
19. F. J. Keller and R. B. Murray, Phys. Rev. 150, 670 (1966).
20. A. W. Sleight, Acta Cryst. B28, 2899 (1972).
21. R. Grasser, E. Pitt, A. Scharmann, and G. Zimmerer, DESY Report SR-75/02, April 1975 and Phys. Stat. Sol. (b) 69, 359 (1975).
22. R. Kebabcioglu and A. Müller, Chem. Phys. Lett. 8, 59 (1971).
23. H. B. Gray, Coord. Chem. Rev. 1, 2 (1966).
24. A. P. Mortola, H. Basch, and J. W. Moskowitz, Intern. J. Quant. Chem. 7, 725 (1973).
25. D. E. Ellis and G. S. Painter, Phys. Rev. B2, 2887 (1970).
26. A. Rauk, T. Ziegler, and D. E. Ellis, Theoret. Chim. Acta (Berl.) 34, 49 (1974).
27. F. A. Kroger, Some Aspects of the Luminescence of Solids, Elsevier, Amsterdam, 1948.
28. M. J. Treadaway and R. C. Powell, J. Chem. Phys. 61, 4003 (1974).
29. R. Grasser and A. Scharmann, J. Lum. 12/13, 473 (1976).
30. R. Grasser and Scharmann, DPG Spring Meeting, Münster, 1979.
31. H. Zeldes and R. Livingston, J. Chem. Phys. 34, 247 (1961).

32. G. K. Born, R. J. Grasser, and A. O. Scharmann, Phys. Stat. Sol. 28, 583 (1968).
33. R. Biederbick, G. Born, A. Hofstaetter, and A. Scharmann, Phys. Stat. Sol. (b) 69, 55 (1975).
34. P. R. Edwards, S. Subramanian, and M. C. R. Symons, J. Chem. Soc. (A), 2985 (1968).
35. M. Böhm, B. Cord, A. Hofstaetter, A. Scharmann, and P. Parot, J. Lum. 17, 291 (1978).
36. A. Hofstaetter, R. Oeder, A. Scharmann, and D. Schwabe, Z. Physik B35, 1 (1979).
37. C. Kikuchi, Bull. Inst. Chem. Res., Kyoto Univ. 48, 10 (1970).
38. A. Hofstaetter, A. Scharmann, D. Schwabe, and B. Vitt, Z. Physik B30, 305 (1978).
39. W. Hayes and J. W. Twidell, Proc. Phys. Soc. 79, 1295 (1962).
40. R. F. Marzke and R. L. Mieher, Phys. Rev. 182, 453 (1969).
41. J. H. Beaumont, W. Hayes, D. L. Kirk, and G. P. Summers, Proc. R. Soc. A315, 69 (1970).
42. M. J. Norgett and A. M. Stoneham, J. Phys. C: Solid State Phys. 6, 229 and 238 (1973).
43. W. Ulrici, Phys. Stat. Sol. (b) 40, 557 (1970).
44. D. J. Gibbons and W. E. Spear, J. Phys. Chem. Solids 27, 1917 (1966).
45. G. B. Street and W. D. Gill, Phys. Stat. Sol. 18, 601 (1966).
46. P. J. Dean and W. J. Choyke, Adv. Phys. 26, 1 (1977).
47. C. H. Henry, J. Lum. 12/13 47 (1976).
48. R. Grasser, G. Roth, and A. Scharmann, Z. Naturforsch. 30a, 1205 (1975).
49. R. Grasser, A. Scharmann, and W. Schwedes, Z. Physik B20, 235 (1975).
50. R. Grasser, K. Hermanns, A. Scharmann, and W. Schwedes, Z. Physik B24, 25 (1976).
51. R. Grasser, A. Scharmann, and W. Schwedes, Z. Physik B28, 247 (1977).
52. G. D. Watkins, Radiation Effects in Semiconductors, Plenum Press, New York, 1968.
53. J. M. Vail, R. J. Brown, and C. K. Ong, J. Physique C 9, 83 (1974).
54. F. Seitz and J. S. Koehler, Solid State Physics 2, 305 (1956).
55. J. D. Weeks, J. C. Tully, and L. C. Kimerling, Phys. Rev. B 12, 3286 (1975).
56. J. C. Bourgoin and J. W. Corbett, Phys. Lett. 38A, 135 (1972).
57. K. Weiser, Phys. Rev. 126, 1427 (1962).
58. G. Kurz, Phys. Stat. Sol. (b) 31, 93 (1969).
59. G. Kurz and S. Susman, Phys. Stat. Sol. (b) 46, K5 (1971).
60. G. Reuter, L. Schwan, and J. M. Spaeth, Phys. Stat. Sol. (b) 53, K29 (1972).
61. J. H. Crawford, Adv. Phys. 17, 93 (1968).
62. F. Seitz, Rev. Mod. Phys. 26, 7 (1954).
63. J. H. O. Varley, Nature 174, 886 (1954) and J. Nucl. Eng. 1, 130 (1954).

64. V. H. Ritz, Phys. Rev. 133, A1452 (1964).
65. C. C. Klick, Phys. Rev. 120, 760 (1960).
66. A. Smakula, Z. Physik 63, 762 (1930).
67. T. P. P. Hall, D. Pooley, W. A. Runciman, and P. T. Wedepohl, Proc. Phys. Soc. 84, 719 (1964).
68. C. B. Lushchik, G. K. Vale, E. R. Ilmas, N. S. Rooze, A. A. Elango, and M. A. Elango, Optics and Spectr. 21, 377 (1966).
69. H. N. Hersh, Phys. Rev. 148, 928 (1966).
70. J. D. Konitzer and H. N. Hersh, J. Phys. Chem. Solids 27, 771 (1966).
71. D. Pooley, Proc. Phys. Soc. 87, 245 (1966).
72. M. N. Kabler, Phys. Rev. 136, A1296 (1964).
73. R. B. Murray and F. J. Keller, Phys. Rev. 137, A942 (1965).
74. D. Pooley, Proc. Phys. Soc. 87, 257 (1966).
75. F. J. Keller and F. W. Patten, Solid State Commun. 7, 1603 (1969).
76. R. G. Fuller, R. T. Williams, and M. N. Kabler, Phys. Rev. Lett. 25, 446 (1970).
77. R. F. Wood, Phys. Rev. 151, 629 (1966).
78. M. N. Kabler and D. A. Patterson, Phys. Rev. Lett. 19, 652 (1967).
79. I. M. Blair, D. Pooley, and D. Smith, J. Phys. C: Solid State Phys. 5, 1537 (1972).
80. M. J. Marrone, F. W. Patten, and M. N. Kabler, Bull. Am. Phys. Soc. 18, 631 (1973).
81. R. T. Williams and M. N. Kabler, Phys. Rev. B9, 1897 (1974).
82. D. Block, A. Wasiela, and Y. Merle D'Aubigné, J. Phys. C: Solid State Phys. 11, 4201 (1978).
83. A. M. Stoneham, J. Phys. C: Solid State Phys. 7, 2476 (1974).
84. K. S. Song, A. M. Stoneham, and A. H. Harker, J. Phys. C: Solid State Phys. 8, 1125 (1975).
85. N. Itoh, A. M. Stoneham, and A. H. Harker, J. Phys. C: Solid State Phys. 10, 4197 (1977).
86. R. T. Williams, M. N. Kabler, and I. Schneider, J. Phys. C: Solid State Phys. 11, 2009 (1978).
87. Y. Kondo, M. Hirai, and M. Ueta, J. Phys. Soc. Japan 33, 151 (1972).
88. J. N. Bradford, R. T. Williams, and W. L. Faust, Phys. Rev. Lett. 35, 300 (1975).
89. N. Itoh and M. Saidoh, J. Physique 34, C9 (1973).
90. I. M. Torrens and L. T. Chadderton, Phys. Rev. 159, 671 (1967).
91. R. Smoluchowski, O. W. Lazareth, R. D. Hatcher, and G. J. Dienes, Phys. Rev. Lett. 27, 1288 (1971).
92. T. P. P. Hall, A. E. Hughes, and D. Pooley, J. Phys. C: Solid State Phys. 9, 439 (1976).
93. Y. Toyozawa, J. Phys. Soc. Japan 44, 482 (1978).
94. C. H. Leung and K. S. Song, Phys. Rev. B 18, 922 (1978).
95. M. N. Kabler and R. T. Williams, Phys. Rev. B 18, 1948 (1978).
96. A. Schmid, P. Bräunlich, and P. K. Rol, Phys. Rev. Lett. 35, 1382 (1975).

97. E. Yablonovitch and N. Bloembergen, Phys. Rev. Lett. 29, 907 (1972).
98. P. Bräunlich, A. Schmid, and P. Kelly, Appl. Phys. Lett. 26, 150 (1975).

THE RELEVANCE OF NONRADIATIVE TRANSITIONS TO SOLID STATE LASERS

L.A. Riseberg

GTE Laboratories, Inc.
Waltham, MA 02154

ABSTRACT

The demonstration of the laser in 1960 (1) in an insulating medium activated by impurity ions, $Al_2O_3:Cr^{3+}$, led to two decades of intensive and revolutionary research in the luminescent properties of insulating solids. This also came at a time when the technology was developed for producing very pure rare earth and transition metal compounds, as well as the means to grow large single crystals. The relationship has been a synergistic one; for, while much important research resulted that has been of great consequence in its own right, the implications of this research for the improvement and development of lasers, have been equally substantive.

Here, we will treat the relationship of nonradiative transitions to the operation of solid state lasers. Since lasers are basically radiative devices, competing nonradiative processes are obviously of great importance. However, the significance goes beyond this. The operation and characteristics of a laser are intrinsically related to the presence of the right combination of radiative and nonradiative processes in the pump level, the lasing level, and the terminal level. In this chapter, we shall explore this relationship.

We shall restrict the discussion here to optically pumped insulator lasers. We exclude for reasons of space the area of semiconductor lasers, which has an entirely different, but equally fascinating pathology. We shall consider primarily the implications of phonon-induced relaxation, although we shall also address the consequences of ion-ion interactions. Our approach will be by example, after the laying of some groundwork in the first sections. Finally, our coverage must of necessity be restrictive, and there is not the space to cover all of the work of all of the contributions to this vast subject.

In Section II, we shall review briefly the basics of laser physics which we shall use in the ensuing discussions. In Section III, we shall review the state of understanding of nonradiative relaxation phenomena. Section IV will study examples from the laser field when nonradiative processes are important. In Section V, we pose some interesting current problems for students of lasers and luminescence.

I. THEORY OF LASER OPERATION

Many texts on laser physics have appeared in recent years. A selected bibliography is included in the references (2-5), and the student is referred to these and other sources for a fuller treatment of this subject.

I.A. Quantum Theory of Radiation

We begin with the consequences of quantum mechanics for an atom interacting with a radiation field. Starting classically with Maxwell's equations, we wish to show that the time-varying electromagnetic waves are quantized as a set of oscillators, where the field excitation is specified in terms of normal modes and number operators.

We begin with Maxwell's equations:

$$\begin{aligned} \nabla\cdot\vec{D} &= \rho \\ \nabla\cdot\vec{B} &= 0 \\ \nabla\times\vec{E} &= -\frac{\partial\vec{B}}{\partial t} \\ \nabla\times\vec{H} &= \vec{j} + \frac{\partial\vec{D}}{\partial t} \quad . \end{aligned} \tag{1}$$

We assume charge-free, isotropic, and homogeneous media and consider the fields $\vec{E}(\vec{r},t)$ and $\vec{H}(\vec{r},t)$ inside a volume V bound by a surface S of perfect conductivity.

$$\left.\begin{aligned} \hat{n}\times\vec{E} &= 0 \\ \hat{n}\cdot\vec{H} &= 0 \end{aligned}\right\} \text{ on S} \quad . \tag{1a}$$

We expand $\vec{E}$ and $\vec{H}$ in terms of two orthogonal sets of vector fields $\vec{E}_a$ and $\vec{H}_a$. These sets, introduced by Slater (6), obey the equations:

$$k_a \vec{E}_a = \nabla \times \vec{H}_a$$

$$k_a \vec{H}_a = \nabla \times \vec{E}_a \quad , \tag{2}$$

where k_a is a constant. Taking the curl of both sides of eq. (2), and using the identity

$$\nabla \times \nabla \times \vec{A} = \nabla(\nabla \cdot \vec{A}) - \nabla^2 \vec{A} \quad , \tag{2a}$$

we find

$$\nabla^2 \vec{E}_a + k_a^2 \vec{E}_a = 0$$

$$\nabla^2 \vec{H}_a + k_a^2 \vec{H}_a = 0 \quad , \tag{3}$$

the familiar wave equations. The fields $\vec{E}(\vec{r},t)$ and $\vec{H}(\vec{r},t)$ can be expressed as a linear superposition of these orthogonal fields

$$\vec{E}(\vec{r},t) = \sum_a \frac{1}{\sqrt{\epsilon}} p_a(t) \vec{E}_a(\vec{r})$$

$$\vec{H}(\vec{r},t) = \sum_a \frac{1}{\sqrt{\mu}} \omega_a q_a(t) \vec{H}_a(\vec{r}) \quad , \tag{4}$$

where $\omega_a = k_a / \sqrt{\mu\varepsilon}$.

Substituting eq. (4) into the third of Maxwell's equations and using eq. (2) we have

$$p_a = \dot{q}_a \quad , \tag{5}$$

and from the fourth of Maxwell's Equations we have

$$\omega_a^2 q_a = -\dot{p}_a \quad , \tag{6}$$

which, with eq. (5), yields

$$\ddot{p}_a + \omega_a^2 p_a = 0 \quad . \tag{7}$$

Thus, $k_a(\mu\epsilon)^{-\frac{1}{2}} = \omega_a$ is the frequency of the ath mode. We thus consider the radiation field formally as a set of quantum oscillators with canonical momentum and coordinate p_a and q_a,

$$H = \sum_a \tfrac{1}{2}(p_a^{\,2} + \omega_a^{\,2} q_a^{\,2}) \quad . \tag{8}$$

We can define creation and annihilation operators a_ν and $a_\nu^\dagger$, such that

$$H = \sum_\nu \hbar\omega_\nu (a_\nu^\dagger a_\nu + \tfrac{1}{2}) \ , \tag{9}$$

and

$$<n_\nu | a_\nu^\dagger a_\nu | n_\nu> = n_\nu \quad , \tag{10}$$

where n_ν is the oscillation number of the νth mode.

I.B. An Atom in a Radiation Field

Let us consider an atom with state vectors $|\Psi_i>$ and energy eigenvalues E_i. The Born-Oppenheimer states describing the system are now

$$|u_{i\nu}> = |\Psi_i>|n_\nu> \ , \tag{11}$$

and the interaction Hamiltonian is

$$V_{int} = -\frac{e}{m}\vec{p}\cdot\vec{A} \quad , \tag{12}$$

where $\vec{p}$ is the electron momentum and $\vec{A}$ is the vector potential associated with the field of eq. (4) by

$$\vec{E} = -\frac{\partial\vec{A}}{\partial t}$$

$$\vec{H} = \nabla\times\vec{A} \quad . \tag{13}$$

We assume, for ease of manipulation that the modes of eq. (4) are plane waves, and we use for the mode ν

$$\vec{A}_\nu = \sqrt{4\pi}\ \hat{n}_\nu\ e^{i\vec{k}_\nu\cdot\vec{r}} \quad . \tag{14}$$

Then by the Golden Rule we have for the transition probability between atomic states $|\Psi_a>$ and $|\Psi_b>$ with emission of light at ω_ν

$$W_{ab} = \frac{e^2}{m^2} \frac{8\pi^2}{\hbar} |P_{ab}|^2 (\bar{n}_\nu + 1)\, \delta\, (E_a - E_b - \hbar\omega_\nu) \quad , \tag{15}$$

in the dipole approximation ($e^{-i\vec{k}_\nu \cdot \vec{r}} = 1$). The $\bar{n}_\nu + 1$ yields a spontaneous and stimulated emission term, while an absorptive transition (i.e., photon annihilation operator) would have only $\bar{n}_\nu$ as the multiplier.

I.C. The Einstein Treatment of Spontaneous and Stimulated Emission

This is a useful classical approach to be used in complement with the results of II. B. Consider a blackbody radiation field whose energy density is given by

$$\rho(\nu) = \frac{8\pi h\nu^3}{c^3} \frac{1}{e^{h\nu/kT} - 1} \quad . \tag{16}$$

For an interacting atom with energy states $|2>$ and $|1>$ the transition from the higher state $|2>$ can be written

$$W_{21} = B_{21}\, \rho(\nu) + A \quad , \tag{17}$$

while the transition rate from 1 to 2 is

$$W_{12} = B_{12} \rho(\nu) \quad . \tag{18}$$

Here A is the spontaneous transition rate and B_{12} and B_{21} are constants to be determined. In thermal equilibrium at a temperature T, the up rate must equal the down rate.

$$N_2 W_{21} = N_1 W_{12} \quad , \tag{19}$$

and the populations are given by the Boltzmann factor

$$\frac{N_2}{N_1} = \frac{g_2}{g_1} e^{-h\nu/kT} \quad , \tag{20}$$

where g_2 and g_1 are the degeneracies. From eqs. (17), (18), and (19) we have

$$\frac{N_2}{N_1} = \frac{B_{12}\rho(\nu)}{B_{21}\rho(\nu) + A} \quad . \tag{21}$$

Using eq. (20) to solve for $\rho(\nu)$ we have

$$\rho(\nu) = \frac{A(g_2/g_1)\ e^{-h\nu/kT}}{B_{12} - B_{21}(g_2/g_1)\ e^{-h\nu/kT}} \quad . \tag{22}$$

For thermal equilibrium, $\rho(\nu)$ must obey eq. (16), so that

$$\frac{8\pi h\nu^3}{c^3}\frac{1}{e^{h\nu/kT}-1} = \frac{A(g_2/g_1)\ e^{-h\nu/kT}}{B_{12} - B_{21}(g_2/g_1)\ e^{-h\nu/kT}} \quad . \tag{23}$$

This last equality requires that

$$B_{12} = B_{21}(g_2/g_1) \quad , \tag{24}$$

and

$$\frac{A}{B_{21}} = \frac{8\pi h\nu^3}{c^3} \quad . \tag{25}$$

Then the stimulated emission rate is

$$(W_{21})_{stim} = \frac{Ac^3}{8\pi h\nu^3}\rho(\nu) \quad . \tag{26}$$

The extension from a blackbody spectrum to any radiation field assumes a constant $\rho(\nu)$ over the width of the transition.

If the transition has a lineshape function $\sigma(\nu'-\nu_o)$, then $(W21)_{stim}$ becomes

$$(W_{21})_{stim} = \int_{-\infty}^{\infty} \frac{Ac^3\rho(\nu')}{8\pi h\nu'^3}\alpha(\nu'-\nu_o)d\nu' \quad . \tag{27}$$

For a monochromatic field $\rho(\nu') = \rho_\nu \delta(\nu'-\nu)$, which yields

$$(W_{21})_{stim} = \frac{c^3 A \rho_\nu}{8\pi h \nu^3} g(\nu-\nu_o) = \frac{Ac^2 I_\nu}{8\pi h \nu^3} g(\nu-\nu_o) \quad , \tag{28}$$

where $I_\nu = c\rho_\nu$ is the energy flux.

The absorption rate is simply

$$W_{12} = \left(\frac{g_2}{g_1}\right) \frac{Ac^2 I_\nu}{8\pi h \nu^3} g(\nu-\nu_o) \quad . \tag{29}$$

We now have the basic tools to address the propagation of light in a resonant medium.

I.D. The Gain Coefficient

For a monochromatic wave propagating in some medium characterized by levels $|1\rangle$ and $|2\rangle$ with populations N_1 and N_2, we have

$$\frac{dI_\nu}{dx} = (N_2 W_{21} - N_1 W_{12}) h\nu \quad , \tag{30}$$

and, substituting eqs. (28) and (29), we have

$$\frac{dI_\nu}{dx} = \left[N_2 - \frac{g_2}{g_1} N_1\right] \frac{Ac^2 g(\nu-\nu_o)}{8\pi \nu^2} I_\nu \quad , \tag{31}$$

or $I_\nu(x) = I_o e^{\alpha(\nu)x}$, where

$$\alpha(\nu) = \left[N_2 - \frac{g_2}{g_1} N_1\right] \frac{Ac^2}{8\pi \nu^2} g(\nu-\nu_o) \quad . \tag{32}$$

If $N_2 < (g_2/g_1)N_1$, i.e., there are more atoms in the lower energy state, then α is a negative quantity and the medium is absorbing. This is the normal equilibrium case. However, if $N_2 > (g_2/g_1)N_1$ then α is positive and the medium has a gain whose spectral distribution is characterized by the line-shape function. This is the condition of population inversion which results in an amplifying medium. With suitable optical feedback supplied to the medium, it becomes possible to achieve, under certain conditions, laser oscillation.

I.E. Optical Resonators

The elegantly simple means of supplying feedback to an amplifying medium is to place the medium in a resonant cavity. For the optical case, which succeeded its microwave analog, this resonator is in the form of a Fabry-Perot cavity. Detailed analyses of the modes of such cavities have been carried out (7), and are beyond the scope of our treatment here. Suffice it to say that for the case of plane or spherical mirrors, there is a sequence of transverse modes, and for each of these there is a set of longitudinal modes, for which the spacing is given by

$$\Delta\nu = \frac{c}{2L} \quad , \tag{33}$$

where L is the length of the cavity; i.e., we require that the round-trip phase shift is an integral multiple of 2π, assuming nearly plane waves. This ensures that the successively fed back light adds coherently in phase after each round-trip.

I.F. The Laser Oscillation Condition

We consider the case of an assembly of ions in a crystal with energy levels E_2 and E_1 with population densities N_2 and N_1. We assume that $g(\nu-\nu_o)$ is the natural lineshape, where $h\nu_o = E_2-E_1$. The gain is then given by eq. (32). We assume that the medium interacts with one mode of the resonator and that the lifetime of a photon in this mode is τ_{photon}. Then the decay of the intensity in this mode is

$$\frac{dI_\nu}{dt} = - \frac{I_\nu}{\tau_{photon}} \quad . \tag{34}$$

The gain in intensity is

$$\frac{dI_\nu}{dt} = c\,\frac{dI_\nu}{dx} = \left[N_2 - \frac{g_2}{g_1} N_1\right] \frac{Ac^3 g(\nu-\nu_o)}{8\pi\nu^2} I_\nu \quad . \tag{35}$$

The start oscillation condition occurs when the gain of eq. (35) just balances the loss of eq. (34), or

$$N_2 - \frac{g_2}{g_1} N_1 = \frac{8\pi\nu^2}{Ac^3 \tau_{photon}\, g(o)} \quad , \tag{36}$$

where it has been assumed that $\nu = \nu_o$. This value of $N_2 - (g_2/g_1)N_1$ is known as the critical inversion density, ΔN_c.

I.G. The Three-Level Laser

In the case of a solid-state laser, the achievement of a population inversion is carried out by the use of optical pumping; i.e., the absorption of intense light, either pulsed or cw, to absorption states either at or above the laser level. There are in general two classes of lasers: the three-level system and the four-level system. We shall treat the dynamics of each of these.

The energy level scheme of the three-level laser is shown in Fig. 1.

Here, τ_2 is the total lifetime of level 2. If we assume for simplicity that $g_2 = g_1$, and that the critical inversion density $\Delta N_c << N$, where N is the total ion density, we have for the oscillation condition that

$$N_2 = N_1 \approx \frac{N}{2} \quad . \tag{37}$$

We define the critical fluorescence power as the total fluorescence power per cm^3 emitted by the upper laser level at threshold. For the three-level case, this is

$$(P_{cf})_{3\text{-level}} = \frac{Nh\nu}{2\tau_2} \quad . \tag{38}$$

For most laser materials, $\tau_2 \simeq 1/A$, and eq. (38) can be written

$$(P_{cf})_{3\text{-level}} = \frac{NAh\nu}{2} \quad . \tag{39}$$

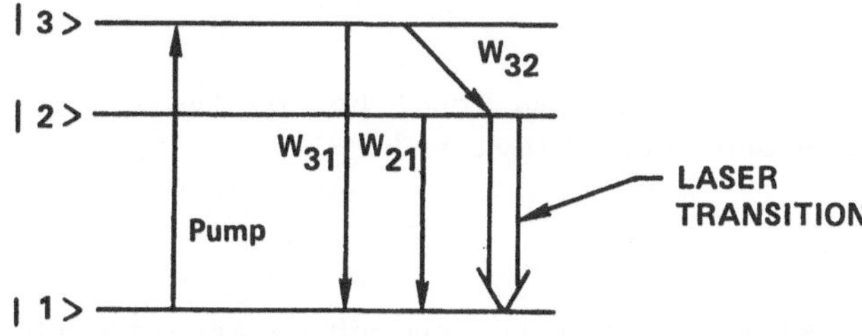

Fig. 1. Energy level scheme for the three-level laser.

The ramifications of having a nonradiative component of τ_2 or significant radiative transitions to other levels complicates the achievement of threshold and we shall treat the details later.

We shall now treat the rate equations for the three level case. We take the pump rate from $|1\rangle$ to $|3\rangle$ as W. The downward relaxation rates between levels are W_{ij} and the stimulated transition rate from $|2\rangle$ to $|1\rangle$ is W_{stim}. There are no other levels in the system, and therefore

$$N = N_1 + N_2 + N_3 \ .$$

The rate equations then can be written

$$\frac{dN_1}{dt} = N_2W_{21} + W_{stim}(N_2-N_1) - W(N_1-N_3) + N_3W_{31}$$

$$\frac{dN_2}{dt} = N_2W_{21} + W_{stim}(N_1-N_2) + N_3W_{32} \tag{40}$$

$$\frac{dN_3}{dt} = W(N_1-N_3) - N_3(W_{32} + W_{31}) \ .$$

Setting the left-side of eq. (40) equal to zero yields the steady-state inversion density

$$N_2-N_1 = N\frac{W[k(W_{32} + W_{31})-W_{21})] - W_{21}(W_{32} + W_{31})}{(W + W_{32} + W_{31})(2W_{stim} + W_{21}) + (W_{stim} + W_{32} + W_{21})} , \tag{41}$$

where $k = W_{32}/W_{32} + W_{31}$ in the branching ratio from level $|3\rangle$ into level $|2\rangle$. If we assume

$$W_{32} + W_{31} >> W_{21} , \tag{42}$$

i.e., the decay from the pump level to the laser level is much faster than its own decay time, and

$$W_{32} >> W \ , \tag{43}$$

the pump intensity is not large enough to saturate the W_{31} transition, then eq. (41) simplifies to

$$N_2 - N_1 = \frac{N(kW - W_{21})}{2W_{stim} + W_{21} + kW} \quad . \tag{44}$$

Below threshold, $W_{stim} = 0$, and the population difference depends linearly on W. Above threshold, the inversion is pinned at ΔN_c, and the stimulated rate is given by

$$W_{stim} = N \frac{kW - W_{21}}{2\Delta N_c} - \frac{kW + W_{21}}{2} \quad . \tag{45}$$

These results describe 3-level laser performance in a general way and will be utilized in our subsequent discussion of the ramifications of the detailed kinetics of the relaxation processes.

I.H. The Four-Level Laser

The four-level laser is described in Fig. 2.

The four-level laser case pertains for an energy level scheme where the lower laser level lies above the ground state, and $E_2 - E_1 >> kT$. Then $N_1 \approx N_o e^{-(E_1 - E_o)/kT} << \Delta N_c$. Here the threshold condition becomes

$$N_2 = \Delta N_c \quad , \tag{46}$$

and the critical fluorescence power is

$$(P_{cf})_{4\text{-level}} = \frac{\Delta N_c h\nu}{\tau_2} \quad . \tag{47}$$

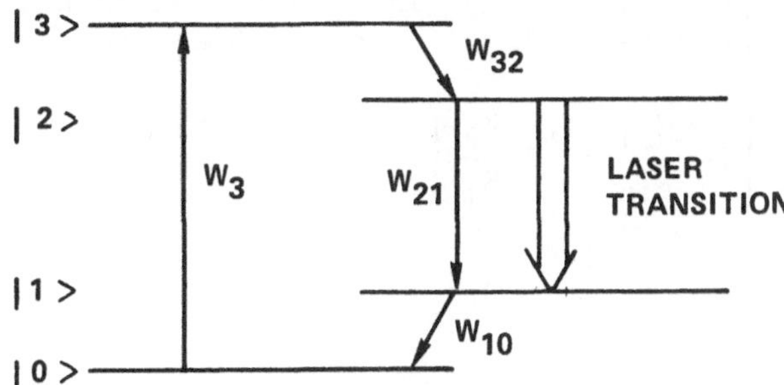

Fig. 2. Energy level scheme for the four-level laser.

For $\tau_2 = 1/A$, we may use eq. (47) and (36) to write

$$(P_{cf})_{4\text{-level}} = \frac{8\pi h\nu^3}{c^3 \tau_{photon} g(0)} , \tag{48}$$

i.e., the critical fluorescence power is independent of the lifetime if $\tau_2 = 1/A$.

In treating the rate equations for the four-level laser, the usual approach is to assume that the pumping and lasing processes are negligible perturbations of the total population. We assume an excitation rate to $|2>$ of R_2. This assumes $N_3 << N_o$, and $W_{32} >> W_{30} + W_{31}$. To account for direct pumping of the terminal laser level, we include an R_1 and then write down a pair of rate equations

$$\begin{aligned} \frac{dN_2}{dt} &= -N_2W_{21} - W_{stim}(N_2-N_1) + R_2 \\ \frac{dN_1}{dt} &= -N_1W_{10} + N_2W_{21} + W_{stim}(N_2-N_1) + R_1 \quad . \end{aligned} \tag{49}$$

The steady-state solution is

$$N_2-N_1 = \frac{R_2\left[1-(W_{21}/W_{10})\ (1 + R_1/R_2)\right]}{(W_{stim} + W_{21})} , \tag{50}$$

and thus the condition for inversion is

$$W_{21}\ (1 + \frac{R_1}{R_2}) < W_{10} \quad . \tag{51}$$

The effective pumping rate R_2' is reduced due to the finite lifetime and pumping rate of level 1 to an effective value

$$R_2' = R_2\left[1- \frac{W_{21}}{W_{10}}\ (1 + \frac{R_2}{R_1})\right] , \tag{52}$$

And eq. (50) becomes

$$N_2-N_1 = \frac{R_2'}{W_{stim} + W_{21}}$$

Equating the left side to ΔN_c the stimulated emission rate is

$$W_{stim} = \frac{R_2'}{\Delta N_c} - W_{21} \quad . \tag{54}$$

From these basic sets of equations, it is possible to discuss specific aspects of laser operation and the ramifications of nonradiative processes.

I.I. Pulsed vs. CW Operation

It can be seen from eqs. (44) and (50) that the achievement of the critical inversion density depends upon the availability of sufficient pump rates W and R_2. In the case of optically pumped solid state lasers such rates are in many cases difficult to sustain continuously. However, with optical flashlamps and pulsed lasers, it is possible to achieve large rates for short periods of time, and build up transient critical inversion and pulsed laser operation.

It behooves us here to discuss briefly the case of Q-switched operation. Normally, the population inversion is pinned at the critical inversion density with additional pumping power coming out as faster W_{stim}; i.e., more power out. Consider, however, the case of a resonator with a variable Q; i.e., different values of τ_{photon}. If the pumping begins at a short τ_{photon}, i.e., a lossy cavity, the inversion density from eq. (36) can be driven to high values. This can be done for periods of the order of τ_2, after which N_2-N_1, will equilibrate. If now the Q of the resonator is suddenly changed, the rate W_{stim} will be more rapid than specified in eqs. (45) or (54) for $N_2-N_1 = \Delta N_c$. The energy will come out in a rapid burst or giant pulse to reestablish steady state equilibrium at $N_2-N_1 = \Delta N_c$. The details of laser kinetics, pulse build-up, etc., are beyond the scope of our presentation here, and the interested student is referred elsewhere for details.

I.J. Lineshapes and Mode Structures

We shall treat the effects on laser characteristics of the line profiles of spectral lines. In solids, two types of broadening are of significance. First of all, there is the natural linewidth due to the lifetimes of the emitting and/or terminal level. For solid state lasers at room temperature, the scattering of phonons is usually the fastest relaxation process and provides the lifetime broadening that is observed. The broadening is Lorentzian with a lineshape given by

$$g(\nu-\nu_o) = \frac{\Delta \upsilon}{2\pi\left[(\nu-\nu_o)^2 + (\Delta\nu/2)^2\right]} \quad , \tag{55}$$

where $\Delta\nu$ is the full-width at half-maximum. This is the lineshape, in a probabilistic sense, for each atom in the entire ensemble; i.e., the broadening is homogeneous.

We now treat the second type of broadening in solids, which is inhomogeneous. Because a crystal is imperfect and there are strains at the ion sites, each level is shifted slightly from site to site in a random way. This is extreme in the case of a glass where none of the sites are identical and the spectral transition is a superposition of contributions from the randomly distributed sites of the ionic ensemble. In the case of such inhomogeneous broadening, the line-shape function is a Gaussian of the form

$$g(\nu-\nu_o) = \frac{2(\ell n\ 2)^{\frac{1}{2}}}{\pi^{\frac{1}{2}}\Delta\nu}\, e^{-4(\ell n\ 2)\ (\nu-\nu_o)^2/(\Delta\nu)^2} \ . \tag{56}$$

Now let us consider such systems in a laser. The gain profile has the form of $g(\nu-\nu_o)$ below threshold. At threshold, for the homogeneous case, the medium will oscillate in the longitudinal mode of eq. (33) close to ν_o. The grain curve, since all ions are equivalent remains of the form of $g(\nu-\nu_o)$ with the gain at ν_o equal to the resonator loss (Fig. 3a). However, for an inhomogeneously broadened line, since it is composed of ions that are not all on "speaking terms" with each other, pinning the gain coefficient at ν_o does not limit the growth of the population of ions which are separated from ν_o. Thus, as shown in Fig. 3b, for an inhomogeneously broadened line, several longitudinal modes within the inhomogeneous linewidth will oscillate simultaneously at these frequencies. Such systems are interesting in their ramifications of resonant energy transfer where there is inhomogeneous broadening.

Fig. 3. Mode structure for (a) homogeneous broadening and (b) inhomogeneous broadening.

II. A REVIEW OF NONRADIATIVE PROCESSES

II.A. General

Although others in this volume have dealt with this subject in great detail, it behooves us to review for our own purposes the chief results of this area for completeness in dealing with our topic. We shall deal with three basic areas here: multiphonon relaxation, ion-ion energy transfer, and vibronic transitions, the latter a combined radiative-nonradiative process.

II.B. Multiphonon Relaxation

The phenomenon of multiphonon relaxation has been subjected to wide study since the late 1960's, when the first systematic experimental results became available. Many review studies have now dealt with this topic.(8-10) The simple diagram shown in Fig. 4 illustrates the process. Here, level $|2\rangle$ relaxes to level $|1\rangle$ nonradiatively, the energy being imparted to the host lattice in the form of two or more phonons of energy ω_i where

$$E_2 - E_1 = \sum_i \hbar\omega_i \quad . \tag{57}$$

The most general and sophisticated approach to this problem is the generalized configurational-coordinate model, developed to a great extent by Struck and Fonger and others.(10,11) The reader is referred elsewhere for this treatment For the present, our needs are well-served by the phenomenological perturbational model, first demonstrated by Riseberg and Moos.(12) This model, though not rigorous, is definitively valid in the weak-coupling limit (as characterized, for example, by rare earths), where the configurational coordinate model gives identical predictions,(11) in view of the nearly zero Franck-Condon offset. We shall therefore review this approach here.

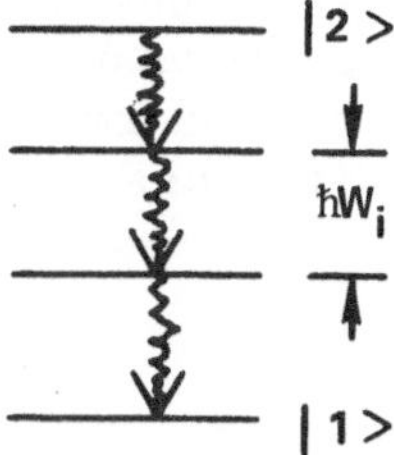

Fig. 4. Schematic representation of multiphonon relaxation process.

We begin with the assumption that the transition probability converges rapidly with the order of the process, and that this can be expressed in terms of a coupling constant; i.e.,

$$\frac{W^{(p)}}{W^{(p-1)}} = \epsilon, \text{ where } \epsilon << 1 \ . \tag{58}$$

Here $W^{(p)}$ is the transition rate for a pth order process, and $W(p-1)$ is that for a (p-1)-th order process. This is a valid approximate treatment in the case of either a perturbation expansion or an expansion of higher-order (anharmonic) terms in the crystal field. The coupling constant contains a density of states, energy denominator, etc. $W(p)$ can then be expressed as

$$W^{(p)} = A\epsilon^{p} \ . \tag{59}$$

We now consider a single frequency model for multiphonon relaxation across an energy gap ΔE. The number of phonons p_i of equal energy $\hbar\omega_i$, required to conserve energy, and hence the order of the process is given by

$$p_i\hbar\omega_i = \Delta E \ . \tag{60}$$

Then, combining eqs. (59) and (60), we have

$$W = c_e^{-\alpha\Delta E}, \text{ where } \alpha > 0 \ . \tag{61}$$

This is the well-known energy-gap dependence law which has been shown to follow for many crystalline (12) and glass (13,14) host matrices where 3<p<9. The individual characteristics of the atomic states and interacting phonon modes are statistically averaged out for transitions involving many phonons. The exponential gap dependence is shown for several lattices containing rare earths in Fig. 5. This is also the form predicted by the single configurational coordinate model in the weak coupling limit. This treatment has been for the spontaneous emission of phonons, and represents the case at low temperatures. Of course, as the temperature is raised, the phonon emission will be enhanced due to the stimulated emission into the thermally populated phonon modes. W then becomes

$$W = W_o \ (\bar{n}_i + 1)^{p_i} \ , \tag{62}$$

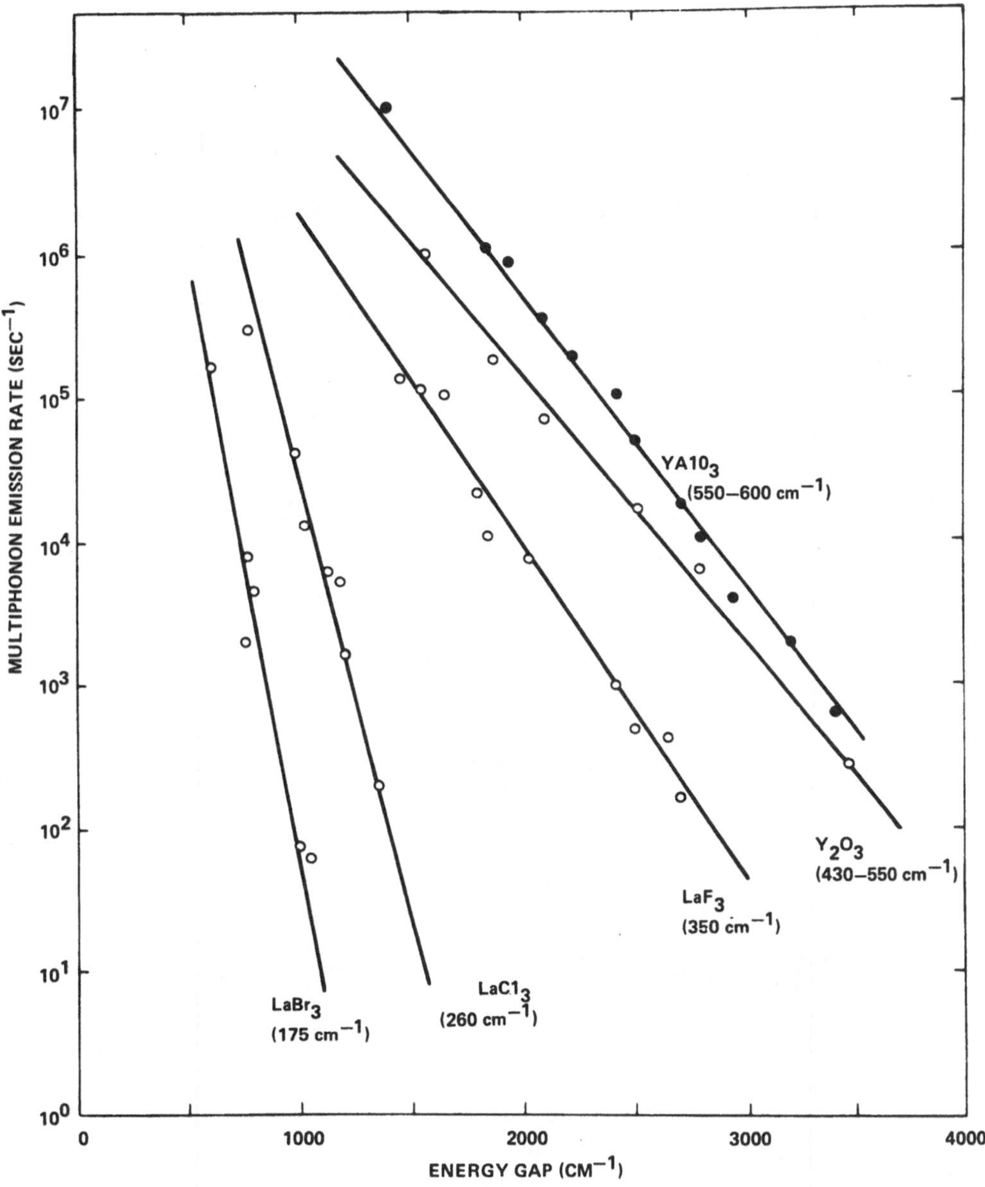

Fig. 5. Energy gap dependence of multiphonon relaxation rate for rare earth levels in several crystal lattices. Maximum phonon energies are in parentheses.

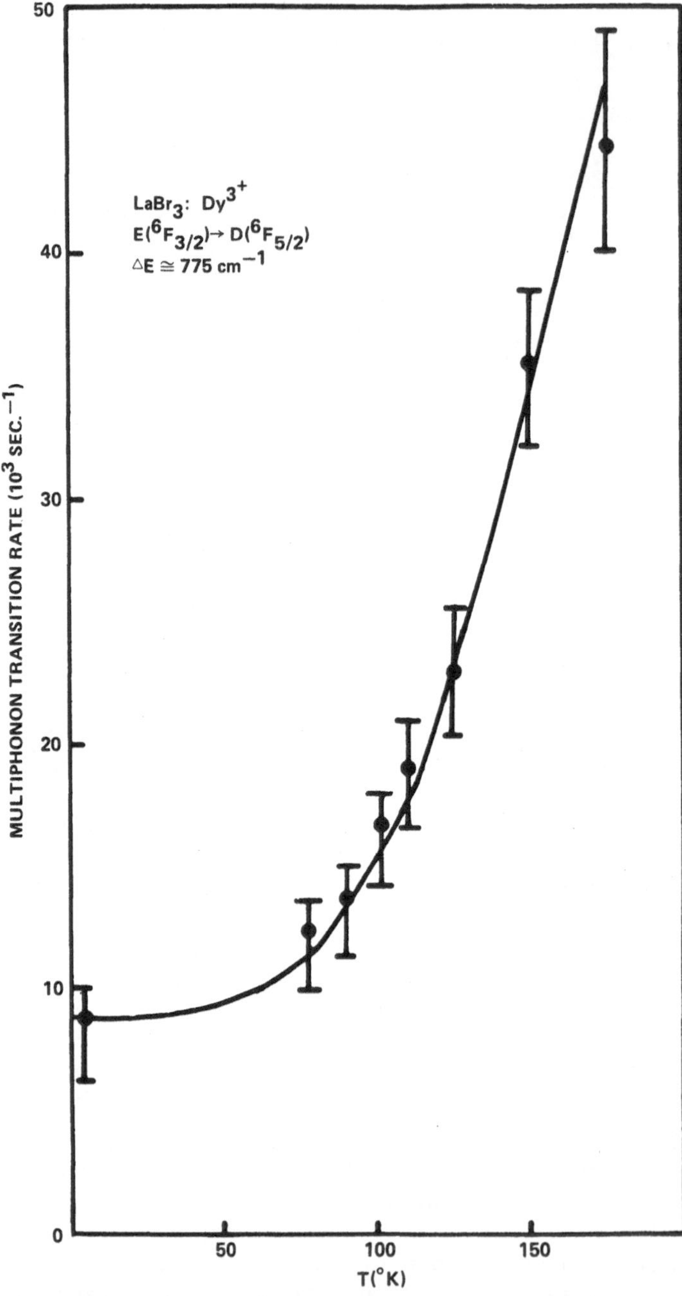

Fig. 6. Temperature dependence of ${}^6F_{3/2} \rightarrow {}^6F_{5/2}$ multiphonon transition rate in $LaBr_3{:}Dy^{3+}$. Theoretical fit is given by eq. (63) with $P_i = 5$.

where $\bar{n}_i$ is the average occupation number of the ith phonon mode. Taking the Bose-Einstein average for $\bar{n}_i$ yields

$$W = W_o \left(\frac{e^{\hbar\omega i/kT}}{e^{\hbar\omega i/kT}-1}\right)^{p_i} \quad . \tag{63}$$

An example of such a temperature dependence is shown in Fig. 6. Many such temperature dependences have been studied in the past decade, and the general applicability of the single frequency model has been categorically demonstrated. The relaxation process is observed to decay in the lowest order consistent with energy conservation and the cut-off in the phonon spectrum. Furthermore, it is the high energy optical phonons which are involved in the process.

The key value of these studies is their utilization for predicting multiphonon transition rates. As can be seen in Fig. 5, one can predict these rates generally to within 50% of the experimental value. Similar predictions can be made, though with less certainty, for the transition ions.

II.C. Ion-Ion Energy Transfer

Another class of radiationless processes of significant consequence is the transfer of energy among ions in a luminescent system by virtue of ion-ion interactions, either static or mediated by the phonon field. There are many processes that have been observed, with two important cases illustrated schematically in Fig. 7.

The processes of Fig. 7 are for like ions; they can be for unlike ions. Those illustrated show resonant energy transfer; the transfer can occur nonresonantly with high probability, where the additional energy can be taken up by the phonon field. Finally, such transfer can be coupled with a radiative transition. There are many important cases of lasers where such processes are of major concern. These will be treated in the next section.

Fig. 7. Schematic illustration of ion-pair processes.

Here we will review briefly the theory behind such processes. Many fine reviews of this area have been published and the reader is referred to the literature for details. (15)

There are two major sources for ion-ion coupling: the exchange interaction between neighboring ions and the electric multipolar coupling arising from the Coulomb interaction between the electron charge clouds. An additional less important mechanism is the magnetic dipole-dipole interaction. First we consider the electrostatic case. Let $\vec{r}_{Ai}$ and $\vec{r}_{Bj}$ be the coordinate vectors of electrons i and j belonging to ions A and B, respectively. The electrostatic interaction is

$$H_{es} = \sum_{i,j} \frac{e^2}{K\,(\vec{r}_{Ai}-\vec{R}-\vec{r}_{Bj})} \quad . \tag{64}$$

The various multipolar terms arise from a power series expansion of the denominator, which can be expressed succinctly as follows, following the formalism of Kushida: (16)

$$H_{es} = \sum_{\substack{k_1 k_2 \\ q_1 q_2}} \frac{e^2}{KR^{(k_1 + k_2 + 1)}} C_{q_1 q_2}^{k_1 k_2} D_{q_1}^{(k_1)}(A)\, D_{q_2}^{(k_2)}(B) \ , \tag{65}$$

where $C_{q_1q_2}^{k_1k_2}$ is a numerical factor dependent on the orientation of the coordinate axes and $D_q^{(k)}$ is a multipole operator defined by

$$D_q^{(k)} = \left(\frac{4\pi}{2K + 1}\right)^{\frac{1}{2}} \sum_i r_i^{\,k}\, Y_{kq}\,(\theta_i, \phi_i) \quad . \tag{66}$$

The leading terms in the expansion in eq. (65) are the electric dipole-dipole (EDD), dipole-quadrupole (EDQ), and quadrupole-quadrupole (EQQ) interactions. These have radial dependences of R^{-3}, R^{-4}, and R^{-5}, respectively. Higher-order terms generally are of negligible importance for the rare-earth and transition metal ions. The matrix elements of H_{ES} are subject to selection rules $\Delta S=0$ and $|\Delta L|$, $|\Delta J| \leqslant k$. These are relaxed for rare-earths because of the large spin-orbit interaction and for the transition-metal ions because of the strong-field quenching of the orbital moment. $\Delta S=0$ remains of importance in the latter, only limited by the degree of spin-orbit mixing.

Next we consider the magnetic dipole-dipole interaction, which is written

$$H_{MDD} = \sum_{i,j} \left[\frac{\vec{\mu}_i \cdot \vec{\mu}_j}{R^3} - \frac{3(\vec{\mu}_i \cdot \vec{R})\ (\vec{\mu}_j \cdot \vec{R})}{R^5} \right] , \tag{67}$$

where $\vec{\mu}_i = \vec{\ell}_i + 2\vec{s}_i$ and $(\vec{\ell}_i, \vec{s}_i)$, $(\vec{\ell}_j, \vec{s}_j)$ are the orbital and spin operators for the ith and j^{th} electrons of ions A and B, respectively. The selection rules $\Delta S, \Delta L, \Delta J = 0, \pm 1$ for intra-configuration transitions are again relaxed as discussed in the electrostatic case. The MDD interaction has the same long-range R^{-3} radial dependence as the EDD interaction. Finally, we consider the exchange interaction given by

$$H_{EX} = \sum_{i,j} J_{ij}\ \vec{S}_i \cdot \vec{S}_j \quad , \tag{68}$$

where J_{ij} represents the isotropic or Heisenberg component of the exchange integral. More generally, J_{ij} is a tensor with additional terms for the anisotropic and asymmetric contributions of the exchange interaction. For rare earths in solids, the large residual orbital angular momentum results in high-rank orbital contributions to J_{ij} expressible in the form

$$J_{ij} = \sum J^{kk'}_{qq'}\ C^{(k)}_q\ (\vec{\ell}_i)\ C^{(k')}_q\ (\vec{\ell}_j) \quad . \tag{69}$$

These terms relax the selection rules for exchange within $4f^N$ to $|\Delta S| \leqslant 1$, $|\Delta L| \leqslant 6$, $|\Delta J| \leqslant 7$. Again, because of the state admixing, exchange is relatively free of selection rule restrictions.

For any of these interactions, the transition probability for energy transfer from ion A to ion B is given by

$$W_{AB} = \frac{2\pi}{\hbar} |\langle \Psi_A(2)\Psi_B(1) | H_{AB} | \Psi_A(1)\Psi_B(2) \rangle|^2 \int F_A(\nu) F_B(\nu) d\nu \quad , \tag{70}$$

where 1 and 2 denote excited and ground states and $F_A(\nu)$ and $F_B(\nu)$ are normalized line-shape functions for the transitions of ions A and B. This denotes a resonance model where the transition rate is dependent upon the degree of overlap of the two line-shape functions, and was first addressed in this way by Dexter. Nonresonant transfer can be brought in by the use of vibronic line-shape functions for $F_A(\nu)$ and $F_B(\nu)$, as was shown by Orbach. (17)

Dexter calculated the dependence on ion-ion separations for the cases of EDD, EDQ, and direct exchange.(18) These result from the squared H_{AB} term and can be inferred from our discussion above. The radial dependencies are R^{-6}, R^{-8}, and R^{-10} for E_{DD}, E_{DQ}, and E_{QQ}, respectively, and R^{-6} for MDD. Direct exchange involves an exponential decrease of the wavefunctions contained in J_{ij} and can be treated in terms of a radial dependence exp $(-R/_L)$, where L is an effective Bohr radius of the ground and excited states under consideration. Thus, W_{AB} has an exp $(-2R/_L)$ radial dependence.

The Dexter theory considered the nearest neighbors as being dominant. Inokuti and Hirayama (19) provided a refined treatment that considered the entire environment of the excited ion. During the past decade and a half, a variety of experimenters have tried, with the aid of these theories, to identify the appropriate multipolar interaction for a specific system. Most concentration dependence studies have been fraught with the complication that different interactions dominate at high concentrations and at low concentrations; i.e., short-range vs. long-range interactions. In general, kinetic studies have been more illuminating. A valuable summary of such work was given by Watts. (15)

At high concentrations, resonant transfer between like ions becomes very fast, and diffusion models are appropriate. In transfer between unlike ions A and B, for sufficiently high concentrations of A, resonant transfer within the A system is sufficiently fast that the excitation may be considered equally shared among all of the ions A. A rate equation approach for A $\rightarrow$ B transfer is then justified.

The foregoing discussion has been relevant primarily to the rare earth materials. However, among the 3d ion systems, the case of ruby stands out as having received a remarkable degree of attention because of its unique spectroscopic properties. In particular, subtle features of resonant energy transfer in the presence of inhomogeneous broadening have recently been addressed experimentally by techniques of laser-induced fluorescence line-narrowing (20,21) and at the same time have been described theoretically by Orbach and co-workers, in terms of Anderson localization theory.(22) Some of these aspects of ruby will be covered below.

II.D. Phonon Broadening of Spectral Lines

For the rare earth ions in crystalline solids as well as for ruby, the inhomogeneous broadening is sufficiently small, that the linewidth observed is the natural width associated with lifetime broadening. Although the radiative lifetimes of these levels are seldom faster than the microsecond regime, linewidths are generally at least several cm^{-1}. This indicates fast nonradiative processes limiting the lifetime of the emitting and/or terminal level. Such processes are, in fact, associated with the emission, absorption, and scattering of phonons. There are several possible schemes for such nonradiative processes. These are shown in Fig. 8. The processes of (a) and (b) represent the direct emission and absorption of a single phonon. The process shown in (c) is simply Raman scattering of phonons, while (d) is the Orbach process, where Raman scattering takes place in resonance with a real state.

The probabilities for such processes can be addressed by simple first- and second-order perturbation theory, and the temperature-dependence for the linewidths delineated. (23) These derivations have been carried out elsewhere and the student is referred to these sources for the details. We summarize here the temperature dependences for each of these kinds of processes.

$$W_{DIRECT} = A\omega^2 T \quad , \tag{71}$$

or a linear dependence on T and an energy squared dependence on the separation of the two levels.

$$\begin{aligned} W_{RAMAN} &= B_1 T^7 \text{ for non-Kramers ions} \\ &= B_2 T^9 \text{ for Kramers ions} \quad , \end{aligned} \tag{72}$$

and finally for the Orbach process

$$W_{ORBACH} = Ce^{-\hbar\omega_1/kT} \quad , \tag{73}$$

all with reference to Fig. 8.

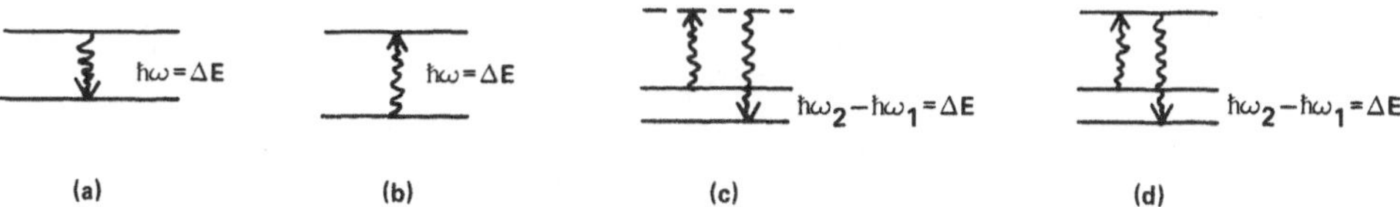

Fig. 8. Phonon-induced relaxation processes.

The results of these processes are to introduce (homogeneous) broadening to the fluorescent line. The temperature-dependent width of a level for all of the phonon-induced relaxation mechanisms can be written

$$\Delta\nu = A\left(\frac{T}{T_D}\right)^7 \int_0^{T_D/T} \frac{x^6 e^x}{(e^x-1)^2}\,dx \tag{74}$$

$$+ \sum_{j<i} B_{ij} \left(\frac{e^{(E_i-E_j)/kT}}{e^{(E_i-E_j)/kT}-1}\right) + \sum_{i<j} B_{ji} \left(\frac{1}{e^{(E_i-E_j)/kT}-1}\right) ,$$

where the B_{ij} are constants for levels above the relevant level j and the B_{ji} for levels below j; and E_i and E_j are the energies for these levels. The first term expresses the level width resulting from Raman processes.

In laser crystals, the width of the levels, and thereby the fluorescent transition, determines the actual peak gain of the medium for a given oscillator strength. Thus, spectral linewidths and temperature dependences are of great consequence in the operation of this class of laser.

II.E. Vibronic Transitions

The vibronic transition is a radiative process where a portion of the energy is transferred nonradiatively to or from the phonon system. This results in a photon whose energy is the difference between the atomic levels with the addition or subtraction of the phonon energy, depending on whether the phonon is emitted or absorbed. The process is shown schematically in Fig. 9. The orbit-lattice interaction $V_{o,}^{(\omega)}{}_{\ell,}$ mixes excited and ground phonon character into the initial and final states. We will not go into the full perturbation treatment here, and the interested student is referred elsewhere for details of this theory. If we take as the perturbed Born-Oppenheimer wavefunctions $|1'\rangle|n_\omega\rangle$ and $|2'\rangle|n_\omega+1\rangle$, then the transition probability for vibronic transitions is

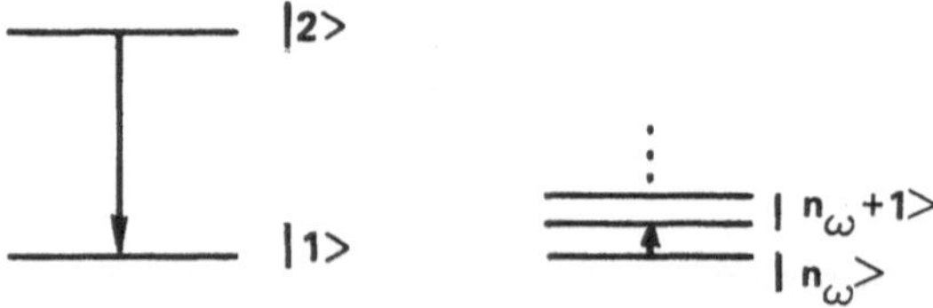

Fig. 9. Schematic diagram of a vibronic transition.

$$W_{vib} = \frac{2\pi}{\hbar} |\langle 2'|\vec{p}|1'\rangle|^2 g(\omega) \quad , \tag{75}$$

where $g(\omega)$ is the density of phonon states. This is a key result, since it predicts that vibronic transitions will have peaks corresponding to the peaks in the phonon density of states. Vibronics often appear as the weak sidebands of no-phonon electronic transitions. If such a transition is forbidden, however, as in the case for intraconfiguration transitions at a site which is a center of inversion, the vibronic emission can dominate.

This presents interesting possibilities for laser operation, since decay to an upper vibrational level of the ground state which is not thermally populated at ambient temperatures has the potential for four-level laser operation. This was the subject of a recent important development, the Cr^{3+}: $A\ell_2BeO_4$ laser, (24) which we shall discuss below.

III. EXAMPLES FROM THE LASER LITERATURE

III.A. General

We now have all of the tools to discuss the ramifications of nonradiative processes in solid state lasers. For pedagogical reasons, we shall take some of the most important cases of lasers and address them in detail, as well as look at some of the unique situations in laser technology that embody these nonradiative phenomena in their design or performance.

III.B. The Ruby Laser

The ruby laser was the first demonstrated solid state laser, and remains an important technological system. Consequently, Al_2O_3:Cr^{3+} has received intensive study both for this reason and for its intrinsic physical interest. An energy level diagram is shown in Fig. 10.

The notation of Fig. 10 is that of strong-field orbitals for the $3d^3$ configuration and reflects octahedral coordination. The laser operates by optical pumping into the strong visible absorption bands $^4T_1$ and $^4T_2$, and fast nonradiative relaxation to the ^2E. The location of the $^2T_2$ is uncertain and may well mediate nonradiative processes. The ^2E is a metastable level because it is spin-forbidden as far as radiative processes are concerned, and the large energy separation to the ground-state implies a many-phonon process and a slow rate. The system is a perfect prototype for the three-level laser, with the ^2E - $^4A_2$ transition at 6943 Å.

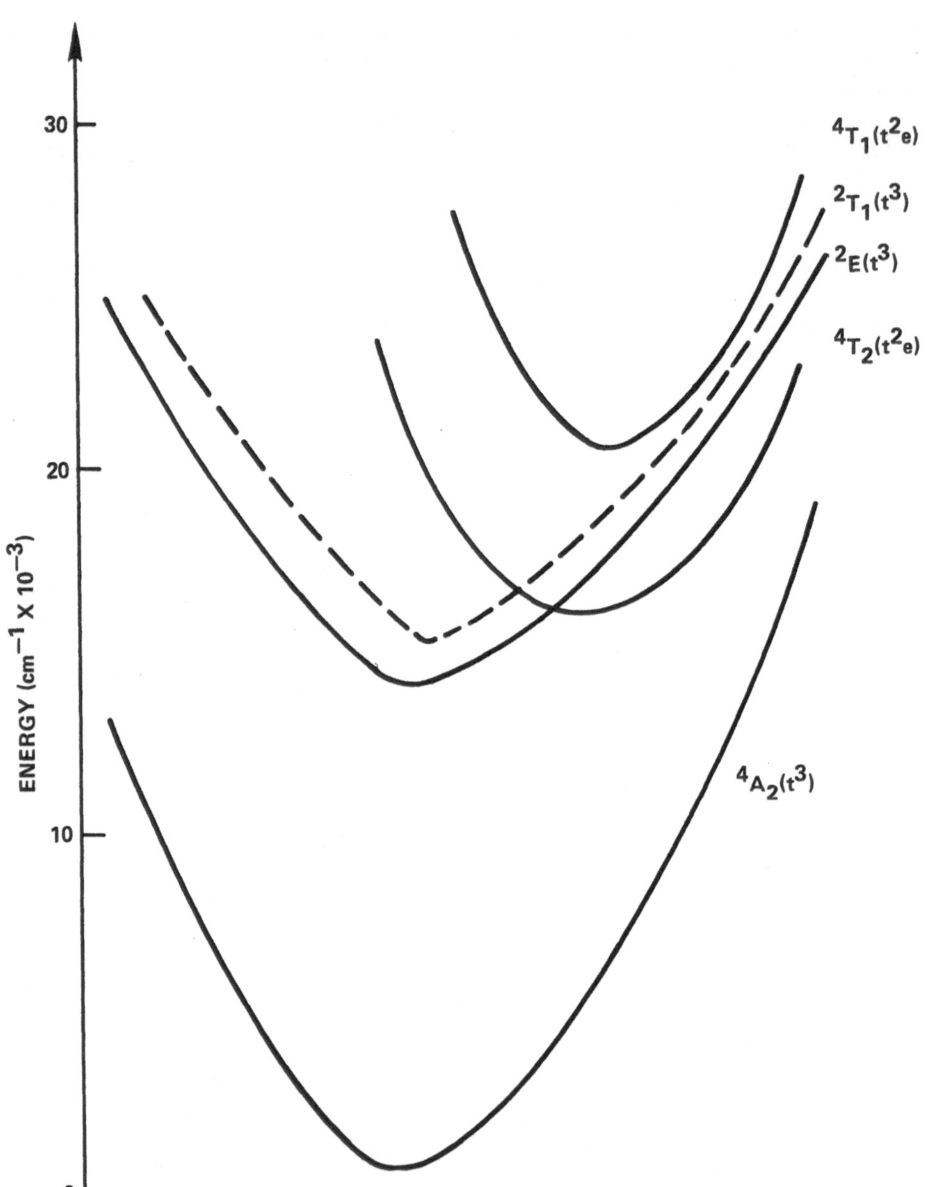

Fig. 10. Energy level scheme for $Al_2O_3:Cr^{3+}$.

Kisliuk and Moore studied the nonradiative properties of ruby in detail. (25) The optically-pumped $^4T_1$ and $^4T_2$ were found to be totally nonradiative by multiphonon relaxation to the ^2E metastable level with a rate in excess of 10^8 sec^{-1}. Since the phonon spectrum in Al_2O_3 goes out to approximately 700-800 cm^{-1}, and the Franck-Condon offset reduces the $^4T_2 \rightarrow {}^2T_1$ energy gap, such as it may be, this nonradiative process should be rapid and efficient. The ^2E for pink ruby (about 0.5% Cr^{3+}) has a decay rate of 300 sec^{-1}, and Kisliuk and Moore determined that the nonradiative contribution $W_{NR} < 50$ sec^{-1}, implying a ^2E radiative quantum efficiency of the order of 80 - 90%. This is compatible with earlier measurements. Since the ^2E and $^4A_2$ belong to the same strong-field orbital, the small Franck-Condon offset leads to a crossing at high energies, and the multiphonon decay is relatively small. It is still contributory, and, as we shall see below for smaller energy gaps in Al_2O_3, the multiphonon decay can dominate. The situation in ruby is indeed fortuitous.

The concentration in ruby is generally chosen to be less than 0.1%. The reason is that at higher concentrations of Cr^{3+}, the ions pair, resulting in transitions with faster lifetimes and lower efficiencies, making laser action harder to achieve. (26) Furthermore, the absorption spectra at these low concentrations are well-matched to the geometrics of flashlamps for the achievement of uniform pumping.

If we consider the critical fluorescence power from eq. (38) for a three-level laser

$$(P_{cf})_{\text{3-level}} = \frac{Nh\nu}{2\tau_2} ,$$

where τ_2 is the lifetime of the laser level, to reach and maintain oscillation, we at least have to pump as fast as the laser level is fluorescing. Thus, the absence of nonradiative decay in $^2E \rightarrow {}^4A_2$ transition and the consequently long lifetime of the ^2E contributes to the relative ease of achieving laser action.

Because the lifetime is relatively long, flashlamp pumping times up to the order of a few milliseconds are effective, prior to reaching a steady-state population of the ^2E. The achievement of CW laser action is possible, though with some difficulty. A further consequence of the long lifetime and low gain is the relative suitability for Q-switching. Much energy can be stored, and the low gain limits the loss of excited ions during the low Q state by amplified spontaneous emission.

A further consequence of nonradiative effects in the ruby laser is the line-broadening. This was studied in detail by McCumber and Sturge, (27) and by Powell et al.,(28) and can be rationalized in terms of the phonon-induced relaxation processes discussed above. The broadening is homogeneous, and ruby exhibits the gain characteristics of the homogeneously broadened system.

III.C. The Nd^{3+} Laser

The single most important class of optically pumped solid state lasers is based on the $^4F_{3/2}$ to $^4I_{11/2}$ transition of the trivalent Nd ion at 1.06μ. The energy level diagram of Nd is shown in Fig. 11, which shows all of the triply-ionized rare earths.

Excitation in a nominal Nd laser system takes place by absorption of optical energy at wavelengths less than 9000 Å to levels lying above $^4F_{3/2}$. In the ideal case this nonradiative decay should be rapid and complete. We can see that this requirement restricts the use of materials for Nd^{3+} lasers. The gaps above $^4F_{3/2}$ are generally 2000 cm^{-1} or less. Referring to our energy gap dependence plots of Fig. 5 or similar such plots for glass matrices, we see that in conventional oxide systems where few phonons are involved, the nonradiative rates are fast enough to dominate the typical radiative rates for rare earth ions in solids of the order of several hundred microseconds. The softer lattices, such as the halides, have optical phonon modes which cut off at only several hundred cm^{-1}, and will have five and six phonon processes for these levels. These can be as slow as the radiative rates, and there can be significant radiative leakage from these levels, reducing the efficiency of $^4F_{3/2}$ pumping.

When the level $^4F_{3/2}$ is populated, it then becomes a question as to the efficiency of radiative transitions from this level. We can see that this 5000 cm^{-1} gap comprises many phonons for virtually all crystalline materials, and in glasses becomes an issue only for borate glasses.(14) Thus, the radiative quantum effeciency of the $^4F_{3/2} \rightarrow {}^4I_{11/2}$ transition can be expected to be very high.

We now come to the terminal level of the laser transition. Regardless of which of the Stark components is the termination of the laser line, the population in the $^4I_{11/2}$ will thermalize rapidly, and decay collectively to the $^4I_{9/2}$. The energy gap to the uppermost level of the ground $^4I_{9/2}$ level is 1500 - 2000 cm^{-1}. If the $^4I_{11/2}$ is not rapidly depleted, the extremely fast fill-up of the level under laser action will bring about an end to the population inversion, and the laser will self-terminate.

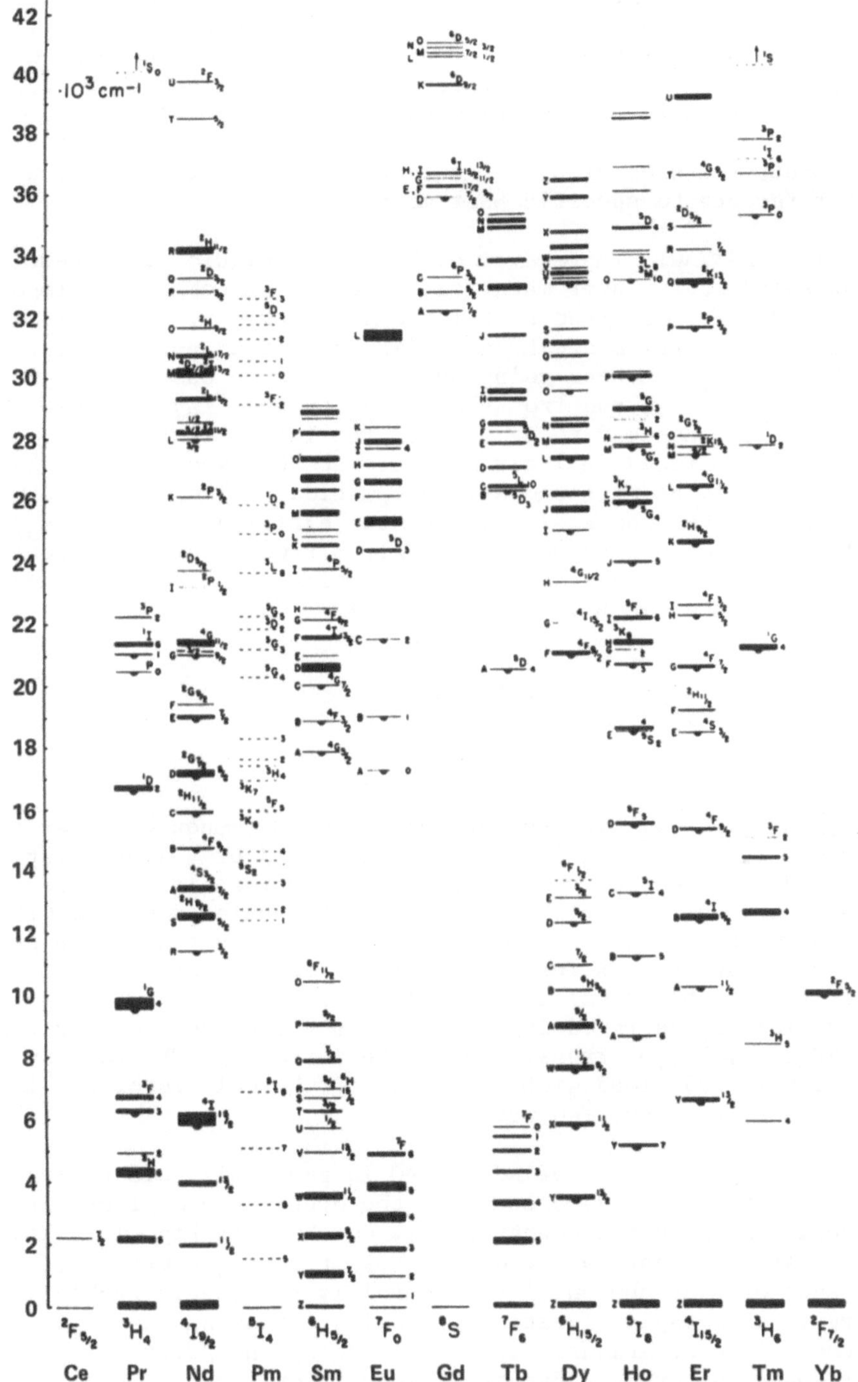

Fig. 11. Energy levels of the triply-ionized rare earths. (Reproduced by permission of H.M. Crosswhite; see also reference (29).)

If we look at our multiphonon gap dependence, we see that we are once again in trouble for the softer hosts. Decays from $^4I_{11/2}$ to $^4I_{9/2}$ can be of the order of a millisecond for these materials, which is incompatible with sustaining 4-level laser characteristics and renders CW operation virtually impossible.

Other features of Nd^{3+} laser action are addressed below, with reference to specific host matrices.

1. Nd:YAG. The system of Nd^{3+} in $Y_3Al_5O_{12}$ is supreme among solid state laser materials in terms of commercial importance. We have observed that for oxide matrices, the desired radiative properties are held to for Nd^{3+} laser performance. YAG has optical phonons going out to the 600 - 800 cm^{-1} region. Decay to the $^4F_{3/2}$ and from the $^4I_{11/2}$ to the $^4I_{9/2}$ can be categorically assumed to be rapid. The lifetime of the $^4F_{3/2}$ is well-established in laser crystals at 260 μsec. The 7-phonon process should be much slower, and Kushida (30) showed in his experiments that the radiative quantum efficiency of the $^4F_{3/2}$ is in excess of ninety percent (90%). If we go back to Section II, for the case when the laser level lifetime τ_2 = the spontaneous transition rate, we have from eq. (48).

$$(P_{cf})_{4\text{-level}} = \frac{8\pi h\nu^3}{c^3 \tau_{photon} g(o)} ,$$

i.e., the critical fluorescence power depends upon τ_{photon}, or the loss rate from the laser. For this kind of 4-level system, low threshold becomes very sensitive to optical loss in the cavity, and the very high optical quality achieved with YAG:Nd, coupled with its high gain, have made it dominant for both pulsed and CW operations. A further important feature of YAG:Nd is its relatively high thermal conductivity. Since the pump energy in the visible is degraded into phonons in the process of exciting the $^4F_{3/2} \rightarrow {}^4I_{11/2}$ transition in the infrared. The ability to handle this heat load with a minimum of thermal distortion is a key feature of the YAG matrix.

The concentration of Nd in YAG is generally held to less than one percent (1%), because of the occurrence of ion-pair quenching at higher concentrations, as shown in Fig. 12. This is primarily a nonresonant decay, but at room temperature, the presence of many phonons leads to fairly reasonable probabilities. There is also rapid migration of excitation among the Nd system at these concentrations, further enhancing the probability of pair quenching or degradation by impurities.

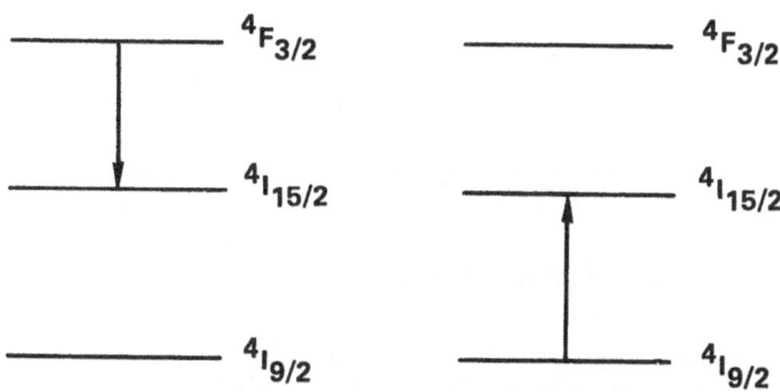

Fig. 12. Ion pair relaxation in YAG:Nd.

The absorption that is achievable at these low concentrations is appropriate for matching to normal macroscopic pump sources. However, to go to ultra-small miniature lasers, the absorption is inadequate for efficient pumping, and one must consider other classes of laser materials, such as the pentaphosphate, to be discussed below.

2. Nd: Glass. The most powerful lasers ever produced have been made from Nd glass. If we refer to the section on multiphonon decay and the results of others on multiphonon relaxation in glasses, we conclude that except for the borate system, glasses have reasonable quantum efficiencies and rapid nonradiative decay to maintain good 4-level laser operation. The poor thermal properties of glasses limit their utilization generally to pulsed, single-shot operation.

The distinctive feature spectroscopically about glasses is that their emission results from the contributions of many different classes of ion sites with different energy levels, radiative decay, and nonradiative decay properties. What has in general been measured only in the past as average characteristics over the entire ensemble of emitting ions has recently been delineated in detail by Brecher, Riseberg, and Weber by means of laser-induced fluorescence line-narrowing techniques.(31) For example, the radiative quantum efficiency for the $^4F_{3/2} \rightarrow {^4I_{11/2}}$ transition in ED-2 laser glass was found to vary from seventy percent (70%) to 100 percent (100%) across the laser transition.

Because of the inhomogeneous broadening, the Nd:Glass system might be expected to oscillate over the entire inhomogeneous width of the laser transition. However, the energy transfer processes discussed above, coupled with substantial homogeneous linewidths lead to rapid cross-relaxation within the laser line, and the inhomogeneity is somewhat suppressed.

In the case of laser fusion experiments, however, when one is concerned with amplification of a single ultra-narrow mode-locked pulse in the subnanosecond regime, the question of cross-relaxation must be examined. Brawer and Weber,(32) in an elegant experiment on nonresonant fluorescence line-narrowing showed that there is indeed a limit to the effective homogeneity of the laser line and an effective ion-ion transfer rate for nonresonant ions of the order of 10^4 sec^{-1} for 3% Nd_2O_3.

3. Nd Pentaphosphate. We mentioned above the difficulty one has in shrinking the dimensions of an Nd:YAG laser because of the limited absorption of the pump radiation for 1% Nd or less. A number of years ago Danielmeyer and Weber (33) demonstrated that NdP_5O_{14}, a stoichiometric Nd compound, had a relatively high radiative quantum efficiency, and with the operation of a laser based on such materials opened a field of major inquiry in solid state luminescence. When Nd is diluted in the diamagnetic analogs LaP_5O_{14} and YP_5O_{14}, very high radiative quantum efficiencies are observed. The implication of this development is the possibility of absorbing much larger quantities of light in a very small volume and obtaining sufficiently high single-pass to build miniature lasers from these materials.

III.D. The Ho^{3+} Energy Transfer Laser

We have thus far highlighted energy transfer as an undesirable phenomenon which limits the concentration achievable for active laser ions in host matrices. However, an important class of lasers utilize energy transfer to have one set of ions strongly absorptive to the pump radiation. These ions then transfer their energy to the lasing ions, enhancing the efficiency over what could be achieved with the absorption bands of the lasing ions alone.

The most important, and among the most interesting of these lasers, is the Ho^{3+} energy transfer laser, first demonstrated in YAG by Johnson, Geusic, and van Uitert.(34) Here the laser line is the $^5I_7 \rightarrow {}^5I_8$ transition at 2.01 μ. In addition to Ho^{3+}, the crystal is doped with Er^{3+}, Tm^{3+}, and Yb^{3+}. The Er^{3+} and Yb^{3+} contribute additional absorption bands, while the Tm^{3+}, in addition, provides a channel to enhance $Er^{3+} \rightarrow Ho^{3+}$ transitions. In particular, the Er^{3+} $^4I_{13/2} \rightarrow Ho^{3+}$ $^5I_7$ transfer can be slow because of the 2000 cm^{-1} gap and the exponential energy gap dependence law for nonresonant ion pair transitions. The Tm^{3+} $^3H_4$ level enhances this decay by serving as an intermediate state lying equally spaced between the Er^{3+} and Ho^{3+} levels. Both processes of Fig. 6 are important in this system, including resonant and nonresonant decays, and the behavior is an extremely complex sequence of interrelaxation steps among all of the ion classes. The most important extension of this work has been to the host matrix $LiYF_4$, (35) which has been the highest efficiency solid state laser ever operated, due to the judicious use of non-radiative energy transfer processes.

III.E. The $LiYF_4:Tb^{3+}$ Laser

One of the prime examples where appropriate nonradiative processes were not present for suitable laser action is the case of Tb^{3+}. Referring to Fig. 10, we see that the $^5D_3 \rightarrow {^5D_4}$ energy gap is approximately 6000 cm^{-1}, while the levels above $^5D_3$ are spaced relatively narrowly. This means that upon optical pumping into these upper energy levels, fast multiphonon decay takes place to the $^5D_3$. This level is not appropriate for four-level laser action to any of the levels of the ^7F ground term, because the wavelength deep in the UV would generally lead to excited state absorption upward from the $^5D_3$, to one of several kinds of levels. These can be either the 5d bands or charge transfer states, which would have absorption cross-sections far greater than any of the possible stimulated emission cross-sections. The desirable laser transition is the $^5D_4 \rightarrow {^7F_5}$ transition in the green, a four-level system at room temperature because of the 2000 cm^{-1} elevation of the $^7F_5$. The problem, however, is that the large energy gap from the $^5D_3$ to the $^5D_4$ prevents multiphonon decay, from our known energy gap law, and it is not possible to pump $^5D_4$ efficiently.

Jenssen, Castleberry, Gabbe, and Linz (35) faced this problem with $LiYF_4:Tb^{3+}$. They solved it by raising the Tb^{3+} concentration to twenty-five percent (25%) to increase the probability for $^5D_3$ to $^5D_4$ decay via ion-pair relaxation involving resonant or near-resonant transitions of the type $^5D_3 \rightarrow {^5D_4}$: $^7F_{0,1}$ or $^5D_3 \rightarrow {^7F_{0,1}}$: $^7F_6 \rightarrow {^5D_4}$. Lasing then occurs from $^5D_4$ to $^7F_5$. Since the $^7F_5$ to $^7F_6$ gap is small, the $^7F_5$ terminal laser level is rapidly depleted by multiphonon decay, thus completing a four-level laser scheme.

Additionally, the large concentration of Tb^{3+} ions enhances the absorption of the relatively fine line system. The upper levels are relatively weakly absorbed, probably as a consequence of the separation of the ^7F term from the rest of the spectrum. Since it is the only septet, the spin-orbit coupling is smaller than it perhaps is in some of the other rare earths, and some residual spin forbiddenness may exist. This is certainly the case for its neighbor Gd^{3+}, where the $^6P_{7/2} \rightarrow {^8S_{7/2}}$ is often magnetic dipole in character, as well as for Eu^{3+} where the $^5D_0 \rightarrow {^7F_1}$ is often magnetic dipole. All of these levels have relatively long radiative rates, usually at least 500 microseconds and often in the millisecond range.

As for the concentration quenching, there is no doubt that at 25% Tb^{3+}, the system is in the rapid energy migration regime. However, with the ultra-high purity materials used for this system, the energy degradation is minimized.

IV. CURRENT PROBLEMS AND OPPORTUNITIES IN SOLID STATE LASER PHYSICS

IV.A. General

The control and utilization of nonradiative processes represent fertile areas for pursuit in the improvement of existing laser materials and in the search for new solid state laser systems. In this section, we shall consider some specific examples of interesting case studies for consideration.

IV.B. Sensitization of Nd

One of the earliest examples of sensitization of an active laser ion was the use of Cr^{3+} in YAG to sensitize the Nd^{3+} laser emission.(36) The sensitization scheme is shown in Fig. 13. Such sensitization was observed and was implemented in laser operation. The broad, strong absorption bands of Cr^{3+} in the octahedral Al^{3+} site of the garnet matrix is ideally suited for optical pumping by visible flashlamp, and efficient energy transfer was indeed observed, and laser action obtained. The difficulty was that the excellent crystal properties of YAG were degraded, however, by the addition of Cr^{3+}, and the gain in pumping was not offset by the increased loss.

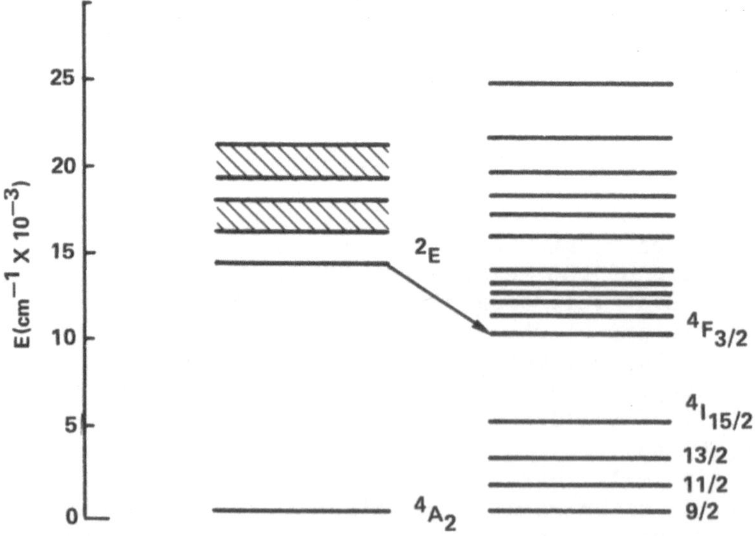

Fig. 13. Sensitization of Nd^{3+} by Cr^{3+} in YAG.

Weber and co-workers undertook an extensive study of the possibility of sensitization by other 3d series ions in both the octahedral and tetrahedral Al^{3+} sites in YAG.(37) No scheme was successful, and the crystal quality remained an obstacle. However, the possibility of sensitization by substitution in the Y^{3+} site remains.

Much more successful was the work of the Raytheon group in $YAlO_3$: Nd^{3+}, where Cr^{3+} substituted easily for Al and an increased efficiency was observed.(38) This is a material that has seen commercial utilization.

There remains a substantial opportunity for impact by the introduction of sensitization by energy transfer in Nd^{3+}-based laser systems. The Nd:glass case, which is of importance to the laser fusion problem, involves a very low efficiency where such an energy transfer process could be of great significance.

IV.C. The Pr^{3+} Laser

One of the most attractive levels in the rare earth series for laser action is the $^3P_0$ level of Pr^{3+}. There are strong transitions in the red and green to the $^3H_5$ and $^3H_6$ levels, which are separated by 2000 cm^{-1} sequentially from the ground $^3H_4$ manifold. The oscillator strengths are generally quite strong, and there is the potential for a high-gain system. Varsanyi (39) in fact observed laser action in $PrCl_3$ at the surface of very small platelets by pumping with a dye laser. The energy gap from the $^3P_0$ to the $^1D_2$ is 3500 cm^{-1} or more, and the level can have a high radiative quantum efficiency in relatively hard host matrices. The absorption band due to the 4f'5d' configuration is sufficiently high that excited state absorption can be avoided. The problem that remains is that there are only a few bands in the blue for pumping. Appropriate sensitization may lead to an extremely efficient laser system.

IV.D. The V^{3+} System

One of the most tantalizing energy level schemes from the viewpoint of laser action is $V^{3+}(3d^2)$ in octahedral coordination. It has been studied in detail only in the Al_2O_3 system.(40,41) The energy level diagram is shown in Fig. 14.

The system is very reminiscent of $Al_2O_3:Cr^{3+}$, with the additional feature that there is a level 1000 cm^{-1} above the ground state, providing semi-4-level operation. The $^3A_2 \rightarrow {}^1E$ has been observed in absorption and has an oscillator strength corresponding to a several millisecond lifetime. This lifetime has been measured at 3 μsec for a very weak fluorescence.(42) It

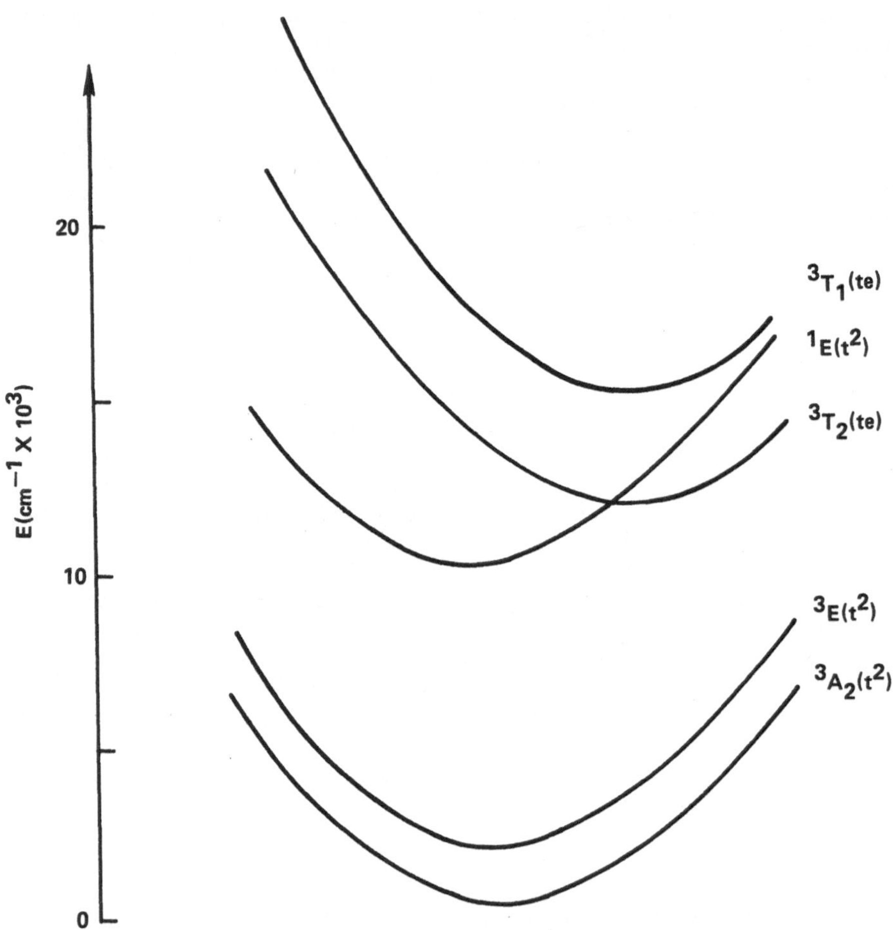

Fig. 14. Energy level diagram of V^{3+} in $A\ell_2O_3$.

is believed that the fluorescence quantum yield is limited by $^1E \rightarrow {}^3E$ multiphonon relaxation. The question to be posed is: are there other host matrices with similar coordination with smaller phonon energies where the quantum yield would be higher?

IV.E. Other Cr^{3+} Lasers

The recent achievement of laser action on the vibrational sideband of the $^2T_1$, ^2E transition in Cr^{3+}-doped alexandrite represents a major milestone in solid state laser technology. The broader implications extend to the case where there is a large Stokes shift for the $^4T_2$ band, and it actually lies below the ^2E. A simplified energy level scheme is shown in Fig. 15. This pattern is the case, for example, in silicate and phosphate glass, (43,44) with the emission occurring at 8000-9000 Å in the near infrared. The lifetime in the glasses is of the order of 10 μsec, and is dominated by multiphonon relaxation. The challenge lies in finding a host matrix with a similar energy level diagram, but with a lower phonon spectrum cut-off to stabilize the system against nonradiative decay. It is obvious that the system is very attractive in terms of 4-level laser operation.

V. CONCLUSION

The reader can no doubt find many other examples from the laser literature where nonradiative processes present a significant feature of the design characteristics of the device. Our presentation was intended to be illustrative rather than exhaustive. Nonradiative processes are intrinsic to every operating solid state laser; the laser researcher is constantly dealing with this issue.

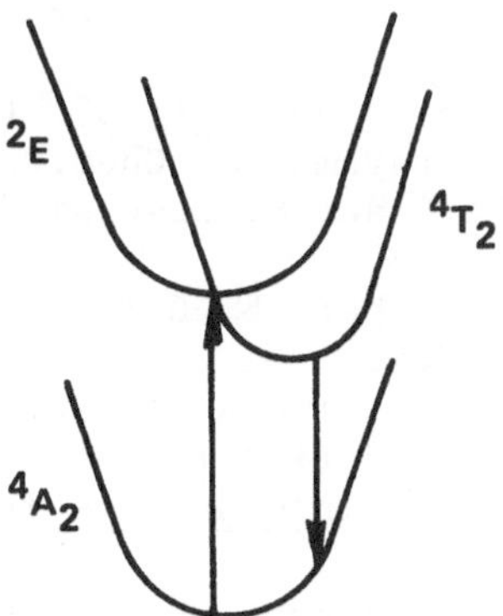

Fig. 15. Schematic energy level diagram for Cr^{3+} in certain hosts.

Furthermore, our last section only touched on some of the systems that are intriguing because of their unique nonradiative characteristics. It is hoped that this will suggest to the student the field of nonradiative phenomena as a fertile one from which to draw direction for future research.

REFERENCES

1. T. H. Maiman, Nature 187, 493 (1960).
2. A. Yariv, Quantum Electronics, Wiley, New York (1967).
3. Laser Handbook, (F. T. Arecchi and E. O. Schulz-Dubois, eds.) North-Holland/American Elsevier, New York (1972).
4. O. Svelto, Principles of Lasers, Plenum, New York (1976).
5. H. E. Puthoff and R. H. Pantell, Fundamentals of Quantum Electronics, Wiley, New York (1969).
6. J. C. Slater, Microwave Electronics, van Nostrand, Princeton (1950).
7. A. G. Fox and T. Li, Bell Syst. Tech. J. 40, 453 (1961).
8. L. A. Riseberg and M. J. Weber, in Progress in Optics, Vol. XIV, (E. Wolf, ed.) North-Holland, Amsterdam (1976).
9. F. Auzel, in Luminescence of Inorganic Solids, (B. Di Bartolo, ed.) Plenum, New York (1978).
10. F. K. Fong, Theory of Molecular Relaxation, Wiley, New York (1975).
11. W. H. Fonger and C. W. Struck, J. Lum. 17, 241 (1978).
12. L. A. Riseberg and H. W. Moos, Phys. Rev. 174, 429 (1968).
13. R. Reisfeld, in Structure and Bonding, Vol. 13, Springer-Verlag, Berlin (1973).
14. C. B. Layne, W. H. Lowdermilk, and M. J. Weber, Phys. Rev. B 16, 10 (1977).
15. R. K. Watts, in Optical Properties of Ions in Solids, (B. Di Bartolo, ed.) Plenum, New York (1975).
16. T. Kushida, J. Phys. Soc. Japan 32, 1059 (1973).
17. R. Orbach, in Optical Properties of Ions in Crystals, (H. M. Crosswhite and H. W. Moos, eds.) Wiley, New York (1967).
18. D. L. Dexter, J. Chem. Phys. 21, 836 (1953).
19. M. Inokuti and F. Hirayama, J. Chem. Phys. 43, 1978 (1965).
20. J. Koo, L. R. Walker, and S. Geschwind, Phys. Rev. Lett. 35, 1669 (1975).
21. P. M. Selzer, D. L. Huber, B. Barnett and W. M. Yen, Phys. Rev. B 17, 4979 (1978).
22. T. Holstein, S. K. Lyo, and R. Orbach, Phys. Rev. B 15, 4693 (1977); ibid. B 16, 934 (1977).
23. B. Di Bartolo, Optical Interactions in Solids, Wiley, New York (1968).
24. J. C. Walling, H. P. Jenssen, R. C. Morris, E. W. O'Dell, and O. G. Peterson, Opt. Lett. 4, 182 (1979).

25. P. Kisliuk and C. A. Moore, Phys. Rev 160, 307 (1967).
26. R. C. Powell, B. Di Bartolo, B. Birang, and C. S. Naiman, in Optical Properties of Ions in Crystals, (H. M. Crosswhite and H. W. Moos, eds.) Wiley, New York (1967).
27. D. E. McCumber and M. D. Sturge, J. Appl. Phys. 34, 1682 (1963).
28. R. C. Powell, B. Di Bartolo, B. Birang, and C. S. Naiman, J. Appl. Phys. 37, 4973 (1966).
29. G. H. Dieke, Spectra and Energy Levels of Rare Earth Ions in Crystals, (H. M. Crosswhite and H. Crosswhite, eds.) Interscience Publishers, New York, 142 (1968).
30. T. Kushida, H. M. Marcos, and J. E. Geusic, Phys. Rev. 167, 289 (1968).
31. C. Brecher, L. A. Riseberg, and M. J. Weber, Phys. Rev. B 18, 5799 (1978).
32. S. Brawer and M. J. Weber, Appl. Phys. Lett. 35, 31 (1979).
33. H. G. Danielmeyer and H. P. Weber, IEEE J. Quant. Elec. 8, 805 (1972).
34. L. F. Johnson, J. E. Geusic, L. G. van Uitert, Appl. Phys. Lett. 8, 200 (1966).
35. H. P. Jenssen, D. Castleberry, D. Gabbe, and A. Linz, IEEE J. Quant. Elec. QU-9, 665 (1973).
36. Z. J. Kiss and R. C. Duncan, Appl. Phys. Lett. 5, 200 (1964).
37. E. Comperchio, M. Weber, R. Monchamp, and L. Riseberg, USAECOM Technical Report DAAB07-69-0227 (1970).
38. M. Bass and M. J. Weber, Appl. Phys. Lett. 17, 395 (1970).
39. F. Varsanyi, Appl. Phys. Lett. 19, 169 (1971).
40. S. Sakatsume and I. Tsujikawa, J. Phys. Soc. Japan 19, 1080 (1964).
41. Z. Goldschmidt, W. Low, and M. Foguel, Phys. Lett. 19, 17 (1965).
42. L. A. Riseberg, H. W. Moos, and W. D. Partlow, IEEE J. Quant. Elec. QE-4, 609 (1968).
43. S. Brawer and W. B. White, J. Chem. Phys. 67, 2043 (1977).
44. E. J. Sharp, J. E. Miller, and M. J. Weber, Phys. Lett. 30A 142 (1969).

RADIATIONLESS PROCESSES IN SEMICONDUCTORS

F. Williams, D. E. Berry and J. E. Bernard

Physics Department
University of Delaware
Newark, Delaware 19711, U.S.A.

ABSTRACT

Radiationless processes in semiconductors are considered in a theoretical framework, with some attention to experimental aspects. The processes generally divide between Auger transitions followed by radiationless relaxation among continuum states and either sequential or simultaneous multiphonon transitions. Radiationless processes which are particularly important for semiconductors include: Auger transitions involving three quasi-electronic particles as occur in electron-hole recombination, transitions involving effective mass states of point charge and associated defects, hot carrier generation and thermalization, decay of electron-hole droplets, and radiationless recombination via interface states. The roles of radiationless processes in semiconductor devices are discussed, including: light emitting diodes, thin film electroluminescent cells, solar energy photovoltaic converters, and semiconducting electrodes in photoelectrolytic cells.

I. INTRODUCTION

Radiationless processes are important in the operation of semiconductor devices. In some devices, radiationless processes are responsible for the primary function of the device. For example, for avalanche diodes, collision ionization (electron-hole pair production) by hot electronic charge carriers is responsible for the primary function of this type of diode, that is, charge carrier multiplication.

In another class of devices, radiationless processes are not the primary goal but are important to satisfy the thermodynamics limiting the operation of the devices. For example, for solar energy photovoltaic converters, illustrated in Fig. 1, the efficiency $\eta = W/L$, where W is the output of electrical work and L is the input of radiant energy, is limited by the first and second laws as follows:

$$L = W + Q; \ -\frac{L}{T_R} + \frac{Q}{T} \geq 0; \therefore \ \eta = (1 - T/T_R), \tag{1}$$

where Q is the heat flow from the solar cell to the heat reservoir at temperature T, and T_R is the temperature characterizing the radiation L, that is, the temperature to which L can heat a black body with ideal optics. Thus, these devices operate near reversible conditions as a consequence of electron-phonon interaction with associated dissipation into heat, at least to the extent specified by thermodynamics. In practical devices, radiationless processes play an even greater role.

In a third class of semiconductor devices radiationless processes are associated neither directly with the primary function nor with thermodynamic limitations but rather are associated with irreversibility in the mechanism of operation and/or parallel dissipative channels of operation. For example, for light emitting diodes consideration of the thermodynamics, essentially the reverse processes of those analyzed for photovoltaics, reveals that efficiencies of conversion from electrical work into light in excess of unity is permitted by the first and second laws, however actual efficiencies are at least an order of magnitude less. As discussed in section IV the processes responsible for these low efficiencies are mainly radiationless recombination processes for

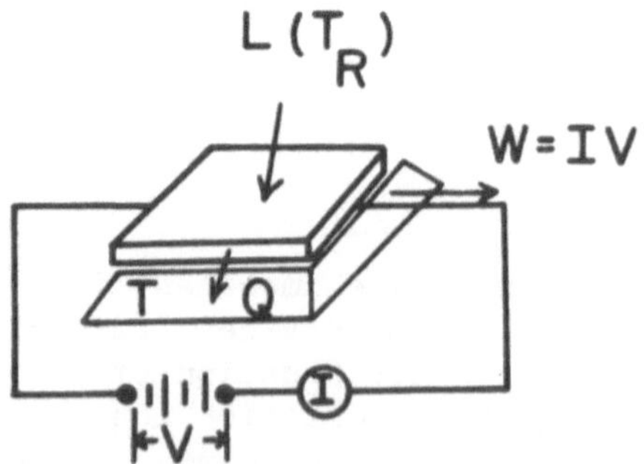

Fig. 1. Solar energy photovoltaic converter.

conduction electrons and valence band holes. Finally, it is to be noted that radiationless relaxation processes are the origin in compound semiconductors, as well as in luminescent materials in general, of the Stokes shift of emission from optical excitation, which results in the interesting and useful property of transparency of these materials to their own emission.

Semiconductors are well suited to both experimental and theoretical research on radiationless processes. Elemental semiconductors, specifically germanium and silicon, have been prepared as large single crystals with low impurity and low defect concentrations and other with controlled dopant concentration. Compound semiconductors, particularly III-V semiconductors such as gallium arsenide, are being prepared, with steadily-advancing quality. The availability of these materials facilitates the separation of intrinsic and extrinsic processes, that is, those characteristic of perfect crystals and those characteristic of defects and dopants.

Semiconductors, particularly Ge and Si, are probably better understood theoretically than any other class of condensed matter. Their electronic band structures have been calculated by many methods, and theory combined with experimental work such as cyclotron resonance measurements has resulted in a detailed understanding of their band structures. Their lattice dynamics are also well-understood. Defects and dopants with shallow electronic states have been thoroughly investigated by effective mass theory; those with deep states are less well-understood and these in fact are most important as regards radiationless recombination. In summary, semiconductors form a class of materials whose technology of preparation is sufficiently advanced and whose electronic states and vibrational levels are sufficiently well-understood so that research on radiationless processes, which depend on electron-vibrational interaction, can be fruitfully pursued.

In addition, there are a number of experimental techniques which have recently become available for investigating radiationless processes, and are especially well-adapted to semiconductors. These include photocapacitance spectroscopy, a specific version of which is illustrated in Fig. 2. The Schottky barrier at the semiconductor-metal junction is excited with band gap laser radiation in order to populate electron and hole states. A tunable second laser then slowly sweeps the spectral range up to the band gap, thus optically changing the occupancy of the charge states of these defects and dopants resulting in capacitance changes, as shown. Another variation involves thermal ionization of these states. By these methods extrinsic states which are important for radiationless recombination can be characterized. Other recent techniques include picosecond time-resolved spectroscopy which has led to the measurement of hot electron relaxation in GaAs.

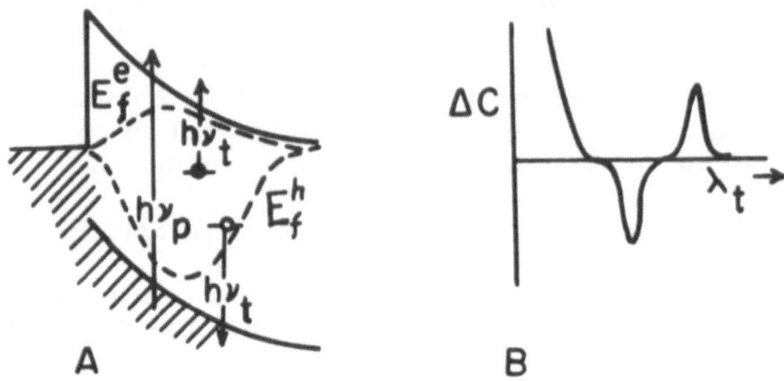

Fig. 2. Photocapacitance spectroscopy: A. Schottky barrier with pump laser $h\nu_p$ splitting Fermi level into two quasi-Fermi levels, E_f^e, E_f^h, and tunable laser $h\nu_t$ photo-ionizing band gap states: and B. Capacitance change ΔC versus wavelength of tunable laser λ_t.

The diversity of types of radiationless processes which occur in semiconductors is illustrated in Fig. 3. These divide into three classes: A. Auger transitions, both intrinsic and extrinsic, the latter involving defect-band edge or intra-impurity or inter-impurity transitions, in all cases a hot conduction electron is created which relaxes non-radiatively by intra-band transitions (similar transitions involving the generation of hot holes in the valence band are evident from symmetry); I. Impact excitation by hot electrons, created for the cases shown by a high applied electrical field but also formed by other methods, for example, with photoexcitation energies in excess of the band gap (collision excitation by hot holes is an obvious variation); and R. Recombination of a conduction electron and a valence band hole involving one or more multiphonon relaxations, either intrinsic or via defect or impurity states.

In section II theoretical research on radiationless processes in semiconductors occurring by Auger transitions will be reviewed; in section III, multiphonon processes will be similarly reviewed; and in section IV, the importance of radiationless processes to specific devices and phenomena will be amplified.

II. THE AUGER EFFECT IN SEMICONDUCTORS

An Auger transition is said to occur when energy is transferred from one electronic particle to another such that in the final state the energy of one of the electronic particles lies in a continuum. In solids these are usually viewed as three particle processes in which an electron and a hole recombine and give up

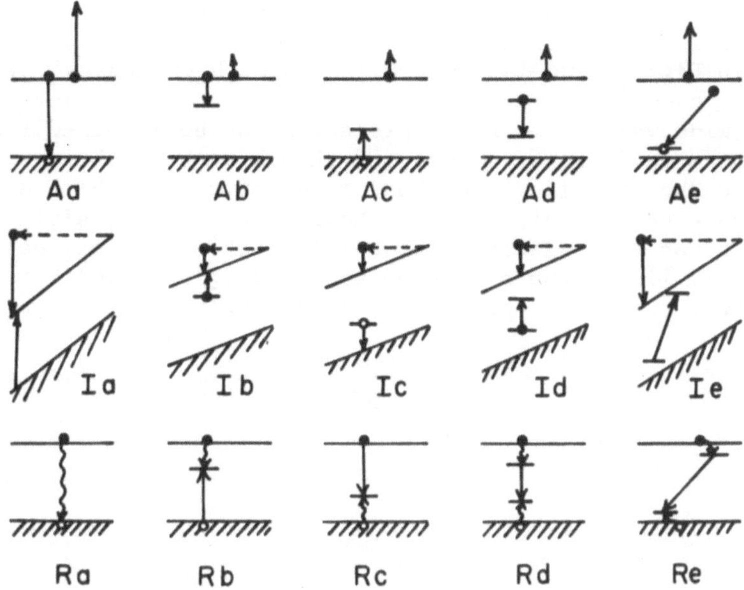

Fig. 3. Radiationless transitions of the following types: A. Auger transitions; I. Impact excitation; and R. Recombination; a involves intrinsic and b, c, d and e involve extrinsic: donor, acceptor, intra-impurity and donor-acceptor pair transitions, respectively.

energy to another electron or hole, as shown in Fig. 3A. Thus, there can be either electron-electron-hole (eeh) or electron-hole-hole (ehh) processes.

Auger processes can be classified according to whether they are intrinsic or extrinsic. Intrinsic processes are those which occur in a pure material, for example, shown in Fig. 3Aa, and extrinsic processes are those which involve electronic states of impurities, with several examples shown in Fig. 3Ab-e. Auger processes can also be classified as either phononless or phonon-assisted. For simplicity, we shall concentrate on phononless processes, and only briefly discuss the changes in the phenomena when phonon assistance is considered. Conservation of quasi-momentum is most evident in plots in wave number vector space, as shown in Fig. 4, and is used to distinguish phononless transitions.

A wide variety of Auger processes can be imagined for semiconductors, and, in fact, many of them have been observed. One of these is the so-called band-band Auger effect in which an electron from one band recombines with a hole from another, giving up energy to another electron (or hole) which may or may not originally lie in the same band as the original electron (or hole). Thus, two, three or four bands may be involved in such a process.

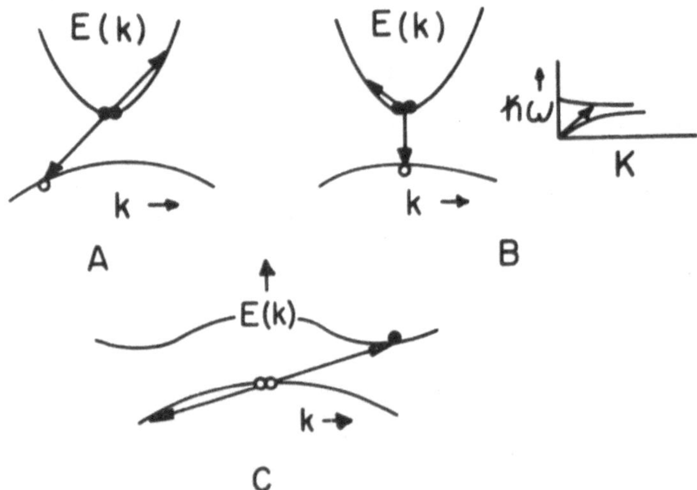

Fig. 4. Auger Transitions in Wavenumber Vector Space, E(k): A. phononless transition for direct gap semiconductor; B. phonon-assisted transition for direct gap semiconductor, with generation of optical phonon of wavevector K; and C. special case of phononless transition at band edges, generating a hot hole.

In Fig. 4A the phononless band-band Auger effect is shown for a direct gap semiconductor. The positive hole which is annihilated in the recombination is not at the valence band edge and was thus thermally activated. It is easy to show that the recombining particles in a direct gap semiconductor cannot participate in the band edge-band edge Auger transition without phonon assistance. The following conditions apply to the quasi-momenta of the recombining band edge electron and hole, k_e and k_h, and to the quasi-momentum change of the Auger particle, k_a:

$$k_e = -k_h = 0, \text{ and } \Delta k_a \neq 0 \tag{2}$$

where the second condition results from energy conservation during the transition and from the Auger particle remaining in allowed states of its band, that is, on its E(k) curve. Thus, phonons must be invoked to conserve total momentum as shown in Fig. 4B. The alternative is the off-the-band edge, phononless transition, as in Fig. 4A, which involves an activation energy for at least one of the recombining particles.

The activation energy for phononless Auger recombination in direct gap materials is related to the energy gap to be jumped by either electronic particle. Hence, it would be expected that such processes would be most important when the gap is not too large compared with kT. In addition, these processes should also be important when the curvature of one band is very small ($\Delta E(k)$ limited to a few kT throughout the reduced zone) and that of the other is large (> energy gap, E_g). Furthermore, phononless Auger transition rates should tend to increase with increasing temperature, as long as competing processes remain constant.

It has been found that band-band Auger transition rates increase faster with increasing carrier concentration than do radiative transition rates. This is easy to understand since, as three-particle processes, Auger transition rates should be proportional to n^2p or p^2n, whereas radiative rates vary in proportion to np. Hence, the ratio of Auger to radiative rates is proportional to n or p, where n is the negative carrier concentration and p is the positive carrier concentration. Note, that this analysis does not apply to degenerate semiconductors, for which the necessity of using Fermi-Dirac statistics in the calculation of the transition rate does not allow such simple forms to be easily obtained.

It is possible for localized centers, or traps, to participate in Auger processes with free carriers. In these, known as band-trap Auger transitions, a free electron recombines with a trapped hole, or a free hole with a trapped electron, and the recombination energy is given to a free carrier. Examples are shown in Fig. 3Ab,c.

In some semiconductors with large impurity concentrations it is possible that the impurity states may form an energy band which is involved in Auger transitions. As a specific example, an exciton bound to an isoelectronic substituent can participate in an Auger process involving an electron from the impurity band. This process is known as an impurity band Auger effect.

Another type of Auger process involves electron transfer between a donor and an acceptor in a semiconductor. Here the energy is given to a conduction band electron or to a valence band hole. For an example see Fig. 3Ae. A similar transition could occur for electron and hole states of a single center, as in Fig. 3Ad.

Other Auger processes involve excitons bound at impurity sites or free excitons. Transitions involving bound excitons are important in radiationless recombination in silicon (Pilkuhn, 1979).

While in direct gap semiconductors phononless Auger transitions require an activation energy, it is theoretically possible that a phononless Auger transition could occur in an indirect gap material with no threshold energy. Such a material would have to have a rather special band structure. A diagram illustrating this type of process is shown in Fig. 4C. Note that the reflection of the valence band across a tangent line at its maximum must intersect the minima of the conduction band. The band structure of germanium is fairly close to that illustrated in Fig. 4C (Huldt, 1971), so that one would expect such an ehh process to be rather probable in that material. The band structure of silicon, however, appears to be unfavorable to this type of process. The related eeh process does seem favorable in both Si and Ge, according to band structure, but it appears to be improbable because of a small matrix element and a small region of k-space for the transition (Pilkuhn, 1979).

The simplest case for which Auger transition probabilities have been calculated is that of phononless, band-band transitions in nondegenerate, direct gap materials. This case was investigated in detail by Beattie and Landsberg (1959), whose calculation has served as a model for much subsequent work on more complicated processes. They used time-dependent perturbation theory to calculate the transition probabilities. The zero-order Hamiltonian for the problem is the Hartree-Fock Hamiltonian. The perturbation causing the transitions is then just the electron-electron interaction

$$H' = \frac{e^2}{\varepsilon r} e^{-\lambda r}, \tag{3}$$

where the dielectric constant ε is included to take care of the effect of screening by inner electrons; the length $1/\lambda$ is used to account for screening by conduction electrons; and r is the separation between the electrons participating in the transition. In practice it is usual to let $\lambda \to 0$ at the end of the calculation, since this introduces no more uncertainty in the result than do certain other approximations which are used. For free electron states it is usual to use Bloch functions as the zero order states.

In order to obtain the net recombination probability, it is necessary to take the difference between the probabilities for the forward (Auger) process and the reverse (impact ionization) process. The latter are shown in Fig. 3I. If energy is conserved, the net probability is zero unless the electrons in the bands of interest are not in equilibrium. Beattie and Landsberg considered the electrons within each band to be in equilibrium among themselves but not with those of the other band. Then they defined quasi-Fermi levels for each band, and it was the difference in these quasi-Fermi levels that was responsible for the existence of a non-zero Auger transition probability. They calculated carrier lifetimes and compared them with experimental results in InSb. Above about 250°K they found reasonable order of magnitude agreement, suggesting that the process they studied is the dominant process in InSb at high temperatures.

In the computation of the matrix elements for the Auger transitions it is necessary to include both direct and exchange terms. It was claimed by Haug and Ekardt (1975) that the exchange term should not be divided by the dielectric constant. Previously, both the direct and exchange terms had been divided by ε. Their modification of the method of calculating transition probabilities should result in an increase of about two orders of magnitude in the transition probabilities previously calculated. While theoretically determined probabilities have generally been significantly lower than experimentally determined values, the increase produced by the Haug-Ekardt modification of the interaction potential is in some cases too great.

Phonon-assisted Auger recombination was studied by Eagles (1961) for direct gap materials, and by Huldt (1976) for indirect gap materials. Phonon-assisted Auger processes are second order processes, and, as such, might be expected to be less likely than phononless processes. However, by invoking phonon assistance it is possible to satisfy momentum conservation conditions more easily. Experimentally, it is observed that Auger transition rates have only weak temperature dependence, which is expected for phonon-assisted processes. Hence, it appears that phonon-assisted processes are quite important in many common semiconductors.

Phononless Auger processes in indirect gap materials were investigated by Huldt (1971) and by Hill and Landsberg (1976), with the result that the recombination coefficients were lower than those determined experimentally.

The case of degenerate semiconductors has been considered by Haug (1977, 1978) for both direct and indirect gap semiconductors. He showed that phononless Auger recombination is highly unlikely in degenerate materials, so both papers concentrate on phonon-assisted processes. In the later paper he applies the theory to the phenomenon of electron-hole-drops and finds fairly good agreement with experiment.

Auger processes involving traps and other localized systems have been reviewed by Landsberg and Adams (1973).

In conclusion, we note that the number and variety of Auger processes in semiconductors are quite large (Landsberg, 1978), but that it appears that the phonon-assisted, band-band processes may be the dominant ones, in both direct and indirect gap materials.

III. MULTIPHONON TRANSITIONS

For multiphonon radiationless transitions, the basic assumption is that electronic energy is directly converted into lattice energy. In the processes considered, the amount of electronic energy converted into phonon energy requires the creation of at least a few and often many phonons. There are several ways in which this conversion can take place. First, we could sequentially create the phonons which is possible only if there exist intervening states to which the electron can relax. This type of multiphonon relaxation is important in intraband relaxation of hot carriers. Lax (1960), in his theory of cascade capture at charged centers, has used the idea of sequential phonon creation. The latter involves as many as a few phonons created in each step of the cascade. A second way n-phonons could be created in a radiationless transition is to have all n-phonons created simultaneously. That is, we assume that the radiationless transition takes place between states, one of which has n more phonons than the other. In the first case, we consider that the transition arises from nth order perturbation theory due to a perturbation that creates a single phonon in first order. However, if the perturbation is small, then the direct transition to the state with n more phonons would be diminished by n factors of this smallness and would therefore be very small. In the second case, in order for the direct creation of n-phonons to be an important radiationless process, the full Hamiltonian must contain terms which in first-order perturbation theory create n-phonons. In this section we consider direct multiphonon radiationless processes.

In the past, several people have worked on the multiphonon radiationless transition, including Huang and Rhys (1950), Kubo and Toyazawa (1955), Lax (1960), Kovarskii and Sinyavskii (see their works, 1962-1969), and Henry and Lang (1977).

Huang and Rhys were the first to develop means of calculating the nuclear overlap factor which is now referred to as the statistical factor. They found the form of this factor and the parameters on which it depended, in the harmonic approximation, in particular, its dependence on S, the Huang-Rhys parameter. Within their analysis no explicit form of the perturbation causing the transition needed to be assumed, however, the Condon approximation had to be used.

Kubo and Toyazawa continued the work on the statistical factor by using the method of generating functions. Once again the Condon approximation was used along with the effective mass theory for charged defects. They determined the parameters for non-radiative capture and applied the theory to point charged defects in germanium and silicon.

Lax considered the general problem of cascade capture of electrons at charged centers. However each cascade step includes a direct multiphonon transition. In this work deformation potential theory is used to describe the interaction of the lattice with the electrons for non-polar materials. This interaction is used as the perturbation coupling the zero-order states whose electronic parts correspond to having the lattice fixed in its equilibrium configuration. The strength of the deformation potential is obtained from experiment.

Kovarskii and Sinyavskii recognized the need to go beyond the Condon approximation. In their works, they have developed equations for radiationless transitions of localized electrons. The derivation of their equations involves the use of electronic wavefunctions that include non-Condon effects. In order to derive these electronic wavefunctions perturbative series were developed where the perturbation used is the change of electron-nuclear interaction with respect to the nuclear coordinates. The chosen perturbation is (H_1 is not the H' for separating electronic and nuclear motions):

$$H_1 = V(r,R) - V(r,R_o) \quad . \tag{4}$$

The main result is the modification of the usual Huang-Rhys

statistical factor. Their results, in general, take the form of multiplicative factors connecting the Condon result with the non-Condon results. An example of their result is the following: The radiationless transition probability between 1s and 2p states of a localized center is found to be of the order of 10^3 larger for the non-Condon calculation and depends on the value of $(E_p-E_s)/\hbar\omega$ and the value of the Huang-Rhys parameter.

Henry and Lang (1977) have presented a model calculation for band-to-band radiationless transitions, in which all details of the calculation can be performed. The model consists of a neutral electron trap which is near enough to the conduction band so that its states are effective mass-like. Upon capture of an electron, the state relaxes so that the relaxed state is also effective mass-like, but is now near the valence band. The relaxed state can then capture a hole. Each capture process is assumed to take place only in those regions of the configurational space in which the adiabatic approximation is invalid. These regions are near the crossing of the extensions of the adiabatic potential curves for the band state and the localized electronic state. In these regions, the adiabatic approximation is replaced by the sudden approximation. In Fig. 5, the region of R-space for electron capture is denoted by δR.

So far we have discussed some of the general aspects of the major works on the direct multiphonon process. We will now go into more detail with respect to the works of Lax, Kovarskii and Sinyavskii, and Henry and Lang.

The problem faced in doing the radiationless transition calculation is two-fold. First a model must be devised which incorporates the primary aspects of the real situation. Here the important characteristic of the different models is the form of the lattice-electron interaction used.

Lax used a deformation potential interaction to model attractive centers in silicon and germanium, and Kovarskii and Sinyavskii also used deformation potential interactions for attractive and neutral centers in Ge, Si, and CdS. In addition, they used a screened Coulomb repulsive barrier for repulsive impurity centers. Henry and Lang considered the case of a neutral impurity in GaP and GaAs and used a potential having the form

$$-[1 + BR]V_o, \tag{5}$$

if $|r| < b$ = nearest-neighbor distance, and zero otherwise.

The second part of the problem faced in deriving numerical results is the form of the approximations to be used. In solving

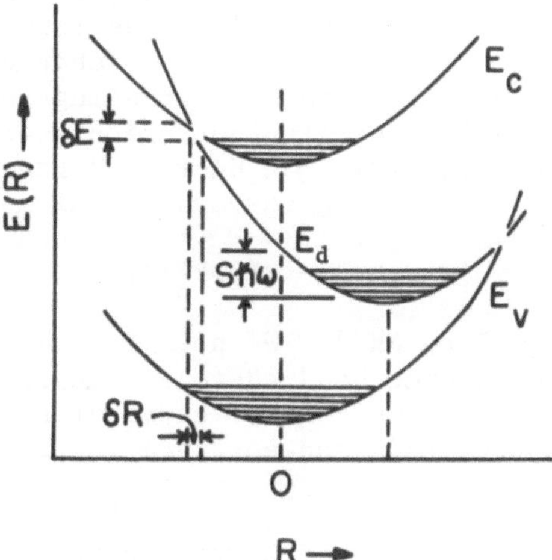

Fig. 5. Configuration coordinate model for a deep center, including adiabatic potential curves for the conduction and valence band edges, E_c and E_v, as well as for the deep center, E_d.

the radiationless transition problem, we assume that a good approximation to the state of the system can be obtained by leaving out some terms in the Schrodinger equation. These terms can be treated as a perturbation which causes radiationless transitions between the states. In the usual adiabatic approximation terms of the form

$$-\frac{\hbar^2}{2M}\left[\chi(R)\frac{\partial^2}{\partial R^2}\phi(r;R) + \frac{\partial}{\partial R}\chi(R)\cdot\frac{\partial}{\partial r}\phi(r;R)\right] \tag{6}$$

are neglected. On the other hand, Lax has used the following approximation: The standard Hamiltonian

$$H = T_N + T_e + V(r,R), \tag{7}$$

can be rewritten in the form

$$H = T_N + T_e + V(r,R_o) + [V(r,R) - V(r,R_o)], \tag{8}$$

where R_o is some fixed value of the nuclear coordinate. The approximation used by Lax is to neglect the last term. The Henry and Lang approximation is to assume the adiabatic approximation in the region of R-space far from the crossing and to assume the same approximation used by Lax in the region near crossing, that is, in the region δR (see Fig. 5). Henry and Lang calculate the transition probability by using the additional approximation

$$H' = [V(r,R_c) - V(r,R_o)], \tag{9}$$

where $|R_c-R_o|$ is equal to $\delta R/2$. We note from Fig. 5 that the conduction band and valence band edges are represented by curved adiabatic potential surfaces. These curvatures arise solely from the elastic energy term and not from any interaction of the electron and lattice.

The problem of using the adiabatic approximation for radiationless transitions as done by Kovarskii and Sinyavskii is that it must include non-Condon effects. These are usually left out of most adiabatic calculations because of difficulty in finding $\phi(r;R)$, that is, the electronic wavefunction as a function of R. These authors overcome this difficulty by using $V(r,R)-V(r,R_o)$ as the perturbation when solving the Schrodingér equation for $\phi(r;R)$, which differs from the Lax approximation.

More recently, Passler (1977) and Ridley (1978) have done further work in this field. Passler uses a static coupling approximation, that is a polaron-like treatment in which the renormalization of the electronic energy by the phonon field is treated exactly by a Platzman transformation. He infers from his results for deep centers and neutral traps that neither the Kovarskii and Sinyavskii nor the Henry and Lang results contain all the terms contained in his results. In particular, he claims that Kovarskii and Sinyavskii have missed a term proportional to $T^{-1/2}$ and that Henry and Lang have missed a term proportional to $T^{1/2}$. This situation has apparently arisen because of a deficiency in the model chosen initially. For example Henry and Lang used a single configurational coordinate model. However, Ridley and other authors have emphasized that the Condon-approximation is not adequate for radiationless transitions. He notes the differences between the results given by (a) Henry and Lang, (b) Kovarskii and Sinyavskii and (c) the standard results of Huang and Rhys. He notes that the Kovarskii and Sinyavskii results are too complicated for general use. He also questions the model for the electronic matrix element used by Henry and Lang; however, Ridley confirms the high temperature semi-classical rate derived by them.

One additional comment to be made here is on the trend towards works on multiphonon processes in which a polaron-like approximation is used. In addition to the work by Passler there are the works of Rebsch (1979) and of Howgate (1969), which use the static coupling approximation. Rebsch presents a theory of Auger processes between polaron states, and Howgate considers the transitions between states of transition metal dopants.

IV. SPECIFIC SEMICONDUCTOR DEVICES AND PHENOMENA

Radiationless processes in some representative semiconductor devices, some of which were introduced in Section I, will now be discussed, utilizing in part the theoretical framework for Auger and multiphonon processes developed in Sections II and III. A major question is the extent to which the radiationless processes occurring in the operation of these devices is intrinsic and to what extent, extrinsic. If extrinsic, then advances in the technology of preparation of materials can lead to improved devices; if intrinsic, then new materials or novel phenomena are alternatives for better devices. Other questions include: (a) the detailed identifications of the intrinsic and extrinsic radiationless processes, for example, if extrinsic whether point defect, associated impurities or interfaces are involved; (b) whether hot or thermal carriers are involved; (c) whether Auger or multiphonon transitions are dominant? In the following we shall consider photovoltaic converters, light emitting diodes, thin film electroluminescent displays, and semiconductor-electrolyte photoelectrochemical converters.

IV.A. Solar Energy Photovoltaic Converters

Although homojunctions, in particular p-n junctions of single crystal, polycrystalline or amorphous silicon, are probably the most important class of photovoltaics for solar energy conversion, we choose a heterojunction, specifically n-CdS/p-Cu_2S, as an example illustrating more diverse possibilities for radiationless transitions. The energy bands for this junction at equilibrium are shown in Fig. 6. The device has the plane-parallel structure shown in Fig. 1 with the junction in the plane. The photovoltage and current are generated by the minority conduction electrons created in the p-Cu_2S where the solar radiation is absorbed. During steady-state operation the Fermi level E_f is split into two quasi-Fermi levels, respectively, for electrons and for positive holes. The compositional change at this junction is quite abrupt. The junction thickness, given by the region with bent band edges, originates from the space charge which takes care of the differences in work functions of n-CdS and p-Cu_2C. As a consequence of lattice mismatch between these two materials, interface states exist in the region of the abrupt compositional change at energies within the band gap. These interface states provide mechanisms for radiationless recombination which competes

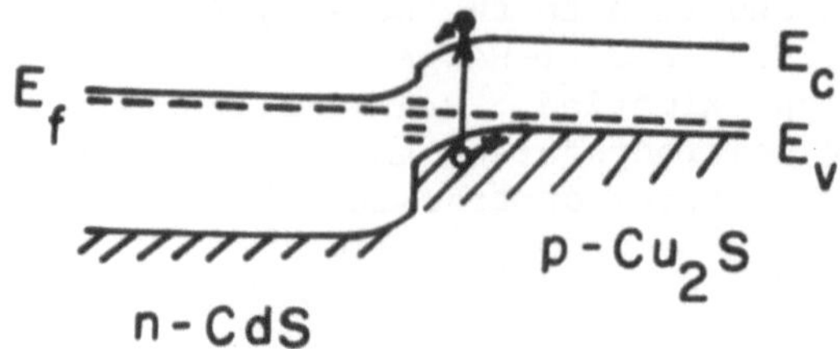

Fig. 6. Solar energy photovoltaic heterojunction cell.

with the photovoltaic current. Recombination via interface states probably occurs mainly by multiphonon transitions.

IV.B. Light Emitting Diodes, LED

Present commercial LEDs are n-p homojunctions of gallium phosphide. Operation in the steady-state with applied voltage V in the forward direction is illustrated in Fig. 7. As briefly discussed in Section I, theoretical power efficiencies in excess of unity are thermodynamically possible, however, measured efficiencies are much less, as a consequence of radiationless recombination competing with radiative recombination. The radiative transition occurs at donor-acceptor pairs, as illustrated in Fig. 7 or at isoelectronic dopants. A decade ago the principal radiationless transition in these devices was thought by some to be an intrinsic Auger transition; the evidence now, in part based on capacitance spectroscopy, is that the dominant radiationless process in these devices is recombination at deep defects or impurity centers by multiphonon transitions.

IV.C. Thin Film Electroluminescent Displays

In addition to the low field, minority charge carrier injection mechanism, just described for LEDs, there is another mechanism for electroluminescence, EL, that is, high field, collision excitation EL. The simplest example is a reverse biased n-p junction, for example of silicon, from which radiation

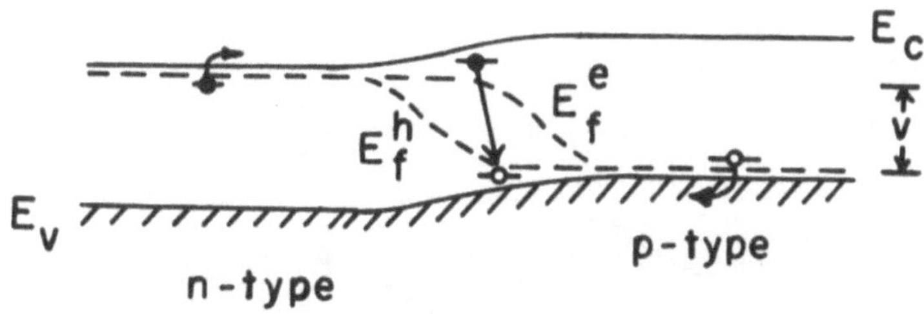

Fig. 7. n-p homojunction light emitting diode, with forward bias V and quasi-Fermi levels E_f^e and E_f^h.

by both intra-band and inter-band transitions are observed, with however the overwhelmingly dominant processes being radiationless.

EL occurring by collision excitation by hot electrons in a high a.c. field was extensively investigated for several decades with powders of zinc sulfide in a dielectric matrix, following the pioneering research of G. Destriau. During the past few years research on this mechanism of EL has experienced a renaissance, in part because of the development by the Sharp Corporation in Japan of stable thin film EL panels capable of information storage, as well as display. The active component is an evaporated film of ZnS:Mn sandwiched between two insulating films of Y_2O_3. The average field is $\sim 10^6$ v/cm so that electronic charge carriers can be accelerated out of equilibrium with the phonons and acquire sufficient kinetic energy for impact excitation of dopants which are capable of luminescent emission. The excitation process is basically as was shown in Fig. 3Id. More details of processes occurring within the ZnS:Mn film are illustrated in Fig. 8: the origin of charge carriers is assumed to be deep donors; the localized electron states of Mn^{2+} are near or in the valence band and thus are stable in the high field; the radiationless relaxation in the region of the Mn^{2+} is also indicated by the Stokes change in transition energy. Competing radiationless processes include: phonon generation during the acceleration, collision ionization of defects and band-to-band transitions, and multiphonon de-excitation of the Mn^{2+}.

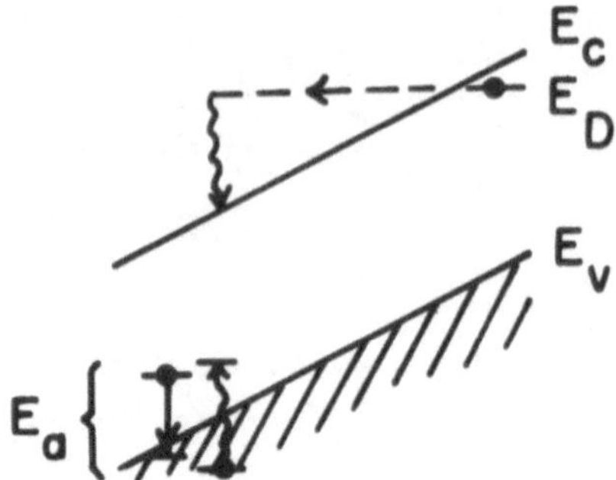

Fig. 8. High field, collision excitation electroluminescence.

IV.D. Photoelectrochemical Cells

Another phenomenon involving semiconductors and radiationless processes is photoelectrolysis, which converts radiant energy such as solar energy directly into chemical energy. The radiation is absorbed in the depletion region of a semiconducting electrode and the photo-created minority carriers are injected into the aqueous electrolyte. This is illustrated in Fig. 9 for an n-type semiconducting anode, such as n-TiO_2. The injected positive holes convert OH^- to OH^o from which O_2 is formed. At the cathode H_2 is evolved, either at an illuminated p-type semiconducting or a metal electrode. Among the radiationless processes are: inter-band recombination in the semiconductor; relaxation of photo-generated hot carriers in the depletion region; and molecular relaxation in the region of the OH^o, which forestalls reverse tunneling. Under some conditions foreward tunneling of hot carriers occurs before thermalization with the phonons. Quantization of the band states by confinement in the depletion region reduces the rate of thermalization.

V. CONCLUSIONS

Semiconductors are well-defined materials with well-understood electronic properties which exhibit a diversity of radia-

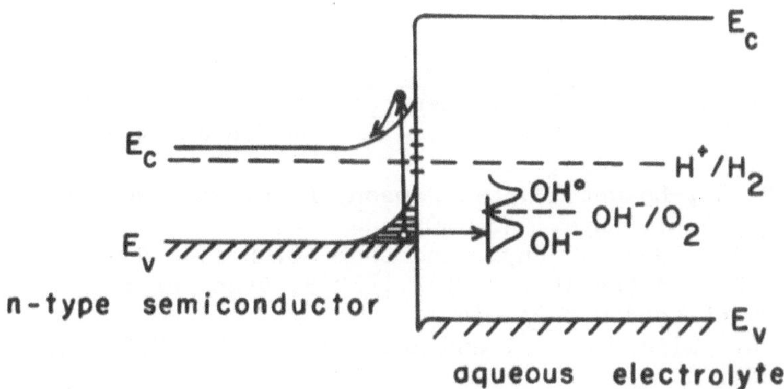

Fig. 9. Photoelectrolysis: positive hole injection from depletion layer of semiconductor into extrinsic state of aqueous electrolyte.

tionless processes, both intrinsic and extrinsic. Auger transitions occur in a variety of types: some phononless; others phonon-assisted, and are probably theoretically understood the best among the radiationless processes in semiconductors. Collision excitation processes are reasonably well understood in general, however, quantitative calculations of specific processes have not significantly advanced. Multiphonon transitions are an active area of current research, and analyses which go beyond the Condon approximation are appropriately being emphasized. Experimental techniques, for example picosecond time-resolution measurements of hot carriers and capacitance spectroscopy of deep recombination centers, are contributing to the clarification of the radiationless processes which occur in semiconductors. These processes are of great importance in the operation of semiconductor devices.

We acknowledge the partial support of the U.S. Army Research Office-Durham during the preparation of this manuscript.

REFERENCES AND RELATED WORKS

Bhargava, R. N., 1970, Proceedings of the 10th International Conference on the Physics of Semiconductors, (USAEC, Washington, D.C.), "Kinetics of Radiative and Non-Radiative Recombination in (Zn,O) Doped GaP," 640.

Bachrach, R. Z., Loumor, O. G., Dawson, L. R., and Wofstern, K. B., 1972, J. Appl. Phys., 43, 5098.

Beattie, A. R., 1962, J. Phys. and Chem. of Solid 24, 1049.

Beattie, A. R. and Landsberg, P. J., 1959, Proc. of the Royal Society of London 249A, 16.

Dexter, D. L., Klick, C. C., and Russell, G. A., 1955, Phys. Rev. 100, 603.

Dishman, J. M., 1971, Phys. Rev. 3B, 2588.

Eagles, D. M., 1961, Proc. Phys. Soc. London 78, 204.

Eliseev, P. G., Zavestovskaya, I. N., and Polvektov, I. A., 1978, 8, 124.

Forbes, L., 1975, Sol. State Elect. (6B) 18, 635.

Frenkel, J., 1931, Phys. Rev. 37, 17.

Frenkel, J., 1931, Phys. Rev. 37, 1276.

Gamurar', V. Ya., 1969, Opt. and Spect. 27, 524.

Gamurar', V. Ya., Perlin, Yu. E. and Tsukerblat, B. S., 1971, Bull. Acad. Sci. USSR, Phys. ser. 35, 1306.

Gebranzig, U., Haug, A. and Rosenthal, W., 1976, J. Lum. 12/13, 547.

Glinchuk, K. D., Prokhorovich, A. V., and Uovnenko, V. I., 1979, Phys. Stat. Sol. 51A, 645.

Grimmeiss, H. G., Monemar, B., and Samuelson, L., 1978, Sol. State Elect. 21, 1505.

Haug, A., 1977, Sol. State Comm. 22, 537.

Haug, A., 1978, Sol. State Comm. 25, 477.

Haug, A., 1978, Sol. State Elec. 21, 1281.

Haug, A., 1979, J. Lum. 20, 173.

Haug, A. and Ekardt, W., 1975, Sol. State Comm. 17, 267.

Henry, C. H., 1973, J. Lum. 7, 127.

Henry, C. H., 1975, J. Elec. Mat. (USA) 4, 1037.

Henry, C. H., 1976, J. Lum. 12/13, 47.

Henry, C. H. and Lang, D. V., 1977, Phys. Rev. 15B, 989.

Henry, C. H. and Logan, R. A., 1977, Appl. Phys. Lett. 31, 203.

Henry, C. H. and Logan, R. A., 1977, J. Appl. Phys. 48, 3962.

Henry, C. H. and Logan, R. A., 1977, IEEE Trans, Elec. Devices ED24, 1215.

Hill, D., 1976, Proc. Royal Soc. of London, 347A, 565.

Hill, D., and Landsberg, P. T., 1976, Proc. Royal Soc. of London, 347A, 547.

Hoshina, T. and Kawai, H., 1976, J. Lum. 12/13, 453.

Howgate, D. W., 1969, Phys. Rev. 177, 1358.

Huang, Kun and Rhys, Avril, 1950, Proc. Roy. Soc. of London, 204A, 406.

Huldt, L., 1971, Phys. Stat. Sol. 8a, 173.

Huldt, L., 1976, Phys. Stat. Sol. 33a, 607.
Iqbal, M. Z. and Northrop, D. C., 1974, J. Phys. D7, L125.
Jaros, M., 1978, Sol. State Comm. 25, 1071.
Johnston, W. D., Jr., 1975, IEEE Trans. on Elec. Devices ED22, 1054.
Kovarskii, V. A., 1962, Sov. Phys.-Sol. State 4, 1200.
Kovarskii, V. A., Chaikovskii, I. A., and Sinyavskii, E. P., 1965, Sov. Phys.-Sol. State 6, 1679.
Kovarskii, V. A., Perel'man, N. F., and Sinyavskii, E'. P., 1973, Sov. Phys.-Sol. State 15, 1207.
Kovarskii, V. A. and Sinyavskii, E'. P., 1963, Sov. Phys.-Sol. State 4, 2345.
Kovarskii, V. A. and Sinyavskii, E'. P., 1969, Sov. Phys.-Sol. State 6, 498.
Kubo, Ryoso, 1952, Phys. Rev. 86, 929.
Kubo, R. and Toyozawa, Y., 1955, Prog. Theor. Phys. 13, 160.
Kusunoki, M., 1979, Phys. Rev. 20B, 2512.
Landsberg, P. T., 1965, Lec. in Theor. Phys. 8a, 313.
Landsberg, P. T., 1970, Phys. Stat. Sol. 41, 457.
Landsberg, P. T. and Adams, M. J., 1973, J. Lum. 7, 3.
Landsberg, P. T. and Beattie, A. R., 1959, J. Phys. and Chem. Sol. 8, 73.
Landsberg, P. T. and Robbins, D. J., 1978, Sol. State Elec. 21, 1289.
Lang, D. V. and Henry, C. H., 1975, Phys. Rev. Lett. 35, 1525.
Lang, D. V. and Logan, R. A., 1977, IEEE Trans. Elec. Devices ED24, 1215.
Lax, Melvin, 1960, Phys. Rev. 119, 1502.
Layne, C. B., Lowdermilk, W. H. and Weber, M. J., 1977, Phys. Rev. 16B, 10.
Lyubchenko, A. V., Fedorov, A. I. and Sheinkman, M. K., 1977, Sov. Phys. Semicond. 11, 562.
Morgan, T. N., 1966, Phys. Rev. 148, 890.
Narayanamurti, V., Logan, R. A. and Chin, M. A., 1978, Phys. Rev. Lett. 40, 63.
Osipov, V. V. and Kholodnov, V. A., 1970, Fiz. & Tekh. Polvprov 4, 2241.
Parrott, J. E., 1976, Sol. State Elec. (GB) 19, 229.
Pässler, R., 1975, Phys. Stat. Sol. 68/69B, 69.
Pässler, R., 1976, Phys. Stat. Sol. 78B, 625.
Pässler, R., 1976, Phys. Stat. Sol. 76B, 647.
Pässler, R., 1977, Phys. Stat. Sol. 83B, K55.
Pässler, R., 1977, Phys. Stat. Sol. 83B, K111.
Pässler, R., 1978, Phys. Stat. Sol. 85B, 203.
Pässler, R., 1978, Phys. Stat. Sol. 86B, K39.
Pässler, R., 1978, Phys. Stat. Sol. 86B, K45.
Peaker, A. R., Hamilton, B., Wight, D. R., Blenkensep, I. D., Harding, W., and Gibb, R., 1977, Proceedings of the 6th International Symposium on Gallium Arsenide and Related Compounds, Part I, (Institute of Physics, London), 326.
Perlin, Yu. E., 1960, Sov. Phys.-Solid State 2, 222.
Perlin, Yu. E., 1964, Sov. Phys. Uspekhi 6, 542.

Perlin, Yu. E., J. Lum. in press.
Perlin, Yu. E., Tsukerblat, B. S. and Perepelitsa, E. I., 1973, Sov. Phys. JETP 35, 1185.
Perlin, Yu. E., Kaminskii, A. A., Enakii, V. N., and Vylegzhanin, D. N., 1979, Phys. Stat. Sol. 92B, 403.
Pilkuhn, M. H., 1979, J. Lum. 18/19, 81.
Queisser, H. J., 1978, Sol. State Elec. 21, 1495.
Rebsch, J.-T., 1979, Sol. State Comm. 31, 377.
Rickayzen, G., 1957, Proc. Roy. Soc. 241A, 480.
Ridley, B. K., 1978, J. Phys. C 11, 2323.
Robbins, D. J. and Dean, P. J., 1978, Adv. Phys. 27, 499.
Rozneritsa, Ya. A. and Prodan, V. D., 1977, Izv. VUZ. Fiz. 10, 16.
Schmid, W., 1978, Sol.-State Elect. 21, 1285.
Sheinkman, M. K., Korsunskaya, N. E., Markevich, I. V. and Torchinskaya, T. V., 1976. Bull. Acad. Sci. USSR, Phys. Ser. 40, 1.
Stoneham, A. M., 1975, Theory of Defects in Solids (Oxford, N.Y.), Chapter 14, p. 477.
Stoneham, A. M., 1977, Phil. Mag. 36, 983.
Stoneham, A. M. and Bartram, R. M., 1978, Sol. State Elec. 21, 1325.
Sinyavskii, E. P. and Kovarskii, V. A., 1967, Sov. Phys.-Sol. State 9, 1142.
Street, R. A., 1977, Sol. State Comm. 24, 363.
Street, R. A., 1978, Phys. Rev. 17B, 3984.
Sturge, M. D., 1973, Phys. Rev. 8B, 6.
Tolpygo, E. I., Tolpygo, K. B., and Sheinkman, M. K., 1974, Sov. Phys.-Semicon. 8, 326.
Toyozawa, Y., 1978, Sol. State Elect. 21, 1313.
Troster, F., 1978, Z. Nat. A 3BA, 1251.
Tsang, W. T., 1978, Appl. Phys. Lett. 33, 245.
Tsukerblat, B. S. and Perlin, Yu. E., 1966, Sov. Phys.-Sol. State 7, 2647.
Vasileff, H. D., 1954, Phys. Rev. 96, 603.
Van Vechten, J. A., 1975, J. Elec. Mat. (USA) 4, 1159.
Weeks, J. D., Tully, J. C., and Kimerling, L. C., 1975, Phys. Rev. 12B, 3286.
Werkhoven, C., VanOpdorp, C., and Vink, A. T., 1977, Proceedings of the 6th International Symposium on Gallium Arsenide and Related Compounds, Part I, (Institute of Physics, London) 317.
Williams, F., Berry, D. E., and Bernard, J. E., 1980, Our earlier chapter.

LONG SEMINARS

PHOTOACOUSTICS AND DEEXCITATION PROCESSES IN CONDENSED MEDIA

A. Rosencwaig

Lawrence Livermore Laboratory
P. O. Box 5508
Livermore, CA 94550

ABSTRACT

In this article, we present a brief introduction to the field of photoacoustics, outlining its principles, methodology and some of its capabilities. We then discuss in some detail the application of this new technique to the study of deexcitation processes in condensed media, both solid and liquid. The deexcitation processes covered include radiationless (heat-producing) deexcitation, radiative deexcitation, photochemical deexcitation, and photovoltaic deexcitation. Photoacoustic experiments involving these different deexcitation modes are described.

I. INTRODUCTION

In its broadest sense, spectroscopy can be defined as the study of the interaction of energy with matter. As such, it is a science encompassing many disciplines and many techniques. The energy used in the oldest form of spectroscopy, optical spectroscopy, exists as photons, with wavelengths ranging from less than one Angstrom in the x-ray region, to more than a hundred microns (10^6 Angstroms) in the far-infrared. Because of its versatility, range and its non-destructive nature, optical spectroscopy remains a widely-used and most important tool for investigating and characterizing the properties of matter.

Conventional optical spectroscopies tend to fall into two major categories. In the first category, one studies the optical photons that are transmitted through the material of interest, that is one studies those photons that have not interacted with the material. In the second category, one studies the light that is scattered or reflected from the material, that is, those photons that have undergone some interaction with the material. Almost all conventional optical methods are variations of these two basic techniques. As such, they are distinguished not only by the fact that optical photons constitute the incident energy beam, but also by the fact that the data is obtained by detecting some of these photons after the beam has interacted with the matter or material under investigation. It should be noted that these optical techniques preclude the detection and analysis of those photons that have undergone an absorption, or annihilation, i.e. interaction with the material - the process that is often of most interest to the investigator.

Optical spectroscopy has been a scientific tool for over a century, and it has proven invaluable in studies on reasonably clear media such as solutions and crystals, and on clean specularly reflective surfaces. There are, however, several instances where conventional spectroscopy is inadequate even for the case of clear, transparent materials. Such a situation arises when one is attempting to measure a very weak absorption, which in turn involves the measurement of a very small change in the intensity of a strong essentially unattenuated transmitted signal. In addition to weakly absorbing materials, there are a great many non-gaseous substances, both organic and inorganic, that are not readily amenable to the conventional transmission or reflection forms of optical spectroscopy. These materials are usually highly light-scattering, such as powders, amorphous solids, tissues, gels, smears and suspensions. Other difficult materials are those that are optically opaque with dimensions that far exceed the penetration depth of the photons. In the former case, the optical signal is composed of a complex combination of specularly-reflected, diffusely-reflected, and transmitted photons, making the analysis of the data extremely difficult. In the latter case, the absorptive properties of the material are difficult, if not impossible to determine since essentially no photons are transmitted. Over the years, several techniques have been developed to permit optical investigation of highly light-scattering and opaque substances. The most common of these are diffuse reflectance, attenuated total reflection or internal reflection spectroscopy, and Raman scattering. All of these techniques have proven to be very useful. Yet, each suffers from serious limitations. In particular, each method is useful for only a relatively small category of materials, has a limited wavelength range, and the data obtained is often difficult to interpret.

During the past few years, another optical technique has been developed to study those materials that are unsuitable for the conventional transmission or reflection methodologies (Rosencwaig,

1973; 1978). The technique, called photoacoustic spectroscopy or PAS, is distinguished from the conventional techniques chiefly by the fact that even though the incident energy is in the form of optical photons, the interaction of these photons with the sample under investigation is studied not through subsequent detection and analysis of some of the photons, but rather through a direct measure of the energy absorbed by the material due to its interaction with the photon beam.

Although more will be said about experimental methodology later in this chapter, a brief description here might be appropriate. In photoacoustic spectroscopy, or PAS, the sample to be studied is usually placed in a closed cell or chamber. For the case of gases and liquids the sample will generally fill the entire chamber. For solids, the sample fills only a portion of the chamber, with the rest of the chamber filled with a nonabsorbing gas such as air. In addition to the sample, the chamber also contains a sensitive microphone. The sample is illuminated with monochromatic light that either passes through an electromechanical chopper or is intensity-modulated in some other fashion. If any of the incident photons are absorbed by the sample, internal energy levels within the sample are excited. Upon subsequent deexcitation of these energy levels, all or part of the absorbed photon energy is then transformed into heat energy through nonradiative deexcitation processes. In a gas this heat energy appears as kinetic energy of the gas molecules, while in a solid or liquid, it appears as vibrational energy of ions or atoms. Since the incident radiation is intensity-modulated, the internal heating of the sample is periodic.

Since photoacoustics measures the internal heating of the sample, it clearly is a form of calorimetry as well as a form of spectroscopy. There are many calorimetric techniques by which one can detect and measure the heat produced during a physical or chemical process. The most obvious approach for the detection of heat production is to employ a conventional calorimeter based upon the usual temperature sensors such as thermistors and thermopiles. These classical techniques, though simple and well-developed, have several inherent disadvantages for photoacoustic spectroscopy in terms of sensitivity, detector risetime, and the speed at which measurements can be made. More suitable calorimetric techniques measure heat production through volume and pressure changes produced in the sample or in an appropriate transducing material in contact with the sample.

In gaseous samples, volume changes can be quite large as a result of internal heating. In these cases, a displacement-sensitive detector such as a capacitor microphone proves to be an excellent heat detector. With present microphones and associated electronics it is possible to detect temperature rises in a gas of 10^{-6}C, or a thermal input of the order of 10^{-9} calories/cm^3-sec. The primary disadvantage with a detector that responds to volume changes is that the response time is limited both by the transit time for a sound wave in the gas within the cell cavity, and by the relatively low frequency response of the microphone. Together, these two factors tend to limit the response time of a gas-microphone system to the order of 100 μsec or longer.

When dealing with liquids, or bulk solid samples it is possible to measure heat production through subsequent pressure or stress variations in the sample itself by means of a piezoelectric detector in intimate contact with the sample. With these detectors, temperature changes of 10^{-7} to 10^{-6}C can be detected, which for typical solids or liquids corresponds to thermal inputs of the order of 10^{-6} calories/cm^3-sec. It should be borne in mind that because the coefficient of volume expansion of liquids and solids is 10-100 times smaller than that of gases, measurement of the heat production in liquids and solids directly with a displacement-sensitive detector such as a microphone would be 10-100 times less sensitive than using a pressure-sensitive device such as a piezoelectric detector.

It is, of course, not always possible to employ a piezoelectric detector, as in the case of a powdered sample, or a smear or gel. In those cases, a gas is used as a transducing medium coupling the sample to a microphone. The periodic heating of the sample from the absorption of the optical radiation results in a periodic heat flow from the sample to the gas, which itself does not absorb the optical radiation. This in turn produces pressure and volume changes in the gas which drive the microphone. This method is not as direct as a contact piezoelectric measurement, but it is quite sensitive for solids with large surface:volume ratios such as powders, and is capable of detecting temperature rises of 10^{-6} to 10^{-5}C in such samples, or thermal inputs of about 10^{-5} to 10^{-6} calories/cm^3-sec.

There are several advantages to photoacoustic spectroscopy. Since <u>absorption</u> of optical or electromagnetic radiation is required before a photoacoustic signal can be generated, light that is transmitted or elastically scattered by the sample is not detected, and hence does not interfere with the inherently absorptive PAS measurements. This is of crucial importance when working with essentially transparent media, such as pollutant-containing gases, that have few absorbing centers. This insensitivity to scattered radiation also permits the investigator to obtain optical absorption data on highly light-scattering materials such as powders, amorphous solids, gels, colloids etc. Another advantage is the capability of obtaining optical absorption spectra on materials that are completely opaque to transmitted light. Coupled with this, is the capability, unique to photoacoustic spectroscopy, of performing non-destructive depth-profile analysis of absorption as a function of depth into a material.

Furthermore, since the sample itself constitutes the electromagnetic radiation detector, no photoelectric device is necessary, and thus studies over a wide range of optical and electromagnetic wavelengths are possible without the need to change detector systems. The only limitations are that the source be sufficiently energetic (at least 10 μwatts/cm^2), and that whatever windows are used in the system be reasonably transparent to the radiation. Finally, the photoacoustic effect results from a radiationless energy-conversion process, and is therefore complementary to radiative and photochemical processes. Thus PAS may itself be used as a sensitive though indirect method for studying the phenomena of fluorescence and photosensitivity in matter.

II. PHOTOACOUSTIC THEORY

The mathematical analysis of a photoacoustic signal is usually fairly laborious and complex. For a description of photoacoustic theory we refer the reader to the review by Rosencwaig (1978) and references therein. Here, we present a simplified photoacoustic theory that can be derived from basic physical insights.

II.A. Gas-Microphone Signals

When a sample absorbs intensity-modulated optical radiation, the sample will undergo periodic heating. This periodic heating results in a periodic heat flow from the sample to the gas at the sample-gas boundary. A thin layer of gas near the boundary is then cyclically heated by this heat flow. The thickness of the boundary layer is determiend by the thermal diffusion length in the gas, μ which is given by

$$\mu = \left(\frac{2\alpha}{\omega}\right)^{\frac{1}{2}} , \tag{1}$$

where ω is the radial frequency at which the light is intensity-modulated and α is the thermal diffusivity defined as

$$\alpha = \kappa/\rho C , \tag{2}$$

where κ is the thermal conductivity, ρ the density and C the specific heat. Generally, the thermal diffusion length for most gases runs in the range of 25-500 μm for the frequencies usually used in photoacoustic spectroscopy (1000 Hz-5 Hz). The thermal diffusion length represents the distance where the temperature rise due to conduction is e^{-1} that at the origin of the heating, which for our case is the sample surface.

The localized heating of a layer of fluid or gas can be thought as producing a localized stress that is then rapidly transmitted through the rest of the enclosed gas in the photoacoustic cell. The local pressure or stress that is generated within a thermal diffusion length of the sample surface can be approximated by the expression

$$p_\mu \simeq B'\alpha_t'(\tfrac{1}{2}\Theta_o), \tag{3}$$

where B' is the bulk modulus and α_t' the volume expansion coefficient of the gas, and where $\frac{1}{2}\Theta_o$ is approximated as the average temperature within this thermal diffusion length, if Θ_o is taken as the temperature at the sample-fluid interface. The pressure at the microphone, a distance ℓ' away, will be given by

$$p = p_\mu\left(\frac{\mu'}{\ell'}\right) = \tfrac{1}{2}B'\alpha_t'\,\Theta_o\left(\frac{\mu'}{\ell'}\right) = \frac{\gamma p_o}{2T_o a'\ell'}\,\Theta_o\ , \tag{4}$$

since for a gas, $B' = \gamma p_o$ where γ is the ratio of specific heats and p_o is the pressure, $\alpha_t' = 1/T_o$ where T_o is the temperature, and $a' = 1/\mu'$.

When the gas or fluid is completely constrained at its borders, then the pressure p is the same everywhere in the cell, as long as the cell dimensions are much smaller than the acoustic wavelength. The temperature Θ_o at the sample-gas interface can be approximated by,

$$\Theta_o \simeq H_{abs}/M_{th}\ , \tag{5}$$

where H_{abs} is the amount of heat absorbed per unit time within the first thermal diffusion length in the sample, and M_{th} is the thermal mass of this region of the sample. For the case where the thermal diffusion length in the sample is smaller than the sample thickness ($\mu < \ell$) and where A is the area illuminated by the light,

$$H_{abs} = \frac{I_o A(1 - e^{-\beta\mu})}{\omega}\ , \tag{6}$$

with I_o being the light intensity, and β the optical absorption coefficient

$$M_{th} = \rho C\mu A\ . \tag{7}$$

This then gives

$$\Theta_o = \frac{I_o(1 - e^{-\beta\mu})}{\rho\ C\ \omega\ \mu} \quad . \tag{8}$$

Similarly, when $\mu > \ell$, then

$$H_{abs} \simeq \frac{I_o(1 - e^{-\beta\ell})A}{\omega} \quad , \tag{9}$$

$$M_{th} \simeq \rho''C''\mu''A \ , \tag{10}$$

where the unprimed symbols represent the sample and the double-primed symbols represent the parameters of the material directly behind the sample.

Thus

$$\Theta_o = \frac{I_o(1 - e^{-\beta\ell})}{\rho''C''\mu''\omega} \quad . \tag{11}$$

Thermal conduction introduces a phase lag in the PAS signal. There is a $\pi/4$ phase lag due to conduction in the gas, and an additional phase lag due to conduction in the sample. The sample phase lag is approximated by $\phi \simeq 1/\beta\mu$. When $\mu < 1/\beta$, ϕ reaches a maximum value of $\pi/4$ as well.

Combining these expressions for the magnitude of the PAS signal with the above remarks on phase, we can reconstruct all six of the special photoacoustic cases treated in the more-detailed Rosencwaig-Gersho theory (1976). We find that aside from a factor of $1/\sqrt{2}$, all six cases are properly given by the above arguments.

One of the most important aspects of photoacoustics is its depth-profiling capability. This unique feature is apparent from eq. (8), which shows that the surface temperature Θ_o is dependent on both the optical and thermal properties of the sample within a thermal diffusion length μ below the surface. Since $\mu \propto \omega^{-\frac{1}{2}}$, as shown in eq. (1), at high frequencies information about only a layer near the surface will be obtained, but as ω is decreased, information from deeper within the sample will become available. It should be noted that this depth-profiling capability exists even for highly opaque samples where $\beta\mu >> 1$, although now information only about the thermal characteristics ($\rho C\mu$) will be available.

II.B. Piezoelectric Signals

The photoacoustic piezoelectric signal can also be derived in a similar manner, by noting that in the near-field, low-frequency case, the photoacoustic stress in a constrained fluid is

$$\begin{aligned} p &= B\alpha_t \bar{\Theta} \\ &= B\alpha_t \frac{I_o(1 - e^{-\beta \ell})}{\rho\, C\, \omega\, \ell}, \end{aligned} \tag{12}$$

where $\bar{\Theta}$ represents the average temperature rise of the entire sample.

At high frequencies, where we are dealing with the far-field case, we can use a variation of White's simple formula (1963).

$$p \simeq B\alpha_t \Theta_o \left(\frac{k^2}{k^2 + \zeta^2} \right), \tag{13}$$

where the surface temperature Θ_o will be given by eq. (5) above, $k = (\omega/c_o)$ is the acoustic wave-vector, with c_o being the sound velocity, and ζ is either $1/\mu$ when $(1/\mu) > \beta$ or $\zeta = \beta$ when $(1/\mu) < \beta$. The term $(k^2/(k^2+\zeta^2))$ can be regarded as giving the extent of coupling between the heat-absorbing region and the acoustic wave.

III. EXPERIMENTAL METHODOLOGY

As in other forms of spectroscopy, a photoacoustic spectrometer is composed of three main parts: a source of incoming radiation, the experimental chamber, and the data acquisition system. A typical photoacoustic spectrometer is shown in block diagram fashion in Fig. 1.

III.A. Radiation Sources

The most common and most versatile sources of optical radiation in the ultraviolet, visible, and infrared regions, are provided by the conventional light sources. These are the arc lamp for the ultra-visible, the incandescent lamp for the visible and near infrared, and the glow-bar for the mid-to far-infrared regions. All three light sources provide strong, broad-band optical radiation, and they must, therefore, be used in conjunction with suitable monochromators. Since the signal-to-noise ratio in photoacoustic spectroscopy increases linearly with the amount of light falling on the sample, one desires an intense light source and a high light throughout (i.e., low f number) monochromator. These light sources generally operate in a continuous mode, and thus a light chopper, usually electromechanical in nature, must be used.

Another source of optical radiation that can be used in photoacoustic spectroscopy of solids is the laser. A laser requires no monochromator and, if operated in a pulsed mode, would also require no chopper. In the visible wavelength region, dye lasers provide an intense highly monochromatic light readily tunable over a fairly large wavelength range.

Dye lasers can also be used with reasonable intensity in the ultraviolet region with the aid of frequency-doubling crystals. In the infrared there are currently no continuously tunable lasers that cover a wide spectral range although if the experiment can be

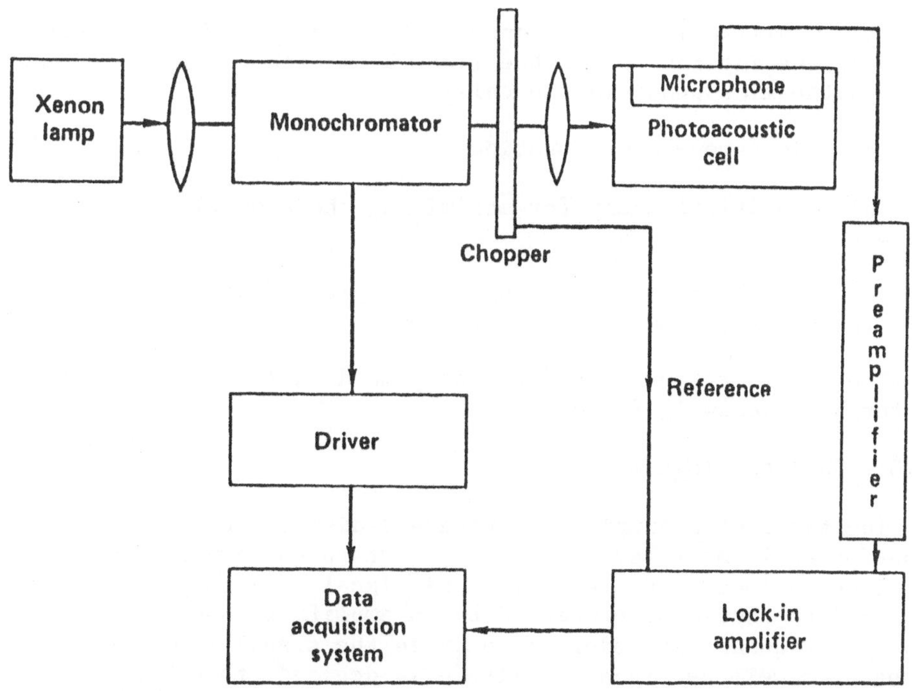

Fig. 1. A block diagram of a typical single-beam photoacoustic spectrometer using a gas-microphone cell. (From Rosencwaig, 1978).

performed over a narrow wavelength range, then a discrete infrared laser (e.g., the CO or CO_2 laser) or a tunable spin-flip Raman laser can be used to great advantage to provide intense, highly monochromatic radiation.

III.B. Experimental Chamber

The experimental chamber is the section containing the photoacoustic cell or cells, and all the required optics. The actual design of this chamber will vary depending on whether one is using a single-beam system employing only one photoacoustic cell, or a double-beam system containing two cells, with appropriate beam-splitting optics. The photoacoustic cells will generally incorporate a suitable microphone with its preamplifier. Both conventional condenser microphones with external biasing, and electret microphones with internal self-biasing provided from a charged electret foil, are good microphones to use.

Some criteria governing the actual design of the photoacoustic cell are:

(a) acoustic isolation from the outside world;

(b) minimization of extraneous photoacoustic signals arising from the interaction of the light beam with the walls, windows and the microphone in the cell;

(c) microphone configuration;

(d) acoustical means for maximizing the acoustic signal within the cell; and

(e) the requirements set by the samples to be studied and the type of experiments to be performed.

The above criteria are considered in detail in the review chapter by Rosencwaig (1978).

III.C. Data Acquisition

The tasks of acquiring, storing and displaying the data can be performed in many ways. However, certain basic procedures should be followed. For example, the signal from the microphone preamplifier should be processed by an amplifier tuned to the chopping frequency in order to maximize the signal-to-noise ratio. If phase as well as signal amplitude is desired, then a phase-sensitive lock-in amplifier should be used.

For a single-beam spectrometer, provision must generally be made to remove from the photoacoustic spectrum, any spectral structure resulting from the lamp, monochromator, and optics of the system. This normalization can be done conveniently by digitizing the analog signal from the tuned amplifier, and then performing a point-by-point normalization (i.e., division) with either a power meter reading or a previously recorded photoacoustic spectrum obtained with a black absorber. In a double-beam spectrometer, normalization can be performed in analog real-time fashion, by dividing the analog output from the tuned amplifier processing the sample signal, by the output derived from a reference signal. This reference output may be from a power meter or from a second photoacoustic cell.

With regard to the storage and display of the data, there are, of course, many possible schemes ranging from the relatively inexpensive chart recorder, to the sophisticated minicomputer.

III.D. Piezoelectric Detection

The gas microphone detection of photoacoustic signals has been found to be very good for many applications, particularly at low modulation frequencies. However, when one is dealing with a large sample or a sample with a small surface/volume ratio, the gas-microphone technique often proves inadequate.

Since photoacoustic detection is primarily the detection of the internal heat produced within a sample by the deexcitation of optical energy levels, it is possible to measure this internal heating by the stresses produced in a piezoelectric detector in intimate contact with the sample.

A piezoelectric transducer is about two orders of magnitude less sensitive than a microphone for a given pressure. However, this decreased sensitivity is often offset if one is dealing with large samples, or samples with low surface/volume ratios, or those in which the heat is generated within the sample. It should be noted that a piezoelectric detection of pressure signals in a gaseous medium would be very inefficient because of the large acoustic impedance mismatch between the gas and the solid transducer. However, a piezoelectric transducer is a suitable detector of thermally-generated pressure or acoustic fluctuations in condensed materials. Furthermore, piezoelectric detectors can operate at much higher frequencies than microphones.

For the study of solids, one can simply attach the piezoelectric transducer to the sample with wax or a suitable cement as depicted in Fig. 2. It is imperative that the coupling agent between the sample and the transducer have a good acoustic impedance match with both the sample and the transducer. At very high frequencies (MHz) it is possible to use a viscous fluid as the coupling agent, but at lower frequencies, where strong shear waves are generated or where strain measurements provide stronger signals, the coupling agent should be a hard solid. In the case of powders one can attach the powder to the transducer with a suitable cement. Care should be taken to prevent any of the incident light from striking the transducer, since this would create a spurious signal arising from the thermally-generated stresses within the transducer, and from the changes in the piezoelectric characteristics with temperature. Although the piezoelectric detector is fairly insensitive to air-borne noise because of the large acoustic impedance between a gas and solid, the sample and transducer should be isolated from vibrational noise.

For the study of liquids the sample is held within a sample chamber, e.g., a cylinder, that is in part or in whole comprised of a piezoelectric material. Windows can be bonded at the ends of the cylinder and suitable filling ports drilled through the walls of the tube.

Except for the preamplifier, which will be a charge or current preamplifier if a piezoelectric ceramic is used as a transducer, the electronic system for these piezoelectric PAS spectrometers are identical to those used in the gas-microphone spectrometers.

IV. DEEXCITATION STUDIES - GENERAL REMARKS

The photoacoustic effect measures the heat-producing deexcitation processes that occur in a system after it has been optically

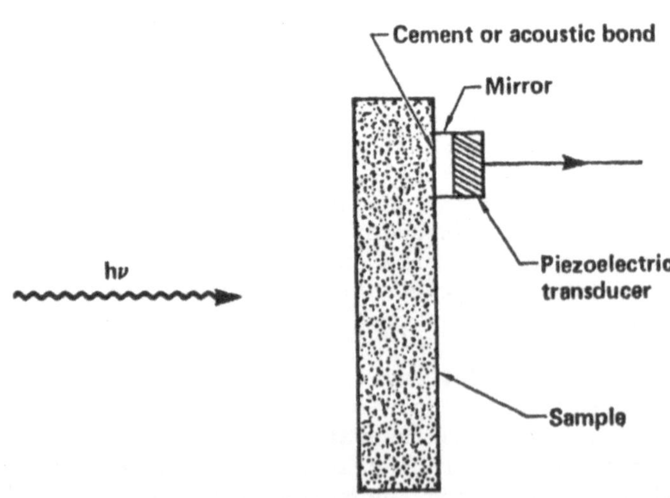

Fig. 2. A piezoelectric transducer mechanically bonded to a sample with wax or cement.

excited, or in more general terms, excited by any electromagnetic radiation. This selective sensitivity of the PAS technique to the heat-producing deexcitation channel can be used to great advantage in the study of fluorescent (or phosphorescent) materials, and in the study of photosensitive substances.

V. FLUORESCENCE STUDIES

When an optically excited energy level decays via fluorescence or phosphorescence, then little or no acoustic signal will be produced in the photoacoustic cell. This is illustrated in Fig. 3, where we consider the case of the fluorescent solid Ho_2O_3 (Rosencwaig, 1975a; 1975b). Several of the trivalent rare earth ions, such as Ho^{3+}, have strongly fluorescent energy levels; that is, levels that tend to deexcite through the emission of a photon, rather than through phonon or heat excitation. The upper PAS spectrum is of some Ho_2O_3 powder containing Co and F impurities. All of the lines present in this spectrum correspond to known Ho^{3+} energy levels, whose positions are designated by the bars below. The dots indicate which of these levels are normally fluorescent. In this material the fluorescence is highly quenched by the presence of the Co and F impurities, and thus both the fluorescent and nonfluorescent lines appear in the PAS spectrum. The lower spectrum is of pure Ho_2O_3. Here all of the fluorescent levels have a greatly diminished relative intensity, since these levels are now deexciting through the emission of a photon rather than through heating of the solid.

A more illustrative example of the potential of photoacoustic spectroscopy in such studies has been reported by Merkle and Powell (1977), who used photoacoustic spectroscopy to study the radiationless decay processes between the excited states of Eu^{2+} ions in KCl crystals.

Fig. 4(A) shows the absorption spectrum of KCl:Eu^{2+} at room temperature. The two strong, broad absorption bands are attributed to transitions from the lowest Stark component of the $^8S_{7/2}(4f^7)$ ground state to the e_g and t_{2g} components of the 4f 5d configuration with the former being at higher energy. The structure on these bands (which is more easily observed at low temperatures) is due to the electrostatic interaction between the d and f electrons and spin-orbit interaction of the latter. The splitting of the bands indicates the cubic crystal field strength for the d electron is about 10Dq = 12000 cm^{-1} whereas the strength of the electrostatic and spin-orbit effects is found from the fine structure to be on the order of $\sim$5000 cm^{-1}. These are dipole allowed transitions and the radiative decay times for the reciprocal emission transitions from these excited states can be calculated (Fowler and Dexter, 1965) to be on the order of 1 μs for both the t_{2g} and e_g excited states.

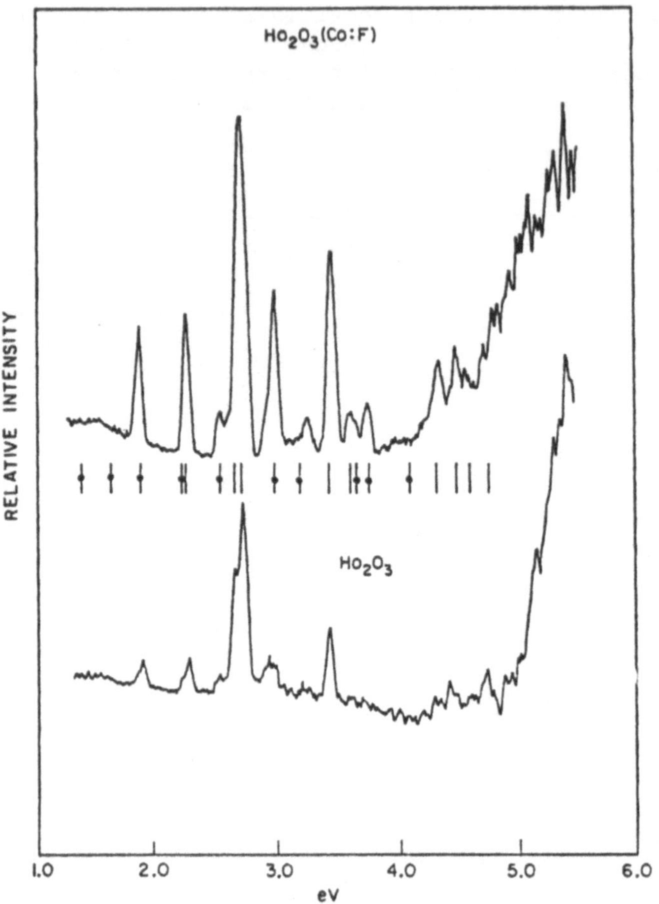

Fig. 3. Photoacoustic study of a fluorescent material. Cobalt and fluorine impurities quench the natural fluorescence of the Ho^{3+} ions in holmium oxide here studied by PAS. The upper spectrum is of doped, and the lower spectrum of undoped, holmium oxide. The fluorescent levels are marked with dots. (From Rosencwaig, 1975a; 1975b.)

The fluorescence spectrum at room temperature consists of a broad band peaked at approximately 23800 cm^{-1} which represents a Stokes shift of ~6000 cm^{-1} from the lowest energy absorption band. The fluorescence decay time at 12 K is found to be about 1.3 μs and the decay pattern is observed to be purely exponential with no measurable rise time. These results are essentially independent of the wavelength of excitation. The measured fluorescence decay time at low temperatures is close to the predicted radiative decay time.

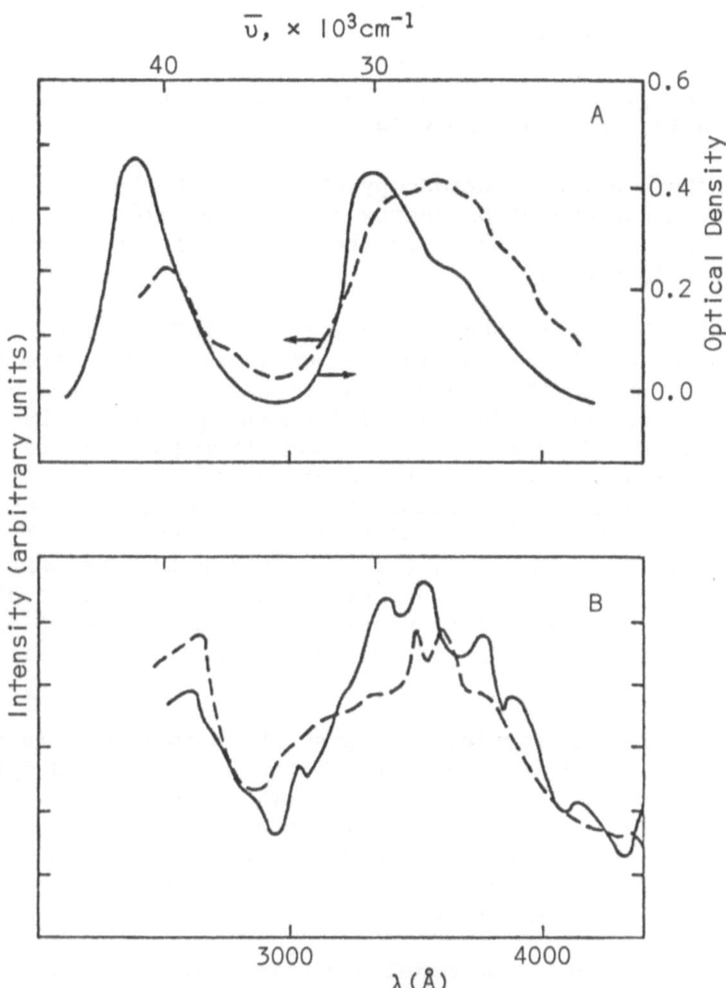

Fig. 4. (A) Optical absorption (solid line) and excitation spectra monitored at 4350 Å (dashed line) for KCl:Eu^{2+} at 300 K. (B) Photoacoustic spectra for two different phase shifts; solid line is 0° phase, dashed line is 45° phase. (From Merkle and Powell, 1977.)

Fig. 4(A) shows the excitation spectrum at room temperature. Both absorption bands appear but their relative intensities are quite different. Excitation in the high energy band leads to a smaller amount of radiative emission than excitation in the low energy band.

The obvious conclusion to be drawn from these optical spectroscopy results is that the e_g level has two different types of decay channels. The dominant one is total radiationless relaxation to the ground state and the secondary one is a multiphonon transition to the t_{2g} level which fluoresces to the ground state. Both of these processes must take place on a time scale much faster than the ~1 μs predicted radiative decay time since no radiative emission is detected from this level.

Photoacoustic spectroscopy provides a method for directly monitoring the amount of energy dissipated through radiationless transitions. Fig. 4(B) shows the results of PAS measurements made on this system.

Both absorption bands appear in the PAS results. The phase angle at which the signal is maximum is related to the lifetime of the state through the expression $\tau = \tan \phi/2\pi\nu_c$ where ν_c is the chopping frequency. The signal from the low energy band is maximum for approximately zero shift in phase from that of the exciting light which indicates that the decay time of the level is much less than 100 μs. This is consistent with the measured fluorescence decay time. The peak intensity of the PAS signal in the high energy band occurs at a phase shift of ∿35° which implies that the lifetime of this state is of the order of milliseconds. It is obvious that this cannot be the radiationless decay time of the e_g level since the radiative decay time is much faster and no radiation is observed from this level.

The model energy level diagram shown in Fig. 5 is proposed to explain the apparent discrepancy between the PAS and optical spectroscopy results. For simplicity, the electrostatic and spin-orbit splittings of the $4f^6 5d$ levels are not shown and only one of the manifold of excited $4f^7$ levels is pictured. The photoacoustic signal at phase angle ϕ is simply the sum over all the relaxation transitions which generate heat.

For the transition model shown in Fig. 5, Merkle and Powell estimate that the t_{2g} PAS signal arises from two sources; those electrons undergoing excited relaxation within the t_{2g} band and those undergoing radiationless and vibronic relaxation to the ground state. The e_g PAS signal is somewhat more complicated. It has a contribution due to all of the electrons relaxing within the e_g band. Another contribution comes from those electrons that relax to the t_{2g} band and subsequently to the ground state. A third contribution arises from those electrons that relax from the e_g band to the $4f^7$ energy level and hence relax radiationlessly to the ground state.

There are thus two excited ground-state radiationless transition times. The one that goes from the t_{2g} to ground (or

$e_g \rightarrow t_{2g} \rightarrow$ ground) is much faster than 0.1 μsec since no radiative emission is detected from this level. The PAS signal that arises from these transitions is at 0° phase shift. The other main relaxation channel is the $e_g \rightarrow 4f^7 \rightarrow$ ground state. Absorption, fluorescence and excitation spectra along with lifetime data can be used to predict the relative PAS intensity ratios $I_{e_g}/I_{t_{2g}}$ for different values of the phase angle. This analysis gives a value of the phase angle for the $4f^7$ ground transition of about 66° in reasonable agreement with experiment. This value implies a $4f^7$ lifetime of 3.6 milliseconds.

The transition model predicted by Merkle and Powell, although rather qualitative, provides an explanation for the apparent discrepancies between optical and photoacoustic data. It thus appears that the dynamics of excited state relaxation in $KCl:Eu^{2+}$ are more complicated than previously thought, and that photoacoustic techniques provide a method for elucidating some of

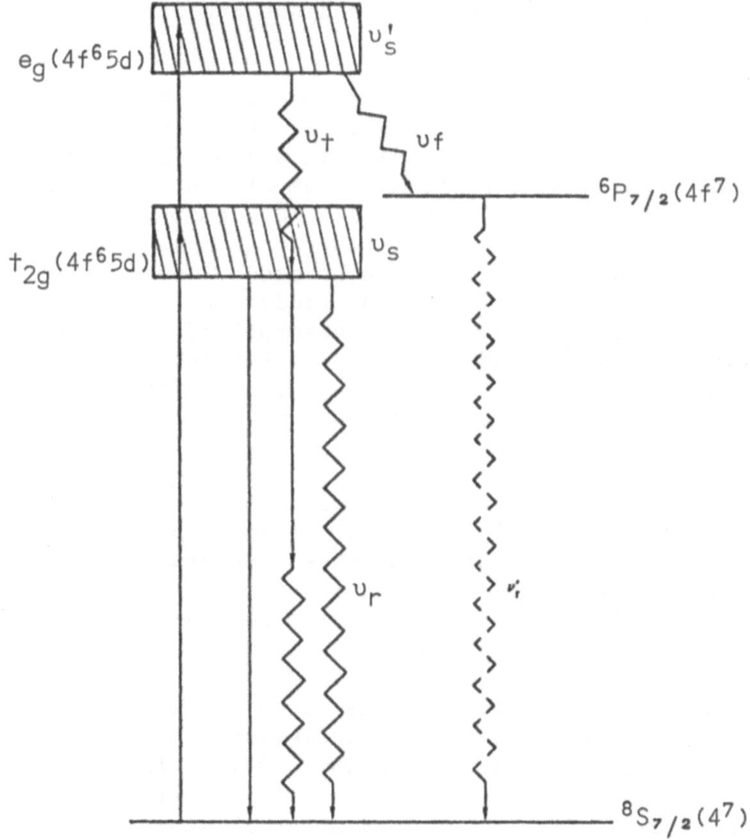

Fig. 5. Proposed model for excited state relaxation of Eu^{2+} in KCl. (From Merkle and Powell, 1977.)

the characteristics of the relaxation processes which cannot be observed by conventional optical means.

In another paper, Peterson and Powell (1976) have reported on a series of PAS studies on various crystals containing Cr^{3+} ions. These crystals included $BaTiO_3$, $SrTiO_3$, MgO and Al_2O_3. Peterson and Powell compared the radiative and non-radiative modes of decay and found, for example, that the dominant decay mode for Cr^{3+} in MgO is quite different from that in Al_2O_3. They attributed this difference in deexcitation modes to the fact that more than half of the Cr^{3+} ions in MgO are in non-cubic sites because of the local defects needed for charge compensation.

VI. QUANTUM EFFICIENCIES

Since a photoacoustic signal is very sensitive to the presence of a radiative mode of deexcitation, measuring, as it were, the nonradiative complement to the fluorescent or phosphorescent signal, it was soon realized that the PAS technique could be used to measure quantum efficiencies.

The precise determination of absolute fluorescence quantum yields by conventional luminescence means has proven to be very difficult (Calvert and Pitts, 1966). In a luminescence measurement, the number of quanta absorbed from a beam of monochromatic light has to be compared with the number of quanta emitted in the polychromatic fluorescent light, whose distribution in space may be geometrically complicated. This can be accomplished by determining a defined fraction of the fluorescent radiation. To that end various corrections (for geometry, reabsorption, reemission, polarization, refractive index) must be taken into account. This is tedious and results in low accuracy; errors exceeding 5-10% are common. Details of this method are reviewed by Demas and Crosby (1971).

Another technique involves the measurement of fluorescence lifetime (Dianov, et al., 1976). Again this method suffers from several experimental difficulties, since a separate measurement of the nonradiative contribution to the lifetime of the state must be made, or alternatively the radiative lifetime must be calculated (Riseberg and Weber, 1976).

There is another method which has been somewhat neglected until recently; to determine the nonradiative part of the absorbed energy by calorimetry (Demas and Crosby, 1971). The temperature rise of an irradiated luminescent sample is compared to the temperature rise of a nonluminescent material showing the same absorption. The main obstacle to this method is the relative insensitivity of common temperature sensors so that strongly absorbing samples usually have to be used.

Photoacoustic spectroscopy is another means for performing the calorimetric determination of radiative quantum yield, and is generally more sensitive than a conventional calorimetric method.

Lahmann and Ludewig (1977) performed such a measurement on rhodamine 6G in water with a liquid cell containing a piezo-electric type of microphone. In this experiment, the PAS signal from the rhodamine 6G solution was compared with that from a non-fluorescent potassium dichromate solution.

For the non-fluorescent sample, the PAS signal is given by

$$q_1 = Sa_1P_1 \quad , \tag{14}$$

where S is the system sensitivity (V/W) and a_1P_1 is the absorbed power with P_1 the incident power. For the fluorescent solution,

$$q_2 = Sa_2P_2\left\{1 - \eta + \eta(\nu_e - \nu_f)/\nu_e\right\} , \tag{15}$$

where η is the quantum efficiency, ν_e is the optical frequency of the exciting radiation, and ν_f the mean frequency of the fluorescent radiation. The first part of eq. (15) denotes the absorption of nonfluorescent molecules, while the second term gives the heat produced by the fluorescent molecules.

Dividing eq. (14) by eq. (15), and rearranging, yields

$$\eta = \frac{\nu_e}{\nu_f}\left\{1 - \frac{a_1q_2P_1}{a_2q_1P_2}\right\} \quad . \tag{16}$$

A similar experiment was performed by Starobogatov (1977), who used a pulsed ruby laser. The samples studied were also dyes, and the PAS signals from these dyes were then compared with appropriate non-fluorescent solutions. In both the Starobogatov and the Lahmann and Ludewig experiments, the measured quantun yields agreed with those obtained by conventional luminescence experiments, and the accuracy appeared to be better.

Adams et al (1977), employed a gas-microphone system to obtain the quantum efficiency of quinine bisulfate in aqueous solution. Rather than use another, non-fluorescent solution as a standard, they measured the quenching effect of the addition of chloride ions to the sample solution. Again an accurate determination of η was obtained. The first experiment of this nature on a solid sample was reported by Murphy and Aamodt (1977). Here the PAS signal of Cr^{3+} in Al_2O_3 was obtained, and in particular a comparison of the $^4T_1$ and $^4T_2$ band intensities was made as a function of Cr^{3+} concentration as shown in Fig. 6.

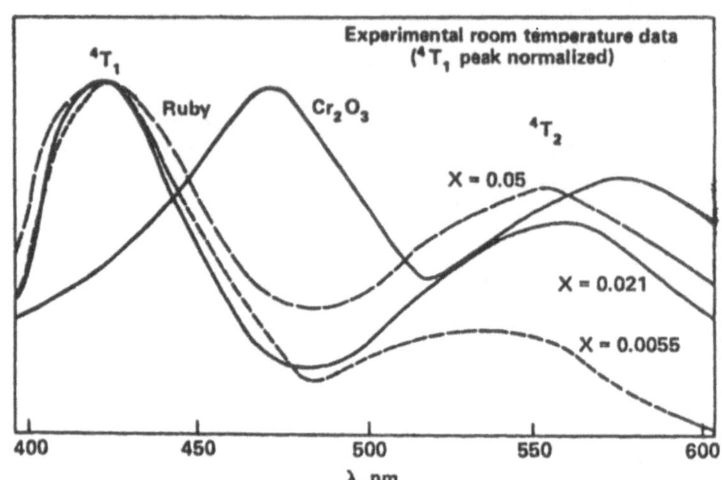

Fig. 6. Room-temperature PAS spectra for three Cr^{3+} concentrations in ruby and for pure Cr_2O_3. For ease of comparison, the $^4T_1$ peak has been normalized to the Cr_2O_3 spectrum. (From Aamodt and Murphy, 1977.)

The level diagram for Cr^{3+} appropriate to ruby is shown in Fig. 7. At low concentrations, light-pumped levels 3 ($^4T_1$) and 4 ($^4T_2$) relax nonradiatively to the single-ion metastable level 2, which relaxes radiatively to ground. With increasing Cr ion concentration, some direct nonradiative relaxation occurs from level 4 to ground, and energy is coupled into level 1 as chromium pairs and higher complexes appear. Level 1 also initially relaxes radiatively, but at higher concentrations this emission is quenched through competition with nonradiative relaxation paths.

Using rate equations for the different transitions allowed in Fig. 7, Murphy and Aamodt were able to relate the ratio of the $^4T_1$ and $^4T_2$ PAS bands to the quantum efficiency η, and obtain data on the quenching effects of increased Cr^{3+} concentration.

Quimby and Yen (1978) also used a gas-microphone method to obtain the quantum efficiency for Nd^{3+} ions in an ED-2 glass matrix. They measured the PAS signal and the fluorescence lifetime for a number of samples as a function of Nd^{3+} concentrations and from these measurements obtained η for Nd^{3+} ions in the glass sample.

The above experiments indicate the type of investigations that can be performed on fluorescent and phosphorescent materials with the photoacoustic technique. A combination of conventional fluo-

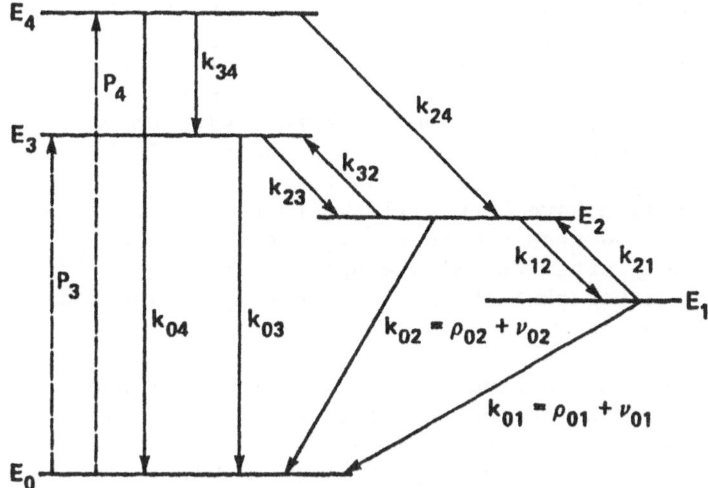

Fig. 7. Five-level energy-level diagram appropriate to ruby. E_i, p_i and k_{ij} indicate energies, pumping rate, and total transition rates respectively. $k_{ij} = \rho_{ij} + \nu_{ij}$, where ρ_{ij} and ν_{ij} are the radiative and nonradiative components of k_{ij} respectively. (From Aamodt and Murphy, 1977.)

rescence spectroscopy and photoacoustic spectroscopy can provide data about both the radiative and nonradiative deexcitation processes within these solids. That is, the complete deexcitation process within these compounds can now be readily studied for the first time. By performing both fluorescence and photoacoustic spectroscopy as a function of temperature and compound composition, one can determine in a straightforward manner, how these two variables affect the efficiencies and rates for the two deexcitation processes. Furthermore, since photoacoustics gives phase as well as amplitude information, one can study exciton processes (random walk, energy level lifetimes, etc.) in these materials as a function of temperature and dopant concentration.

VII. PHOTOCHEMISTRY

Another channel of deexcitation for absorbed light energy in some compounds is through photochemistry. Photoacoustics offers a unique tool for the study of photochemical process in solids. An illustration of this, shown in Fig. 8, gives the results of a PAS experiment on the photosensitive material Cooper blue (2,2-dimethyl-4-phenyl-6-p-nitrophenyl-1, 3-diazabicyclo (3.1.0) hex-3-ene) (Rosencwaig, 1975a). This compound is colorless in the dark but turns a strong blue when exposed to light of short wavelength.

The bottom spectrum of Fig. 8 is the PAS spectrum of dark-adapted Cooper blue. There is substantial UV absorption but little visible absorption. The middle spectrum was obtained immediately after the first, and is quite different, showing two strong absorption bands in the visible. These are the bands that give Cooper blue its blue color. These bands arise from a photochemical change in Cooper blue wherein some of the photons absorbed in the short wavelength region have been utilized to break a ring in the Cooper blue molecule and thus create a new compound. The upper spectrum, run immediately after the middle spectrum shows yet further changes, reflecting further photochemical and even photoinduced thermochemical processes.

Not only can one readily see the effects of photochemistry by means of photoacoustic spectroscopy, but one can also establish the activation spectrum for the photochemical process directly, by simply comparing the PAS spectrum with a conventional absorption spectrum. Information about the activation spectrum of photosensitive materials is at present quite difficult to obtain by other means. In addition, one can obtain, from the phase measurements of the photoacoustic signal, data about photochemical reaction rates, and even distinguish between true photochemical events and photoinduced thermochemical events. Photoacoustic studies on photosensitive materials will not only provide valuable basic information about the physical and chemical processes in these materials, but can also be of great benefit in the understanding of technologically important compounds such as photoresists, and in the study of photoinduced physical and chemical changes in polymers, plastics and pigments.

VIII. PHOTOSYNTHESIS

In biology, one of the most important manifestations of photochemistry is the process of photosynthesis both in green plants and in certain bacterial organisms. As in the case of radiative deexcitation, photochemical processes such as photosynthesis compete with the photoacoustic process. The normalized photoacoustic signal q can be written as

$$q = a\left\{1 - \sum (\eta'_i \Delta E_i / Nh\nu) - (\eta\nu_f/\nu_e)\right\} \tag{17}$$

Here a is the fraction of light absorbed by the sample; η'_i is the quantum yield for photochemical reactions; ΔE_i is the molar internal energy change per product formation in this reaction, η is the quantum yield for luminescence; ν_f and ν_e are the frequencies of radiated and absorbed light respectively, N is Avogadro's number and h is Planck's constant.

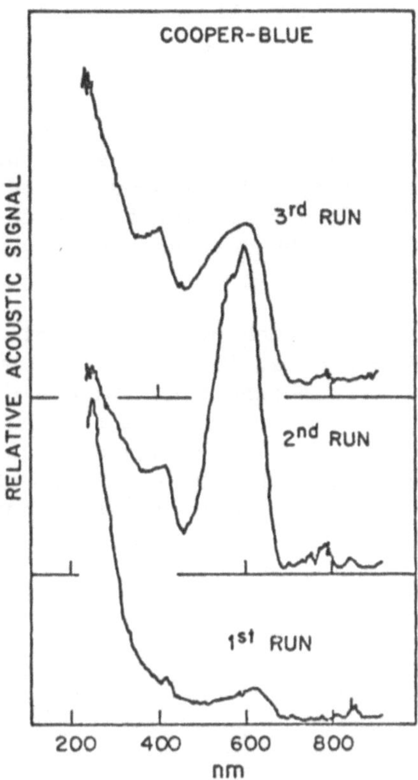

Fig. 8. Photoacoustic study of a photosensitive material. The Cooper-blue was dark-adapted before the lower spectrum was run. The middle spectrum was run afterward and then the upper spectrum, with only a few minutes separating the different runs. (From Rosencwaig, 1975a.)

If we neglect the role of radiative deexcitation, it is clear that any photochemical process will result in a decrease of the PAS signal with increasing quantum yield for photochemistry. Furthermore, by varying the modulation frequency and analyzing the phase of the PAS signal, information can be obtained in kinetic parameters and on the energy content of intermediates.

Cahen et al. (1978b) have studied photosynthesis in lettuce chloroplast membranes by means of photoacoustics. Since they did not use a piezoelectric cell for the chloroplast suspension, they found that the best results were obtained when the chloroplast suspension was abosrbed on cotton wool. Fig. 9 shows the PAS spectra obtained for photosynthetically active membranes, and for DCMU-poisoned (DCMU = 3-(3,4-dichlorophenyl) -1,

1-dimetheylurea, an electron transport inhibitor), chloroplast membranes. The spectra are normalized at 440 nm where little photosynthetic activity is to be expected. At the 680 nm chlorophyl band, the DCMU - poisoned chloroplasts give a signal that is 10% stronger than that of the active chloroplasts.

As the modulation frequency is decreased, the difference between the spectra of normal and poisoned membranes increases. Cahen et al. explained this observation by hypothesizing that the inhibitory activity of the DCMU does not reach completion at high modulation frequencies.

Another experiment performed by Cahen et al. (1978a) investigated the photosynthetic process in the purple membrane of Halobacterium halobium. This membrane contains a single protein, bacteriorhodopsin, covalently bonded to a retinal molecule. Absorption of light by the retinal brings about a cyclic photochemical process which drives the translocation of proteins from one side of the membrane to the other through conformational changes of the protein.

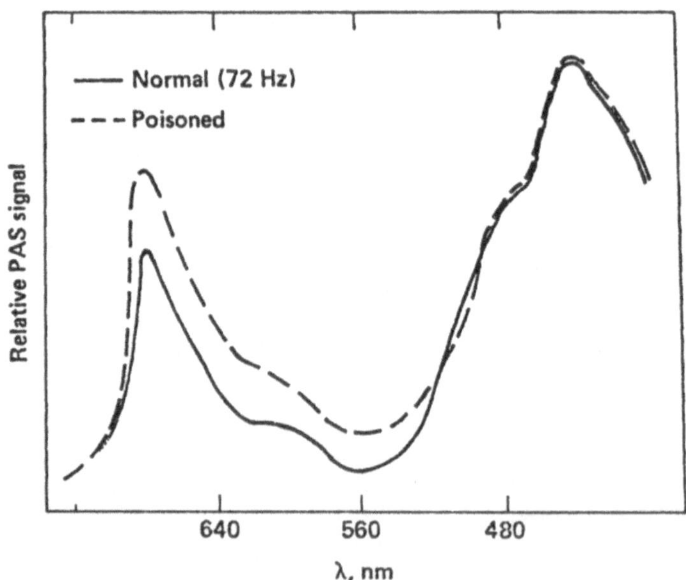

Fig. 9. Photoacoustic spectra of photosynthetically active (solid line) and of photosynthetically inactive (dashed line) lettuce chloroplasts. The photosynthetically inactive sample was immersed in a DCMU-saturated methanol bath and allowed to dry. DCMU is 3-(3,4-dichlorophenyl)-1, 1-dimethylurea, an electron transport inhibitor. (From Cahen, et al., 1978b.)

In Fig. 10 is shown the PAS spectra of dried purple membrane fragments with and without strong side illumination. The side illumination causes an accumulation of the photointermediates absorbing at 415 nm and 660 nm, and a decrease of the population absorbing at 565 nm.

In lyophillized purple membranes, the absorbed light drives the photocycle only, while in whole cells, part of the absorbed energy is stored as ATP and ion gradients. From eq. (17), the photoacoustic signal normalized to the absorbed energy is given by

$$q/a = 1 - \frac{\eta' \Delta E \lambda}{\text{const.}} \quad . \tag{18}$$

If $\eta'\Delta E = 0$, as would be the case if there was no photochemical reaction ($\eta' = 0$), or if there were a cyclic reaction with $\Delta E = 0$ then q/a will be independent of λ. However, if some of the absorbed energy is stored in the products sensed by PAS, a valley-shaped

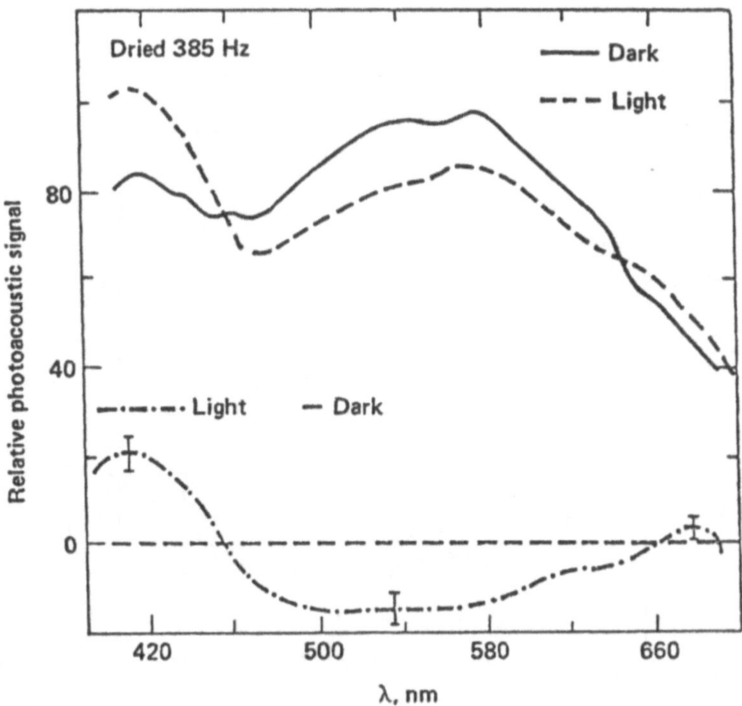

Fig. 10. PAS of dried purple membrane fragments of Halobacterium halobium, with and without $200 mW/cm^2$ continuous side illumination. The bottom spectrum is the difference spectrum (light minus dark). (From Cahen, et al., 1978a.)

curve would be obtained, with the lowest point occurring in the region of highest energy storage.

The results of Cahen et al. appear to demonstrate this situation as seen in Fig. 11. In the freeze-dried purple membrane fragments, where presumably no energy is stored (on the time scale of the experiment) in the steady state, q/a is constant within experimental error. However, in intact cells q/a passes through a minimum in the 540-620 nm region.

These results indicate that photoacoustics appears to be a sensitive tool for the study of photosynthesis. Furthermore, PAS can provide information on the energetics of the process, by clearly distinguishing between samples in which part of the absorbed energy is stored or used to do work, and those samples in which all of the absorbed energy is dissipated in the photocycle as heat.

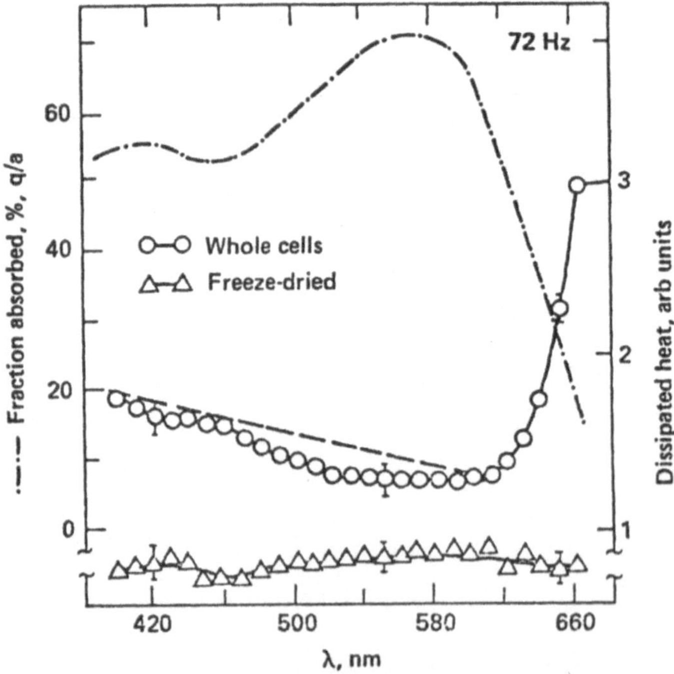

Fig. 11. Absorption and PAS spectra of whole cells, and PAS spectrum of lyophilized purple membrane fragments. (From Cahen, et al., 1978a.)

IX. NON-SPECTROSCOPIC STUDIES OF PHOTOCHEMISTRY

Gray and Bard (1978) have used photoacoustics to study photochemical reactions where the PAS signals were, in the main, attributable to gas evolution or consumption. They studied the oxygen consumption in the photo-oxidation of rubrene, wnere singlet oxygen is formed which then attacks the rubrene.

Fig. 12 shows the diffuse reflectance and the PAS spectrum of 5 wt.% rubrene on MgO. If the rubrene from this sample is redissolved in a small quantity of benzene which is allowed to evaporate in air to apparent dryness, enough benzene is retained in the sample to solvate the rubrene and, on exposure to light of wavelengths shorter than 580 nm in the presence of oxygen, the endoperoxidation reaction proceeds. The first scan from long wavelengths in the PAS spectrum of such a sample (Fig. 12c) shows a large negative-going transient due to oxygen uptake from the gas boundary layer, and this transient, is larger than the conventional PAS signal at this wavelength. At shorter wavelengths, the absorbance increases, and the negative O_2-uptake signal is overtaken by the thermal signal. Scans D-F were recorded immediately after

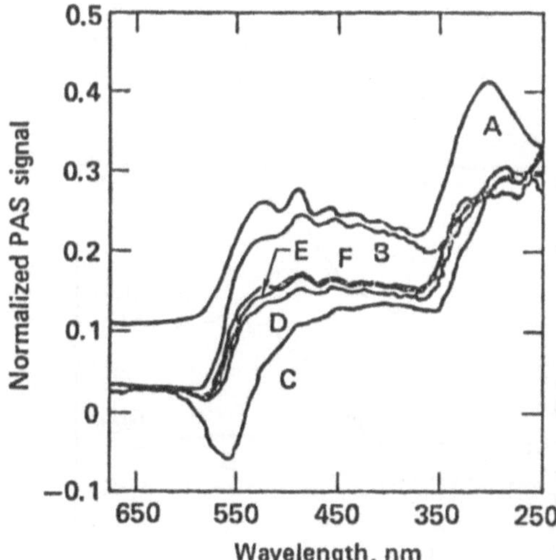

Fig. 12. PAS of rubrene supported on MgO powder. (A) reflectance spectrum of 5 wt. % rubrene in MgO, dried in Rota-vap (ordinate in arbitrary units). (B) PAS spectrum of sample in (A). (C) PAS spectrum of sample in (B) solvated with benzene, first scan. (D)-(F) are successive scans following (C). (From Gray and Bard, 1978.)

scan C. Note that the intial negative-going signal is smaller on the second scan D and has disappeared by scan F. Note also the shoulder that grows at 300 nm, the absorbance peak of the photoproduct. The final scan F looks much like the original scan except for a diminished overall intensity in the rubrene band and the presence of the photoproduct band at 300 nm.

An example of gas evolution was obtained in a study of heterogeneous photocatalytic oxidation of acetic acid to methane and CO_2 at a platinized TiO_2 catalyst. In Fig. 13, curve A shows the PAS spectrum of the Pt/TiO_2 catalyst in dry powdered form. When the sample is wet with benzonitrile, which does not undergo photodecomposition, the signal level drops due to the presence of the liquid surface layer (curve B). However, when the catalyst is wet with acetic acid, the signal level in the region of TiO_2 absorption is enhanced considerably as a result of the release of gas from the sample.

These examples demonstrate the utility of PAS in the study of photochemical reactions involving gas evolution and consumption, with a sensitivity of 10^{-13}–10^{-11} moles of gas evolved or consumed per second.

X. PHOTOCONDUCTIVITY

In materials where photovoltaic processes can occur, another deexcitation mode competes with the photoacoustic effect.

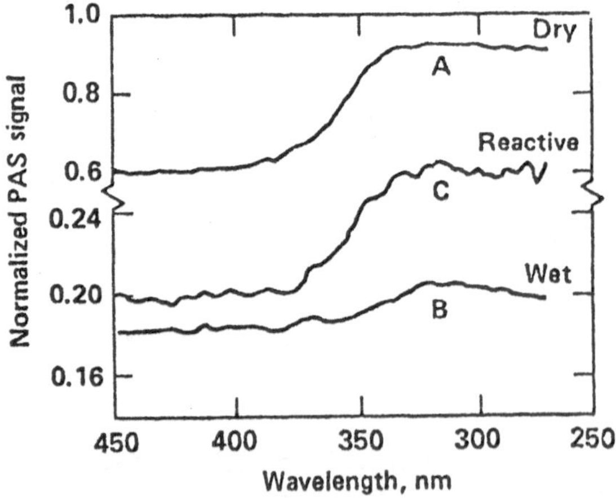

Fig. 13. Effect of gas evolution on the PAS signals for platinized doped TiO_2 (anatase) in absence and presence of acetic acid. (A) dry TiO_2 powder. (B) TiO_2 powder wet with benzonitrile. (C) sample in (B) with 10 μL of acetic acid added. (From Gray and Bard, 1978.)

(Cahen, 1978). Eq. (17) above can be expanded to include this possibility.

$$q = a \left\{ 1 - \sum (\eta_i' \Delta E_i / Nh\nu) - (\eta \nu_f / \nu_e) - \gamma' \right\}, \tag{19}$$

where γ' is the energy-conversion efficiency for the photovoltaic process.

In a photovoltaic device, the efficiency γ' is a function of the electrical load on the device. Thus if only the photovoltaic process is present as for instance in silicon, eq. (19) can be written for a specific resistance load R as

$$q(R) = a \left\{ 1 - \gamma'(R) \right\}, \tag{20}$$

where $\gamma'(R)$ can be considered to be the photovoltaic loss. From eq. (20), the photovoltaic conversion efficiency is given by

$$\gamma'(R) = \left\{ a - q(R) \right\} / a. \tag{21}$$

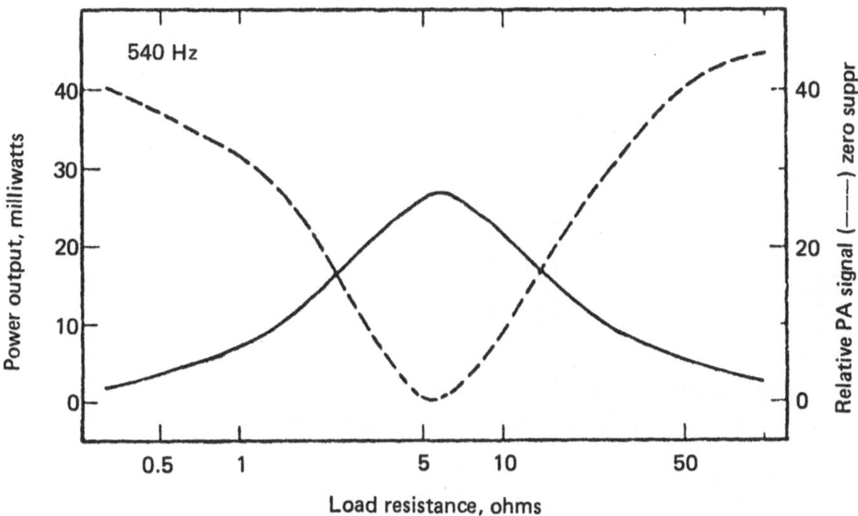

Fig. 14. Electrical output power (solid line) and zero-suppressed PAS signal (dashed line) as function of the resistance load on Si photovoltaic cell. (From Cahen, 1978.)

To solve eq. (21) for $\gamma'(R)$ requires a knowledge of a, the energy absorbed by the sample, a sometimes difficult task. A photovoltaic device will, however, have zero energy-conversion efficiency at open-circuit (OC) conditions. Thus, q(oc) = a, and

$$\gamma'(R) = \left\{ q(oc) - q(R) \right\} / q(oc). \tag{22}$$

Fig. 14 shows Cahen's results on a Si solar cell, where both the PAS signal and the measured electrical power are shown as a function of the load resistance. As expected from eq. (22), the PAS signal is at a minimum when the output electrical power is at a maximum. From this a maximum efficiency of 17.5% is obtained for the wavelengths used. Finally, Cahen notes that the photovoltaic efficiency can be readily measured as a function of wavelength as well as load resistance with photoacoustic spectroscopy.

A nonspectroscopic experiment on photoconductivity has been reported by Ghizoni, et al., (1978). In this experiment, the periodic heating of a semiconductor due the electron transport processes in an electric field is studied. Under a DC electric field, the free carriers of a semiconductor, gaining energy from the field, will ultimately establish a steady state where the energy gained from the field equals the energy lost to the lattice,

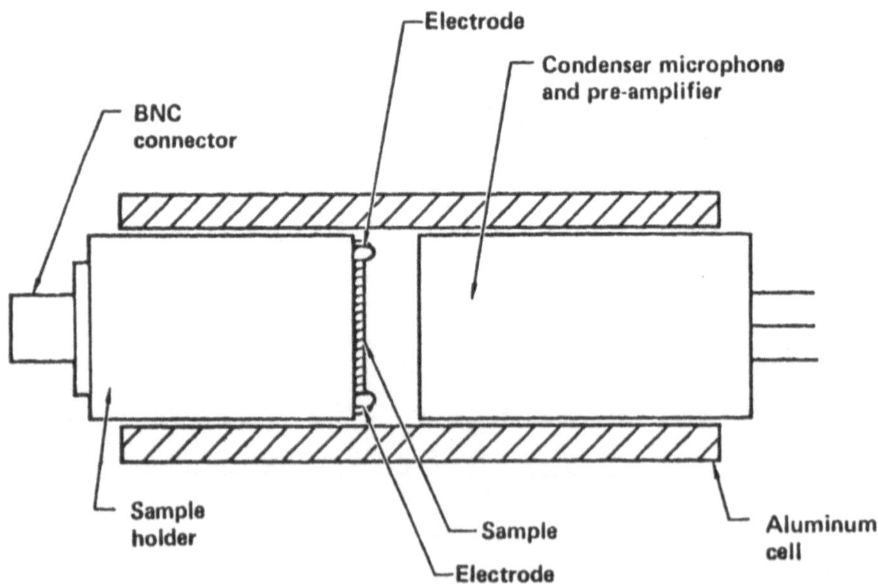

Fig. 15. Acoustic cell used for investigating the transport properties in semiconductors. (From Ghizoni, et al., 1978.)

via electron-phonon interactions. Hence, by pulsing a DC voltage in a semiconductor, mounted in a PAS cell as shown in Fig. 15, this periodic heating can be detected by the microphone.

In Fig. 16 is shown the acoustic signal as a function of pulse amplitude, V_p, for two different sample thicknesses and different values of pulse duration τ_p. In Fig. 17 is shown the variation of the acoustic signal as a function of pulse duration. These data indicate that at low values of the electric field, $E = V_p/d$, the acoustic signal is simply given by resistive Joule heating, that is $q \sim E^2\tau_p$. However, at high fields, the signal increases exponentially, such that $q \sim e^{E\tau_p}$.

Ghizoni et al., explained these results on the basis that at sufficiently high electric fields, phonon-stimulated emission dominates, that is, the phonon production rate due to electron-phonon collisions, becomes larger than the inverse of the phonon

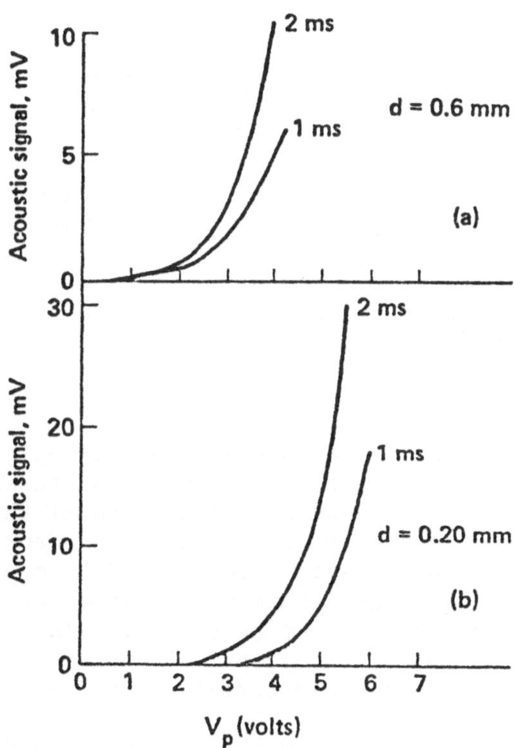

Fig. 16. Acoustic signal versus the applied peak voltage (V_p) for various pulse durations (τ_p). (a) Sample thickness d = 0.6 mm. (b) d = 0.2 mm. (From Ghizoni, et al., 1978.)

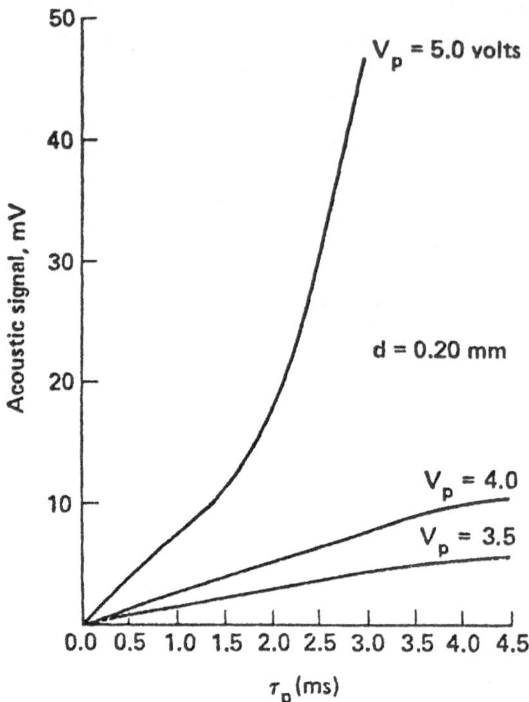

Fig. 17. Acoustic signal versus pulse duration for different values of the amplitude V_p. (From Ghizoni, et al., 1978.)

relaxation time. Under such conditions, there is phonon gain in the medium, and the photoacoustic signal increases exponentially.

XI. CONCLUSIONS

Photoacoustics and photoacoustic spectroscopy are still in their formative stages, yet their potential as research and analytical tools appears almost boundless. Until the development of these new techniques, many materials, both natural and synthetic, could not be readily investigated by conventional optical methodologies, since these materials occur in the form of powders, or amorphous solids, or as smears, gels, oils, suspensions, and so on. With photoacoustics, optical absorption data on virtually any solid or liquid material can now be obtained.

In addition, since photoacoustics provides a direct measure of the nonradiative deexcitation of optical energy, this new technique can be used to study deexcitation phenomena in any medium. As we showed in this Chapter, photoacoustics can be used to investigate radiative deexcitation, photochemical deexcitation, photovoltaic deexcitation, as well as the more ubiquitous thermal deexcitation of optical energy levels. The next few years promise to be an exciting period of growth for the science of photo-acoustics.

REFERENCES

Adams, M. J., Highfield, J. G. and Kirkbright, G. F., 1977, Anal. Chem., 49, 1850.

Cahen, D., 1978, Appl. Phys. Lett. 33, 810.

Cahen, D., Garty, H. and Caplan, S. R., 1978a, FEBS Lett. 91, 131.

Cahen, D., Malkin, S. and Lerner, E. I., 1978b, FEBS Lett. 91, 339.

Calvert, J. G. and Pitts, J. N., Jr., 1966, Photochemistry, Wiley, New York.

Demas, J. N. and Crosby, G. A., 1971, J. Phys. Chem. 75, 991.

Dianov, E. M., Karasik, A. Ya., Neustruev, V. B., Prokhorov, A. M. and Shcherbakov, I. A., 1976, Sov. Phys. Dokl. 20, 622.

Fowler, W. B. and Dexter, D. L., 1965, J. Chem. Phys. 43, 1968.

Ghizoni, C. C., Sigueira, M. A. A., Vargas, H. and Miranda, L. C. M., 1978, Appl. Phys. Lett. 32, 544.

Gray, R. C. and Bard, A. J., 1978, Anal. Chem. 50, 1262.

Lahmann, W. and Ludewig, H. J., 1977, Chem. Phys. Lett. 45, 177.

Merkle, L. D. and Powell, R. C., 1977, Chem. Phys. Lett. 46, 303.

Murphy, J. C. and Aamodt, L. C., 1977, J. Appl. Phys. 48, 3502.

Peterson, R. G. and Powell, R. C., 1978, Chem. Phys. Lett. 53, 366.

Quimby, R. S. and Yen, W. M., 1978, Opt. Lett. 3, 181.

Riseberg, L. A. and Weber, M. J., 1976, in Progress in Optics, Vol. XV (E. Wolf, ed.), North-Holland, New York.

Rosencwaig, A., 1973, Opt. Commun. 7, 305.

Rosencwaig, A., 1975a, Anal. Chem. 47, 592A.

Rosencwaig, A., 1975b, Phys. Today 28 (9), 23.

Rosencwaig, A., 1978, in Advances in Electronics and Electron Physics, Vol. 46 (L. Marton, ed.), Academic Press, New York.

Rosencwaig, A. and Gersho, A., 1976, J. Appl. Phys. 47, 64.

Starobagatov, I. O., 1977, Opt. Spect. 42, 172.

White, R. M., 1963, J. Appl. Phys. 34, 3559.

RADIATIONLESS DECAY OF EXCITONS

J. Singh

Research School of Chemistry, Australian National University, Box, 4, P.O. Canberra A.C.T. 2600, Australia

ABSTRACT

Radiationless decay of excitons plays a very important role in the generation of charge carriers in semiconductors and insulators, and consequently in the photoconducting properties of such materials. Besides the generation of charge carriers, the nonradiative migration of the exciton energy from one region of the crystal to the other can cause a decrease in the photoluminescence of the crystal when the excitation energy of the host material is imparted to impurities or other charge carriers and excitons.

We shall, first of all, review briefly the general theory of excitons in order to form a common background. For example, we shall discuss the following points. What are excitons? What is the important perturbation responsible for creating exciton states in crystals? We shall then consider the various possibilities and probabilities of the nonradiative decay of excitons namely, (a) exciton-exciton collisions (1,2), (b) exciton-impurity (donor or acceptor) interactions, (c) exciton-charge carrier interactions (3) etc. These processes will be considered in both organic and inorganic materials.

I. EXCITONS

Excited but nonconducting electronic states created by incident photons in single crystals of semiconductors and insulators are called excitons. In such crystals the conduction band is usually unoccupied and separated from the fully occupied valence band by a large energy gap. The excitation of an electron from the valence band can, therefore, be considered as a simultaneous generation of an electron in the conduction band and a positively charged vacancy (hole) in the valence band. The attractive interaction between the excited electron and the hole, thus created, builds up a bound neutral excited state which carries a net linear momentum and moves throughout the crystal.

There are two representations for excitons; i) Wannier or the large radii orbital excitons and ii) Frenkel or the molecular excitons. The former describes excitons in inorganic solids in which the one electron wavefunctions are best represented by Bloch type functions, whereas the latter is applicable to the study of excitons in organic crystals where one electron wavefunctions are written in terms of the molecular electronic functions. The kinetic energy of Wannier excitons is contributed by the individual kinetic energies of the electron and hole whereas the motion of Frenkel excitons originates from the translational symmetry of the crystal where any one of the atoms/molecules stands an equal chance of being excited by the incident photon.

The microscopic view of the motion of a Wannier exciton (in inorganic materials) is that the excited electron in the conduction band and the hole in the valence band remain in their respective bands, due to the Coulomb interaction, during the motion of the exciton from one region of the Brillouin zone to the other. The exchange interaction, between the electron and the hole, which exchanges the charge carriers between the conduction and valence bands is zero for triplet, and is neglected for singlet Wannier excitons because the exchange interaction is known to be of short range. For a singlet exciton the theory is therefore applicable strictly for large inter-charge carrier separation and such an exciton is called a large radii orbital exciton.

In the case of Frenkel excitons or molecular excitons, however, the singlet exciton moves predominantly due to the Coulomb interaction between the electron and hole, and the excited electron and hole recombine on one molecule and thus the liberated energy is transferred to another molecule to excite another electron and hole pair. This is called the hopping transport or the coherent transport of molecular excitons in organic crystals. The triplet molecular exciton cannot move like a singlet exciton because the above type of interaction between electron and hole vanishes for triplet excitons. However, the exchange interaction between the electron and hole, which transfers the excited electron from one molecule to the other in its excited state, and the hole from the ground state of one molecule to that of the other is nonzero for triplet excitons. This interaction can be called carrier transfer interaction, whereas the Coulomb interaction responsible for the motion of singlet excitons is an energy transfer type of interaction. In organic crystals usually the overlap of intermolecular wavefunctions is relatively small and, therefore, the carrier transfer interaction is smaller than the energy transfer interaction. In inorganic materials, however, the situation is the other way around. Once can, therefore, undoubtedly conclude that the motion of singlet molecular excitons is faster than that of triplet molecular excitons. In other words the crystal volume encountered by the singlet molecular exciton in a particular time is larger than that encountered by a triplet exciton in the same amount of time. Such conclusions, however, cannot obviously be drawn for the motion of singlet and triplet Wannier excitons.

The energy of an exciton in a crystal is derived as:

$$E(K) = E_o + E_{ex} + \hbar^2K^2/2M^* \quad , \qquad (1)$$

where E_o represents the total electronic energy of the crystal before being excited. E_{ex} is the exciton state energy which is less than the energy of the band gap by the binding energy of exciton, E_b, as shown in Fig. 1. For molecular excitons E_{ex} is usually taken to be equivalent to the excited state energy of molecules in the crystal. The last term of eq. (1) represents the kinetic energy of an exciton with wavevector $\vec{K}$ and the effective mass M*.

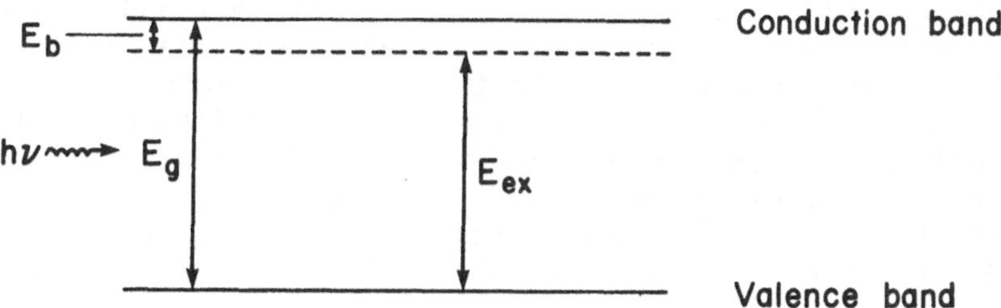

Fig. 1. Exciton state created by an incident photon in a semiconductor crystal.

II. RADIATIONLESS PROCESSES

There are several possible channels in which an exciton can lose its energy nonradiatively in the crystal. This can be due to the interaction with lattice vibrations as well as to the presence of impurities in the crystal such as donors and acceptors. An exciton can also lose its energy due to the interaction with other excitons and the excess charge carriers present in the crystal. Normally such excitonic reactions result in the generation of charge carriers and consequently the photoconductivity of the crystal is enhanced. The interaction of an exciton with another exciton, however, can lead to the formation of some other exciton states. A well-known example of this is the fission of a singlet exciton into two triplets (6) in tetracene crystals, and fusion of two triplets into one singlet exciton in anthracene crystals (7).

The interaction of an exciton with an excess charge carrier is another interesting example of the nonradiative decay of excitons. The excess charge carrier can cause annihilation of the exciton by generating another charge carrier of high kinetic energy and thus increase the carrier's mobility in the crystal. Annihilation of the exciton due to an excess charge carrier can also take place by exciting the excess charge carrier to a higher conducting state. Such Auger-type processes have recently been observed in several experimental works (5). Moreover, the observed decrease in the fluorescence of anthracene in the presence of an excess charge carrier can also be explained in terms of the formation of a complex charge carrier's bound state (8,9). Such bound states of an exciton and an excess electron can be formed from the free and trapped excitons and excess electrons. The binding energy of such a bound state is obtained in terms of the interacting potential U(R) derived as:

$$U(\underset{\sim}{R}) = -\frac{e\underset{\sim}{R}}{R^3} \cdot (\underset{\sim}{\mu}^f(o)-\underset{\sim}{\mu}^g(o)) + (\alpha^{ff'}-\alpha^{gf'})\underset{\sim}{E}(\underset{\sim}{R}) \; ; \; \underset{\sim}{E}(R) = e\underset{\sim}{R}/\varepsilon R^3 \quad , (2)$$

where $|R|>o$ and represents the separation between the exciton and excess electron in the crystal. $\underset{\sim}{\mu}^f(o)$ and $\underset{\sim}{\mu}^g(o)$ are the permanent dipole moments of a molecule in its excited state f and ground state g. $\alpha^{ff'}$ and $\alpha^{gf'}$ represent the static polarizability tensors of a molecule in its excited state f and ground state g, respectively, induced by an excess electron, in the excited state f' of another molecule. $\underset{\sim}{E}(\underset{\sim}{R})$ is the electrostatic field at a distance $\underset{\sim}{R}$ induced by the excess electron, and ε is the dielectric constant of the crystal. Usually $\mu^f(o) \geq \mu^g(o)$, depending on the symmetry of the crystal and molecules, and $\alpha^{ff'}-\alpha^{gf'} > o$. The potential of interaction (eq. (2)) thus becomes attractive and can possibly form a bound state. Agranovich and Zakhidov (8) have calculated $U(\underset{\sim}{R})$ for centrosymmetric crystals of anthracene where $\underset{\sim}{\mu}^f(o) = \underset{\sim}{\mu}^g(o) = 0$, and thus at R = 10 Å and 1 Å they have obtained

$U(R) \sim -100\ cm^{-1}$ and $-1500\ cm^{-1}$. These are, however, only a rough estimate of $U(\tilde{R})$.

Some of the other calculated rates of the nonradiative decay of excitons are shown in Table 1. Most of these rates are obtained from time-dependent perturbation theory assuming that the crystal undergoes a quantum transition from its intitial electronic state to the final one. We have shown in Table 1 the corresponding reaction of the decay of the exciton as well.

TABLE 1. SOME OF THE RATES OF THE EXCITON'S DECAY

Crystal	Reaction	Derived rates	Magnitude ($cm^3\ sec^{-1}$)
G_e	$e + h \rightarrow \overset{e}{\underset{h}{\vert}}$	$R_b = \alpha_b n_h$ (a)	$\alpha_b \sim 10^{-7}$
G_e	$\overset{e}{\underset{h}{\vert}} \rightarrow e + h$	$R_d = \alpha_d n_{ex}^2$ (a)	$\alpha_d \sim 10^{-6}$
Benzophenone	$\overset{e}{\underset{h}{\vert}} + \overset{e}{\underset{h}{\vert}} \rightarrow e + h$	$R = \alpha n_{ex}^2$ (b)	$\alpha \sim 10^{-10}$
GaP	$\overset{e}{\underset{h}{\vert}}$ + ionized donor		
	$\rightarrow$ h + ionized donor	$R = \beta_d n_d$ (c)	$\beta_d \sim 10^{-4}$

$\overset{e}{\underset{h}{|}}$ represents an exciton, e, an electron and h a hole. n_e, n_h and n_{ex} are respectively the concentrations of electrons, holes and excitons in the crystal. (a) Reference 10, (b) Reference 2 and (c) Reference 4.

REFERENCES

1. J.B. Webb and D.F. Williams, J. Phys. (C) 11, 3245 (1978).
2. J. Singh, to be published.
3. M. Trlifaj, Czech. J. Phys. B14, 227 (1964) and B15, 780 (1965).
4. J. Singh and P.T. Landsberg, J. Phys. (C) 9, 3627 (1976).
5. M. Pope, J. Burgos and N. Wotherspoon, Chem. Phys. Lett. 12, 140 (1971), and P. Schlotter and J. Baessler, Chem. Phys. 19, 353 (1977).
6. J. Singh, J. Phys. Chem: Solids 39, 1207 (1978).
7. P. Avakian and R.E. Merrifield, Mol. Cryst. 5, 37 (1968) and P. Avakian and E. Abramson, J. Chem. Phys. 43, 821 (1965).
8. N.M. Agranovich and A.A. Zakhidov, to be published.
9. J. Singh, to be published.
10. A.A. Lipnik, Sov. Phys. (Solid State) 2, 1835 (1961).

PHONON COUPLING TO RADIATIVE, NONRADIATIVE, AND ENERGY-TRANSFER TRANSITIONS

W. H. Fonger and C. W. Struck

RCA Laboratories
Princeton, New Jersey 08540, U.S.A.

ABSTRACT

The model of phonon coupling to transitions based on Condon's quantum-mechanical treatment of the Franck-Condon principle and Manneback's recursion formulas for overlap integrals is reviewed. The model is valid for arbitrary Franck-Condon offset and temperature. For equal force constants, the results are expressible in Huang-Rhys-Pekar W_p functions. The model is illustrated by Yb^{+3} in Y_2O_2S. The Yb^{+3} 4f states ($^2F_{7/2,5/2}$) are simultaneously common to small- and large-offset transitions. Radiative, nonradiative, and energy-transfer transitions from the charge-transfer state are comparable in strength and compete where quantum-mechanical formulas are needed.

I. INTRODUCTION

The Franck-Condon principle, as stated quantum mechanically by Condon (1) in his 1928 treatment of absorption in diatomic molecules, is that the transition rate is proportional to the squared vibrational overlap integral $\langle u_n|v_m\rangle^2$. This "Franck-Condon factor" gives the transition's dependence on coupling to the phonons; v_m and u_n are the initial- and final-state vibrational wavefunctions. For parabolic energy potentials, v_m and u_n are harmonic-oscillator wavefunctions, the $\langle u_n|v_m\rangle$ integrals are given by the Manneback recursion formulas (2), and the initial-v-state thermal weights are $(1-r_v)r_v^m$, where r_v is the Boltzmann factor $\exp(-\hbar\omega_v/kT)$ for the initial-state phonon energy $\hbar\omega_v$.

The model has been applied to phonon coupling in radiative, nonradiative, and energy-transfer transitions for arbitrary

Franck-Condon offset and temperature (3). The model has been worked out for multiple nuclear coordinates and for the linear and derivative nuclear operators in place of the Condon operator 1. The calculations can be carried out numerically in any case and, when all phonons have the same energy, have been carried out analytically for any number of nuclear coordinates (4). For brevity, only one coordinate and the operator 1 are treated here.

II. THE MODEL

The $v_m \rightarrow u_n$ radiative and nonradiative transitions are diagramed in Fig. 1. For the radiative transition, the energy mismatch between the m and n energy levels is taken up by the photon energy $h\nu_p$; for the nonradiative, the m and n energy levels match. The net phonon energy generated in the transition is

$$n\hbar\omega_u - m\hbar\omega_v \equiv p\hbar\omega_0 , \tag{1}$$

where $\hbar\omega_0$ is an arbitrarily selected energy unit, and p is the number of these units. The transition weight of all $v_m \rightarrow u_n$ transitions in which p units of phonon energy are generated is

$$U_p = \sum_{m=0}^{\infty} (1-r_v)r_v^{\ m} \langle u_n|v_m\rangle^2 , \tag{2}$$

where, in the sum over m, n is coupled to m by eq. (1). The index p has the integer values 0, ±1, ±2, ... If the energy difference on the left side of eq. (1) is a non-integer multiple of $\hbar\omega_0$ bracketed by $p\hbar\omega_0$ and $(p+1)\hbar\omega_0$, the $v_m \rightarrow u_n$ weight $(1-r_v)r_v^m \times \langle u_n|v_m\rangle^2$ is broken up into two parts, in some reasonable manner, and assigned one part to U_p, the other to U_{p+1}. As thus defined, U_p is normalized: $\Sigma_p U_p = 1$.

For the inverse $u \rightarrow v$ transition, there is an analogous weight V_p defined as in eqs. (1) and (2) with u,n and v,m interchanged.

For the $v \rightarrow u$ emission in Fig. 1, U_p is the weight of the emission at photon energy $h\nu_p$ given by

$$h\nu_p = h\nu_{zp} - p\hbar\omega_0 , \tag{3}$$

where $h\nu_{zp}$ is the zero-phonon energy. That is, the U_p distribution is the normalized emission bandshape. In an analogous way, V_p is the weight of the $u \rightarrow v$ absorption transition at the photon energy

$$h\nu_p = h\nu_{zp} + p\hbar\omega_0 . \tag{4}$$

The rate of the nonradiative transition in Fig. 1 is

$$\text{Rate}_{v\rightarrow u} = N\ U_p , \tag{5}$$

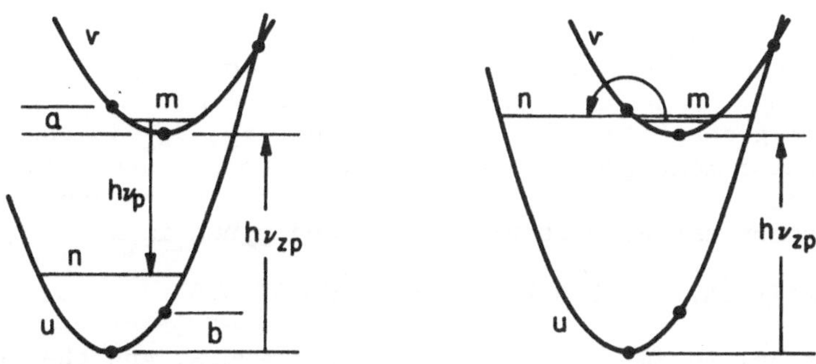

Fig. 1. The v → u radiative (left) and nonradiative (right) transitions; $h\nu_{zp}$ is the transition zero-phonon energy.

where the "electronic factor" N is, empirically, 10^{12} - 10^{14} sec^{-1}, and p has the particular value given by

$$p\hbar\omega_0 = h\nu_{zp}. \tag{6}$$

The factor U_p is called the nuclear factor or the phonon-coupling factor or the thermal-Franck-Condon weight.

The $v_1 \to u_1$, $u_2 \to v_2$ energy-transfer transition between two centers 1 and 2 is shown in Fig. 2. The dashed line indicates that both transitions occur together. The rate of this transition is

$$\text{Rate}_{v_1 \to u_1, u_2 \to v_2} = N \sum_{p_1=-\infty}^{+\infty} U_{p_1} V_{p_2} , \tag{7}$$

where the electronic factor N is empirically 10^5 - 10^8 sec^{-1}, and the indices p_1, p_2 are constrained by

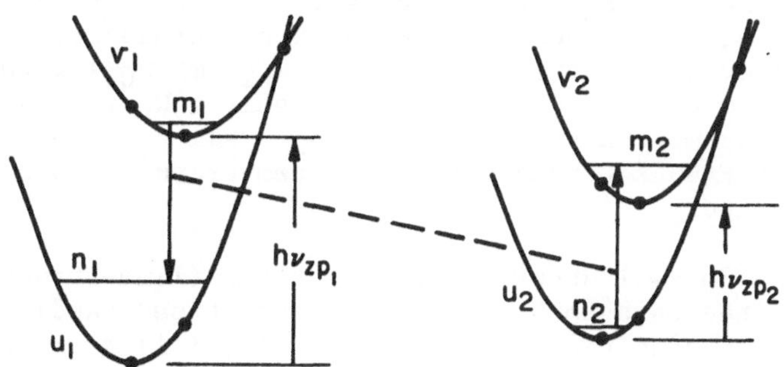

Fig. 2. The $v_1 \to u_1$, $u_2 \to v_2$ energy-transfer transition between two centers 1 and 2.

$$p_1 + p_2 = p, \qquad p\hbar\omega_0 = h\nu_{zp1} - h\nu_{zp2} . \tag{8}$$

This constraint is needed for energy to be conserved in the transition. Eq. (7) is evident since U_{p1} is the weight of $v_1 \to u_1$ transitions generating $p_1\hbar\omega_0$ in phonon energy; V_{p2} is the weight of $u_2 \to v_2$ transitions generating $p_2\hbar\omega_0$ in phonon energy, and $U_{p1}\,V_{p2}$ is the joint weight of $v_1 \to u_1$, $u_2 \to v_2$ transitions generating $p\hbar\omega_0$ in phonon energy with $p_1\hbar\omega_0$ in 1 and $p_2\hbar\omega_0$ in 2.

III. SIMPLIFICATIONS FOR EQUAL FORCE CONSTANTS

For equal force constants, $\hbar\omega_u = \hbar\omega_v = \hbar\omega_0$, and the U_p sums (2) add up to familiar Huang-Rhys-Pekar W_p functions (5,6). The computational work is then smaller. W_p functions depend on two parameters, $\langle m\rangle$ and S_0, and are defined as the normalized ($\Sigma_p W_p = 1$) solution of the recursion formula (7)

$$S_0\langle m\rangle W_{p+1} + p\,W_p - S_0\langle 1+m\rangle W_{p-1} = 0 ; \tag{9}$$

S_0 is the Huang-Rhys-Pekar dimensionless measure of Franck-Condon offset, and $\langle m\rangle = [\exp(\hbar\omega_0/kT) - 1]^{-1}$ is Planck's average-thermal-occupancy measure of the temperature. For equal force constants, $S_0\hbar\omega_0$ is the common value of the energies a and b in Fig. 1

For equal force constants, eqs. (3,4,6,8) are unchanged, and eqs. (1,2,5,7) reduce to

$$n - m = p , \tag{1'}$$

$$U_p = W_p(\langle m\rangle, S_0) , \tag{2'}$$

$$\text{Rate}_{v\to u} = N\,W_p(\langle m\rangle, S_0) , \tag{5'}$$

$$\text{Rate}_{v_1\to u_1, u_2\to v_2} = N\,W_p(\langle m\rangle, S_{01} + S_{02}) . \tag{7'}$$

In eq. (7'), S_{01} and S_{02} are the offset parameters of centers 1 and 2. This equation uses the W_p-function reproductive property that the sum over products $W_{p1}(\langle m\rangle, S_{01})\, W_{p2}(\langle m\rangle, S_{02})$ with the indices constrained by $p_1 + p_2 = p$ is itself a W_p function with the index p and parameters $\langle m\rangle$ and $S_{01} + S_{02}$. All equations are valid for arbitrary Franck-Condon offset and temperature, including S_{01} and S_{02} separately.

The case $S_0 \ll 1$ describes small-offset (line-structured) transitions, the case $S_0 \gg 1$ large-offset (broad-band) transitions. For "Poisson b_1 conditions" $S_0^2\langle 1+m\rangle\langle m\rangle/(|p| + 1) \ll 1$, W_p reduces to

$$W_p \sim \exp(-S_0\langle 2m+1\rangle)\,(S_0\langle 1+m\rangle)^p / p! , \tag{10}$$

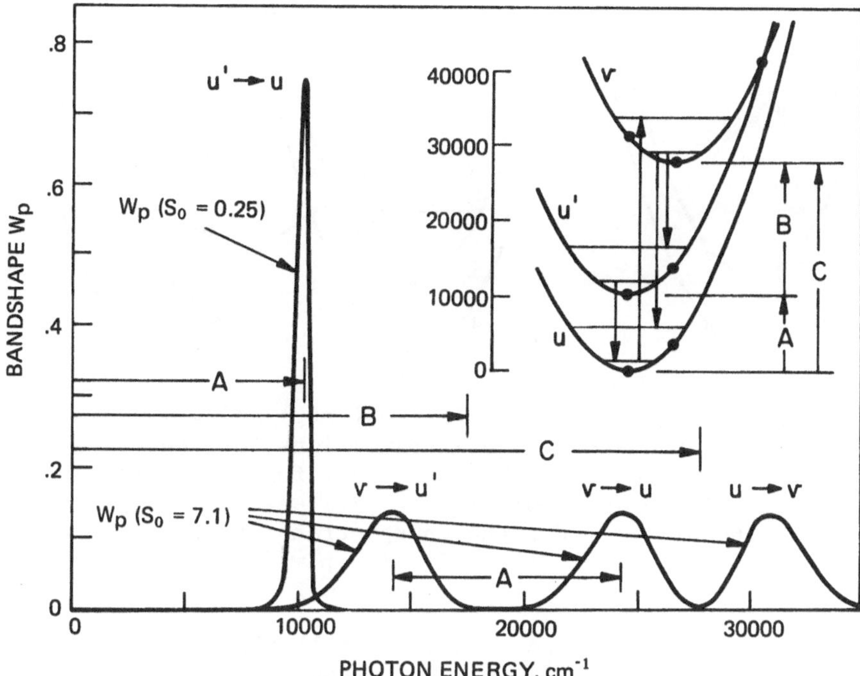

Fig. 3. The Yb^{+3} optical transitions in Y_2O_2S at 300°K for the equal-force-constants model and $\hbar\omega_0 = 500\ cm^{-1}$. A, B, and C are the transition zero-phonon energies.

for $p \geq 0$. For $p < 0$, the same formula holds with $\langle 1+m \rangle \to \langle m \rangle$, $p \to |p|$. For S_0 small, W_p's temperature dependence is substantially $W_p \propto \langle 1+m \rangle^p$. This "Kiel" $\langle 1+m \rangle^p$ dependence has been widely used (8,9) to describe the temperature dependence of nonradiative relaxation between rare-earth +3 4f levels. A description by W_p functions is equivalent.

For large-offset transitions, the nonradiative rates (5,5') replace the Mott rate $N \exp(-E_x/kT)$ previously used for crossovers (10). The Mott rate poorly approximates rates (5,5') for downward transitions ($p > 0$) but is better for upward transitions.

IV. APPLICATION TO Yb^{+3} TRANSITIONS IN Y_2O_2S

Yb^{+3} transitions in oxysulfides have been studied by Buchanan, Nakazawa (11), and ourselves. They provide a good example for the model. The $Y_2O_2S:Yb^{+3}$ states and optical transitions are diagramed in Fig. 3. For brevity, the two small-offset 4f states ($^2F_{7/2,5/2}$) are called u and u'. The large-offset v state is the charge-transfer

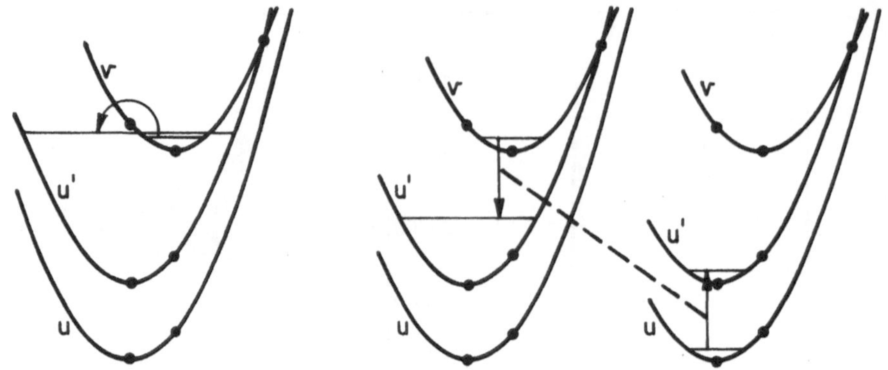

Fig. 4. The Yb^{+3} $v \to u'$ nonradiative (left) and $v \to u'$, $u \to u'$ energy-transfer (right) transitions for the equal-force-constants model. The model rate for the nonradiative transition is $\sim 10^{14}$ $W_p(\langle m \rangle, 7.1)$ sec^{-1} with $p = B/\hbar\omega_0 = 37.6-.013$ kT. For the energy transfer, the rate is $\sim 10^8$ $W_p(\langle m \rangle, 7.35)$ sec^{-1} with $p = (B-A)/\hbar\omega_0 = 17.2-.013$ kT.

state. The u,u' states are common to small-offset $u \rightleftarrows u'$ transitions and large-offset $v \rightleftarrows u,u'$ transitions. The nonradiative and energy-transfer transitions from v are diagramed in Fig. 4.

The three zero-phonon energies are called A, B, and C. The A energy is $\sim$ 10200 cm^{-1}. The v state moves downward with increasing temperature, and its C energy is $\sim$ 29000 cm^{-1}-6.5kT. This downshift, common for many offset states, is due to thermal lattice expansion, which is not otherwise handled in the model. The B energy is C-A. Because of this constraint, the two broad $v \to u,u'$ emission bands are spaced by A, the energy of $u \rightleftarrows u'$ transitions.

Figs. 3 and 4 were constructed to give the best fit to the Yb^{+3} optical bandshapes for equal force constants, $\hbar\omega_0$ = 500 cm^{-1}, and 300°K. The offset parameters were S_0 = 0.25 for small-offset $u \rightleftarrows u'$ transitions, S_0 = 7.1 for large-offset $v \rightleftarrows u,u'$ transitions. The model optical bandshapes are therefore $W_p(\langle m \rangle, 0.25)$ for $u \rightleftarrows u'$ and $W_p(\langle m \rangle, 7.1)$ for $v \rightleftarrows u,u'$.

The model rates for the nonradiative and energy-transfer transitions from v are given in the Fig. 4 caption. For the energy transfer, the W_p-function S_0 parameter is the sum 7.35 of the individual 7.1 and 0.25 S_0 parameters for $v \to u'$ and $u \to u'$, respectively. Compared to the energy-transfer rate, the model nonradiative rate will increase much more rapidly with temperature because its p number ($\sim$35 vs. $\sim$15) is further out in the temperature-sensitive wing of the W_p distribution. [The region of the W_p maximum is near $p = S_0$ ($\sim$ 7 here).]

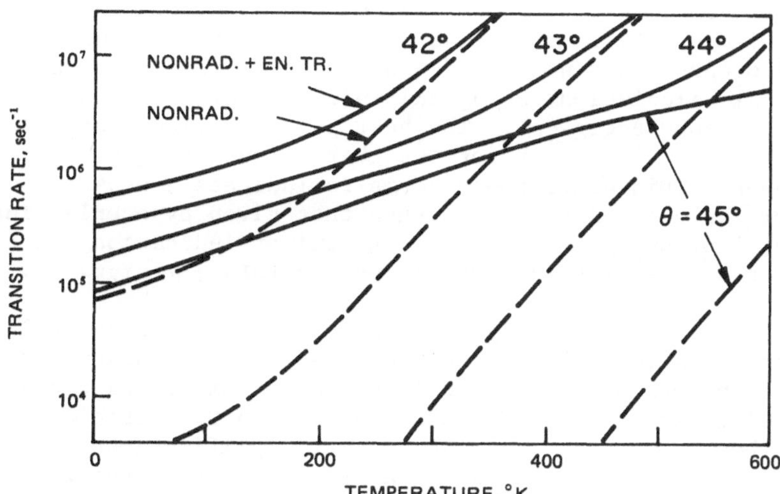

Fig. 5. Calculated nonradiative and energy-transfer rates from the Yb^{+3} offset v state. The parameter θ is the Manneback angle. The competing emission rate from v is ∿2.5x10^6 sec^{-1}.

The calculated rates are plotted in Fig. 5—together with analogous rates for unequal force constants ($k_u = k_{u'} \neq k_v$) calculated from eqs. (5,7). The dashed curves give the $v \rightarrow u'$ nonradiative rate, the solid curves the sum of this rate and the $v \rightarrow u'$, $u \rightarrow u'$ energy-transfer rate. The Manneback angle θ used as a parameter gives the force-constants ratio through $k_v/k_{u'} = \tan^4\theta$. As expected, the energy-transfer rates are fairly flat, the nonradiative rates increase strongly with temperature. The nonradiative rates also change sharply with θ.

These relaxation rates compete with the double-broad-band emission from v. Empirically, the v emission rate is ∿ 2.5 x 10^6 sec^{-1}. The v emissions quench thermally near 250°K. At 77°K, their efficiency is ∿ 1 for 0.05% Yb^{+3} and decreases to ∿ 3/4 for 1% Yb^{+3}, the decrease being manifested by a 3:1 change in the ($v \rightarrow u,u'$)/($u' \rightarrow u$) emission ratio.

These observations are explained by the model results for θ = 42°. The v-state concentration quenching at 77°K is accounted for by θ = 42° $v \rightarrow u'$, $u \rightarrow u'$ energy transfer; the v-state thermal quenching near 250°K by the strong increase of θ = 42° $v \rightarrow u'$ nonradiative transitions. By contrast, for θ = 45° (equal force constants), $v \rightarrow u'$ nonradiative transitions are completely negligible to 600°K, and energy transfer would produce only a mild v-state quenching above 400°K.

The observed $v \rightarrow u,u'$ optical bandshapes (not shown) are best fitted with $\theta \sim 44°$, not the 42° wanted for thermal quenching. This discrepancy was not avoided by going to more elaborate models (numerical calculations using multiple nuclear coordinates and multiple phonon energies). The best parabola placements were always somewhat different for the thermal quenching vs. the optical bandshapes. This discrepancy seems to indicate a weaker p dependence for the nonradiative rate than that given by overlap integrals ($<u_{p+m}|v_m>^2$ or $<u_{p+m}|\partial/\partial z|v_m>^2$). A similar indication had been found for small-offset 4f transitions in LaF_3 and Y_2O_3 (3).

These difficulties might be corrected with non-parabolic potentials. Nevertheless, the model as presently described has handled a number of small- and large-offset cases and probably will be useful in many more. See the examples in G. Blasse's lectures in this book. Despite the model's oversimplifications, its combined ease and breadth recommend it as a useful first overview of phonon coupling.

REFERENCES

1. E. U. Condon, Phys. Rev. 32, 858 (1928).
2. C. Manneback, Physica 17, 1001 (1951).
3. C. W. Struck and W. H. Fonger, J. of Lum. 10, 1 (1975); ibid. 17, 241 (1978).
4. C. W. Struck and W. H. Fonger, J. of Lum. 18/19, 101 (1979); Phys. Rev. B 19, 4400 (1979).
5. K. Huang and A. Rhys, Proc. Roy. Soc. Lond. A 204, 406 (1950).
6. S. I. Pekar, Zh. Eksp. Teor. Fiz. 20, 510 (1950); Usp. Fiz. Nauk 50, 197 (1953); "Untersuchungen über die Elecktronentheorie der Kristalle," Akademie, Berlin (1954), Chs. 5,6.
7. C. W. Struck and W. H. Fonger, J. Chem. Phys. 60, 1988 (1974).
8. H. W. Moos, J. of Lum. 1/2, 106 (1970).
9. L. A. Riseberg and M. J. Weber, in Progress in Optics (E. Wolfe, ed.), North Holland, Amsterdam, Vol. 14, 89 (1977).
10. N. F. Mott, Proc. Roy. Soc. Lond. A 167, 384 (1938); R. W. Gurney and N. F. Mott, Trans. Faraday Soc. 35, 69 (1939).
11. E. Nakazawa, J. of Lum. 18/19, 272 (1979).

ON THE EVALUATION OF DECAY RATES IN SOLIDS

R. Englman

Soreq Nuclear Research Centre
Yavne, Israel.

ABSTRACT

A rederivation of the exponential energy-gap law for the non-radiative decay rate of impurities in solids is presented, using the saddle point method. Assuming realistic dispersions of the vibrational frequencies one finds that only the principal saddle point is significant and that the nature of the electron-phonon coupling near the maximum frequency determines the decay rate.

I. INTRODUCTION

In the decay of electronic excitations on an impurity, energy-excesses ΔE of the order of 5-50,000 cm^{-1} get transferred to the vibrational modes of the crystal. The energy $\hbar\omega_i$ of these modes is characteristically a few-hundred wavenumbers, so that the electronic decay is necessarily a multi-phonon process. All pathways of dissipation must satisfy the energy conservation law

$$\Delta E = \sum_i n_i \hbar\omega_i, \tag{1}$$

where i runs over the normal modes of the lattice (including the impurity) and n_i are integers. But not all pathways satisfying eq. (1) are equally efficient. Customarily the most efficient pathways are recognized by a saddle-point integration of the expression for the decay-rate. It then turns out that, as in many statistical-mechanical sums, the most efficient terms overwhelmingly dominate the process.

For electronic relaxation the saddle-point method appears to have been first implemented by Jortner and the present author (1), to derive energy-gap laws. It was shown that (in the regime of the exponential energy-gap law) the highest-frequency (ω_M) modes of the system formed the efficient pathway for relaxation. Although the derivation in (1) was aimed at large molecular systems, its results were taken over to interpret experimental results in solid-state impurity systems. In particular, rare-earth ions in crystalline (2,3) and glassy (4) hosts constituted systems in which the decay rate showed exponential dependence on the energy-gap and confirmed the predominant importance of the highest frequency lattice modes.

In contrast to its experimental successes, the saddle-point method has run into several theoretical difficulties. These have been described at some length in Chapter 15 of a recent work (5) and will not be reproduced here in detail. Only two conclusions arrived at in (5) will be rederived here in a more elaborate and pedagogically more systematic manner, supplemented by numerical estimates. These conclusions refer to the uses of the principal (k=0) saddle point and of the effective density of states ($f(\omega) \equiv \frac{1}{2}\rho(\omega)\Delta^2(\omega)$) for electronic relaxation of impurities in solids (both terms will be explained in the text) and can be stated as follows.

(i) The principal saddle-point gives the dominant contribution to the rate (or, equivalently, the subsidiary saddle-points may be neglected).

(ii) The value of the effective density of states $f(\omega)$ near the maximum lattice-frequency is alone relevant to the rate.

Both conclusions hold for large energy gaps only, but our numerical work indicates that it is sufficient to have for the scaled energy gap

$$\frac{\Delta E}{\hbar\omega_M} > 1.5\times S,$$

where S (the Huang-Rhys number) is a measure of the electron-vibration coupling strength. The results are also somewhat different if there is a localised mode above the vibrational continuum, due to the impurity in the lattice (Case B in §3.3.2 of (5)).

II. HISTORICAL NOTE ON THE SADDLE-POINT METHOD

In equilibrium statistical-mechanics the names of C.G. Darwin and R.H. Fowler are associated with this method (6). In a purely mathematical framework a rigorous statement of the conditions for the validity of the method are provided in (7). Its use for the asymptotic evaluation of the integral representation of a Bessel function, $K_p(z)$ when z is large, was developed by P. Debye (8) and is reproduced on p.24 of (7). In this case the subsidiary saddle points do not contribute to the integral.

In the subject of small polaron diffusion, that is analogous to non-radiative decay with a negligible energy gap, Holstein (9) criticized an earlier work (10) for what is essentially the neglect of subsidiary saddle-points. In radiationless decay proper, Fischer (11) and Gelbart et al. (12) pointed out the presence of the subsidiary saddle-points for molecular systems. Fain (13) and Medvedev et al. (14) further discussed their contributions, with conclusions differing from ours.

III. ANALYSIS

Our starting point will be the Fourier-transform for the rate at zero temperatures (for displaced parabolic potential surfaces)

$$\text{rate} = 2\pi\hbar^{-2}K^2 \int_{-\infty}^{\infty} dt \, \exp[-i\Delta E_{-p}t/\hbar + \int_0^{\omega_M} d\omega f(\omega) e^{i\omega t}], \tag{2}$$

where K is the matrix-element of the electron-transition (non-adiabatic) operator including also the so-called promoting modes (p.36 (5)). The quanta of these modes $\hbar\omega_p$ appear also in the exponent through

$$\Delta E_{-p} = \Delta E - \hbar\omega_p(\sim\Delta E), \tag{3}$$

with ΔE being the purely electronic energy gap. For a more complete treatment of the promoting modes the reader is referred to Chapter 8 in (5). In the exponent we also have the effective density of states

$$f(\omega) = \tfrac{1}{2}\rho(\omega)\Delta^2(\omega), \tag{4}$$

$\rho(\omega)$ being the density of states of phonons and $\Delta(\omega)$ the normalised displacement of the potentials for the modes with frequency ω. ω_M is the maximum frequency in the crystal. It is convenient to introduce a number G by writing

$$\Delta E_{-p} = \hbar\omega_M e^G. \tag{5}$$

For our results to hold the energy gap is supposed to be large. Our numerical work indicates that

$$G > 0.4$$

is already sufficient for the validity of our approximations. Let us denote the exponent in (1) by Q(t), so that

$$Q(t) = -i\omega_M t\, e^G + \int_0^{\omega_M} d\omega\, f(\omega) e^{i\omega t}$$
$$= -i\omega_M t\, e^G + e^{i\omega_M t}\, q(t) - p(t). \tag{6}$$

The introduction of the functions q(t) and p(t) is a crucial step in the development of our theory. As defined q and p do not depend exponentially on $i\omega_M t$ and in the development of the saddle-point method (where $|e^{i\omega_M t}|$ is a large quantity of order e^G) the term p(t) will be found to be negligible. This is significant, since q(t) depends only on the function $f(\omega)$ near $\omega = \omega_M$. In fact, through integration by parts one finds that formally

$$q(t) = (it)^{-1} \left[\left(1 + \frac{1}{it}\frac{\partial}{\partial\omega}\right)^{-1} f(\omega)\right]_{\omega=\omega_M} . \tag{7}$$

A few examples may be helpful at this stage. Supposing

(i) $f(\omega) \equiv \frac{1}{2}\rho(\omega)\Delta^2(\omega) = S^{(1)}/\omega_M$, a constant,

one derives

$$q(t) = S^{(1)}(i\omega_M t)^{-1} = p(t); \tag{8}$$

(ii) $f(\omega) = S^{(2)}(\omega_M - \omega)\omega_M^{-2}$, vanishing at ω_M,

$$q(t) = S^{(2)}(i\omega_M t)^{-2},\; p(t) = S^{(2)}(1+i\omega_M t)(i\omega_M t)^{-2} . \tag{9}$$

In the above expressions $S^{(1)}$ and $S^{(2)}$ are measures of the electron-phonon coupling and are in fact analogous to the Huang-Rhys number S. More generally than (i) and (ii), if $f(\omega)$ and its first, second, up to r-2 derivatives vanish at $\omega=\omega_M$

$$q(t), p(t) \sim S^{(r)}(i\omega_M t)^{-r} . \tag{10}$$

Thus eq. (8) represents r=1. A discrete mode has r=0.

The saddle point of the integrand in eq. (2) will be at $\dot{Q}(t)=0$, where the dot stands for differentiation with respect to G. Hence

$$e^G = e^{i\omega_M t}[q+\dot{q}/i\omega_M] - \dot{p}/i\omega_M . \tag{11}$$

After rearrangement we get the saddle points in the t-plane given by

$$i\omega_M t_k \equiv \mu_k = G + 2\pi ik + \ln\{[1+e^{-G}\dot{p}/i\omega_M]/[q+\dot{q}/i\omega_M]\}, \text{ where} \tag{12}$$

$(k = 0,\pm 1,\ldots)$, $q \equiv q(t_k)$, $p \equiv p(t_k)$.

The original and deformed contours are seen in Fig. 1. The real solution μ_o is the principal saddle-point, the solutions μ_k with $k\neq 0$ are the subsidiary points. The second derivative near these points

$$\ddot{Q} = (i\omega_M)^2\, e^{\mu_k}[q+2\dot{q}/i\omega_M + \ddot{q}/(i\omega_M)^2] - \ddot{p} \tag{13}$$

$$\sim -\omega_M^2\, e^G,$$

so that the integration along the deformed contour (Fig. 1) near the saddle-point converges (and does so fast).

Then the integral in eq. (2) works out to be

$$\sqrt{\pi}\,\omega_M^{-1}\, e^{-G/2} \sum_{k=-\infty}^{\infty} e^{Q(t_k)}, \tag{14}$$

where

$$Q(t_k) = -e^G\{\mu_k - (1 + \frac{\dot{q}}{i\omega_M q})^{-1} - e^{-G}[\frac{\dot{p}}{i\omega_M}(1 + \frac{\dot{q}}{i\omega_M q})^{-1} - p]\}. \tag{15}$$

It may be shown algebraically that for large G all subsidiary saddle-points make a negligible contribution in eq. (14). This follows since the roots of eq. (12) are of the form

$$\mu_k = GA_k(G) + 2\pi ik\, B_k(G), \tag{16}$$

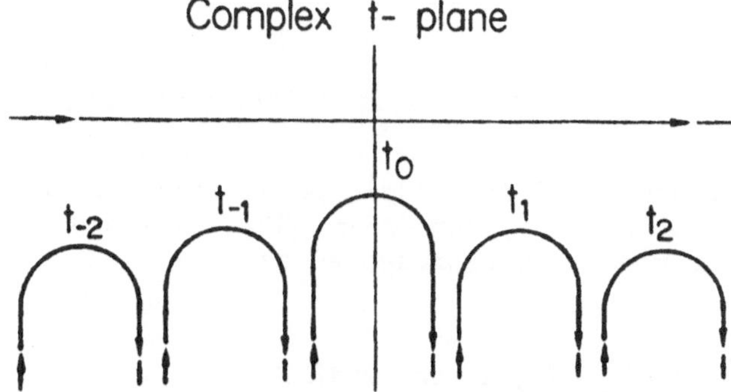

Fig. 1. The integration paths of eq. (2) in the t-plane, with saddle-points on the deformed path shown.

where A_k and B_k are real, and

$$A_o(G) < A_{k\neq 0}(G) \sim 1 + \alpha(\ln|G|/G) + \beta(k^2/G^3) \ . \tag{17}$$

For large G the first term in the curly brackets in eq. (15) is dominant. In consequence the contribution of the subsidiary saddle-points relative to that of the principal one (k=0) is as

$$e^{-e^G \beta k^2/G^2} \sim e^{-\beta \frac{\Delta E}{\hbar\omega_M} k^2 (\ln \frac{\Delta E}{\hbar\omega_M})^{-2}} << 1 \ . \tag{18}$$

The values of the saddle-points as found by numerical solution of eq. (12) are shown in Fig. 2. One notes that

$$A_{k+1}(G) > A_k(G), \ B_{k+1}(G) < B_k(G) \ . \tag{19}$$

The contributions of the terms in eq. (14) with $k\neq 0$ relative to that of the principal term (k=0) are plotted in Fig. 3 as functions of $\Delta E/\hbar\omega_M$. It is seen that already at $\Delta E/\hbar\omega_M = 1.25S$,

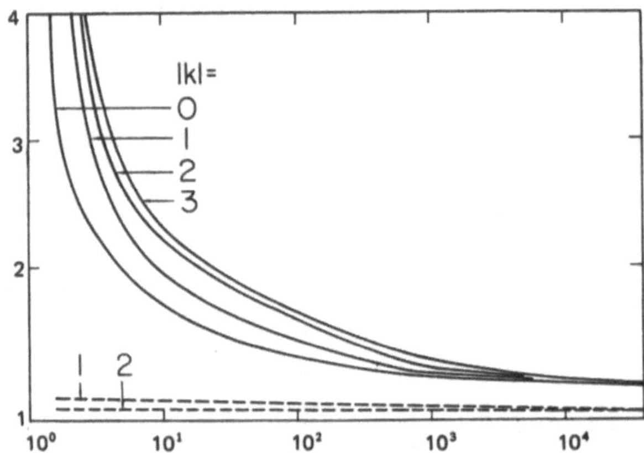

Fig. 2. The roots μ_k of the saddle-point equation (eq. (12)) vs. the scaled energy-gap ($\Delta E/\hbar\omega_M S$). Real and imaginary parts of the roots are represented as

$$\ln(\Delta E_{-p}/\hbar\omega_M)A_k \text{ and } 2\pi ik\ B_k$$

and the factors A_k (full lines) and B_k (broken lines) are plotted for a few values of the integer $|k|$.($B_o\equiv 0$). The constant effective density of state [eq. (8):r=1 and $S=S^{(1)}=1$] is used for the computation (throughout).

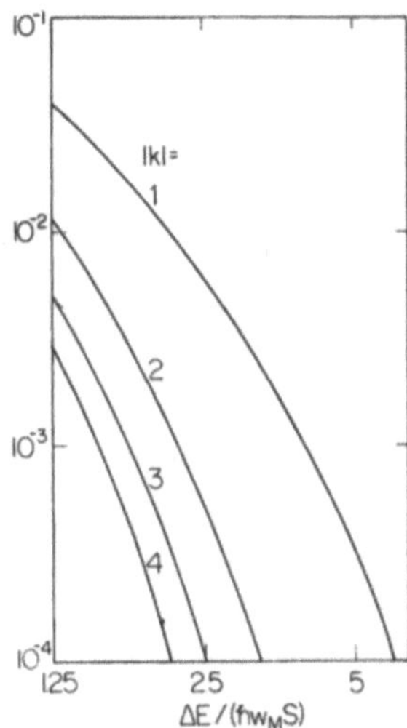

Fig. 3. The contribution of the subsidiary saddle points to the rate, relative to the principal saddle point (k=0).

the relative contribution of the subsidiary saddle points adds up to little more than 1/10. This justifies the common practice in the literature (e.g. (1) or (5)) of using the principal saddle point only, already at moderately large values of $\Delta E/h\omega$. Let it be remarked that strictly speaking for molecules with line-like, rather than continuous, vibrational spectra the subsidiary saddle points must be kept for all k's. Keeping these terms in eq. (14) ensures conservation between electronic and vibrational energies.

IV. OFF-PEAK FREQUENCY MODES

It remains to investigate the effects on the decay rate of the modes away from the maximum frequency ω_M. (Let it be recalled that in intra-molecular relaxation (1) only $\omega=\omega_M$ enters, though this exclusivity was queried (15).) We are in fact assessing the importance of the p-term in eq. (6), since by eq. (7), q comes from $f(\omega\to\omega_M)$. From eq. (12) we see that the contribution of p to the saddle point is of the order of $e^{-G}|G+2\pi ik|^{-r-1}$ and from eq. (15) we find that the contribution of p to the exponentials in eq. (14) is of the order of $|G+2\pi ik|^{-r}$, and therefore negligible as $G \sim \ln(\Delta E/\hbar\omega)\to\infty$. From this we derive the important conclusion that for large energy gaps the highest frequency modes in the electron-phonon coupling dominate. Summarizing our work we obtain the following expression of the energy-gap law:

$$\text{rate} = 2\pi\hbar^{-2}K^2 \frac{\sqrt{\pi}}{\sqrt{\Delta E_{-p}\hbar\omega_M}} \exp\left\{-\frac{\Delta E_{-p}}{\hbar\omega_M} < \ln\left[\frac{\Delta E_{-p}}{\hbar\omega_M q}\left(1+\frac{\dot{q}}{iq\omega_M}\right)-1\right]>\right\}, \quad (20)$$

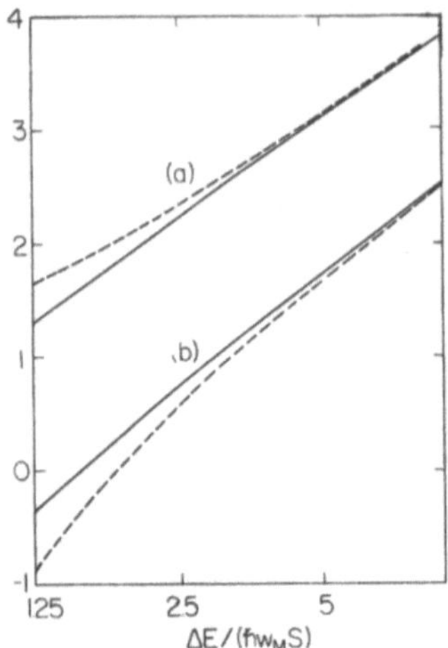

Fig. 4. The principal saddle point, k=0. Curves (a) show the values of the root μ_o and curves (b) the exponent $Q(\mu_o)$, that appears in eq. (14) and (15) multiplied by $\hbar\omega_M/\Delta E$. The full and broken lines compare the results based on the complete eq. (6) and on the neglect of the off-peak frequency term p(t).

q and $\dot{q}$ being evaluated at the principal saddle point $\mu_o \sim G$.

We have also investigated the neglect of the p-term numerically again using the model (i) in III. Fig. 4 shows the saddle point values and the associated exponent $Q(\mu_o)$, arising both from the full expression (eq. (6)) and from that in which the off-peak frequency term p is neglected. It is clear from Fig. 4 that the difference between these cases fades away even for moderately large $\Delta E/\hbar\omega$. For the subsidiary saddle-points the difference vanishes even faster.

V. CONCLUSION

Several problems arising from the saddle-point method for radiationless decay in solids have been satisfactorily resolved. The extension of this theory to non-zero temperatures is indicated in (5), p. 36. It is alluring to contemplate experiments (involving, e.g., selective probing of excited vibrational levels) in which the saddle points would be directly observed!

REFERENCES

1. R. Englman and J. Jortner, Mol. Phys. 18, 145 (1970).
2. L.A. Riseberg and H.W. Moos, Phys. Rev. 174, 429 (1968).
3. M.J. Weber, Phys. Rev. 171, 283 (1968).
4. R. Reisfeld and Y. Eckstein, J. Chem. Phys. 63, 4001 (1975).
5. R. Englman, Non-radiative Decay of Ions and Molecules in Solids North-Holland, Amsterdam (1979).
6. R.H. Fowler and E.A. Guggenheim, Statistical Thermodynamics, University Press, Cambridge (1956).
7. A. Erdélyi, Higher Transcendental Functions, Vol. II, McGraw-Hill, New York (1953).
8. P. Debye, Math. Ann. 67, 535 (1909).
9. T. Holstein, Ann. Phys. 8, 343 (1959).
10. I. Yamashita and T. Kurosawa, J. Phys. Chem. Solids 5, 34 (1958).
11. S. Fischer, J. Chem. Phys. 53, 3195 (1970).
12. W.M. Gelbart, K.F. Freed and S.A. Rice, J. Chem. Phys. 52, 2460 (1970).
13. V.M. Fain, J. Chem. Phys. 65, 1854 (1976).
14. E.S. Medvedev, V.I. Osherov and V.M. Pschenichnikov, Chem. Phys. 23, 397 (1977).
15. R. Englman and J. Jortner, J. Lum. 1/2, 134 (1970).

MULTIPHONON RELAXATION IN GLASSES

R. Reisfeld

Department of Inorganic and Analytical Chemistry
The Hebrew University of Jerusalem
Jerusalem, Israel

ABSTRACT

Experimental data of multiphonon relaxation between electronic levels of rare earth ions in glasses are presented. These are based on measurements of the decay times or quantum efficiencies of the specific levels and subtracting the radiative transition probabilities in oxide, chalcogenide and fluoride glasses. It is shown that the nonradiative transfer in all glasses depends mainly on the energy gap between the emitting and next lower level of the ion incorporated in the glass and that the high energy phonons of the network formers are responsible for the relaxation permitting the lowest order process. The phenomenological parameters α, β and ε appearing in the formulae of multiphonon transfer are computed and compared to these values in crystals. It is also shown that the Huang-Rhys number S being a measure of the electron-phonon coupling strengths is smaller than 0.1 for the oxide glasses as predicted by the theory of multiphonon relaxation. However it achieves a large value of about 2 in the chalcogenide glasses. The discrepancy in this high value is explained by the covalency of the rare earths incorporated in the chalcogenide glasses with the surrounding sulfur ions, thus the rare earth ions cannot be treated as isolated centers in these glasses. Experimental findings of phonon-assisted energy transfer between uranyl and rare earth ions in phosphate glasses are presented. It is shown that the experimental dependence between multiphonon relaxtion rates and the energy gap is that obtained for similar to the multiphonon relaxation in a single ion. The coupling constant is higher in the case of the energy transfer due to the stronger coupling of uranyl to the glass.

I. INTRODUCTION

The theory of multiphonon relaxation of ions in crystals, organic molecules and in semiconductors has attracted the interest of many researchers during the last two decades. A number of reviews have been written recently (1-6) dealing with these important phenomena. A unified treatment of multiphonon relaxation in both organic and inorganic systems as well as a critical review of the existing theories can be found in an excellent book by Englman (7). Excited electronic levels of rare earths (RE) in solids decay nonradiatively by exciting lattice vibrations (phonons). When the energy gap between the excited level and the next electronic level is larger than the phonon energy several lattice phonons are emitted in order to bridge the energy gap. It was recognized that the most energetic vibrations are responsible for the non-radiative decay since such a process can conserve energy in the lowest order. In glasses the most energetic vibrations are the stretching vibrations of the glass network polyhedron and it was shown that these distinct vibrations are active in the multiphonon process (8) rather than the less energetic vibrations of the bond between the RE and its surrounding ligands. Later (9) it was demonstrated that these less energetic vibrations may participate in cases when the energy gap is not bridged totally by the high energy vibrations. The experimental results reveal that the logarithm of the multiphonon decay rate decreases linearly with the energy gap, or the number of phonons bridging the gap. The theory of nonradiative decay by multiphonon mechanism was first proposed by Kubo and Toyozawa (10) who proposed that the basic mechanism allowing such transitions is the correction in the Born-Oppenheimer (BO) approximation due to vibrational motion of ions which admixes the electronic wave function and causes transitions that represent stationary states in the zero order BO approximation. For multiphonon processes one has to proceed to higher order being $\Delta E/h\omega$ to get real transitions between the electronic states (7). It should be noted that in contrast to the smallness of the radiative processes of high order, the non-radiative high order processes are quite high.

From the point of view of theoretical treatment, it is extremely difficult to calculate accurately the perturbation of a high order. However a considerable part of ΔH_{vibr} can be eliminated as a perturbation by including it exactly in the wave function by a "renormalization." The part of ΔH_{vibr} which still remains after renormalization is the nonadiabacity of the BO correction operator. At the end of the calculation an approximation is made of the interaction with only one phonon mode (1,5,11-13). For small coupling and low temperature a Poisson-like function is obtained (12) for the distribution of the multiphonon relaxation rate with the number of phonons

$$W_p \sim (\exp - S)(S^{p/p:}) \quad , \tag{1}$$

where S is the Huang-Rhys-Pekar number (7), and

$$S(T=0) = \frac{1}{2} \Delta^2 \ . \tag{2}$$

[S is the measure of electron phonon coupling strength.]

The displacement Δ measures the horizontal shift of the electronic state potentials in units of the zero point amplitude $\sqrt{(h/M\omega)^{\frac{1}{2}}}$ where M is the mass of the vibrator. In the case of an isolated RE(S<<0.0), S can be incorporated in the exponential formula of Dexter:

$$W = \beta \exp(-\alpha\Delta E) \text{ with} \tag{3}$$

$$\alpha \quad (h\omega)^{-1} \left[\ln(p/S) - 1\right] \quad \text{for low temperature and} \tag{4}$$

$$\alpha = (h\omega)^{-1} \left[\ln(p/S)(n+1) - 1\right] \quad \text{for } T>0 \ ; \tag{5}$$

n being the phonon occupancy number or

$$n = (\exp(-h\omega/kT - 1)^{-1} \text{as explained in ref. 6.} \tag{6}$$

Application of the multiphonon theory to glasses requires the knowledge of the structural units forming the glass. Similar to the electronic spectra in glasses, the vibrational fequencies show inhomogeneous broadening due to the variation of sites. Table 1 shows the average frequencies of the network formers. The vibrations involving the network formers are lower by a factor of 2 to 4. The lack of symmetry in a glass and the molecular character of the high-energy vibration were taken into account by Layne et al (14) in developing the theory of multiphonon relaxation in glass by using higher-order terms in the perturbation theory. The dependence of a multiphonon rate on the energy gap to be bridged results then from the ratio of p-phonon process to that for a p-1 phonon decay. Assuming the average matrix elements to be the same for p and p-1 order processes, the ratio of W_p to W_{p-1} is

$$W_p/W_{p-1} = (h/2M\omega)(n+1)4m^2|\langle a|V^1|b\rangle|^2/h\omega^2 \ . \tag{7}$$

Since the perturbation is weak

$$W_p/W_{p-1} = \varepsilon \ll 1 \ . \tag{8}$$

This result leads to the following exponential dependence of the rate on the energy gap:

$$W_p = W_0\varepsilon^p = W \exp\left[\ln(\varepsilon)/h\omega\right]\Delta E \ . \tag{8}$$

Considering the dependence of W_p on the phonon occupation number from eq. (6), the rate for a p-order multiphonon decay is at temperature T>0

$$W_p = W_0\{n(T)+1\}^p \exp(-\alpha\Delta E) \text{ where} \quad (10)$$

$$\alpha = -\ln(\varepsilon)/h\omega \quad (11)$$

and $W_0 = \beta$ of eq. (3) are dependent on the host but independent of the specific electronic level of RE from which the decay occurs.

By comparing eq. (4) with eq. (11) one obtains the connection

$$\ln S/P = \ln\varepsilon - 1. \quad (12)$$

TABLE 1. PARAMETERS OF NONRADIATIVE RELAXATIONS

The host	β sec^{-1}	α cm	$h\omega$ cm^{-1}	ε
GLS, ALS[x]	$1\text{x}10^6$	$2.9\text{x}10^{-3}$	350	0.36
Tellurite[#]	$6.3\text{x}10^{10}$	$4.7\text{x}10^{-3}$	700	0.037
Germanate	$3.4\text{x}10^{10}$	$4.9\text{x}10^{-3}$	900	0.013
Phosphate	$5.4\text{x}10^{12}$	$4.7\text{x}10^{-3}$	1200	0.0037
Borate	$2.9\text{x}10^{12}$	$3.8\text{x}10^{-3}$	1400	0.0049
BeF_2[xx]	$9\text{x}10^{11}$	$6.3\text{x}10^{-3}$	500	0.042
$LaCl_3$	$1.5\text{x}10^{10}$	$13.0\text{x}10^{-3}$	260	0.037
$LaBr_3$	$1.2\text{x}10^{10}$	$19.0\text{x}10^{-3}$	175	0.037
LaF_3	$6.6\text{x}10^8$	$5.6\text{x}10^{-3}$	350	0.14
Y_2O_3	$2.7\text{x}10^8$	$3.8\text{x}10^{-3}$	550	0.12
SrF_2	$3.1\text{x}10^8$	$4.0\text{x}10^{-3}$	360	0.20
$Y_3Al_5O_{12}$	$9.7\text{x}10^7$	$3.1\text{x}10^{-3}$	700	0.045
$YAlO_3$	$5\text{x}10^9$	$4.6\text{x}10^{-3}$	600	0.063
$LiYF_4$	$3.5\text{x}10^7$	$3.8\text{x}10^{-3}$		
BaY_2F_8				

[x]Ref. 18; #R. Reisfeld, L. Boehm and N. Spector, The Rare Earths in Modern Science and Technology, (G.J. McCarthy and J. J. Rhyne, eds.), Plenum Press, N.Y., 513 (1978); [xx]C.B. Layne and M. J. Weber, Phys. Rev. B/,16, 3259 (1977); Ref. 2., p. 89.

II. ANALYSIS OF THE EXPERIMENTAL RESULTS

The multiphonon relaxation rates of RE in oxide glasses were studied by us (3) and by Weber and his group (14). The results show an exponential behavior of W on the energy gap. The low temperature results may be presented by a linear dependence of ln W versus ΔE or ln W versus p as predicted by eqs. (10 and 11).

The spectral behavior of RE in chalcogenide glasses of the composition $3Ga_2S_3.La_2S_3$(GLS) and $3Al_2S_3.La_2S_3$(ALS) were studied by A. Bornstein and the author (15-19). The multiphonon relaxation rates for a number of levels of Ho^{3+} were obtained from lifetime measurements subtracting the radiative transition probabilities using the formula

$$1/\tau_{meas} = W + \sum A, \qquad (13)$$

where the radiative transition probabilities A were calculated by the Judd-Ofelt formula.

In cases where energy differences were higher than $2500cm^{-1}$ and nonradiative probabilities small, the values were obtained from measurements of quantum efficiencies using the formula

$$\text{quantum yield} = \frac{\sum A}{\sum A + W} . \qquad (14)$$

Specifically, the latter formula was applied to the non-radiative decays for the levels $^4S_{3/2}$ and $^4F_{9/2}$ of Er^{3+}. The low temperature values of multiphonon relaxations in the chalcogenide glasses versus energy gap are presented in Fig. 1. Nonradiative rates for some levels of Nd^{3+} and Er^{3+} in fluoride glasses of the composition $49BeF_2.27KF.14CaF_2.10AlF_3$ at room temperature were studied by Layne and Weber (20) by measurement of the transient fluorescence followed by pulsed laser excitation of selected levels. Their results can also be presented by an exponential behavior.

The multiphonon decay rate from a given level to the next lower level decreases with the lowering of the energy of the stretching frequencies of the glass former since a large number of phonons is needed in fluoride glasses and more so in chalcogenide glasses in order to reach the same energy gap. Dependence of the multiphonon transition rate on the number of phonons from the emitting to the next lower level for a number of glasses and crystal hosts are presented in Fig. 2.

The parameters α and β of eq. (8) and ε of eq. (11) are presented in Table 1 together with the phonon frequences. The

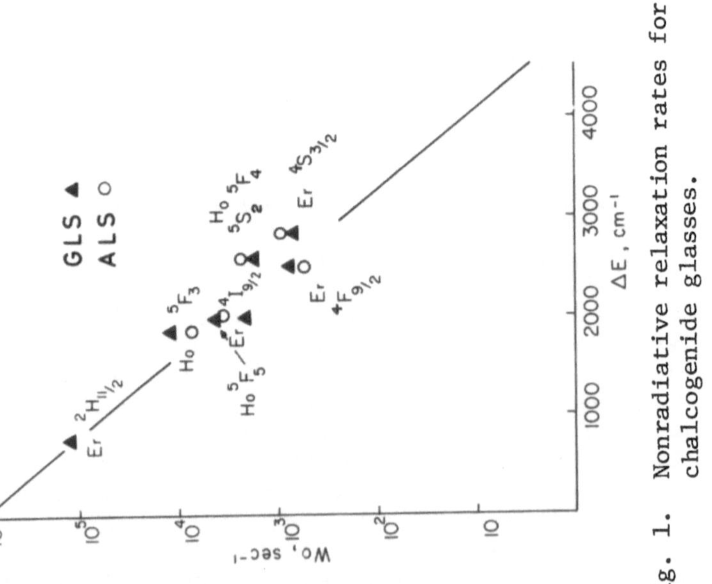

Fig. 1. Nonradiative relaxation rates for chalcogenide glasses.

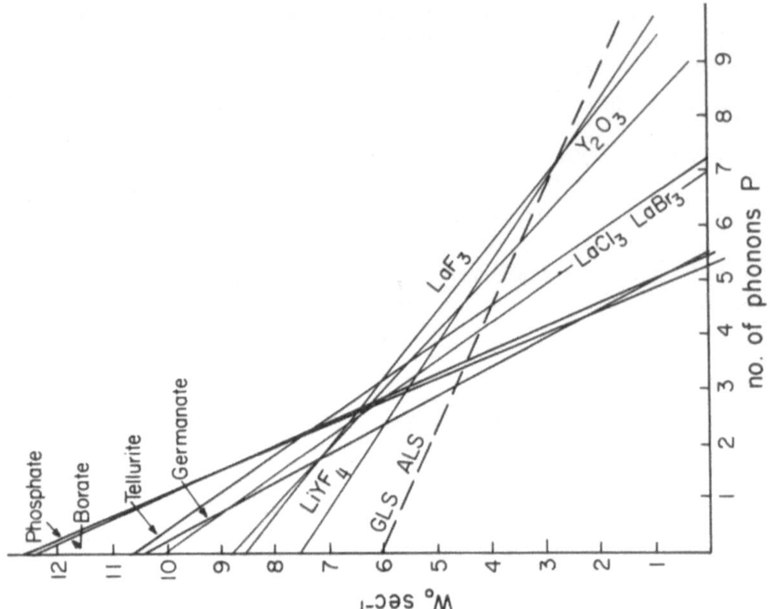

Fig. 2. Nonradiative relaxation; W_0 versus number of phonons p.

TABLE 2. VALUES OF S FOR 4 AND 5 PHONONS

Host	GLS	Tell	Ger	Phos	Borate	BeF_2	$LaCl_3$
ε	0.36	0.037	0.013	0.0037	0.0049	0.042	0.037
S(4p)	0.53	0.054	0.019	0.005	0.007	0.06	0.054
S(5p)	0.66	0.067	0.024	0.007	0.009	0.08	0.067

4p and 5p refer to 4 and 5 phonons respectively

Huang-Rhys-Pekar (HRP) number S was calculated using eq. (12) for 4 and 5 phonons. Its values are presented in Table 2. Our calculation differs slightly from that of Fonger and Struck (13) who used a different formula.

It should be noted that the measurement of lifetime and quantum efficiency from which the decay rates are calculated are reflecting an average of sites and Stark splittings of the specific levels, thus the parameters presented in the Table reveal a gross behavior of variation of the level energies. When the glasses are selectively excited by line-narrowing techniques, differences in multiphonon rates are observed (21) but even those cannot be attributed to specific sites having accidentally similar energies (22). Also when excited by a lines source, various rare earth ions, having 0→0 transitions, with the exception of Eu^{3+}, transfer energy between them by multiphonon processes; and both the lifetime and steady-state fluorescence observed, reflect the emission from a number of sites. Thus the numbers presented in the Table while not specific to a given site are still characteristic for a specific host. Low coupling constants in the oxide glasses are in accordance with the theory assuming weak coupling between the RE ion and its surroundings. A similar behavior is observed in the fluoride glasses. The chalcogenide glasses exhibit a much higher coupling constant while still showing exponential behavior between ln W_0 and the number of phonons. A plot of eq. (1) versus the number of phonons shows that an exponential behavior is observed only for $S \leqq 0.1$. Higher values of S show deviation from exponentiality. From this it can be concluded that the multiphonon relaxation theory as finally reflected by eq. (1) is a good approximation for glasses in which mainly an ionic bond is formed between the RE ion and its surrounding ligands. In the chalcogenide glass, the bond between the RE ion and the surrounding S ions is of a covalent type as reflected by the nephelauxetic effect (15,23). The covalency in the chalcogenide glasses is further reflected in the hypersensitive transitions in this glass (24).

Dr. W. H. Fonger has kindly sent us his computation of the orders of change of the function appearing in eq. (1) between 4 and 9 phonons as a function of values of S. From this calculation it is evident that the change in the oxide and fluoride glasses being of 10 orders of magnitude can be expressed as values smaller than 0.1. On the other hand, the changes in the chalcogenide glasses are of 2 orders of magnitude corresponding to the high number of S=2. As mentioned above the high number in the chalcogenide glass may be due to the covalency and inability to look on the RE incorporated in the chalcogenide glasses as a point charge.

Energy transfer from the UO_2^{2+} ion to various RE ions which are not in resonance with the electronic emitting state of the UO_2^{2+} ion (25) are presented as a function of energy gap between the uranyl level and the levels of the RE ions in which energy transfer is taking place. Here again an exponential behavior is observed according with the Dexter-Miyakawa theory (25) which predicts that the phonon-assisted energy transfer may be described by formula

$$W_{PAT}(\Delta E) = W_{PAT}(0)e^{-\beta\Delta E} , \qquad (15)$$

where ΔE is the energy gap between the levels of donor and acceptor ions and β is a parameter determined by the strength of electron-lattice coupling as well as by the nature of the phonon involved.

In reference (13) it is proposed that energy transfer probabilities should have the same functional dependence as eq. (1) where the HRP function is the sum of the donor and acceptor S values. Phonon-assisted energy transfer between trivalent RI ions in yttrium oxide was studied by Yamada, Shionoya and Kushida (27). An exponential behavior of the energy transfer rate on the energy gap resulted in $\beta=2.5\times10^{-3}$. β values obtained from Fig. 3 for energy transfer from a UO_2^{2+} ion to a RE ion in a phosphate glass equal 2.3×10^{-4}. The coupling constants obtained from these values using eq. (11) are $\hat{\varepsilon}=0.3$ for yttrium oxide and $\hat{\varepsilon}=0.76$ in phosphate glass. Since $\hat{\varepsilon}$ in this case reveals the sum of the coupling constant of the donor and acceptor ions (28), it is not surprising that when the acceptor ion is a uranyl which is much more strongly coupled to the host, the coupling constant is higher.

In conclusion, multiphonon processes in glasses can be expressed by an exponential behavior with the host-dependent parameters α and β. The coupling constants ε and S obtained from α reveal a weak interaction in the oxide and fluoride glasses and a much stronger interaction in the chalcogenide glasses being a result of the covalency between the RE's in the chalcogenide host.

The author is deeply grateful to Professors R. Englman and C. K. Jørgenson, Dr. Fonger and Mr. A. Bornstein for valuable

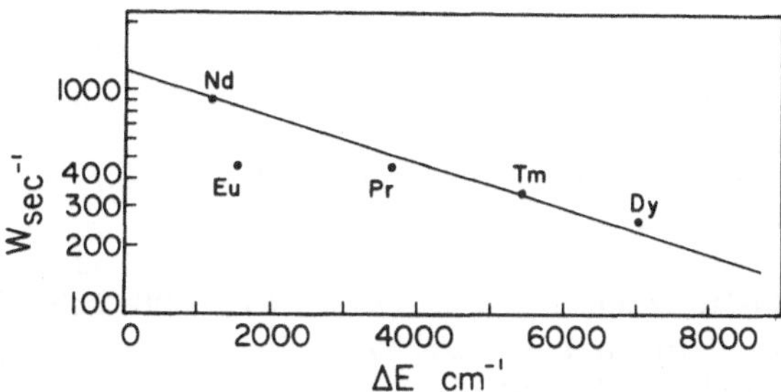

Fig. 3. Energy transfer probability from UO_2^{2+} to the rare earths.

discussions and to Mr. E. Greenberg for her help in preparation of the manuscript.

REFERENCES

1. Y.E. Perlin, Sov. Phys. Uspekhi 6, 542 (1963).
2. L.A. Riseberg and M.J. Weber in Progress in Optics, Vol. XIV (E. Wolf, ed.), North Holland (1976).
3. R. Reisfeld, Structure and Bonding 22, 123 (1975).
4. K.F. Freed in Radiationless Processes in Molecules and Condensed Phases (F.K. Fong, ed.), Springer-Verlag (1976).
5. F.K. Fong, Theory of Molecular Relaxation, John Wiley & Sons (1975).
6. F. Auzel in Luminescence of Inorganic Solids (B. Di Bartolo, ed.) Plenum Pres, N.Y., 1978.
7. R. Englman, Nonradiative Decay of Ions and Molecules in Solids, North Holland Pub. Co., Amsterdam, New York, Oxford (1979).
8. R. Reisfeld and Y. Eckstein, J. Chem. Phys. 63, 4001 (1975).
9. R. Reisfeld, J. Hormadaly and A. Muranevich, Chem. Phys. Lett. 38, 188 (1976); J. Hormadaly and R. Reisfeld, ibid 45, 436 (1977).
10. Y. Toyozawa in Dynamical Processes in Solid State Optics (R. Kubo, ed.), Benjamin, 90 (1976).
11. R. Englman and J. Jortner, Mol. Phys. 18, 145 (1970).
12. T. Miyakawa and D.L. Dexter, Phys. Rev. B 1, 2961 (1970).
13. W.H. Fonger and C.W. Struck, J. of Lum. 17, 241 (1978).
14. C.B. Layne, W.H. Lowdermilk and M.J. Weber, Phys. Rev. B 16, 10 (1977).

15. A. Bornstein, J. Flahaut, M. Guittard, S. Jaulmes, A.M. Loizeau-Lazac'h, G. Lucazeau and R. Reisfeld, The Rare Earths in Modern Science and Technology (McCarthy and Rhyne, eds.) Plenum Press, 599 (1978).
16. R. Reisfeld and A. Bornstein, J. Bodenheimer and J. Flahaut, J. Luminescence 18/19, 253 (1979).
17. R. Reisfeld, A. Bornstein, J. Flahaut, M. Guittard and A.M. Loireau-Lazac'h, Chem. Phys. Lett. 47, 408 (1977).
18. A. Bornstein, PhD Thesis (1979).
19. R. Reisfeld and A. Bornstein, J. Noncryst. Solids 27, 143 (1978).
20. C.B. Layne and M.J. Weber, Phys. Rev. B 16, 3259 (1977).
21. C. Brecher, L.A. Riseberg and M.J. Weber, Phys. Rev. B. 18, 5799 (1978).
22. Y. Kalisky, R. Reisfeld and Y. Haas, Chem. Phys. Lett. 61, 19 (1979).
23. R. Reisfeld and C.K. Jørgensen, Lasers and Excited States of Rare Earths, Springer-Verlag, Heidelberg (1977).
24. R. Reisfeld, A. Bornstein and J. Flahaut, The Rare Earths in Modern Science and Technology, Vol. 2, (McCarthy and Rhyne, eds.) Plenum Press (1979).
25. C.K. Jørgensen and R. Reisfeld, Chem. Phys. Lett. 35, 441 (1975).
26. R. Reisfeld and N. Lieblich-Sofer, "Phonon-Assisted Energy Transfer in Glasses" in Dynamical Processes in the Excited States of Ions and Molecules in Solids, DPC Wisconsin, Madison (1979).
27. N. Yamada, S. Shionoya and T. Kushida, J. Phys. Soc. Jap. 32, 1577 (1972).
28. N. Lieblich-Sofer, R. Reisfeld and C.K. Jørgensen, Inorg. Chim. Act 30, 259 (1978).

STIMULATED EMISSION IN THE PRESENCE OF STRONG NONRADIATIVE DECAY IN CRYSTALS

A. Kaminskii

Institute of Crystallography
Academy of Sciences of USSR
Leninsky pr. 59
117333 Moscow B-333, USSR

ABSTRACT

This article deals with the current state of the physics of laser crystals which present stimulated emission at three microns or longer wavelengths. In the beginning a short history and a review are given. Then some specific properties of self-saturating three-micron transitions of Er^{3+} ions and Ho^{3+} ions in insulating crystals are considered. Special attention is paid to the new functional schemes of three micron solid state lasers, in particular to those involving sensitization, deactivation, and to the so-called feed-flowing schemes. Finally, some problems concerning the prospects and possible applications of three-micron and longer wavelength crystal lasers are briefly discussed.

I. INTRODUCTION

The response of a condensed medium to the action of an external radiative excitation includes very different phenomena and processes. Some of these processes result in the change of the properties of the medium itself, and some others in the reemission of the exciting radiation with a change of wavelength. In the latter process, the medium acts as a "converter." In activated media, such as crystals or glasses, the efficiency of the conversion processes depends on many factors, among which a special place is occupied by nonradiative processes, in particular multiphonon nonradiative transitions.

This short contribution deals with the problem of obtaining 3-micron and longer wavelength stimulated emission of crystals doped with trivalent rare-earth ions. During the previous meeting of the School of Atomic and Molecular Spectroscopy in 1977, I dealt briefly with this problem (1). In the past few years, some new results in this field of physics and spectroscopy of laser crystals were obtained in a number of laboratories, and these results make us look at the problem of stimulated emission from a new point of view.

II. THE PROBLEM OF THE 3-MICRON LASER ACTION

When trying to obtain stimulated emission in the visible region or in a region close to it, one problem must be solved: how to make a strongly luminescing quantum system to generate laser emission. In our days this is not an experimental difficulty (2). When trying to obtain stimulated emission in the 3-micron or longer wavelength region, the problem presents some different kinds of difficulties. In many cases here, one should not call these systems luminescing, but nonradiatively decaying: the problem of 3-micron and longer wavelength crystal lasers is a problem of nonradiative transitions.

Many of us still remember the case of ruby, when Maiman, convinced of this crystal's high quantum yield, made experiments which lead to the creation of the first laser. In glasses doped with trivalent rare-earth ions, in particular with neodymium, there has been a long search for higher quantum yield. It is not an exaggeration to say that the first years of the crystal and glasses laser development went under the banner of search for lasing compounds with high luminescence quantum yield. Some rare experiments on stimulated emission excitation in activated crystals from initial levels characterized by low quantum yield, did not attract the attention of specialists and the results were considered not interesting from the practical point of view.

In the following years the notion about the acceptable value of the luminescence quantum yield of laser crystals began to change. Thus, at about the middle of the seventies, the investigations of a large number of self-activated neodymium crystals with weak luminescence concentration quenching and the development of miniature lasers for optical fiber communication systems and optoelectronics, showed that for the solution of some problems, a quantum yield of $\sim$35% is quite suitable.

Laser experiments, carried out in some laboratories in the Soviet Union and in the United States, showed that the creation of a crystal laser with a wavelength in the 3-micron region is connected with transitions of low luminescence quantum yield (3,4). The estimates are that the quantum yield of the known 3-micron laser levels is at best about only some percents. When concerned with 4- and 5-micron generation of activated crystals, experimenters should come across quantum yields of $\sim$ a fraction of a percent.

The first 3-micron laser was created in 1967 by Robinson and Devor from "Hughes Research Labs" on the base of mixed fluoride crystal with Er^{3+} ions (5). Because of a number of reasons, this work did not attract the attention of the specialists at the time. In the following years, the development of tunable parametric lasers generated a search for effective pumping sources for them. For this reason, a room-temperature laser based on a fluoride crystal with erbium ions and an output energy of 4 Joules, was used in the Lebedev Physical Institute in 1973 (6). In the same year, Johnson and Guggenheim produced a 3-micron laser using a fluoride crystal with Dy^{3+} ions, but because of the low position of the terminal laser level, this system worked only at liquid nitrogen temperature (7). Soon after, in 1974, 3-micron generation by Er^{3+} ions in a yttrium aluminum garnet host was obtained at the Lebedev Physical Institute (8) and in a year, the same was achieved in our Institute, but using Ho^{3+} ions (9).

Further progress in the 3-micron laser field was achieved after some new properties of the multiphonon nonradiative processes in crystals doped with rare-earth ions were understood. It was clear for instance, that the activity of these processes in a number of structurally complex crystals, is not equally determined by all the lattice vibrations of these crystals (2). For instance, the lattice vibration spectrum of yttrium aluminum garnet is stretching to ≃ 860 cm^{-1}, but the most active vibrational modes in the electron-phonon interaction are the ones with energy 350-500 cm^{-1}; this is illustrated by the curve of the effective phonon density of this crystal (see Fig. 1). Such facts have shown, that, under special pumping conditions, 3-micron and longer wavelength generation can be achieved in various crystals doped with trivalent rare-earth ions.

Recently Esterowitz, Eckhardt, and Allen of the "Naval Research Lab," using the principle of cascade generation in activated crystals, suggested by me in 1971 (10), have obtained laser action at 3.9-micron wavelength by using Ho^{3+} ions in $LiYF_4$.

Figure 2 presents the simplified level schemes of Dy^{3+}, Ho^{3+}, and Er^{3+} ions. It follows, from this figure, that 3-micron generation is connected with the transitions $^5I_6 \to {}^5I_7$ of Ho^{3+} ions, $^4I_{11/2} \to {}^4I_{13/2}$ of Er^{3+} ions and $^6H_{13/2} \to {}^6H_{15/2}$ of Dy^{3+} ions. These transitions are indicated by double arrows. The 4-micron transition of Ho^{3+} ion in the channel $^5I_5 \to {}^5I_6$ is marked with a thick arrow. All lasers, except dysprosium, have 4-level operation schemes with high terminal levels and can operate at room and higher temperatures.

III. LASER ACTION AT 3 MICRON

A search carried out lately resulted in the finding that more then 20 crystals are now known which can lase in the 3-micron retion at room temperature. The list of such crystals is given

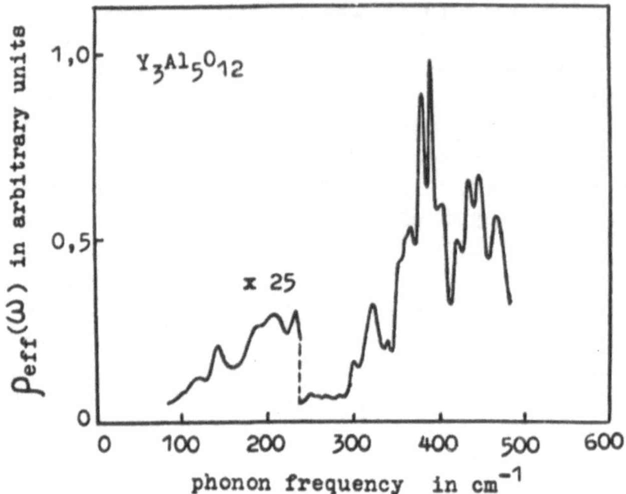

Fig. 1. Effective phonon density of the $Y_3Al_5O_{12}$ crystal.

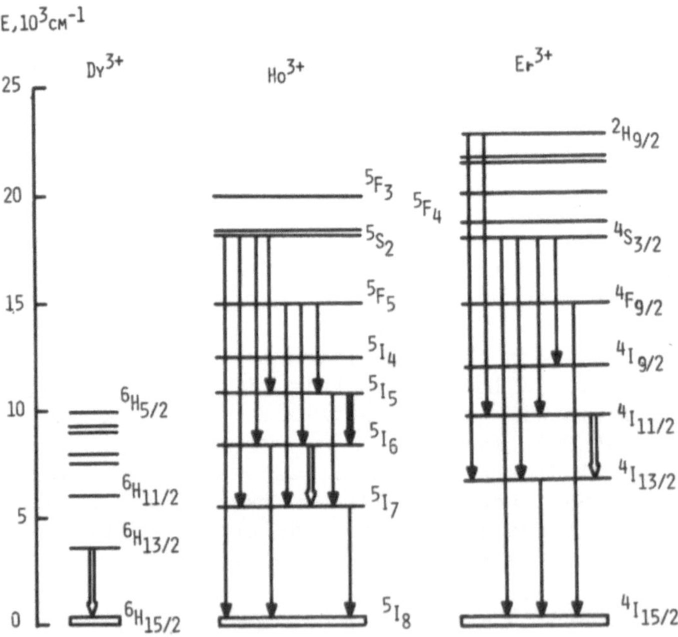

Fig. 2. Laser channels of the Dy^{3+}, Ho^{3+}, and Er^{3+} ions.

in Tables 1 and 2. The excitation thresholds of 3-micron lasers are only few tens of Joules.

The 3-micron laser transitions are self-saturating transitions, namely, transitions in which the lifetime of the initial laser level is smaller than that of the terminal laser level. Here, we have a picture which is the reverse of that of conventional neodymium laser schemes, where the lifetime of the initial state is thousands or more times longer than that of the terminal one. As an example, let me consider yttrium aluminium garnet doped with Er^{3+} (see Fig. 3). Here the 3-micron laser transition $3 \rightarrow 2$ is characterized by a small luminescence branching ratio β_{32} (about 15%) and a relatively high nonradiative transition probability W_{32} (about 10^4 sec^{-1}) Therefore, the quantum yield from this level is just a little more than 1%. The other 3-micron erbium laser crystals have nearly the same characteristics. These data show, that in order to obtain stimulated emission in self-saturated $3 \rightarrow 2$ transition, pulse pumping is demanded. It is necessary to create a population inversion between 3 and 2 levels for such time interval, during which nonradiative transition $3 \rightarrow 2$ will not deactivate the excited laser level 3. The laser excitation condition in these transitions is outlined in Fig. 4. In this figure, the positions of the three lowest levels of the Er^{3+} ions in yttrium aluminium garnet are shown. From the figure, it follows that the initial laser level $^4I_{11/2}$ is practically free from concentration quenching. This is due to the absence of cross-relaxation channels responsible for the concentration quenching of the 3-rd level in this system. The possibility opens for the generation of 3-micron emission by using crystals with 100% concentration of Er^{3+} ions and the laser experiments confirm this conclusion. Now, already two erbium self-activated crystals lasing in the 3-micron transition at room temperature, are known. They are $Er_3Al_3O_{12}$ and $KEr(WO_4)_2$. In Fig. 4 it can be seen that the lasing ions concentration is higher than that of Nd^{3+} in crystals of the pentaphosphate type. The slight concentration dependence of the $^4I_{11/2}$ state's decay time seen in Fig. 4 may be due to undesirable quenching impurity. The laser threshold of self-activated erbium crystals, as seen from Table 2, are practically the same as in the crystals with low Er^{3+} ion concentration.

It is interesting to consider how the self-saturation effect is displaying itself in the 3-micron $^4I_{11/2} \rightarrow {}^4I_{13/2}$ transition of Er^{3+} ions in $Y_3A\ell_5O_{12}$. In the left part of Fig. 5, we show the crystal-field splitting scheme of the $^4I_{11/2}$ and $^4I_{13/2}$ levels with the indication of level energy (in cm^{-1}) and the laser transitions (in μm) for room and liquid nitrogen temperatures. In the right part, we show the usually observed spiking picture of the 3-micron stimulated emission in correspondence to the different lines of the self-saturating transition. At liquid nitrogen temperature and with excitation energy several times

Table 1. 3- AND 4- MICRON HOLMIUM CRYSTAL LASERS

Crystal	Ho^{3+} Ion Concentration at. %	Laser Wavelength µm	Laser Threshold J	Reference
$LiYF_4$	2	2.850 2.95	25-40	
	1	3.914	laser pump	
$BaYb_2F_8$	5	2.9054	6	present work
$KY(WO_4)_2$	3	2.9395	14	11
$KGd(WO_4)_2$	3-5	2.9342	25	12
$KLu(WO_4)_2$	3	2.9445	13	
$YAlO_3$	2-10	2.9180 3.0132	7.5 50	13
$Y_3Al_5O_{12}$	5-10	2.9403	25	9,13
$LaNbO_4$	2	2.8510	50	14
$Gd_3Ga_5O_{12}$	10	≈ 2.9	115	15
$Lu_3Al_5O_{12}$	5-10	2.9460	40	9,13

TABLE 2. 3-MICRON ROOM TEMPERATURE ERBIUM CRYSTAL LASERS

Crystal	Er^{3+} Ion Concentration at. %	Laser Wavelength µm	Laser Threshold J	Reference
$LiYF_4$	2	2.81 2.870	- 130	 16
CaF_2	10	2.7307	400	6
CaF_2-TmF_3	12.5	2.69	10	5
$KY(WO_4)_2$	3-50	2.8070	30	17
$KGd(WO_4)_2$	3-30	2.7222 2.7990	50 15	12
$KEr(WO_4)_2$	100	2.8070	20	18
$KLu(WO_4)_2$	25	2.8092	15	19
$YAlO_3$	2	2.7309	40	9,13
$Y_3Al_5O_{12}$	3-50	2.7953 2.8302 2.9365	10 10 10	present work 20 8
$Er_3Al_5O_{12}$	100	2.9367	12	20
$Gd_3Ga_5O_{12}$	10	2.8218	125	15
$Lu_3Al_5O_{12}$	3-50	2.7987 2.8289 2.9395	10 10 10	present work 9,21

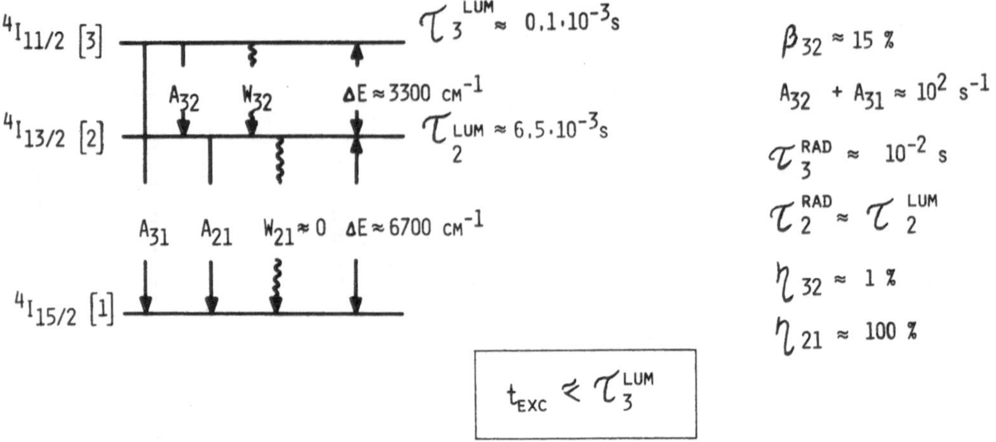

Fig.3. $Y_3Al_5O_{12}:Er^{3+}$ (at low concentration)

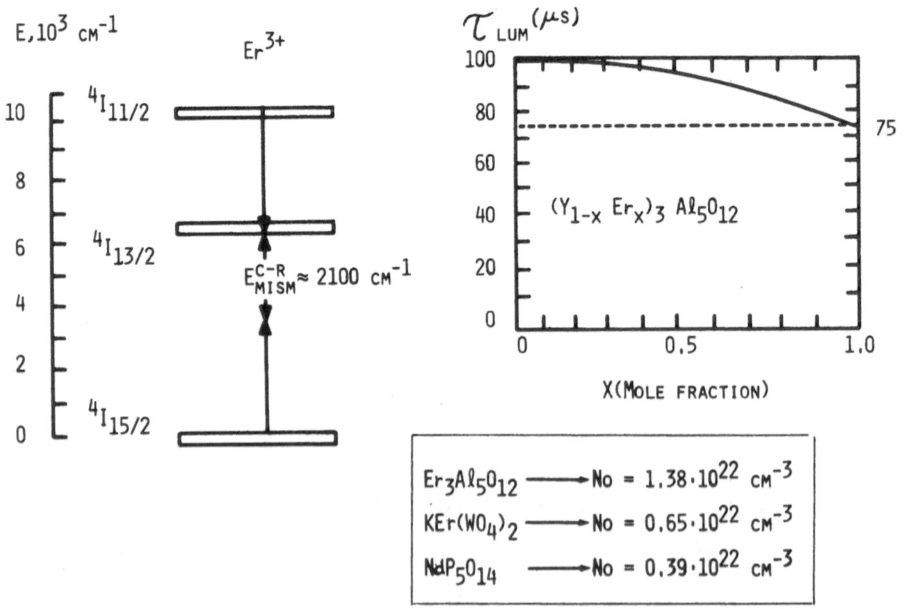

Fig. 4. Concentration quenching in the $(Y_{1-x}Er_x)_3Al_5O_{12}$ system

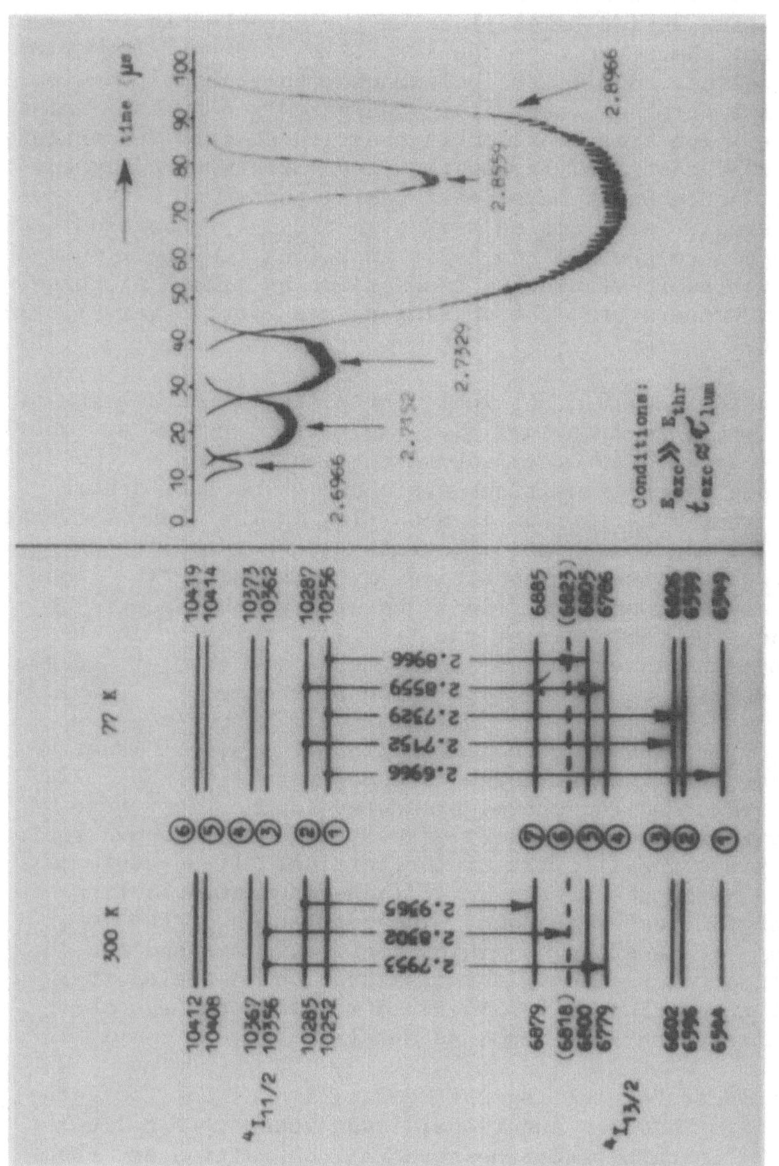

Fig. 5. (Left) Crystal field splitting of the $^4I_{11/2}$ and $^4I_{13/2}$ manifolds and three micron laser transitions of a $Y_3A\ell_5O_{12}:Er^{3+}$ crystal at 77 and 300K. (Right) Time evolution of a 3-micron laser action of the self-saturating $^4I_{11/2}\rightarrow{}^4I_{13/2}$ channel in a $Y_3A\ell_5O_{12}:Er^{3+}$ crystal at ∿77K.

higher than threshold with duration $t_{exc} \simeq \tau_{lum}$, it is seen that laser action takes place first at the short-wavelength transition 1→1 (2.6966 μm). This transition is then rapidly saturated. Laser action changes at once to the transition 2→3 (2.7152 μm), then to the transition 1→2 (2.7329 μm) and only after about 30 microseconds, laser action takes place at the "stable" long-wavelength lines. Here, we have the so-called "red shifting" effect of laser lines, due to the Boltzmann population of the low Stark levels of a terminal laser state characterized by long value of its lifetime. The same behavior is observed at room temperature. Naturally, in this case, only transition connected with a higher group of $^4I_{13/2}$ state Stark levels are taking part: at first, for short time interval, laser action arises in the 3→4 transition, and then it is changed to the 3→6 and 2→7 transitions. Here it must be noted that in the short-wavelength transitions at liquid nitrogen, as well as room temperature, the spiking pulses carry a very little part of the total energy.

Let us consider now Fig. 6, where three simplified schemes of the Er^{3+} ions levels in luthetium aluminum garnet crystal at room temperature are shown. In order to prevent overloading of the figure, only some laser transitions are shown. The left scheme shows the case when laser action is occurring in the long-wavelength transitions. The middle scheme illustrates the possibility of improvement of the excitation conditions by introducing Yb^{3+} ions into the crystal as a sensitizer ion. Naturally, the spectral composition and time character of the Er^{3+} ions emission in the presence of the sensitizer ions is the same as for the left scheme. Sensitization here is occurring by a resonant scheme.

Another way of improving the excitation of 3-micron laser systems is connected with the usage of the crystals with deactivator ions according to the scheme presented in the right part of Fig. 6. In this case, the deactivators are Tm^{3+} ions and their role is to decrease the lifetime of the terminal $^4I_{13/2}$ state of the Er^{3+} ions. Here, the mechanism of (non-resonant) electron excitation energy transfer is used. The population of the lower Stark levels of $^4I_{13/2}$ state of erbium ions decreases and the 3-micron generation takes place without saturation in the short-wavelength 1→1 transition at 2.6996 μm. Cr^{3+} ions can also be used as sensitizers and Ho^{3+} ions, as deactivators.

Fig. 7 demonstrates the logical combination of the two ideas considered above. This new functional laser scheme was called "feed-flowing." In this case, the spectral composition and the generation kinetics will be the same as in the case of the scheme with deactivator ions.

All the schemes considered here have been experimentally realized: the relevant data are shown in Table 3. Then, it is

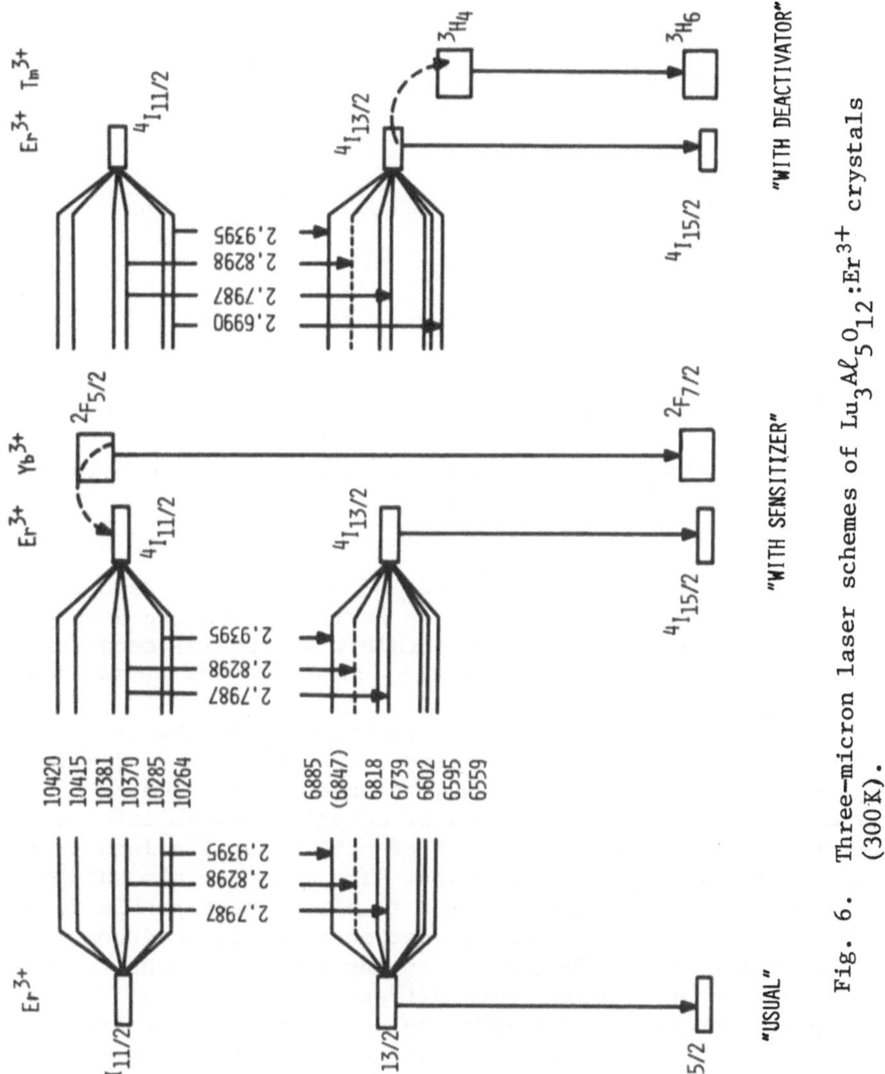

Fig. 6. Three-micron laser schemes of $Lu_3Al_5O_{12}:Er^{3+}$ crystals (300 K).

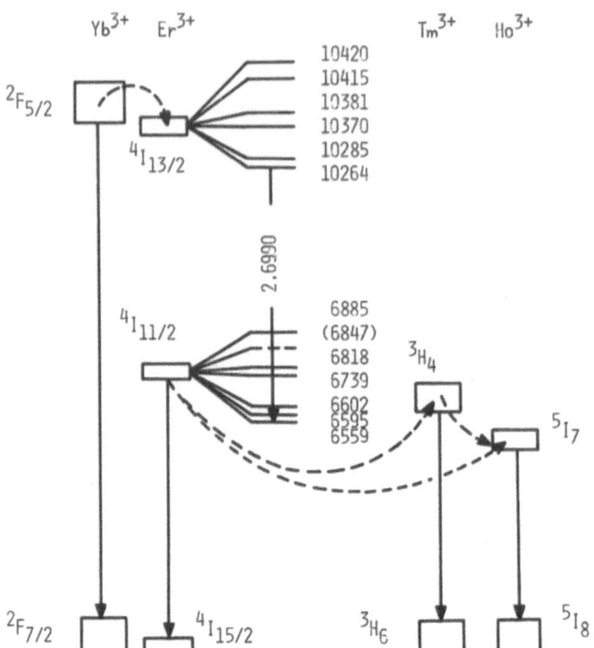

Fig. 7. Feed-flowing laser scheme ($Lu_3A\ell_5O_{12}$:Ho^{3+},Tm^{3+},Yb^{3+}-Er^{3+}, 300K).

seen that the excitation thresholds of some crystals are quite acceptable for their development as devices. One can assume that new functional schemes in the crystals with optimal concentrations of coactivator ions will permit the achievement of better parameters.

We shall now examine briefly the problem of the optimization of the compositions and coactivators concentration in 3-micron laser crystals. In our investigations, we have faced a number of difficulties. Among them the absence of detailed enough information about the spectroscopic behaviour of both separated ions pairs and group of ions in any crystal hosts. The measurements showed that what is optimal for a pair of ions is often unacceptable for a triplet of ions, and, of course, for more complicated combinations of coactivator ions. Therefore, in our investigations, we needed a large amount of crystals with different compositions and of perfect optical quality. For instance, experiments with erbium and its coactivators in luthetium aluminum garnet needed more than fifty samples. But still, we were able only to define the region of optimal concentrations of coactivator ions: I wish to underline only the region. Such series of crystals are difficult not only to grow, but also to investigate. The insufficient chemical purity of the starting materials used for the crystal growth is also a serious barrier to such investigations.

Table 3. IMPROVED ROOM TEMPERATURE 3-MICRON CRYSTAL LASERS

Crystal	Coactivator Ion			Laser Wavelength µm	Laser Threshold J	Reference
	Laser	Sensitizator	Deactivator			
$BaYb_2F_8$	Ho^{3+}	Yb^{3+}	-	2.9054	6	present work
$Lu_3Al_5O_{12}$	Ho^{3+}	Yb^{3+}, Cr^{3+}	-	2.9460	10	22
CaF_2-TmF_3	Er^{3+}	-	Tm^{3+}(?)	2.69	10	5
$KY(WO_4)_2$	Er^{3+}	-	Ho^{3+}, Tm^{3+}	2.6887	10	23
$Y_3Al_5O_{12}$	Er^{3+}	Yb^{3+}	-	2.7953	6	22
				2.8302	6	
				2.9365	6	
		Yb^{3+}, Cr^{3+}	Tm^{3+}	2.6975	5	3,4,22
				2.7953	5	
				2.8302	5	
				2.9365	15	
		-	Ho^{3+}, Tm^{3+}	2.6975	16	
$Lu_3Al_5O_{12}$	Er^{3+}	Yb^{3+}, Cr^{3+}	-	2.7987	6	22
				2.8298	6	
				2.9395	6	
		Yb^{3+}, Cr^{3+}	Tm^{3+}	2.6990	6.5	3,4,22
				2.7987	6.5	
				2.8298	6.5	
				2.9395	60	
		Yb^{3+}	Ho^{3+}, Tm^{3+}	2.6990	≈ 10	

Investigating the spectral and laser properties of the Er^{3+}-Yb^{3+} and Er^{3+}-Ho^{3+} pairs, it was found that the increase in concentration of sensitizer and deactivator ions not always results in desirable results. Also, investigations on electron-excitation energy transfer show that it is more advantageous to work at higher concentrations of the assisted ions. In particular, our measurements show that at Yb^{3+} ions concentration of more than 10 at. % the excitation threshold of the Er^{3+} ions 3-micron emission increases sharply. It was found that one of the causes of this undesirable phenomenon can be the influence of the "screening" action of an intense absorption band of the Yb^{3+} ions, hampering the direct excitation of the initial $^4I_{11/2}$ state of the Er^{3+} ions. Measurements also showed that, for some crystals, Auzel's mechanisms are playing a role, when the excitation from resonantly connected levels $^4I_{11/2}$ of Er^{3+} and $^2F_{5/2}$ of Yb^{3+} is transferring to the higher levels of the Er^{3+} ions. In the Er-Ho pair at high concentrations, an essential role is played by the absorption at the laser wavelength from the excited Ho^{3+} ions state $^5I_7$ with long lifetime, because the Ho^{3+} ions also have a 3-micron channel connected with this state. These are only some of the difficulties encountered during the search for ways of improving the parameters of 3-micron lasers. Besides the problem of multiphonon nonradiative transitions, which decrease the luminescence quantum yield from the initial state in these crystals, there are evidently some other problems related to various manifestations of the electron-phonon interaction, namely, other nonradiative processes such as energy transfer, up-conversion, concentration quenching, etc.

IV. APPLICATIONS OF 3-MICRON CRYSTAL LASERS

The future development of 3-micron crystal lasers will depend on a lot of factors, the main of which is their suitability to applications. Here, of course, a serious difficulty arises connected with the fact that the 3-micron emission, or the emission at 2.7-2.9 micron is falling into the absorption band of water molecules (H_2O) and of carbon dioxide gas (CO_2). Therefore, for such lasers some special fields of application have to be found.

Here we can enumerate some scientific fields, where 3-micron laser may be useful: molecular spectroscopy, photochemistry physics and chemistry of surfaces, physics of semiconductors with narrow energy gap, and laser isotope separation. Recently I received from Poland a book on laser applications in mining. The physicist who sent it to me called in his letter my attention to the chapter where laser methanometers (for detecting CH_4) were described. Now, in such devices He-Ne lasers at 3.39 micron are used. But gas lasers have some shortcomings when compared with crystal lasers (compactness, Q-switching regime, etc.)

It seems to me that the 3-micron laser transitions of the Ho^{3+} and Er^{3+} ions can also help in the construction of effective crystal lasers with the emission in the region of 1.6 - 2 micron by using the cascade principle of stimulated emission in the transitions $^5I_6 \rightarrow {}^5I_7 \rightarrow {}^5I_8$ of the Ho^{3+} ions and $^4I_{11/2} \rightarrow {}^4I_{13/2} \rightarrow {}^4I_{15/2}$ of the Er^{3+} ions (see Fig. 8). We have already obtained some positive results with such schemes. It should be noted that the second laser transition in such schemes improves the 3-micron laser action because here a prompt deactivation of the terminal 3-micron level takes place.

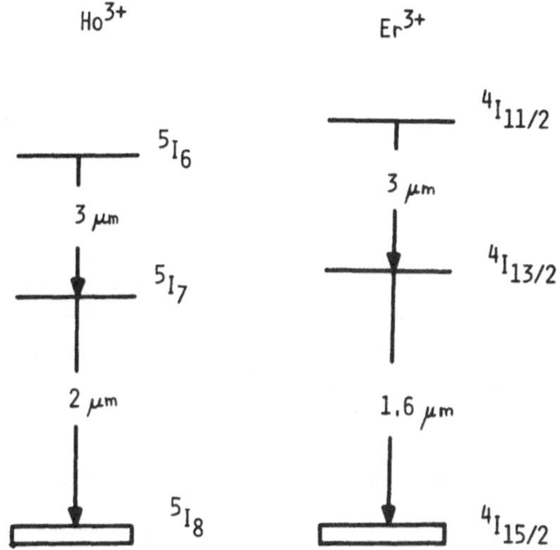

Fig. 8. Cascase laser schemes of the Ho^{3+} and Er^{3+} ions in garnet crystals

The investigations of 3-micron laser transitions of crystals with rare-earth ions undoubtfully discredits the fear of specialists regarding low luminescence quantum yields. The number of laboratories involved in such investigations is increasing. A step can be made forward into the 4-micron region. I like to add that everything I have reported here has been possible due to the fact that many physicists, several of whom are contributing to this volume, have actively investigated the different manifestations of nonradiative processes in condensed media.

REFERENCES

1. A. A. Kaminskii, in Luminescence of Inorganic Solids, B. Di Bartolo, ed., Plenum Press, New York and London, 1978, p. 511.
2. A. A. Kaminskii, Laser Crystals - Their Physics and Properties, Springer, Verlag, Berlin, Heidelberg, New York, 1979.
3. A. A. Kaminskii, Izv.Akad.Nauk SSSR, Ser.neorgan.Mater. 15, 1028 (1979).
4. A. A. Kaminskii, Izv.Akad.Nauk SSSR, Ser.fizicheskaya 43, 1169 (1979).
5. M. Robinson and Devor, Appl. Phys. Lett. 10, 167 (1967).
6. S. Ch. Batygov, L. A. Kulevskii, S. A. Lavrykhin, et al., Kurzfassungen, Internationale Tagung "Laser and ihre Anwendungen" Teil 2, Dresden, 1973, S.K97.
7. L. F. Johnson and Guggenheim, Appl. Phys. Lett. 23, 96 (1973).
8. E. V. Zharikov, V. I. Zhekov, L. A. Kulevskii, et al., Kvantovaya Elektron. 1, 1867 (1974).
9. A. A. Kaminskii, T. I. Butaeva, A. M. Kevorkov, et al., Izv.Akad. Nauk SSSR, Ser.neorgan.Mater. 12, 1508 (1976).
10. A. A. Kaminskii, Izv.Akad.Nauk SSSR, Ser.neorgan.Mater. 7, 904 (171).
11. A. A. Kaminskii, A. A. Pavlyuk, Tran Ngoc et al., Dokl.Akad.Nauk SSSR 245, 575 (1979).
12. A. A. Kaminskii, A.A. Pavlyuk, T. I. Butaeva et al., Izv. Akad. Nauk SSSR, Ser. Neorgan. Mater. 13, 1541 (1977).
13. A. A. Kaminskii, T. I. Butaeva, A. O. Ivanov et al., Zh.Tekhn. Fiz. Pis'ma 2, 787 (1976).
14. A. A. Kaminskii, V. A. Fedorov and Tran Ngoc, Izv.Akad.Nauk SSSR, Ser.neorgan.Mater. 14, 1357 (1978).
15. A. A. Kaminskii, V. A. Fedorov, S. E. Sarkisov et al., Phys. Stat. Sol. (a), 53, K219 (1979).
16. A. M. Tkachyuk and M. V. Petrov, Optika i Spektroskopiya 45, 147 (1978).
17. A. A. Kaminskii, A. A. Pavlyuk, I. F. Balashov et al., Izv. Akad. Nauk SSSR, Ser.neorgan.Mater. 14, 2256 (1978).
18. A. A. Kaminskii, A. A. Pavlyuk, T. I. Butaeva et al., Izv.Akad. Nauk SSSR, Ser.Neorgan.Mater.15, 549 (1979).
19. A. A. Kaminskii, A. A. Pavlyuk, N. R. Agamalyan et al., Izv. Akad.Nauk SSSR, Ser.neorgan.Mater., 15, 1495 (1979).
20. A. M. Prokhorov, A. A. Kaminskii, V. V. Osiko et al., Phys. Stat. Sol. (a) 40, K69 (1977).
21. A. A. Kaminskii, T. I. Butaeva, V. A. Fedorov et al., Phys. Stat.Sol. (a), 39, 541 (1977).
22. A. A. Kaminskii and S. G. Petrosyan, Dokl.Akad.Nauk SSSR 246, 63 (1979).
23. A. A. Kaminskii, A. A. Pavlyuk, A. M. Kevorkov et al., Izv. Akad.Nauk SSSR, Ser.neorgan.Mater. 13, 582 (1977).

THE ROLE OF POLARONS IN RADIATIONLESS PROCESSES

R. Evrard

Institut de Physique, Université de Liège
B-4000 Sart Tilman/Liège 1, Belgium

ABSTRACT

The theory of electron-phonon interactions in semiconductors is reviewed. The effective Hamiltonian is derived for phonons with large wavelengths in three cases: the deformation-potential interaction with acoustic phonons, the Fröhlich polar interaction with longitudinal-optical phonons and the piezoelectric coupling with acoustic phonons. Some possible applications of polaron theory to radiationless processes are briefly discussed.

I. INTRODUCTION

The word polaron has been introduced to refer to an electron interacting with the electric polarization in an ionic crystal. The reason for using a new word is to emphasize the difference in behaviour between a "bare" electron in a rigid lattice and the composite system formed by the electron and the associated polarization. Indeed an electron slowly moving in an ionic crystal is accompanied by a lattice distortion due to the Coulomb interaction with the crystal ions. The system has an effective mass larger than the band mass since the distortion cloud contributes to the inertia and the electron has an extra self-energy in the polarization field. This justifies calling this composite particle by a different name: a polaron.

Polarons by themselves have a rather academic interest. Indeed the band energies and band masses are not precisely known, so that the corrections to these quantities due to the electron-polarization interaction are not directly measurable. However, polaron theory is just an aspect of the physics of electron-phonon (e-ph) interactions. Obviously, the latter is of primary importance in understanding many properties of electrons in solids and more specially the radiationless processes. Therefore, in this seminar, I would like to briefly describe the mechanisms responsible for the e-ph interaction in semiconductors and show how the concept of polarons can bring an interesting contribution to the study of several radiationless processes.

II. ELECTRON-PHONON INTERACTIONS IN SEMICONDUCTORS

In principle, one could start from the ion cores and valence electrons (as well as conduction electrons) and study the motion of all these particles in interaction. However, this seems an impossible task in practice due to the large number of particles involved in the problem. Moreover, the conduction band for the electrons and the valence band for the holes form continuous energy spectra so that the electronic particles can undergo low energy transitions, corresponding to slow motions. Therefore, the Born-Oppenheimer approximation is not applicable to these particles.

Thus, it is better for practical purposes to assume that the problem of the intrinsic semiconductor at $T = 0^{\circ}K$ has first been solved and that its band structure as well as its vibratory modes are known. The conduction electrons or valence holes are then introduced and their interactions with the ion displacements are calculated. This procedure seems valid for non-degenerate semiconductors where the particle concentration is low enough for the vibrations to remain unaffected, except possibly for the existence of local modes due to the presence of impurities or defects.

The derivation of the e-ph interaction is very simple in principle. The electronic particle (electron or hole; from now on, we will simply say electron) moves in a potential $V(\vec{r},\vec{R})$ due to the Coulomb forces between the ion cores and the electrons. Here $\vec{r}$ denotes the position of the electron and $\vec{R}$ is a collective notation for the position of the ion cores. In using such a type of potential, we neglect the correlations with the position of the other electrons, i.e. we use a one-electron approximation which is probably valid for non-degenerate semiconductors. $\vec{R}_o$ is the equilibrium position. Due to the lattice vibrations, the ions are displaced by $\vec{u}$ so that their position $\vec{R}$ is

$$\vec{R} = \vec{R}_o + \vec{u} , \tag{1}$$

and the potential is changed from $V(\vec{r},\vec{R}_o)$ to $V(\vec{r},\vec{R}_o + \vec{u})$ so that the e-ph interaction potential can be written as

$$V_{e-ph}(\vec{r},\vec{u}) = V(\vec{r},\vec{R}_o + \vec{u}) - V(\vec{r},\vec{R}_o) . \tag{2}$$

To calculate this interaction potential is not an easy task. Indeed, the ion oscillations not only change the direct Coulomb interactions with the conduction electron, but also modify the valence bands and the electronic charge distribution associated with them. Therefore, the electron-electron Coulomb and exchange forces give an important contribution to the e-ph interaction. This means that the complete band structure should in principle be calculated for all the ion displacements $\vec{u}$ of importance in the problem, which is almost impossible. Therefore one restricts oneself to the case where the displacement amplitudes are small enough for an expansion of $V(\vec{r},\vec{R}_o,\vec{u})$ restricted to the order linear in $\vec{u}$ to be valid.

In this linear approximation, the interaction potential is the sum of independent contributions coming from all the possible lattice modes λ. To describe these modes, we use annihilation and creation operators a_λ and a_λ^+. The linear expansion for the potential acting on the electron gives

$$V(\vec{r},\vec{R}) = V(\vec{r},\vec{R}_o) + \sum_\lambda [v_\lambda(\vec{r})a_\lambda + v_\lambda^+(\vec{r})a_\lambda^+] \ . \qquad (3)$$

The function $v_\lambda(\vec{r})$, which we call <u>interaction function</u>, describes the strength of the interaction with the mode λ. It can be calculated by allowing static ion displacements corresponding to small a_λ's and calculating $V(\vec{r},\vec{R})$ by some method of band theory, the pseudo-potential method for instance. Knowing the rigid band potential $V(\vec{r},\vec{R}_o)$ one deduces the interaction function by subtraction according to eq. (3). However, a band calculation for a distorted lattice is not an easy task.

The main goal of this seminar is to describe how to obtain the interaction function in a simpler way for some particular cases. But, before coming to this point, let us introduce the band-mass approximation and show what the e-ph becomes with this approximation. From now on, we discard the local modes and focus attention on the case of phonons.

III. EFFECTIVE HAMILTONIAN FOR e-ph INTERACTION.

Due to the translational invariance of the lattice, the e-ph interaction potential must have the following form

$$V_{e\text{-}ph} = \sum_{\vec{q}} [v_{\vec{q}}(\vec{r})a_{\vec{q}}e^{i\vec{q}\cdot\vec{r}} + v_{\vec{q}}^+(\vec{r})a_{\vec{q}}^+ e^{-i\vec{q}\cdot\vec{r}}], \qquad (4)$$

where $v_{\vec{q}}(\vec{r})$ has the same periodicity as the lattice. In this expression, $\vec{q}$ denotes the phonon wave vector and the sum is taken not only over all the values of $\vec{q}$ in the first Brillouin zone, but also over the different types of phonons (acoustic and optical, longitudinal and transverse).

The total Hamiltonian for the electron and the phonons in interaction is

$$H = \frac{p^2}{2m_o} + V(\vec{r},\vec{R}_o) + \sum_{\vec{q}} h\omega_{\vec{q}}\, a^+_{\vec{q}} a_{\vec{q}} + V_{e-ph} \,. \tag{5}$$

In this expression, $\vec{p}$ is the electron momentum operator, m_o its mass in vacuum, $\omega_{\vec{q}}$ the frequency of the phonon with wave vector $\vec{q}$ and $V(\vec{r},\vec{R}_o)$ the potential energy of the electron in the rigid crystal.

In semiconductor physics, effective Hamiltonians are introduced whenever possible. The basic idea is to replace the actual Hamiltonian acting on the total Bloch state

$$\psi^{(i)}_{\vec{k}}(\vec{r}) = V^{-\frac{1}{2}}\, u_{i,\vec{k}}(\vec{r})\, e^{i\vec{k}\cdot\vec{r}} \,, \tag{6}$$

by an effective operator acting only on the plane wave part of this wavefunction, namely exp $(i\vec{k}\cdot\vec{r})$. In the Block state of eq. (6), $\vec{k}$ is the electron wave vector and i the band to which it pertains. V is used for the volume of the crystal and $u_{i,\vec{k}}(\vec{r})$ has the periodicity of the lattice.

To obtain the effective Hamiltonian, take a matrix element of eq. (5) between two different Bloch waves with $\vec{k}$,i and $\vec{k}'$,j as respective indices. The first two terms correspond to the motion of the electron in the rigid lattice, whose solutions are given by the Bloch states themselves. Therefore

$$\langle i,\vec{k}\left|\frac{p^2}{2m_o} + V(\vec{r},\vec{R}_o)\right|j,\vec{k}'\rangle = \varepsilon_i(\vec{k})\,\delta_{i,j}\,\delta_{\vec{k},\vec{k}'} \,, \tag{7}$$

where $\varepsilon_i(\vec{k})$ is the energy of the state $\vec{k}$ in the band i. Close to a minimum or a maximum of the band, a parabolic approximation can be used, i.e.

$$\varepsilon_i(\vec{k}) = \varepsilon_i(o) + \frac{h^2k^2}{2m} \,, \tag{8}$$

where $\varepsilon_i(o)$ is the energy at the extremum, m the band mass and $\vec{k}$ is measured with respect to the position of the extremum in the Brillouin zone. We suppose that the band is isotropic about $\vec{k}$ = o. Moreover, we restrict ourselves to the case of a single extremum of interest in each band (no degeneracy). The extension to a more general situation is straightforward.

The matrix elements appearing in the calculation of the e-ph interaction are of the following type

$$<i,\vec{k}|v_{\vec{q}}(\vec{r})e^{i\vec{q}\cdot\vec{r}}|j,\vec{k}'> = V^{-1}\int d^3r e^{i(\vec{k}'+\vec{q}-\vec{k})\cdot\vec{r}}$$

$$\times\ u^*_{i,\vec{k}}(\vec{r})v_{\vec{q}}(\vec{r})u_{j,\vec{k}'}(\vec{r}) \quad . \tag{9}$$

To perform this integration over the whole crystal, we first sum over all the cells and then integrate within each cell. This leads us to write $\vec{r} = \vec{a} + \vec{\rho}$ where $\vec{a}$ is any lattice translation and $\vec{\rho}$ remains within the boundaries of the cell centered around $\vec{a}$. Then

$$<i,\vec{k}|v_{\vec{q}}(\vec{r})e^{i\vec{q}\cdot\vec{r}}|j,k'> = V^{-1} \sum_{\vec{a}} e^{i(\vec{k}'+\vec{q}-\vec{k})\cdot\vec{a}}$$

$$\times \int d^3\rho e^{i(\vec{k}'+\vec{q}-\vec{k})\cdot\vec{\rho}} u^*_{i,\vec{k}}(\vec{\rho})v_{\vec{q}}(\vec{\rho})u_{j,\vec{k}'}(\vec{\rho}) \quad , \tag{10}$$

since $u_{i,\vec{k}}$ and $v_{\vec{q}}$ are invariant under the lattice translations.

The sum over the lattice translations is zero except when

$$\vec{k} = \vec{k}' + \vec{q} + \vec{g} \quad , \tag{11}$$

where $\vec{g}$ is either zero or a reciprocal vector.

In what follows, we restrict ourselves to phonons with large wavelengths and vectors $\vec{k}$ and $\vec{k}'$ near the same band extremum. Then $\vec{g} = 0$ and

$$<i,\vec{k}|v_q(\vec{r})e^{i\vec{q}\cdot\vec{r}}|j,\vec{k}'> = v^{-1}\delta_{\vec{k},\vec{k}'+\vec{q}}\int d^3\rho u^*_{i,\vec{k}'+\vec{q}}(\rho)u_{j,\vec{k}'}(\vec{\rho}), \tag{12}$$

where v is the volume of the elementary cell. Finally, focusing on intraband matrix elements and assuming that the u's are smooth functions of $\vec{k}$, one obtains

$$<i,\vec{k}|v_q(\vec{r})e^{i\vec{q}\cdot\vec{r}}|j,\vec{k}'> = v^{-1}\delta_{\vec{k},\vec{k}'+\vec{q}}\int d^3\rho |u_{i,o}(\vec{\rho})|^2 v_{\vec{q}}(\vec{\rho}) \quad . \tag{13}$$

Together with eqs. (7) and (8), this shows that the intraband matrix elements are exactly similar to those obtained for the effective Hamiltonian

$$H_{eff} = \frac{p^2}{2m} + \sum_{\vec{q}} h\omega_{\vec{q}} a^+_{\vec{q}} a_{\vec{q}} + \sum_{\vec{q}} (V_{\vec{q}} a_{\vec{q}} e^{i\vec{q}\cdot\vec{r}} + V^*_{\vec{q}} a^+_{\vec{q}} e^{-i\vec{q}\cdot\vec{r}}), \tag{14}$$

using plane waves as wave functions. The interaction coefficients $V_{\vec{q}}$ are given by

$$V_{\vec{q}} = v^{-1}\int d^3\rho |u_{i,o}(\vec{\rho})|^2 v_{\vec{q}}(\vec{\rho}) \quad , \tag{15}$$

and the zero of energy has been chosen at the minimum ($\varepsilon_i(o) = 0$). Eq. (15) is the general expression of a polaron Hamiltonian.

Let us now calculate the interaction coefficient in three simple cases.

IV. DEFORMATION POTENTIAL FOR LONG-WAVE ACOUSTIC PHONONS

In the case of a nonpolar crystal (e.g. the elemental semiconductors like Si or Ge), it is possible to relate the interaction coefficient $V_{\vec{q}}$ to the shift in the band energy produced by a homogeneous and constant strain resulting from an external stress (see ref. (1)). For example, consider a longitudinal acoustic phonon with the q-vector along a symmetry axis taken as 0_x. The displacement of an ion whose equilibrium position is $\vec{R}_o$ is given by

$$u(\vec{R}_o) = \sqrt{\frac{\hbar}{2NM\omega_{\vec{q}}}} \left(a_{\vec{q}} e^{i\vec{q}\cdot\vec{R}_o} + a_{\vec{q}}^{+} e^{-i\vec{q}\cdot\vec{R}_o}\right) , \tag{16}$$

where N is the number of elementary cells in the crystal and M their total mass. The strain associated with this lattice mode of vibration is

$$\varepsilon_{xx}(\vec{R}_o) = \frac{\partial u_x}{\partial R_{ox}} = i\sqrt{\frac{\hbar}{2NM\omega_{\vec{q}}}}\, q\left(a_{\vec{q}} e^{i\vec{q}\cdot\vec{R}_o} - a_{\vec{q}}^{+} e^{-i\vec{q}\cdot\vec{R}_o}\right) . \tag{17}$$

Now, for large wavelengths ($qa \ll 1$), the strain is homogeneous over large distances. Therefore, it produces a change in the electron energy similar to that obtained with a strain homogeneous in the whole crystal and due to an external stress. Obviously, this energy shift is the e-ph interaction potential. It is proportional to the strain, so that it can be written as

$$V = \Xi\, \varepsilon_{xx}(\vec{r}) = \sum_{\vec{q}} \left(V_{\vec{q}}^{d} a_{\vec{q}} e^{i\vec{q}\cdot\vec{r}} + V_{\vec{q}}^{d*} a_{\vec{q}}^{+} e^{-i\vec{q}\cdot\vec{r}}\right) \tag{18}$$

with

$$V_q^d = i\Xi \sqrt{\frac{\hbar}{2\rho V v_s}}\, \sqrt{q} , \tag{19}$$

where the deformation potential Ξ can be measured experimentally in studying the effects of an external stress on the electron energy. The notation ρ is used for the specific mass ($\rho = NM/V$) and v_s denotes the sound velocity ($v_s = \omega_q/q$). The derivation of the interaction coefficients is readily extended to the case of anisotropic bands (2).

V. POLAR INTERACTION WITH LONGITUDINAL-OPTICAL MODES

In ionic crystals, the longitudinal-optical (L-O) phonons with large wavelengths are the source of a macroscopic polarization. Indeed, for this type of vibrations, the ions of opposite sign move in opposite directions. The e-ph interaction studied in this section is simply due to the long range effect of the polarization on the electron. More precisely, the interaction potential (eq. (4)) is the electric potential induced at the position of the electron by the ion displacements. This type of interaction leads to the well known Hamiltonian for Fröhlich's polarons (3,4).

If one assumes that the conduction electron is distributed evenly in the elementary cell, i.e.

$$|u_{i,o}(\vec{\rho})|^2 = 1 \quad , \tag{20}$$

then the integration for obtaining the interaction coefficient (eq. (15)) leads to averaging the electric field over the elementary cell. Thus, the problem is reduced to a problem of macroscopic electricity, namely to find the macroscopic electric field produced by the L-O vibrations in an ionic crystal.

Consider a crystal with two ions of opposite sign per cell. As already stated above, for the long-wave L-O modes, the ions of opposite sign move in opposite directions and the dipoles add up to give a macroscopic polarization. The relative displacement of the ions in the cell located at $\vec{R}_o$, for a vibration with wave vector $\vec{q}$ is

$$u(\vec{R}_o) = \sqrt{\frac{h}{2N\mu\omega}} \; (a_{\vec{q}}\, e^{i\vec{q}\cdot\vec{R}_o} + a^+_{\vec{q}}\, e^{-i\vec{q}\cdot\vec{R}_o}) \, , \tag{21}$$

where N is the number of cellsin the crystal, μ the reduced mass of the cell and ω the frequency of the long-wave L-O phonons. This frequency can be taken as independent of $\vec{q}$. The polarization associated with this displacement can be written as

$$\vec{P}(\vec{R}_o) = \frac{N}{V} \; [Ze\; \vec{u}(\vec{R}_o) + \alpha\vec{E}_\ell(\vec{R}_o)] \, , \tag{22}$$

where Ze is an effective charge making the electric dipole directly related to the displacement $\vec{u}$, α the total poarizability of the ions in a cell and $\vec{E}_\ell(\vec{R}_o)$ the local field at a lattice site in the cell located at $\vec{R}_o$.

For free oscillations, the polarization has no external source, so that $\vec{D} = 0$ and therefore

$$\vec{E} = -4\pi\vec{P} \quad . \tag{23}$$

The local field is

$$\vec{E}_\ell = \vec{E} + \frac{4\pi}{3}\vec{P} = -\frac{8\pi}{3}\vec{P} \quad . \tag{24}$$

Eqs. (22) and (24) allow one to relate the polarization to the displacement, giving

$$(1 + \frac{8\pi}{3}\frac{N}{V}\alpha)\vec{P}(\vec{R}_o) = \frac{N}{V} Ze\, \vec{u}(\vec{R}_o) , \tag{25}$$

and

$$(1 + \frac{8\pi}{3}\frac{N}{V}\alpha)\vec{E}(\vec{R}_o) = -4\pi \frac{N}{V} Ze\, \vec{u}(\vec{R}_o) \quad . \tag{26}$$

The potential $\Phi(\vec{R}_o)$, related to $\vec{E}(\vec{R}_o)$ by

$$\vec{E}(\vec{R}_o) = -\vec{\nabla}\phi(\vec{R}_o) \quad , \tag{27}$$

is obtained by integration. This gives the e-ph interaction, which is simply $e\phi(\vec{r})$.

The last difficulty is to relate the parameters in eq. (26) to measurable properties of the crystal. For this purpose suppose that some external static sources are present, giving rise to an electric displacement $\vec{D}(\vec{R}_o)$. When the ions move in the presence of this field they bring about an electrostatic energy density equal to

$$-\vec{P}_i(\vec{R}_o)\cdot\vec{D}(\vec{R}_o) = -\frac{\frac{N}{V}Ze}{1 + \frac{8\pi}{3}\frac{N}{V}\alpha}\vec{u}(\vec{R}_o)\cdot\vec{D}(\vec{R}_o), \tag{28}$$

where $\vec{P}_i$ denotes the polarization produced by the ion displacements. To obtain the total polarization, one has to add the electronic polarization $\vec{P}_e$ directly induced by the external sources, so that

$$\vec{P}(\vec{R}_o) = \vec{P}_i(\vec{R}_o) + \vec{P}_e(\vec{R}_o)$$

$$= \frac{\frac{N}{V}Ze}{1 + \frac{8\pi}{3}\frac{N}{V}\alpha}\vec{u}(\vec{R}_o) + \frac{1}{4\pi}(1 - \frac{1}{\varepsilon_\infty})\vec{D}(\vec{R}_o) \quad . \tag{29}$$

In this relation, ε_∞ denotes the electronic (high frequency) dielectric constant, i.e. the square of the refraction index. To obtain the ion displacement $\vec{u}(\vec{R}_o)$ forced by $\vec{D}(\vec{R}_o)$, one minimizes the total energy density which is the sum of the electrostatic contribution (28) and the elastic term

$$\frac{1}{2}\frac{N}{V}\mu\omega^2\ u^2(\vec{R}_o)\ . \tag{30}$$

This gives

$$\mu\omega^2\vec{u}(\vec{R}_o) = \frac{Ze}{1+\frac{8\pi}{3}\frac{N}{V}\alpha}\ \vec{D}(\vec{R}_o)\ . \tag{31}$$

Introducing this result into eq. (29), noting that

$$P(\vec{R}_o) = \frac{1}{4\pi}\ (1 - \frac{1}{\varepsilon_o})D(\vec{R}_o)\ , \tag{32}$$

where ε_o is the static dielectric constant, leads immediately to

$$\left(\frac{\frac{N}{V}\ Ze}{1+\frac{8\pi}{3}\frac{N}{V}\alpha}\right)^2 = \frac{N}{V}\frac{\mu\omega^2}{4\pi}\ (\frac{1}{\varepsilon_\infty} - \frac{1}{\varepsilon_o})\ . \tag{33}$$

One is now able to obtain the e-ph interaction potential by using the procedure described above. One has in this way

$$\vec{E}(\vec{R}_o) = -\sqrt{4\pi\frac{N}{V}\ \mu\omega^2(\frac{1}{\varepsilon_\infty} - \frac{1}{\varepsilon_o})}\vec{u}(\vec{R}_o)\ , \tag{34}$$

and

$$V_{e-ph} = e\phi(\vec{r})\ , \tag{35}$$

with $\phi(\vec{r})$ given by

$$-\vec{\nabla}\phi(\vec{r}) = \vec{E}(\vec{r})\ . \tag{36}$$

The final result is

$$V_{e=ph} = \sum_{\vec{q}}\ (V_{\vec{q}}a_{\vec{q}}e^{i\vec{q}\cdot\vec{r}} + V^*_{\vec{q}}a^+_{\vec{q}}e^{-i\vec{q}\cdot\vec{r}}) \tag{37}$$

with

$$V_{\vec{q}} = -i\ \frac{e}{q}\ \sqrt{2\pi\ \frac{h\omega}{V}\ (\frac{1}{\varepsilon_\infty} - \frac{1}{\varepsilon_o})}\ . \tag{38}$$

Following Fröhlich, we introduce the following coupling constant

$$\alpha = \frac{e^2}{2h\omega\sqrt{h/2m\omega}}\ (\frac{1}{\varepsilon_\infty} - \frac{1}{\varepsilon_o})\ , \tag{39}$$

which gives the average number of virtual phonons in the distortion cloud. In terms of α, the interaction coefficient, eq. (38) can be written as

$$V_{\vec{q}} = -\frac{i}{q}(4\pi/V)^{\frac{1}{2}}\alpha^{\frac{1}{2}}h\omega(h/2m\omega)^{\frac{1}{4}} \quad . \tag{40}$$

VI. PIEZOELECTRIC COUPLING

We have seen in section IV that an acoustic phonon with a small wave vector $\vec{q}$ (large wavelength) is the cause of a strain

$$\varepsilon_{ij}(\vec{R}_o) = \frac{\partial\mu_i(R_o)}{\partial R_{oj}}$$

$$= i\sum_{\vec{q}}\sqrt{\frac{h}{2NM\omega_{\vec{q}}}}\,\eta_i q_j(a_q e^{i\vec{q}\cdot\vec{R}_o} - a_q^+ e^{-i\vec{q}\cdot\vec{R}_o}), \tag{41}$$

where $\vec{\eta}$ is a unit vector defining the direction of polarization of the wave. To describe the strain, the quantities S_{ij} are usually preferred to the ε_{ij}'s. They are defined as

$$S_{ij} = \tfrac{1}{2}(\varepsilon_{ij} + \varepsilon_{ji}) \quad . \tag{42}$$

In an ionic crystal that lacks a center of inversion, the strain produces an electric field by piezoelectricity. The components of this field are

$$E_i = -\frac{4\pi}{\varepsilon}\sum_{j,k=1}^{3} e_{ijk}\,S_{jk}, \tag{43}$$

where e_{ijk} are components of the piezoelectric e tensor and ε, the dielectric constant. The e-ph coupling potential is again given by

$$V_{e-ph} = e\phi(\vec{r}), \tag{44}$$

with $\phi(\vec{r})$ defined by the relation

$$\vec{E}(\vec{r}) = -\vec{\nabla}\phi(\vec{r}) \quad . \tag{45}$$

This leads to

$$E_i = -i\,\frac{4\pi}{\varepsilon}\sum_{\vec{q}}\sqrt{\frac{h}{2NM\omega_{\vec{q}}}}\,(a_{\vec{q}}e^{i\vec{q}\cdot\vec{R}_o} - a_{\vec{q}}^+e^{-i\vec{q}\cdot\vec{R}_o})$$

$$\times \sum_{j,k} e_{ijk}\,\frac{\eta_j q_k + \eta_k q_j}{2}, \tag{46}$$

and

$$V_{e-ph} = \sum_{\vec{q}} (V_{\vec{q}} a_{\vec{q}} e^{i\vec{q}\cdot\vec{r}} + V^*_{\vec{q}} a^+_{\vec{q}} e^{-i\vec{q}\cdot\vec{r}}), \tag{47}$$

with the interaction coefficient $V_{\vec{q}}$ given by

$$V_{\vec{q}} = 4\pi\frac{e}{\varepsilon} \frac{h}{2\rho V\omega_{\vec{q}}} \sum_{j,k} e_{ijk} \frac{\eta_j q_k + \eta_k q_i}{2} \quad . \tag{48}$$

Now, for acoustic phonons

$$\omega_{\vec{q}} = qv_S , \tag{49}$$

with

$$v_s = \sqrt{\frac{C}{\rho}} , \tag{50}$$

where C is an appropriate elastic constant. Usually (5), an average is taken over all the directions of propagation and of polarization. This leads to

$$V_q = \frac{hv_S}{\sqrt{q}} \left(\frac{4\rho\alpha_p}{V}\right)^{\frac{1}{2}} , \tag{51}$$

where α_p is the coupling constant for the piezoelectric coupling and defined as

$$\alpha_p = \frac{1}{2} \frac{e^2}{hv_S} \left\langle \frac{e^2_{ijk}}{\varepsilon^2 C} \right\rangle . \tag{52}$$

However, it is more correct to explicitly take the angular dependence into account (6,7).

VII. APPLICATIONS TO RADIATIONLESS PROCESSES

Let us first recall that the different forms of Hamiltonians described in this paper are effective Hamiltonians based on the band-mass approximation. Therefore, they are appropriate to the description of large polarons. For small polarons, one should use a different approach. Holstein's molecular model is often used. We refer the reader interested in this point to the original paper by Holstein (8) as well as the Emin's papers (9).

Let us just point out that, in the cases where self-trapped particles are unambiguously observed, the self-trapping is due to the formation of a kind of molecular complex that involves a lattice distortion much too strong for the linear form of the e-ph interaction potential used in polaron theory to be valid.

The theory of large polarons has been used or can be used to study different radiationless processes in semiconductors, for instance

- the scattering of charge carriers by phonons and its effect on the electrical conductivity in the ohmic regime as well as at high electric field (10,11,12);

- the relaxation of hot carriers (12) after injection into the band for instance by Auger transitions or at semiconductor junctions; or

- the radiationless capture of charge carriers by shallow or moderately deep impurity centers, and the probability of transitions between states of bound polarons.

The processes can be either cascade captures or multiphonon transitions. For application of polaron theory to this question, see the papers by Pässler (13). In my opinion, an important difficulty is the following: the excited states one deals with are usually broad resonances with short lifetimes. Therefore, they are not eigenstates of the Hamiltonian and the golden rule must be used with caution.

To conclude, polaron theory provides a convenient model to study the radiationless transitions of free charge carriers or carriers bound to shallow or moderately deep impurities in semiconductors. Since the interaction potential is known, "ab initio" calculations are possible, contrarily to the case of the configuration-coordinate model where unknown parameters remain in the calculations and are determined at the end by fitting to the experimental data. On the other hand, the polaron model is not adapted to the study of deep centers, since the band-mass approximation and sometimes the assumption of an e-ph coupling linear in the phonon variables are no longer valid in this case.

ACKNOWLEDGEMENT

These notes are based on lectures given at the Physics Department of the University of Delaware when the author was Unidel visiting professor. The author thanks the University of Delaware and the staff of the Physics Department for their kind hospitality.

REFERENCES

1. J. Bardeen, W. Shockley, Phys. Rev. 80, 72 (1950).

2. C. Herring, E. Vogt, Phys. Rev. 101, 944 (1956).
3. H. Fröhlich, H. Pelzer, S. Zienau, Phil. Mag. 41, 221 (1950).
4. Polarons in Ionic Crystals and Polar Semiconductors (J.T. Devreese, ed.), North-Holland, Amsterdam (1972).
5. A. R. Hutson, J. Appl. Phys. 32, 2287 (1961).
6. G. D. Mahan, J. J. Hopfield, Phys. Rev. Letters 12, 241 (1964).
7. J. J. Licari, G. Whitfield, Phys. Rev. B 9, 1432 (1974).
8. T. Holstein, Ann. Phys. 8, 325 (1959).
9. D. Emin, Adv. Phys. 22, 57 (1973) and ref. therein.
10. J. T. Devreese, R. Evrard, Phys. Stat. Sol. (b) 78, 85 (1976).
11. E. Kartheuser, J. T. Devreese, R. Evrard, Phys. Rev. B 19, 546 (1979).
12. Solid-State Electronics, Vol. 21, Pergamon Press, London (1978).

QUANTUM MECHANICAL CALCULATION OF THE RELATION BETWEEN THE RADIATIONLESS AND RADIATIVE DECAY RATES WITH APPLICATION TO MOLECULES IN THE GAS PHASE AND IONS IN THE SOLID STATE AND THE CONSEQUENCES FOR TEMPERATURE QUENCHING (Abstract Only)

J.M.F. van Dijk

Philips Research Laboratories

Eindhoven, The Netherlands

ABSTRACT

A general relation between the radiationless and radiative decay rates is derived, whereby we not only consider the vibrational overlap integrals for the two processes, but also the electronic factors that cause these processes. This proves possible, because the operators responsible for these processes, the electric dipole moment and the non-adiabatic Born-Oppenheimer coupling, turn out to be related to each other.

In the first part of the paper, we treat the case of non-crossing potential energy curves. Then the energy gaps for which the radiationless rate can compete with the radiative rate turn out to be rather small for ions in the solid state (i.e., about one third of the energy gap encountered in visual emission processes). For molecules, not enough data are as yet available to warrant the same conclusion, but we suspect that the situation will not differ very much from that of ions in the solid state. Also the temperature dependence of the radiationless decay is examined.

In the second part of the paper, we consider systems with (pseudo-) crossing potential energy curves. It follows that the energy gaps, for which radiationless decay is competitive with radiative decay, are much larger and can thus explain the "quenching" of visible and UV-emission. Also, the temperature dependence of the radiationless decay is examined, which on application to ions in the solid state gives qualitative rules for the "temperature quenching" of these ions; these rules are similar to Blasse's empirical rules for this phenomenon.

SHORT SEMINARS (ABSTRACTS)

VIBRATIONAL STRUCTURE OF THE ABSORPTION BAND OF A JAHN-TELLER CENTER IN A NON CONDUCTING CRYSTAL

N. Terzi

Instituto di Fisica dell'Università di Milano
Gruppo Nazionale di Struttura della Materia del CNR

The phonon processes involved in the optical transitions of an electron bound to a center in a non conducting crystal are discussed. The lineshape $I(\omega)$ of the absorption band is deduced for a dispersive Jahn-Teller (JT) system, i.e. in a theoretical framework in which one takes into account at the same time two important characteristics of the center, whose cumulative effect is not usually considered in the most popular models:

1) The interaction of the optic electron with the continuum of the crystal normal modes (i.e. the dispersion), and

2) The orbital degeneracy of the electronic levels involved in the optical transition (i.e. the JT effect).

The assumptions made are those usually adopted for an impurity in a crystal:

i) The level structure of the electron consists of a non degenerate ground state and only one degenerate excited state.

ii) The lattice dynamics are harmonic.

iii) The electron interacts with the phonons via a linear electron-phonon (EP) interaction.

In the framework of the second quantization formalism and by using the diagram technique, one finds that the phonon processes contributing to $I(\omega)$ can be summed up to all the orders in a compact expression[1]. Explicit and analytical expressions are then deduced in special limits.

First, I(ω) is evaluated both in the strong and in the weak EP interaction limits.

In the strong interaction limit, the well-known symmetric Toyozawa-Inoue I(ω) is found[2] in the lowest-order approximation. The next-order contributions are seen to be responsible for an asymmetry of I(ω), which then arises in absence of a quadratic EP interaction.

In the weak interaction limit and at T = 0 K, the sideband of the zero-phonon line is studied. The sideband is found to be composed by a JT-modified phonon density of states and its convolutions. The structures and the singularities of such a phonon density are discussed and the result of a numerical evaluation is shown.[3]

Finally, the usual Pekarian function for the lineshape is obtained, when only a normal mode is assumed to interact with an electron whose electronic levels are non degenerate.

1. E. Mulazzi and N. Terzi, Phys. Rev. B. 10, 3552 (1974).
2. Y. Toyozawa and M. Inoue, J. Phys. Soc. Japan 21, 1633 (1966).
3. N. Terzi and G. Brivio, J. Phys. C. (Paris) (in press).

ELECTRON TRANSFER PROCESSES IN PHOTOSYNTHETIC BIOLOGICAL SYSTEMS

E. Buhks

Institute of Chemistry
Tel-Aviv University
Ramat-Aviv, Tel-Aviv, Israel

This seminar presents a conceptual model of the sequence of primary light induced electron transfer (ET) steps in photosynthetic bacteria. The temperature dependence of some of these redox reactions, like ET process between cytochrome and bacteriochlorophyll in Chromatium, is characterized by a temperature-independent rate at low temperatures and exhibits the Arrhenius-type dependence at high temperatures. The other primary ET processes, like an ET reaction between bacteriopheophytin and Fe-quinone complex in Rps. spheroides, are temperature-independent in the broad range of 4-300K. The third type of ET processes, exemplified by back ET reactions between Fe-quinone and bacteriochlorophyll in Rhs. rubrum, exhibits negative activation energy at high temperatures.

The theoretical approach, describing the primary ET processes in photosynthesis, is based on the non-adiabatic multiphonon ET theory, which incorporates both a continuous distribution of optical phonons in a polar solvent and discrete intramolecular vibrational modes. The last two types of the redox reactions are attributed to activationless ET processes which play an essential role in highly efficient charge separation in primary photosynthetic processes. The transition temperature, separating the tunneling region from the activated region indicates the range of phonon frequencies involved in the ET process. Comparing the low-temperature rates with calculated Franck-Condon factors one can determine the value of the electron-exchange matrix element, which in turn provides a rough estimate of the distance scale between a donor and an acceptor in the primary ET events.

THEORETICAL CALCULATIONS OF THE RADIATIVE AND NONRADIATIVE TRANSITION PROBABILITIES OF F-CENTERS IN IONIC CRYSTALS

I. Rojas Hernández

IIM de la Universidad Michoacana
Morelia, Mich/Mexico

Although the radiative and radiationless processes of the F-centers pose a very old problem, there are questions that the theory cannot answer satisfactorily. A lot of theoretical models have been developed by physicists in order to explain the fundamental experimental results obtained in the laboratories.

In this paper we have calculated analytically the radiative and nonradiative transition probabilities (rates) on the basis of the microscopic model developed by Stampf et al.[1] The numerical results, which we obtained for the F-center in NaCl have been compared with the experimental results measured by Honda and Tomura.[2] In order to do this we needed the radiative, nonradiative and ionization transition rates in the formula

$$\eta = \frac{W^s}{W^s + W^k + W^i} ,$$

which represents the quantum yield for the luminescence for the above mentioned color centers. In this formula,

W^s is the radiative transition rate

W^k is the nonradiative transition rate

W^i is the ionization transition rate.

1. H. Stumpf, Phys. Kondeus. Materie 13, 101 (1971); Phys. Kondeus. Materie 18, 217 (1974); W. Heinzel, Dissertation, Tübingen 1973, Phys. Kondeus. Materie 17, 99 (1974).
2. S. Honda, M. Tomura, J. Phys. Soc. Japan 33, 1003 (1972).

ENERGY TRANSFER IN RANDOM SYSTEMS

K. Godzik

Weizmann-Institute of Science
Department of Chemical Physics
Israel

A theoretical study of incoherent electronic energy transfer in an impurity band of substitutionally disordered molecular crystals is presented. It is assumed that the exciton is well described within a localized representation and that migration between randomly distributed impurity sites is due to resonance or phonon-assisted energy transfer. In the case of phonon-assisted transfer we shall confine ourselves to transfer processes which do not depend on the energy mismatch between two sites.

The time evolution of the microscopic probability of excitation at lattice site n, $p_n(t)$, is then described by a master equation. Due to the randomness of the impurity system, the transition from the microscopic description of an individual configuration to a macroscopic average probability of excitation, $P(\underline{r},t)$ involves an averaging process over all possible configurations. The way this averaging process is performed, determines various kinds of models. The pertinent experimental observables, i.e., the initial site occupation probability $P_o(t)$, the mean square displacment $\Sigma^2(t)$ and the time dependent diffusion coefficient D(t) will be expressed within the scope of these models. The time dependence of these quantities will be discussed and also their dependence on various kinds of interactions.

ENERGY TRANSFER PROCESSES IN "PURE" Mn^{2+} FLUORIDE SYSTEMS

R. Moncorgé and B. Jacquier

University of Lyon ER °10 C.N.R.S.
43 Bd du 11 Nov. 1918
69621 Villeurbanne, France

At very low temperatures (below the Néel point T_N), the optical spectra of manganese fluoride systems such as MnF_2 and $AMnF_3$ (A = K, Rb, Cs) exhibit fine structures which are related to electronic, magnetic and lattice collective excitations, namely, exciton emissions and their magnon/phonon sidebands. These transitions are intrinsic if they occur at unperturbed magnanese sites. However, small amounts of impurities, typically a few parts per million of Mg, Zn, and Ca, are inevitably present in the crystals. These impurities may perturb the nearest neighbor Mn ions, inducing in the lattice impurity perturbed manganese sites which are populated via the migration of the intrinsic exciton and phonon assisted energy transfer or trapping. The trapped excitons then release their energy giving rise to a number of extrinsic exciton lines and their magnon/phonon sidebands which are observed in the fluorescence spectra.

Our attention here has been focused on the radiative and non-radiative characteristics of the intrinsic and impurity-induced fluorescence lines. Initially, we shall consider the thermally-induced intensity and decay time variations of the trapped excitons, and secondly, the relationship between the intrinsic exciton decay and the trap fluorescence rise observed at a short time delay, with respect to the relative importance of diffusion and direct phonon/magnon assisted energy transfer. Finally, the effect of a possible biexciton process at very low temperatures and under high pumping power conditions is discussed.

TEMPERATURE BEHAVIOR OF THE BLUE FLUORESCENCE OF ZnS:Tm^{3+} CRYSTALS

J. Mareš, L. Jastrabik and S. Pačesová

Institute of Physics
Czechoslovak Academy of Sciences
Na Slovance 2
180 40 Prague 8, Czechoslovakia

The temperature behavior of the fluorescence spectra of Tm^{3+} impurity ions in cubic ZnS crystals in the blue range was studied either under selective, high-density excitation by nitrogen and dye laser beams or under non-selective excitation by a deuterium lamp in the temperature range from 4.2K to room temperature. At room temperature a structureless non-selectively excited fluorescence band ranging from 470 to 485 nm was observed, while at liquid nitrogen temperature and lower we observed more than 30 narrow fluorescence lines, the half widths of which were about 0.14 nm. The blue strongly temperature dependent fluorescence depends on the energy of excitation. This means that this fluorescence is either host-sensitized by an energy of excitation above the band gap energy of ZnS (E(cub.) = 3.78 eV at T = 77 K), or impurity excited by an energy of excitation below the band gap energy of ZnS.

The condition of preparation of ZnS:Tm^{3+} crystals and the observed temperature behavior of the selectively or non-selectively excited fluorescence spectra indicate that Tm^{3+} ions can create complexes of the type Tm^{3+}-V_{Zn} , where V_{Zn} is a zinc vacancy located at one of the nearest neighbors of the Tm^{3+} ion which substitutionally replaces Zn^{2+}. We conclude from a comparison of the ZnS:Tm^{3+} fluorescence spectra with those of several other II-VI compounds that the Tm^{3+}-V_{Zn} complexes have C_2 symmetry. Most of the fluorescence lines can be ascribed as $G_4 \rightarrow {}^3H_6$ and ${}^1I_6 \rightarrow {}^3F_4$ transitions among energy levels of Tm^{3+}, while further observed lines can be either phonon-assisted transitions or transitions from other Tm^{3+} sites in ZnS.

ION-PHONON INTERACTIONS IN SEMICONDUCTORS

A. Radliński

Institute of Experimental Physics
Ul. Hoza 69
Warszawa, Poland

A short review of the experimental results of the infrared (~3 μm) luminescence of Co^{2+} impurity centers in II-VI semiconducting compounds will be given. These systems show an intermediate ion-phonon interaction ($1 \lesssim S \lesssim 3$, S being the Huang-Rhys number), which makes it possible to observe both no-phonon and one-phonon structure. Due to extreme simplicity of the electronic structure the interaction of the ionic system with the whole phonon density of states (as a function of temperature) can be interpreted quantitatively.

LIST OF CONTRIBUTORS

F. Auzel, Centre National d'Etudes des Télécommunications, 196, Rue de Paris, 92220 Bagneux, France

J. E. Bernard, Department of Physics, University of Delaware, Newark, Delaware 19711, U.S.A.

D. E. Berry, Department of Physics, University of Delaware, Newark, Delaware 19711, U.S.A.

G. Blasse, Physical Laboratory, State University, P.O. Box 80.000, 3508 TA Utrecht, The Netherlands

E. Buhks, Department of Chemistry, Tel-Aviv University, Tel-Aviv, Israel

B. Di Bartolo, Department of Physics, Boston College, Chestnut Hill, Massachusetts 02167, U.S.A.

R. Englman, Soreq Nuclear Research Center, Yavne, Israel

R. Evrard, Institut de Physique, Université de Liège, Sart Tilman, Liège, Belgium

W. H. Fonger, RCA Laboratories, David Sarnoff Research Center, Princeton, New Jersey 08540, U.S.A.

K. Godzik, The Weizmann Institute of Science, Chemical Physics Department, Rehovot, Israel

R. Grasser, I. Physikalisches Institut de Justus-Liebig-Universität, 6300 Giessen, West Germany

B. Jacquier, Université de Lyon, E.R. No. 10 C.N.R.S., 43 Bd. du 11 Novembre 1918, 69621 Villeurbanne, France

L. Jastrabik, Institute of Physics, Czechoslovak Academy of Sciences, Na Slovance 2, 18040 Prague 8-Liben, Czechoslovakia

J. Jortner, Department of Chemistry, Tel-Aviv University, Tel-Aviv, Israel

A. Kaminskii, Institute of Crystallography, Academy of Sciences of the U.S.S.R., Leninsky pr. 59, 117333 Moscow B-333, U.S.S.R.

J. Mareš, Institute of Physics, Czechoslovak Academy of Sciences, Na Slovance 2, 18040 Prague 8-Liben, Czechoslovakia

R. Moncorgé, Université de Lyon, E.R. No. 10 C.N.R.S., 43 Bd. du 11 Novembre 1918, 69621 Villeurbanne, France

S. Pačesová, Institute of Physics, Czechoslovak Academy of Sciences, Na Slovance 2, 18040 Prague 8-Liben, Czechoslovakia

R. Pappalardo, GTE Laboratories, 40 Sylvan Road, Waltham, Massachusetts 02154, U.S.A.

A. Radliński, Institute of Experimental Physics, Ul. Hoza 69, Warzawa, Poland

D. A. Ramsay, Herzberg Institute of Astrophysics, National Research Council of Canada, 100 Sussex Drive, Ottawa, Ontario, Canada K1A 0R6

R. Reisfeld, Department of Inorganic and Analytical Chemistry, The Hebrew University of Jerusalem, Jerusalem. Israel

L. A. Riseberg, GTE Laboratories, 40 Sylvan Road, Waltham, Massachusetts 02154, U.S.A.

I. Rojas Hernandez, IIM de la Universidad Michoacana, Morelia, Michoacan, Mexico

A. Rosencwaig, Lawrence Livermore Laboratory, P.O. Box 5508, Livermore, California 94550, U.S.A.

A. Scharmann, I. Physikalisches Institut der Justus-Liebig-Universität, 6300 Giessen, West Germany

J. Singh, Research School of Chemistry, The Australian National University, Box 4 P.O., Canberra A.C.T. 2600, Australia

C. W. Struck, RCA Laboratories, David Sarnoff Research Center, Princeton, New Jersey 08540, U.S.A.

N. Terzi, Istituto de Fisica dell'Università di Milano, Gruppo Nazionale di Struttura della Materia del CNR, Via Celoria 16, 20133 Milano, Italy

J. M. F. van Dijk, Philips Research Laboratories, 5600 MD Eindhoven, The Netherlands

F. Williams, Department of Physics, University of Delaware, Newark, Delaware 19711, U.S.A.

SUBJECT INDEX